THE CULTURAL GEOGRAPHY READER

There has ... with the so-called
"cultural ... been no generally
accessible ... cultural geography
and relate ...

The Cu ... abridged readings,
including ... ham, J.B. Jackson,
Gillian Ro ... yn Longhurst, Don
Mitchell, ... g Culture; Cultural
Geograph ... bal Context; Home
and Away ... pe of the discipline
and its ke ... nd empirical focus,
allowing s ... rough the grounded
research ... t discusses the key
concepts, ... her disciplines and
practices. ... own focused introductions,

The re ... pieces from the US
and UK ... audience of undergraduate ... itions in the US and
UK, as w ... ny. In doing so, The
Cultural C ... enduring benchmark
text.

Timothy ... His research focuses
on China ... he author of Tourism
and Mod ... slocal China (2006).

Patricia ... , USA. Her research
focuses ... sity. She is the author
of Dry Pl ... an Mosaic: A Thematic
Introduct...

The Cultural Geography Reader

Edited by

Timothy S. Oakes

and

Patricia L. Price

Routledge
Taylor & Francis Group

LONDON AND NEW YORK

First published 2008
by Routledge
2 Park Square, Milton Park, Abingdon, Oxon, OX14 4RN

Simultaneously published in the USA and Canada
by Routledge
270 Madison Avenue, New York, NY 10016

Routledge is an imprint of the Taylor & Francis Group, an informa business

Typeset in Amasis MT Lt and Akzidenz Grotesk by Graphicraft Limited, Hong Kong
Printed and bound in Great Britain by MPG Books Ltd, Bodmin

British Library Cataloguing in Publication Data
A catalogue record for this book is available from the British Library

Library of Congress Cataloging in Publication Data
The cultural geography reader / edited by Timothy S. Oakes and Patricia L. Price.
p. cm.
Includes bibliographical references and index.
1. Human geography. I. Oakes, Tim. II. Price, Patricia Lynn, 1965–
GF43.C85 2008
304.2—dc22

2007032651

ISBN10: 0-415-41873-9 (hbk)
ISBN10: 0-415-41874-7 (pbk)
ISBN10: 0-203-93195-5 (ebk)

ISBN13: 978-0-415-41873-7 (hbk)
ISBN13: 978-0-415-41874-4 (pbk)
ISBN13: 978-0-203-93195-0 (ebk)

To our students

Contents

Acknowledgments

We wanted to write a *Cultural Geography Reader* mostly because we liked the *City Reader* (edited by LeGates and Stout, Routledge), and wanted a similar text for students in our cultural geography courses. When we discussed the project with Andrew Mould – who would become our editor at Routledge – we had no idea what sort of journey we would embark upon, that cold afternoon deep in the book display area of the 2004 Association of American Geographers' meeting in Pittsburgh. And though the quest has been a long one, fraught at time with the perils of securing permissions and the tedium of optical character recognition software, we have also made new friends and renewed old acquaintances. In addition, the editors were able to add a new chapter to their long friendship, which was itself born of a shared journey through graduate school together (but that's another story).

Among the many, many individuals who have helped us along the way are Andrew Mould, our fearless editor at Routledge, and his fair assistants, Zoe Kruze and Jennifer Page. Andrew provided the encouragement we needed to set out on this path in the first place, and intervened at key moments in the process to keep us on track. One of the many helpful suggestions he made was to aim for a readership that spans the Atlantic and what can be quite different understandings of cultural geography. By engaging patient, careful, and smart reviewers, Andrew put together a team that helped us to craft the project that at least in spirit has as one of its primary goals to speak to diverse traditions in a rich discipline. Zoe and Jennifer patiently worked with us to navigate the daunting waters of the minute, and not-so-minute, details of assembling the manuscript, securing permissions, sourcing images, and general troubleshooting. Production editor Jodie Tierney crossed vast oceans over email, slaying grammatical dragons and otherwise helping us shape the manuscript into its final form. Alex Dorfsman, Mathias Woo, and Brian Taylor graciously granted permission to use their images, which breathe beautiful life into the cover of this book and the spaces between sections.

Of course, this *Reader* would never have come into existence were it not for the original writings gathered between these covers. We are thus hugely indebted to all of the scholars whose works constitute the jewels showcased here. Some of these individuals have long since passed on. Their insights form the substrate from subsequent contributions are forged. Fortunately, most of the contributors are very much alive and, we hope, well. We are lucky to count many of them among our colleagues and friends. We are particularly grateful to those who gave their personal permission as part of the process involved in securing rights to reproduce their work from the original publishers. Of course, all errors of abridgement, editing, and interpretation rest squarely on our shoulders.

Tim would like to thank Julie, Eva, Angus, and Sydney for putting up with him, as usual. And he of course thanks Patricia for convincing him to take up the project and for humoring his crankiness.

Patricia would like to thank Tim (who should know better by now) for agreeing to partner up with her on this project in the first place. She has long relied on Tim's friendship to help her look much smarter and faster than she really is. She would also like to thank Ari, Nina, and Daniel for their unflagging good humor over the three years she spent with noise-cancelling headphones on grumbling "Be quiet, I'm trying to write!" You can turn the television on now.

T.S.O.
P.L.P.

INTRODUCTION

Is Hong Kong a Good City?

On the cover of this book are sixteen stills from the film *A Very Good City*, by Mathias Woo. Shot in 1998, one year after Hong Kong's reunification with China, Woo's eighty-minute film gazes on the Hong Kong one would see from the front seat of a double-decker bus or tramcar. As the city's streetscapes fly by with increasing speed, a Cantonese voiceover narrates the difficulties of growing up in a city whose story is best told by "telling its pace." Passages of the voiceover flash across the screen in seemingly isolated and random Chinese phrases, while English "subtitles" offer not a translation of the narrative at all but a reworking of T.S. Eliot's "The Love Song of J. Alfred Prufrock." Clearly, urban alienation is on display in Woo's work. *A Very Good City* – which draws its name from Kevin Lynch's classic treatise on urban planning *A Theory of Good City Form* (1981) – leaves the viewer with a sense that the utopian promise of modern urban planning has been a colossal failure, that Hong Kong, like other world cities, is a chaotic and uninhabitable landscape of mass transit. As Woo himself puts it, "to tell a story about a city is just about telling the lines in it; to tell a story about a story may just be telling lines and lines." And everyone knows that a line is not a space. One cannot live on a line.

Why have we chosen Woo's images of Hong Kong for the cover of *The Cultural Geography Reader*? Is it to suggest that our world has become chaotic and uninhabitable? Is it to suggest that we live always in transit, always commuting but never dwelling in place? We invite the reader to wonder about these images. Imagine yourself visiting Hong Kong for a day, seeing the city on a bus. Imagine doing this as a visitor in your own city or town. Collect in your mind for a moment the images of your own city that you might see passing by. What does your city look like? What is it like to live and work there? Is it a good city? If you collected these thoughts and sights in your mind and wrote them down, you would be doing the sort of thing that a cultural geographer does.

A Very Good City can be viewed as a cultural geography of Hong Kong; it raises questions about how we live in, experience, and shape a particular environment, about what living in and reshaping that environment means to us, about how that environment (and thus our relationship to it) is changing in various ways. These are basic themes of cultural geography. And they are basic themes in the lives of all humans.

As an academic discipline, cultural geography draws on these themes as the basis for in-depth ana-lyses of how we live in and act upon the world, and what that world means to us. Viewed as a cultural geography of Hong Kong, Mathias Woo's film suggests a rich set of issues and questions that go far beyond the rather well worn theme of urban alienation. One might thus view *A Very Good City* not simply as another version of "life out of balance" (though Woo's film is reminiscent in some ways of Godfrey Reggio's 1982 film *Koyaanisqatsi*, which is Hopi for "life out of balance") but as an ongoing *search for place in a world of rapid change*. Woo's film implicitly raises questions about living in Hong Kong in the context of its recent transition from British colony to Chinese Special Administrative Region: a city prospering on the precarious brink between powerful states. The betweenness of Hong Kong – long serving as a gateway between China and "the West" – has always made it a space of transit and transition, and of all kinds of cultural displacement. Mobility, then, has always been a major part of people's lives in Hong Kong. And now, with a new Disneyland recently opened and an ongoing government campaign to emphasize tourism, mobility continues to define Hong Kong in fundamental ways. Touristic transience has become a way of life in Hong Kong, with residents being asked to present their city to visitors as a consumable landscape. And they are themselves encouraged by the government of Hong Kong to consume their own city as tourists would. Woo's view from the bus, then, is the view both of a resident and of a visitor, an insider and outsider, a local tourist. The film struggles to make sense of this oxymoron, of Hong Kong as a kind of paradox, an illegible landscape for sightseers and inhabitants alike. How does one live in a place like Hong Kong? How does one carve a space of identity and meaning out of the transience, transition, displacement, and mobility that seems to define the experience of being there? Ultimately, these are questions that could be asked of many places around the world, not just Hong Kong.

The point of viewing *A Very Good City* as cultural geography, however, is not to suggest that *all* cultural geography raises such existential questions about what it means to inhabit places in an increasingly globalized world of transition and change. Rather, it is to offer Woo's film as just one example of how people experience their world, how they represent that experience, and how they make that experience and its representation meaningful. There is certainly no *one* cultural geography of Hong Kong, and for all Woo's questions about alienation, mobility, and transience there are undoubtedly many others living in Hong Kong who have reflected upon their city in very different ways. Indeed, the fact that our experiences, knowledge, and meanings of the world are so very different generates a great deal of debate, conflict, and even violence. Cultural geographies can be highly contested, for the very simple reason that what makes a good city for one group of people can in turn make that city alienating or dangerous for another. For some, it is the very transience of Hong Kong that makes it a good city, and not a failed landscape of what French anthropologist Marc Augé has termed *non-places*.[1] The political, then, is never very far from the cultural. Our focus here on the politics of culture is meant to convey not simply the idea that "the culture is political," but also the opposite: "politics is cultural." By this we mean to suggest a rather broad conception of politics, one defined as power relations expressed in many different ways, and not merely limited to electoral behaviors or the social relations of class.

While it is obvious that there are seemingly infinite ways individuals and groups experience and shape their world, and that there may be many cultural geographies that represent, reflect on, or analyze these experiences and changes and their meanings, there *are* several themes that tend to reappear with frequency in cultural geography. A number of them are apparent in Mathias Woo's film. Most cultural geographies focus in one way or another on *landscape*, a term which typically refers to the appearance of a particular section of land, but which is more thoroughly discussed in Part Three of the *Reader*. In addition, Woo's film is about *place*, that is, the space with which one identifies and feels at home, and which carries meanings and memories that individuals and groups share about a particular environment. Place is discussed in more detail in Part Five of the *Reader*. Both of these terms – landscape and place – direct our attention to the ways people know, fashion, and come to understand their place in space. How we construct *knowledge* of our world, then, is a fundamental subject of cultural geography.

There are other central themes of cultural geography. A great deal of our experience of the world brings us into contact with the non-human. How we define, understand, and engage *nature*, then, is another fundamental theme, though one which is much less apparent in Woo's film, except perhaps in its striking *absence*. Nature is the focus of Part Six of the *Reader*. There is also another constellation of themes which are not dealt with directly in Woo's film at all. These focus most explicitly on the above claim that the political is never very far from the cultural. This is not simply because there are many different ways in which people know and experience their world, but because that knowledge and experience is shaped by power. People do not share equal access to the world, and this inequality manifests itself in many ways. Social patterns of discrimination, exploitation, and patriarchy shape cultural geographies in fundamental ways. Some people are highly mobile, while others are relatively fixed. Some consume the world at high levels, while some consume hardly at all. Some are more responsible than others in determining the dominant ways we interact with and shape the world. These are topics that are dealt with specifically in Part Seven of the *Reader*. Below we explore some of these more political aspects of cultural geography before moving on to a general discussion of cultural geography as an academic discipline.

L'Affair Foulard, or, Did My Culture Make Me Do It?

In his film, Mathias Woo asked whether Hong Kong – a global city of transience, transition, and travel – was a good city. Here we turn to another question raised by globalization and mobility: How do we reconcile strong religious identities and convictions within states that profess religious tolerance as long as religious practices are kept out of public life? What role does culture play in the ways that minority groups seek to integrate themselves (or not) into dominant societies? How do the cultural geographies of multiculturalism and globalization help us understand some of the seemingly intractable issues of cultural diversity being faced by states around the world today? Certainly there are many examples from around the world that we could draw upon to discuss these issues. Here, we turn to a brief account of the so-called *affair foulard*, or the headscarf issue in contemporary France, to provide a sense of how cultural geographers might understand such questions.

On October 19 1989, Ernest Chenier, the headmaster of the College Gabriel-Havez of Creil, France, expelled three Muslim girls – Fatima, Leila, and Samira – for refusing to remove their headscarves, or *foulards*, while attending school. Although the scarf is an expression of a particular religious identity that is protected by France's commitment to religious freedom, for Chenier it was also a symbol of beliefs that directly challenged the very idea upon which France's principle of religious freedom was based. That idea is *laïcité*, a term which, though difficult to translate into English, refers generally to the concept of a secular state in which freedom of religion exists, but exists in a distinctly private realm that does not interfere with the public sphere in which citizenship exists. Indeed, it was perhaps a sincere belief in religious freedom that impelled Chenier to act in the first place, assuming as he might have that the girls were being *required* to wear the scarf in public, presumably against their will. This, however, was not the case. School officials and the parents of the girls had already reached an agreement in which they were to attend class without their heads covered. But Fatima, Leila, and Samira went to class covered anyway. In this way, their act took on a *deliberate* character, a gesture of both identification *and* defiance, which was thus explicitly political. Two weeks later, the Minister of Education, Lionel Jospin, took the matter to the Conseil d'Etat, France's high court, which delivered an ambiguous interpretation of how *laïcité* applied to the *foulard*. The Conseil ruled that the wearing of religious signs by students was not incompatible with *laïcité*, but the wearing of such signs as an act of "pressure, provocation, proselytizing, or propaganda" or in a way that would disturb the normal function of public education, violated the basic principles of French law. The Conseil left it to school officials to interpret this distinction between the scarf as a sign of religious devotion or identification and the scarf as an act of political provocation or public disturbance. In 1994 the Ministry of Education clarified the ruling by declaring explicitly that whereas students were free to wear religious symbols discreetly, the *foulard* could not be worn with any discretion and was thus forbidden in French state schools.

Cultural geography is central to understanding why Chenier responded to the actions of Fatima, Leila, and Samira with expulsion, and why the issue was taken all the way to the French high court for resolution. There are at least two ways in which we can view the issue through the analytic lens of cultural geography. First, *laïcité* depends upon a clear boundary between two distinct spaces: the private (where religion is said to belong) and the public sphere (where all citizens are equal under the law). The girls' actions clearly challenged this boundary by projecting into the public sphere an article of clothing that the state regarded as private. But from the perspective of cultural geography – that is, by paying closer attention to how people actually interact with their environment, and how they make that interaction meaningful – the abstract spaces of private and public don't always represent the ways people actually live their lives. While the boundary perhaps makes sense at the scale of the nation, an *imagined community* where citizenship is defined, actual cultural practices are carried out at much more local scales.[2] Indeed, the extremely local scale of *the body* is perhaps the most important scale of all in cultural geography. Our bodies travel between private and public spaces all the time. Often, how we dress and act with our bodies depends upon whether we are in private or public space, and there are social norms that govern such dress and actions, norms that vary according to gender, socio-economic class, ethnicity, place, time of day, and so forth. But some bodies may fail to abide by these norms. As the selection by Peter Jackson (see p. 413) makes clear, the cultural politics of how we dress our bodies is an important field within cultural geography. Paying attention to culture at the scale of the body makes clear that the division between private and public is not a natural fact but a socially constituted, and quite unstable, norm. As different bodily practices begin to change society, those norms are also challenged.

Second, the meaning of *identity* has shifted from an emphasis on sameness (this being the general sense of identity assumed under the concept of citizenship) to one of recognition, in which one claims difference from others based on recognizable, or identifying, traits. These are not of course mutually exclusive approaches to identity, but they do reflect a shift in which markers and practices of cultural difference are now central to claims of inclusion, exclusion, entitlement, and disenfranchisement. This shift has come about partly as a result of the increased scale of mobility around the world since the mid-twentieth century, bringing previously distant groups into daily contact with each other to an unprecedented degree. *Culture* has become a general term for the practices, symbols, and meanings that different groups refer to in claiming rights of recognition. Such claims of cultural citizenship – the rights and entitlements afforded to groups in recognition of their cultural identity – differ significantly from the abstract notions of citizenship in the public sphere upon which *laïcité* is based. For one thing, claims of cultural recognition derive from *practices* that are often very local in scale, embodied, and have historical and geographical origins beyond the national space in which abstract citizenship is defined. To paraphrase an observation by the anthropologist Talal Asad: the spatialities of many tradition-rooted practices cannot be translated into the abstract space of the nation.[3]

Did culture make Fatima, Leila, and Samira wear the *foulard* in deliberate provocation of *laïcité*? While many people in France might believe so, culture is not a thing with causal powers, but a way of understanding how we experience the world and what that experience means to us. What we can learn here from cultural geography is that culture does not, by itself, explain behavior without an understanding of the different scales and contexts within which people do things. And our objective here is not to present the truth behind the girls' actions, but to convey what a cultural geography of *l'affair foulard* might look like. On one level, a cultural geography of *l'affaire foulard* might simply point out that covering one's head as a sign of religious devotion is certainly not a practice restricted to Islam. There are people all over the world who cover their heads for religious or spiritual reasons. Indeed, many Christian and Jewish acts of devotion involve head covering of some kind. So the issue here is not simply one of majority societies accommodating the cultural practices of minorities. Thus, on another level, whether we view their actions as deliberate and political or not, we must consider the *embodied* nature of cultural practice, the fact that embodied practices do not necessarily translate into abstract spaces like private and public. We must also consider the socio-cultural norms that define French citizenship and understand how those are challenged by claims of cultural recognition by minority groups. These questions of scale and identity, then, are also important issues that cultural geography brings to bear on our understanding of cultural politics.

The example of *l'affair foulard* highlights several aspects of cultural geography emphasized in some of the later sections of this *Reader*. As already mentioned above, Part Seven focuses on questions of *difference*, while Part Five examines issues of *identity*, and the place-based contexts and scales within which different identities are worked out. Part Six considers *mobility*, and Part Seven looks at some of the ways culture has become a *resource* (for example, in making claims of recognition). All of these themes are at play in the case of Fatima, Leila, and Samira. Taken as a whole, however, all eight parts of the book raise questions about the *politics* of culture, exploring the various ways in which the political is never far from the cultural. Certainly that is a central message of *l'affair foulard*. Power is thus a consistently fundamental theme in cultural geography. Culture, as numerous contributors to this *Reader* observe, is laden with power. And vice versa: power is almost always encoded, transmitted, negotiated, and contested through – at least in part – cultural practices. The spaces that shape and are shaped by people's experience of their world are not neutral, but are always socially constituted and thus always subject to political practices.

Cultural Geography as an Academic Discipline

What follows is a brief outline of some of the intellectual lineage of cultural geography. We present this not as a conclusive history of the field – those histories exist in publication already, and several are quite comprehensive[4] – but rather as a means of conveying to the reader the general understanding of the field that the editors have relied upon in assembling the *Reader*. A more detailed discussion of the genealogy of cultural geography will be offered in the introduction to Part Two. In addition, specific moments in the history of cultural geography that pertain to the section themes in this *Reader* are discussed in further detail in the introductions to those parts.

Cultural geography as a broadly understood practice far pre-dates the modern academic sub-field of geography we introduce in this *Reader*. Indeed, for as long as humans have lived in groups and been aware of other human groups that share the earth with them, we can speculate that we have been interested in one another's customs and practices, and the differences in how different people interact with their environments. You might argue that some of the earliest scholarly concerns published in fact constitute the first research and writings on cultural geography. Among these would be the ancient Greek Herodotus's (fifth century BC) account of the ancient Persian empire, the Roman Strabo's (first century AD) seventeen-volume *Geographia* in which he details peoples and places of the Mediterranean world that he visited; and the writings of the Moroccan-born Ibn Battuta (mid-fourteenth century AD) recounting his travels through the Muslim empire at the time, which stretched from North Africa through India to Southeast Asia and China.

Whether this interest in other peoples inhabiting other places arises from some intrinsic curiosity leading to wanderlust, a desire to conquer (or avoid being conquered), or some combination of the two, is certainly open to debate. However, it does point to the long-standing affinity between geography and anthropology, as well as archeology. Geography, anthropology, and archeology all have their roots as established academic disciplines in the latter half of the nineteenth century, a period of intense colonial activity on the part of their main practitioners: Germany, France, and Britain. As discussed in relation to many of the selections in Part Two, the discipline of geography has been central to projects of nation building, colonization, imperialism, fascism, and just about any other political development that has involved the systematization of knowledge about people and places. In order to conquer others it is necessary first to know about them: What are their habits? How do they live? Where are they located? Thus all three disciplines share a keen interest in the material culture of human civilizations, and – at least early on in their development – an extensive effort to collect and catalog those cultural differences deemed significant.

As discussed in greater detail in Part Two, early modern cultural geography, from the nineteenth century onward, was largely descriptive. Attempts to understand and map national character, travel accounts, and descriptions of the relationship between the conditions of physical world and human societies formed the backbone of early modern cultural geography. It is not until the early twentieth century in the Anglophone world that attention began to shift from the descriptive to the analytical. Yet there remained strong currents of cross-fertilization between the different national versions of geography, primarily between America, Germany, Britain, and France. An example is the rise of the so-called Berkeley school, centered on the figure of Carl O. Sauer and his students at the University of California at Berkeley from the 1920s to the 1950s (see p. 96). Sauer's legacy was to draw German developments in landscape studies together with the French style of regional monographs, and bring these into concert with American interests in patterns of material culture. Another example would be the development of a culture–region approach by Welsh and Irish geographers which drew on French regional approaches, but worked them into new kinds of analyses. Thus, while cultural geographers shared many basic interests in the relationships between culture, nature, region, landscape, and so on, they developed these interests in very different ways.

Other key developments in cultural geography have happened more recently, beginning in the 1970s and extending to scholarly debates among cultural geographers and within wider academic ambits today. These questions have focused in part on the concept of culture itself. What exactly *is* culture? Is culture a *thing that acts* upon humans and human societies? Is culture an *independent realm* of power relations, an arena of contention, power, and inequality, as with any other social construction? Or is there in fact "no such thing as culture"?[5] These questions will be examined at length in Part One of the *Reader*. However, it is worth considering in this Introduction that, extending from the proposition that culture is political – a proposition with which the editors wholeheartedly agree – isn't there, *shouldn't* there, be more to the study of cultural geography than descriptions of visible features on the landscape? Indeed, does the study of culture not carry with it the imperative to study, even to participate in, the major struggles of society? As more and more women and non-white geographers have attained positions of influence in academia, the practitioners of cultural geography themselves have changed. In many instances, this has opened the door to issues of race, gender, and sexuality as legitimate topics of study in cultural geography. This transformation is examined in depth in Part Seven of the *Reader*. Finally, within geography, *cultural* geographers have often been at the forefront of questioning the discipline's fascination with the rather dehumanized processes of model building, computer-driven spatial analysis, and quantification. A more humanist approach by some cultural, as well as other, geographers has kept alive long-standing concerns with the lived experience of place, literary and philosophical inquiry into our place on earth, and the role of human creativity in making meaning of the human condition.

These larger questions have tended to recur among the practitioners of cultural geography. Of course, there have also been noticeable temporal trends that have shaped much of the work that scholars generate in cultural geography at any given time. For instance, in the 1980s there was a significant emphasis on issues of representation, reflexivity, and on social movements that were mobilizing around new clusters of cultural identities. In the 1990s came many inquiries into the nature of subjectivity – a concept whose subtle difference with identity owes much to the work of the French philosopher and historian Michel Foucault – as well as many challenges to the centrality of representation in our analyses of social and cultural practice. Examinations of social movements were in some sense replaced by studies of everyday life, and this trend has in some ways accelerated of late. In the 2000s, British cultural geographers in particular have tried to go beyond the focus on representation that was so important in the previous decade. The so-called non-representational geographies emerging from this tradition bring the experiential, the other-than-visual, and the contested divide between human and non-human into the cultural geography conversation.

What these developments point to is not so much an evolutionary course of cultural geography along a particular historical trajectory as an on-going process of questioning the content and boundaries of the field. As the discussions above of Mathias Woo's film and *l'affair foulard* in France demonstrate, cultural geography is a contested field both in terms of the broader politics of culture, place, and identity as well as in terms of scholarly inquiry into these practices. But, as argued above, it is not our intention to write a comprehensive history of cultural geography here, nor is it our intention to impose a vision of where we believe cultural geography is or should be heading. What we will say here, and discuss in greater detail in individual part introductions, is that cultural geography is best thought of in the plural, that it is probably misguided to divide cultural geography neatly into traditional and new halves, and that any attempt to impose a strong coherence on to the field will necessarily leave out a good deal of important work. Cultural *geographies* have provided a rich and dynamic field of study that distills some of the principal intellectual and social concerns of the times. With the growing recognition across the social sciences more broadly that culture is a vital arena of investigation, and with the heightened emphasis on the spatial dimension in disciplines other than geography, the field of cultural geography is front and center in academia today. Rather than highlighting the constant change in the field and underscoring a seemingly endless succession of "new" cultural geographies, the perspective of the editors of this *Reader* tends to side more with that of geographers Nigel Thrift and Sarah Whatmore, who emphasized the remarkable continuities across time in the concerns of practicing cultural geographers, by noting the "almost obsessive *return* to preoccupations that have never really gone away, preoccupations with the stuff of life; with what constitutes personhood; with the legacy of empire; with the sense of belonging."[6]

Some Final Thoughts on this Reader

As we have hoped to illustrate in this brief Introduction, cultural geography is a wide-ranging field of inquiry that touches on many different aspects of our social and personal lives. We have tried to convey this breadth of topics in the *Reader* by including many selections by non-geographers who have nevertheless worked on cultural geography topics, or whose work cultural geographers have built upon. Along with academic geographers, the *Reader* includes contributions by historians, sociologists, anthropologists, cultural studies theorists, as well as several public intellectuals. The academic eclecticism represented in the *Reader* is both a recognition that cultural geography is a diverse and in many ways interdisciplinary field, and a recognition that culture has become an increasingly important variable in the broader social sciences today. Once the purview of anthropology only, the study of culture is now taken up by scholars in fields as diverse as political science (e.g. Seyla Benhabib's *The Claims of Culture*, 2002), sociology (e.g. Sharon Zukin's *The Cultures of Cities*, 1995; see p. 431 of the *Reader*), and history (e.g. the contribution by William Sewell in Part One of the *Reader*, see p. 40).

Because of this so-called *cultural turn* in the social sciences, there is a need to explore the concept of culture that is at the heart of the multiple cultural geographies explored in these pages. As Part One discusses in further detail, there is a tremendous amount of academic ambiguity surrounding the term "culture." This is partly because it is a concept that travels widely across disciplinary and more popular realms of knowledge. We do not set out to resolve the contested interpretations of "the cultural" in this *Reader*, either for the social sciences in general or for cultural geography more specifically. But we do believe that whereas culture can be an awkward, ambiguous, obfuscating, and indeed sloppy concept in much academic writing, this does not justify the temptation to jettison the concept from scholarly analyses. In other words, we do not feel – as some have suggested – that we live in a post-cultural era, or that to invoke culture is to tread the dangerous ground of making excuses for behaviors we would otherwise denounce as contemptuous of human dignity regardless of their context. Instead, we present the detailed discussions of culture in Part One as part of a general conviction that geographers must be clear about how they are conceiving the cultural, and what intellectual communities beyond geography they should be engaged with as part of their approach to culture. Thus, while we do not offer up our own favored definition of culture in the *Reader*, and while we do not wish to police the cultural content of geography, we do believe that geographers should be committed to a theoretically informed deployment of culture in their scholarship.

Beyond this, our editorial goals in the *Reader* have been to present a series of original, accessible, and relevant works by scholars past and present who have worked across the many different topics in cultural geography. Several of our original choices were precluded from reproduction here due to copyright restrictions, costs, or difficulties in adapting the original material to a length or level appropriate for a student readership. Of course, any selection of the most "original, accessible, and relevant" works in cultural geography will entail choices that will certainly be contested by others in the field. Indeed, we have seen a spate of handbooks, readers, textbooks, and the like, all offering a particular angle on cultural geography. With these other resources in mind, we have approached this *Reader* with the goal of representing the diverse breadth of the field in both American and Anglo contexts, in the most accessible and introductory way possible. To that end, each selection has been abridged for length and clarity, and each selection has been provided with an introductory overview that situates the key arguments and contributions of the selection within a broader scholarly context, as well as providing key background information on the ideas and the authors. Selections have been grouped into eight parts, which do not seek to provide a firm-and-fast set of subdivisions for cultural geography; rather, the parts help beginning students of cultural geography to make sense of the wide variety of topics that cultural geographers have made use of in their work. Each section is prefaced by a brief introduction, in which the editors gather the overarching threads running through the individual selections, and provide discussion of the more challenging terms where needed.

Perhaps one of the most daunting tasks in assembling such a reader is making it useful for the many different types of students who will use it in their courses. Both of the editors, Tim Oakes and Patricia Price, were trained in the United States, but as part of the general goal of conveying the breadth of cultural geography we have compiled a volume that represents cultural geography in both its American and British inflections. We discuss some of these differences, and the occasional misunderstandings, in the introduction to Part Two of the *Reader*. For now, it is simply important to point out that while there has been significant cross-fertilization across the Atlantic, some distinct traditions and approaches remain. Regardless of your location – whether you are studying in Britain, the United States, Hong Kong, Canada, Australia, Singapore, New Zealand, or anywhere elsewhere in the world for that matter, we trust this *Reader* will help you gain an appreciation for the varied contributions to cultural geography, for its enduring themes, and – most important – for the windows that cultural geography can open on to your world.

NOTES

1 M. Augé, *Non-places: Introduction to an Anthropology of Supermodernity*, trans. J. Howe (1995).
2 The term "imagined community" comes from Benedict Anderson, *Imagined Communities: Reflections on the Origins and Spread of Nationalism* (1983).
3 T. Asad, *Formations of the Secular* (2003), p. 179.
4 A very partial selection of these includes the first chapter of Peter Jackson *Maps of Meaning: An Introduction to Cultural Geography* (1989); Part I of James Duncan, Nuala Johnson, and Richard Schein (eds.) *A Companion to Cultural Geography* (2004); and the "Introduction" to Nigel Thrift and Sarah Whatmore (eds.) *Cultural Geography: Critical Concepts in the Social Sciences*, Volume I, *Mapping Culture* (2004), pp. 1–17.
5 This phrase is from Don Mitchell "There is no such thing as culture: towards a reconceptualization of the idea of culture in geography," *Transactions of the Institute of British Geographers* 19 (1995): 102–16. See also p. 159 of this *Reader*.
6 Thrift and Whatmore, *Cultural Geography*, p. 4, emphasis in original.

PART ONE

Approaching Culture

Courtesy of Alex Dorfsman

INTRODUCTION TO PART ONE

Noting the many discussions, debates, and "near tantrums" concerning the term culture, Katharyne Mitchell quips: "the lack of specificity as to what culture actually is and means drives many scholars to distraction."[1] Indeed, our task in this part is less to nail down such a slippery concept than to develop an appreciation for its complexity and for some of the most influential approaches to culture within and, in particular, outside of geography. As the various selections that follow indicate, "culture" is one of the most bedeviling and complicated words in the English language. In one of the following selections, for instance, Clifford Geertz notes that a definitive textbook on the subject, Clyde Kluckhohn's *Mirror for Man* (1949), devoted some twenty-seven pages to the term, and settled on no fewer than eleven distinct uses or definitions.

"Culture," as Raymond Williams observes in one of the following selections, has been equated with *civilization* (that is, a process of intellectual, spiritual and aesthetic advancement), has meant a *whole way of life* (an approach initiated by German philosopher Johann Gottfried von Herder), and has also been associated with intellectual and artistic works and practices (the kind of culture that appears in "Arts" sections of major newspapers like *The New York Times* or *The Guardian*). But, there is even more to culture than this. The term enjoys a kind of sticky plasticity that has allowed it to adhere to anything that has a symbolic dimension, or anything that might involve a distinct way of doing things, or simply a unique way of looking, acting, or presenting oneself. Stuart Hall has noted this plasticity, observing that just about any aspect of social life these days can have a "culture" attached to it: "the culture of corporate enterprise, the culture of the workplace, the growth of an enterprise culture . . . , the culture of masculinity, the cultures of motherhood and the family, a culture of home decoration and shopping, a culture of deregulation, even a culture of the fit, and – even more disturbingly – a culture of the thin body."[2]

One wonders whether culture means anything at all! At the least, it should be clear that culture is less a *thing* than a highly malleable *category* of social relations and practices. But in fact there is significant debate among scholars not only regarding the definition of and approach to culture, but also whether culture is in any way analytically useful. Don Mitchell's suggestion along these lines ("there's no such thing as culture") has many echoes within the broader field of cultural studies.[3] Many scholars regard the term with outright suspicion, wrapping it in scare quotes ("culture"), or avoiding it altogether. Some might even consider themselves "*post*culturalists." There are of course several reasons for this. For instance, the plasticity of the term makes it analytically sloppy. It can obfuscate more than it reveals, and there may be more precise ways of explaining practice, belief, ritual, behavior, and so on, than attributing them to something as vague and ill defined as culture.

Along these lines, culture can be a neutralizing or naturalizing mask for group differences that have more disturbing social determinations. Saying that someone does something because of "his culture" can have the effect of diverting analytical attention away from the broader social structures that may play a role in conditioning people's behavioral decisions or choices. In the example of France's headscarf issue raised in the *Reader*'s Introduction, for example, it was suggested that to attribute the behavior of Fatima, Leila, and Samira simply to culture missed the inherently political nature of their actions, which were structured by specific place-based social relations. In general terms, such structures may

include, for example, institutionalized racism or various kinds of ethnic, gender, and/or sexual discriminations. This point is suggested by Lila Abu-Lughod, who, in a selection later in this selection, argues
that culture carries too much colonial and imperial baggage to be used as a neutral category of social
inquiry. For her, scholars should be doing what they can to dismantle culture as an explanatory category
of human behavior or practice.

For the most part, however, most scholars are not willing to jettison the concept of culture altogether.
Hall, for instance, argues that the point is not that culture has expanded into every nook and cranny of
social life and therefore lost its ability to identify a distinct slice of social life, but rather that "every social
practice depends on and relates to meaning," and that "culture is one of the constitutive conditions of
existence of that practice, that every social practice has a cultural dimension. Not that there is nothing
but discourse, but that every social practice *has a discursive character*."[4] By "discursive character," Hall
is referencing the idea that social practices are not the automatic result of, say, economic "laws", but
of conscious decisions in which *meaning* also plays a part. James and Nancy Duncan make a related
appeal for the continuing relevance of culture.[5] Noting that culture is inherently *unstable* and that there is
no such thing as a *pure* culture, they observe that the norm is now hybrid cultures, borderland cultures,
blurred cultures, shifting cultures. This is a point that echoes arguments made by Akhil Gupta and James
Ferguson, in one of the following selections. Gupta and Ferguson argue that while culture was traditionally associated with fixed and bounded regions or territories, it is now much more useful to think of
culture in terms of spaces of mixture, borderland, and even mobility (see also selections by Cresswell
and Clifford, pp. 325 and 316). For the Duncans, then, the analytical power of culture lies in determining how relatively stable cultural formations come about in the first place: "If change, process, fluidity,
heterogeneity, and transformation are our basic starting ontological assumptions then what becomes
remarkable are those things that are relatively stable and coherent such as organizations and institutions that become entrenched over time and which generally hold their shape and content through time
and across space."[6]

Overall, and as will also be clear from the selections in Part Two of the *Reader*, we can note that
culture has traveled a path from describing the superficial outcome of more basic social (or even environmental) determinants to serving as an increasingly important variable *explaining* behavior. The emergence of culture as an explanatory variable came about because culture represented an alternative to
simplistic approaches which saw humans responding to external influences (such as "environment," or
"relations of production") in predictable, almost mechanical, ways. Such approaches, in other words,
failed to recognize Hall's "discursive character," mentioned above. Over time, Hall continues, culture
has come to be seen as "a constitutive condition of existence of social life, rather than a dependent
variable."[7] As a result, culture has become increasingly central to explanation throughout the social
sciences, and, with this centrality, definitions and approaches to the term have become increasingly
diffuse and confusing.

The selections that follow do not offer a comprehensive accounting of this emergence of culture as
an explanatory variable. But they do offer a useful set of distinctive approaches to culture, all of which
have and continue to play an important role in cultural geography. Given that geography (and social science in broader terms) has experienced something of a *cultural turn* since the 1980s, it is important to
understand that not everyone in the discipline understands culture in the same way. This causes a great
deal of perhaps unnecessary debate and confusion. As Raymond Williams observes in his selection,
the important point about culture is not that it escapes a single definition, but that it captures an ongoing conviction among human scientists that understanding behavior necessarily involves accounting for
both "the material" and "the symbolic." These two dimensions are held in tension throughout social science inquiry. Human behavior always has its discursive character – meanings which are typically marked
by signs and symbols – *and* its more determined character – that is, our responses to the constraints
of our material lives: the basic needs of production, reproduction, and consumption. Debates over the
meaning and usefulness of culture are largely debates over how best to resolve the tension between
these material and symbolic dimensions of human behavior.

This tension has its roots, perhaps, in the fundamental problem of Western metaphysics: how do we grasp or comprehend the world? How can we be sure our representations of the world are accurate? How do we test the reliability of knowledge given the role that subjectivity, perception, and representation necessarily play in the formation of that knowledge? Philosophers from Aristotle to Descartes to Kant and Heidegger have of course wrestled famously with this problem. Restated in terms of culture, the problem asks how we can recognize the ways we perceive, experience, and represent the world, symbolically and with meaning, without losing some sense of the world's "external and objective reality". Given the fact that our understanding of how the world *actually is* depends upon *how we know* the world in the first place, knowledge can become frustratingly circular, reflecting as much the subjectivity of knowing as the objectivity of what is known. Debates over culture are essentially debates generated by this frustration. Raymond Williams, then, suggests that we approach culture not simply as the symbolic or subjective side of this tension, but a way of making sense of the tension itself. That seems to be as good a starting point as any for an exploration of the central concept in cultural geography.

NOTES

1 Page 667 in K. Mitchell, "What's culture got to do with it?" *Urban Geography* 20, 7 (1999): 667–677.
2 S. Hall, "The Centrality of Culture" in K. Thompson (ed.) *Media and Cultural Regulation* Sage (1997), pp. 207–238. See also pp. XX of this *Reader*.
3 Don Mitchell, "There's no such thing as culture: towards a reconceptualization of the idea of culture in geography," *Transactions of the Institute of British Geographers* 20, 1 (1995): 102–116. See also pp. XX of this *Reader*. For similar discussions outside of geography, see R. Brightman, "Forget culture: replacement, transcendence, relexification," *Cultural Anthropology* 10 (1995): 509–546.
4 Hall, "Centrality of culture," 225–226.
5 J. Duncan and N. Duncan, "Culture unbound," *Environment and Planning A* 36 (2004): 391–403.
6 *Ibid.*, p. 397.
7 Hall, "Centrality of culture," p. 220.

"Culture"

from *Keywords: A Vocabulary of Culture and Society*, revised edition (1983)

Raymond Williams

Editors' introduction

In the following brief etymology of culture, Raymond Williams explores the lineage of "one of the two or three most complicated words in the English language." But his account should not be mistaken for the kind of entry one might expect to find in an encyclopedia or dictionary. *Keywords* is meant to be an inquiry into the shared meanings that form the basis of English culture and society. For years, scholars have turned to *Keywords* not just for definitions or historical summaries of important English words and concepts, but more for clues to the *relationships between those words and broader patterns of social and cultural change*. Thus, the dominant impression one gets from the book as a whole is the *dynamic* quality of meaning, that meanings change in relation to social changes that are also occurring. Throughout *Keywords*, Williams insists that language does not simply reflect social change and historical process, but that these changes and processes themselves occur *within* language. In the following account, for instance, Williams traces the ways the emergence of culture as an independent noun helped frame nineteenth century intellectual and social movements such as Romanticism. In such movements, problems, meanings, and relationships are worked out in the confusions and ambiguities of language itself. Culture and society are in a continuous process of change, and that change occurs most fundamentally at the level of language. Nor is change a straightforward process of the old giving way to the new. Old meanings linger in language, just as they do in other aspects of our everyday lives.

What makes culture so complicated is that – like language more generally – such a great range of meanings are simultaneously wrapped up in the term. Some of these are quite old and continue to linger in the use of the term, while others are quite new. Williams ultimately identifies three broad uses of "culture". First, as a noun describing a general process of intellectual, spiritual, and aesthetic development since the eighteenth century (similar to the term "civilization"); second, a noun indicating a particular way of life (what we might call an "anthropological" sense of the term); and third, a noun describing works and practices of intellectual and especially artistic activity (that is, a more "elite" sense of the term).

In addition to tracing the etymology of "culture", Williams also offers a way of thinking through the complexities of the term without surrendering to the desire for a final, simple and reliable definition that will resolve ambiguity. This is an extremely important, yet subtle, message. While noting that it is important for any discipline – such as anthropology or geography – to clarify its terminology, Williams argues that "in general it is the range and overlap of meanings that is significant." The confusion of meanings inherent in culture, in other words, offers insight into the complex *relationship between our material and symbolic worlds*. Indeed, this relationship between the material production and symbolic meanings of culture formed the basis of Williams's approach to culture as the outcome of the meanings we produce out of our ordinary, daily lives.

Raymond Williams (1921–1987) was Professor of Drama at Jesus College, Cambridge, and was a wide-ranging literary and media critic, political analyst, dramatist, novelist, and social historian. The author of over twenty books, Williams is perhaps best known for *Culture and Society* (1958), *The Long Revolution* (1961), and *Marxism and Literature* (1977). Perhaps his most geographical work of non-fiction was *The Country and the City* (1973), but Williams's short stories and novels – such as *Border Country* (1960) – are also rich in geographical themes. In these and many other works, Williams explored the social history of the ideas, practices, and meanings that together make up culture. His most well known contribution to cultural theory was perhaps the concept of *structure of feeling*, which he defined as "a particular quality of social experience and relationship" that gave a certain historical period its distinctiveness (see *Marxism and Literature*, p. 131). While Williams argued that there were definite social and material structures that limited the range of this "experience and relationship," he sought to focus attention on experience itself as an often overlooked variable in social analysis. He countered the crude Marxist view that culture was determined by the economic base of society by showing how culture was an active part of a broader process of social change, rather than the mere expression or illustration of that change. Culture itself was, therefore, a terrain of social struggle, a field in which social relations worked themselves out. Culture was also decidedly "ordinary" in this approach – part of our everyday lives – rather than merely the elite realm of high art and literature.

Raymond Williams's approach to culture – typically referred to as *cultural materialism* – was central to the development of cultural studies, beginning in the 1970s, and also relates to cultural geography in several ways. By emphasizing the relationship between material production and the symbolic systems of signification, Williams provided an approach to culture that helped radicalize cultural geography in the early 1980s. Cultural materialism, for instance, forms the conceptual centerpiece of Peter Jackson's critique of cultural geography in *Maps of Meaning* (1989) and helps shape Don Mitchell's approach to culture in *Cultural Geography: A Critical Introduction* (2000). It also helped inspire Denis Cosgrove's project of linking cultural landscapes to modes of production in *Social Formation and Symbolic Landscape* (1985). A concise overview relating Williams's work to cultural geography can be found in Longhurst's "Raymond Williams and Local Cultures" (*Environment and Planning A* 23, 1991: 229–238). Conceiving culture as a terrain of struggle has helped inform cultural geography as a field examining the ways material relations get worked out in place-based cultural politics. Williams's approach, in other words, would insist that an understanding of people's place-based experiences – a structure of feeling – is crucial to understanding processes of social change occurring at broader scales of space and over longer periods of time. Such understanding has come to shape the research agendas of many contemporary cultural geographers.

Culture is one of the two or three most complicated words in the English language. This is so partly because of its intricate historical development, in several European languages, but mainly because it has now come to be used for important concepts in several distinct intellectual disciplines and in several distinct and incompatible systems of thought.

The immediate forerunner is *cultura* [Latin], from the Latin root word *colere*. *Colere* had a range of meanings: inhabit, cultivate, protect, honour with worship. Some of these meanings eventually separated, though still with occasional overlapping, in the derived nouns. Thus 'inhabit' developed through *colonus* [Latin], to *colony*. 'Honour with worship' developed through *cultus* [Latin], to *cult*. *Cultura* took on the main meaning of cultivation or tending, including, as in Cicero, *cultura animi*, though with subsidiary medieval meanings of honour and worship (cf. in English **culture** as 'worship' in Caxton (1483)). The French forms of *cultura* were *couture* [Old French], which has since developed its own specialized meaning, and later *culture*, which by the early fifteenth century had passed into English. The primary meaning was then in husbandry, the tending of natural growth.

Culture in all its early uses was a noun of process: the tending *of* something, basically crops or animals. The subsidiary *coulter* – ploughshare, had travelled by a different linguistic route, from *culter* [Latin], – ploughshare, *culter* [Old English], to the variant English spellings *culter, colter, coulter* and as late as the early seventeenth century **culture** (Webster, *Duchess of Malfi*, III, ii: 'hot burning cultures'). This provided a further basis for the important next stage of meaning, by metaphor. From the early sixteenth century the tending of natural growth was extended to a process of human development, and this, alongside the original meaning in husbandry, was the main sense until the late eighteenth and early nineteenth centuries. Thus More: 'to the culture and profit of their minds', Bacon 'the culture and manurance of minds' (1605); Hobbes: 'a culture of their minds' (1651); Johnson: 'she neglected the culture of her understanding' (1759). At various points in this development two crucial changes occurred: first, a degree of habituation to the metaphor, which made the sense of human tending direct; second, an extension of particular processes to a general process, which the word could abstractly carry. It is of course from the latter development that the independent noun **culture** began its complicated modern history, but the process of change is so intricate, and the latencies of meaning are at times so close, that it is not possible to give any definite date. **Culture** as an independent noun, an abstract process or the product of such a process, is not important before the late eighteenth century and is not common before mid nineteenth century. But the early stages of this development were not sudden. There is an interesting use in Milton, in the second (revised) edition of *The Readie and Easie Way to Establish a Free Commonwealth* (1660): 'spread much more Knowledg and Civility, yea, Religion, through all parts of the Land, by communicating the natural heat of Government and Culture more distributively to all extreme parts, which now lie num and neglected'. Here the metaphorical sense ('natural heat') still appears to be present, and *civility* is still written where in the nineteenth century we would normally expect **culture**. Yet we can also read 'government and culture' in a quite modern sense. Milton, from the tenor of his whole argument, is writing about a general social process, and this is a definite stage of development. In eighteenth

century England this general process acquired definite class associations though **cultivation** and **cultivated** were more commonly used for this. But there is a letter of 1730 (Bishop of Killala, to Mrs Clayton; *cit* Plumb, *England in the Eighteenth Century*) which has this clear sense: 'it has not been customary for persons of either birth or culture to breed up their children to the Church'. Akenside (*Pleasures of Imagination*, 1744) wrote: '. . . nor purple state nor culture can bestow'. Wordsworth wrote 'where grace of culture hath been utterly unknown' (1805), and Jane Austen (*Emma*, 1816) 'every advantage of discipline and culture'.

It is thus clear that culture was developing in English towards some of its modern senses before the decisive effects of a new social and intellectual movement. But to follow the development through this movement, in the late eighteenth and early nineteenth centuries, we have to look also at developments in other languages and especially in German.

In French, until the eighteenth century, **culture** was always accompanied by a grammatical form indicating the matter being cultivated, as in the English usage already noted. Its occasional use as an independent noun dates from the mid eighteenth century, rather later than similar occasional uses in English. The independent noun *civilization* also emerged in the mid eighteenth century; its relationship to **culture** has since been very complicated. There was at this point an important development in German: the word was borrowed from French, spelled first (late eighteenth century) *Cultur* and from the nineteenth century *Kultur*. Its main use was still as a synonym for *civilization*: first in the abstract sense of a general process of becoming 'civilized' or 'cultivated'; second, in the sense which had already been established for *civilization* by the historians of the Enlightenment, in the popular eighteenth century form of the universal histories, as a description of the secular process of human development. There was then a decisive change of use in Herder. In his unfinished *Ideas on the Philosophy of the History of Mankind* (1784–91) he wrote of *Cultur*: 'nothing is more indeterminate than this word, and nothing more deceptive than its application to all nations and periods'. He attacked the assumption of the universal histories that 'civilization' or 'culture' – the historical self-development of humanity – was what we would now call a unilinear process, leading to the high and dominant point of eighteenth century European culture. Indeed he attacked what he called European subjugation and domination of the four quarters of the globe, and wrote:

> Men of all the quarters of the globe, who have perished over the ages, you have not lived solely to manure the earth with your ashes, so that at the end of time your posterity should be made happy by European culture. The very thought of a superior European culture is a blatant insult to the majesty of Nature.

It is then necessary, he argued, in a decisive innovation, to speak of 'cultures' in the plural: the specific and variable cultures of different nations and periods, but also the specific and variable cultures of social and economic groups within a nation. This sense was widely developed, in the Romantic movement, as an alternative to the orthodox and dominant '*civilization*'. It was first used to emphasize national and traditional cultures, including the new concept of **folk-culture**. It was later used to attack what was seen as the mechanical character of the new civilization then emerging: both for its abstract rationalism and for the 'inhumanity' of current industrial development. It was used to distinguish between 'human' and 'material' development. Politically, as so often in this period, it veered between radicalism and reaction and very often, in the confusion of major social change, fused elements of both. (It should also be noted, though it adds to the real complication, that the same kind of distinction, especially between 'material' and 'spiritual' development, was made by von Humboldt and others, until as late as 1900, with a reversal of the terms, **culture** being material and *civilization* spiritual. In general, however, the opposite distinction was dominant.)

On the other hand, from the 1840s in Germany, *Kultur* was being used in very much the sense in which *civilization* had been used in eighteenth century universal histories. The decisive innovation is G.F. Klemm's *Allgemeine Kulturgeschichte der Menschheit* – 'General Cultural History of Mankind' (1843–52) – which traced human development from savagery through domestication to freedom. Although the American anthropologist Morgan, tracing comparable stages, used 'Ancient *Society*', with a culmination in *Civilization*, Klemm's sense was

sustained, and was directly followed in English by Tylor in *Primitive Culture* (1870). It is along this line of reference that the dominant sense in modern social sciences has to be traced.

The complexity of the modern development of the word, and of its modern usage, can then be appreciated. We can easily distinguish the sense which depends on a literal continuity of physical process as now in 'sugar-beet culture' or, in the specialized physical application in bacteriology since the 1880s, 'germ culture'. But once we go beyond the physical reference, we have to recognize three broad active categories of usage. The sources of two of these we have already discussed: (i) the independent and abstract noun which describes a general process of intellectual, spiritual and aesthetic development, from the eighteenth century; (ii) the independent noun, whether used generally or specifically, which indicates a particular way of life, whether of a people, a period, a group, or humanity in general, from Herder and Klemm. But we have also to recognize (iii) the independent and abstract noun which describes the works and practices of intellectual and especially artistic activity. This seems often now the most widespread use: **culture** is music, literature, painting and sculpture, theatre and film. A **Ministry of Culture** refers to these specific activities, sometimes with the addition of philosophy, scholarship, history. This use, (iii), is in fact relatively late. It is difficult to date precisely because it is in origin an applied form of sense (i): the idea of a general process of intellectual, spiritual and aesthetic development was applied and effectively transferred to the works and practices which represent and sustain it. But it also developed from the earlier sense of process; cf. 'progressive culture of fine arts', Millar, *Historical View of the English Government*, IV, 314 (1812). In English (i) and (iii) are still close; at times, for internal reasons, they are indistinguishable, as in Arnold, *Culture and Anarchy* (1867); while sense (ii) was decisively introduced into English by Tylor, *Primitive Culture* (1870), following Klemm. The decisive development of sense (iii) in English was in the late nineteenth and early twentieth centuries.

Faced by this complex and still active history of the word, it is easy to react by selecting one 'true' or 'proper' or 'scientific' sense and dismissing other senses as loose or confused. There is evidence of this reaction even in the excellent study by

Kroeber and Kluckhohn, *Culture: a Critical Review of Concepts and Definitions*, where usage in North American anthropology is in effect taken as a norm. It is clear that, within a discipline, conceptual usage has to be clarified. But in general it is the range and overlap of meanings that is significant. The complex of senses indicates a complex argument about the relations between general human development and a particular way of life, and between both and the works and practices of art and intelligence. It is especially interesting that in archaeology and in *cultural anthropology* the reference to **culture** or **a culture** is primarily to *material* production, while in history and *cultural studies* the reference is primarily to *signifying* or *symbolic* systems. This often confuses but even more often conceals the central question of the relations between 'material' and 'symbolic' production, which, in some recent argument – cf. my own *Culture* – have always to be related rather than contrasted. Within this complex argument there are fundamentally opposed as well as effectively overlapping positions; there are also, understandably, many unresolved questions and confused answers. But these arguments and questions cannot be resolved by reducing the complexity of actual usage. This point is relevant also to uses of forms of the word in languages other than English, where there is considerable variation. The anthropological use is common in the German, Scandinavian and Slavonic language groups, but it is distinctly subordinate to the senses of art and learning, or of a general process of human development, in Italian and French. Between languages, as within a language, the range and complexity of sense and reference indicate both difference of intellectual position and some blurring or overlapping. These variations, of whatever kind, necessarily involve alternative views of the activities, relationships and processes which this complex word indicates. The complexity, that is to say, is not finally in the word but in the problems which its variations of use significantly indicate.

It is necessary to look also at some associated and derived words. Cultivation and cultivated went through the same metaphorical extension from a physical to a social or educational sense in the seventeenth century, and were especially significant words in the eighteenth century. Coleridge, making a classical early nineteenth century distinction between civilization and culture, wrote (1830): 'the permanent distinction, and occasional contrast, between cultivation and civilization'. The noun in this sense has effectively disappeared but the adjective is still quite common, especially in relation to manners and tastes. The important adjective **cultural** appears to date from the 1870s; it became common by the 1890s. The word is only available, in its modern sense, when the independent noun, in the artistic and intellectual or anthropological senses, has become familiar. Hostility to the word **culture** in English appears to date from the controversy around Arnold's views. It gathered force in the late nineteenth and early twentieth centuries, in association with a comparable hostility to *aesthete* and aesthetic. Its association with class distinction produced the mime-word *culchah*. There was also an area of hostility associated with anti-German feeling, during and after the 1914–18 War, in relation to propaganda about *Kultur*. The central area of hostility has lasted, and one element of it has been emphasized by the recent American phrase **culture-vulture**. It is significant that virtually all the hostility (with the sole exception of the temporary anti-German association) has been connected with uses involving claims to superior knowledge (cf. the noun intellectual), refinement (*culchah*) and distinctions between 'high' art (**culture**) and popular art and entertainment. It thus records a real social history and a very difficult and confused phase of social and cultural development. It is interesting that the steadily extending social and anthropological use of **culture** and **cultural** and such formations as **sub-culture** (the culture of a distinguishable smaller group) has, except in certain areas (notably popular entertainment), either bypassed or effectively diminished the hostility and its associated unease and embarrassment. The recent use of *culturalism*, to indicate a methodological contrast with *structuralism* in social analysis, retains many of the earlier difficulties, and does not always bypass the hostility.

"Community"

from *The Making of the English Working Class* (1963)

E.P. Thompson

Editors' introduction

This selection from E.P. Thompson's most influential book begins with the curious observation that "the passing of old England" evades analysis. The "passing" to which Thompson refers here is the transition from England as a predominantly agrarian to an industrial society: a transition about which much was already known when Thompson wrote *The Making of the English Working Class*. Why, then, does he argue that it continues to "evade analysis"? A careful reading of the following selection will perhaps suggest that such a transition cannot be fully understood without an appreciation for *the daily experiences* of the people who lived that transition. Ultimately, Thompson's work is an appeal for the inclusion of experience – alongside more quantifiable variables, such as levels of production and consumption – in our evaluation of the industrial revolution. And Thompson's point is that while standard quantifiable variables might indicate that industrialization brought about an improved quality of life for most people, an account of those people's experiences will yield a much more complicated and ambivalent picture of the daily costs of such improvements.

Published in 1963, *The Making of the English Working Class* set out, as Thompson famously wrote in his preface, "to rescue the poor stockinger, the Luddite cropper, the 'obsolete' hand-loom weaver, the 'Utopian' artisan, and even the deluded follower of Joanna Southcott,[1] from the enormous condescension of posterity." Thompson's is a history from below, focusing on the daily lives, beliefs, attitudes and practices of working people who experienced industrialization. But it is important to understand that rescuing the worker from the "condescension of posterity" involved much more than merely illuminating her daily life experiences. As a Marxist, Thompson sought to also rescue the working class from the narrow economic determinism that so often resulted from the historical materialist method. Thus, Thompson also wrote a history of class formation in which the working class were their own agents in bringing themselves into being. It was, in this sense, that the working class was a product of "conscious working class endeavor." And while material relations of production provided the structure of immiseration within which the working class formed, it was the everyday practices and experiences of the workers that really brought the working class into a state of self-awareness. There were great losses that working people experienced on account of industrialization. And to give us a sense of that experience, Thompson cites not the statistics of poverty or disease, but Blake's poetry. Clearly there was something of the experience of workers that poetry had captured better than numbers. Thompson set out to bring experience into the analysis and found that amid the experience of immiseration the working people did build their own community and their own culture.

In this selection, Thompson focuses on those practices and experiences by which workers built a sense of collective community. Values such as self-discipline, "decency and regularity," and the mutual aid institutions of trade unions and friendly societies promoting the "code of the self-respecting artisan," all played a role in building this community. Part of Thompson's point here is to garner some respect for the working class, to show that they too had ideals of sobriety and decency and regularity which were thought to only be the terrain of the upper classes. Already disciplined by new work regimes, workers disciplined themselves as part of their sense of collective identity.

There are two messages here about culture. One is that working class culture didn't simply reflect the material base of industrialism, but was actively crafted by workers in the ways they worked out the constraints of their daily lives. The other message is that the working class had a distinctive way of life that can be called a culture of their own; they created it. Both messages feature the agency of the worker.

Edward Palmer Thompson (1924–1993) was an English historian, journalist, and essayist. Other well known works included his biography *William Morris* (1955), *The Moral Economy of the English Crowd in the Eighteenth Century* (1971), and the posthumously published *Witness against the Beast: William Blake and the Moral Law* (1993). After leaving the Communist Party in disgust over the Soviet invasion of Hungary in 1956, Thompson started the *New Reasoner*, an important journal of what was to become known as the *New Left*. (It later joined with another journal to become the *New Left Review*.) Along with Raymond Williams, Stuart Hall, and Richard Hoggart, Thompson's work is also credited with helping to initiate the field of cultural studies in Britain (see also Hall, p. 264).

While Thompson did not devote the attention of Raymond Williams to theorizing culture, his work is important to cultural geography in several ways. First, he provided a model whereby everyday experience could be considered "culture." In ascribing a culture to the working class, Thompson was implicitly arguing for a view of culture as a "way of life" that was not restricted to the high art and literature of the upper classes. Second, Thompson's focus on working class culture meant that one could analytically link daily experience to much broader historical processes. For cultural geographers, this meant relating place-based experience to social processes operating at broader scales. Geographers who have explicitly engaged Thompson's work include Paul Glennie and Nigel Thrift ("Reworking E.P. Thompson's 'Time, work–discipline and industrial capitalism'", *Time and Society* 5, 3, 1996) and Derek Gregory ("Human agency and human geography," *Transactions of the Institute of British Geographers* n.s. 6, 1981).

NOTE

1   Joanna Southcott was a self-described religious prophetess of early nineteenth century England.

COMMUNITY

ii. The rituals of mutuality

Again and again the "passing of old England" evades analysis. We may see the lines of change more clearly if we recall that the Industrial Revolution was not a settled social context but a phase of transition between two ways of life. And we must see, not one "typical" community (Middleton or Pudsey), but many different communities, coexisting with each other. In south-east Lancashire alone there were to be found, within a few miles of each other, the cosmopolitan city of Manchester upon which migrants converged from every point in the kingdom; pit-villages (like the Duke of Bridgewater's collieries) emerging from semi-feudalism; paternal model villages (like Turton); new mill-towns (like Bolton); and older weaving hamlets. In all of these communities there were a number of converging influences at work, all making towards discipline and the growth in working-class consciousness.

The working-class community of' the early nineteenth century was the product, neither of paternalism nor of Methodism, but in a high degree of conscious working-class endeavour. In Manchester or Newcastle the traditions of the trade union and the friendly society, with their emphasis upon self-discipline and community purpose, reach far back into the eighteenth century. Rules which survive of the Manchester smallware weavers in the 1750s show already meticulous attention to procedure and to institutional etiquette. The committee members must sit in a certain order. The doors must be kept locked. There are careful regulations for the safe-keeping of the "box." Members are reminded that "Intemperance, Animosity and Profaneness are the Pest and Vermin that gnaw out the very Vitals of all Society."

> If we consider this Society, not as a Company of Men met to regale themselves with Ale and Tobacco, and talk indifferently on all Subjects: but rather as a Society sitting to Protect the Rights and Privileges of a Trade by which some hundreds of People . . . subsist . . . how awkward does it look to see its Members jumbled promiscuously one amongst another, talking indifferently on all Subjects . . .

"Decency and Regularity" are the watchwords; it is even hoped that when "Gentlemen and Magistrates" observe such order "they will rather revere than punish such a Society."

This represents the code of the self-respecting artisan, although the hope that such sobriety would win the favour of the authorities was to be largely disappointed. It was in a similar school that such men as Hardy and Place received their education in London. But as the Industrial Revolution advanced, it was this code (sometimes in the form of model rules) which was extended to ever-wider sections of working people. Small tradesmen, artisans, labourers – all sought to insure themselves against sickness, unemployment, or funeral expenses, through membership of "box clubs" or friendly societies. But the discipline essential for the safe-keeping of funds, the orderly conduct of meetings and the determination of disputed cases, involved an effort of self-rule as great as the new disciplines of work. An examination of rules and orders of friendly societies in existence in Newcastle and district during the Napoleonic Wars gives us a list of fines and penalties more exacting than those of a Bolton cotton-master. A General Society imposed fines for any member "reflecting upon" another member in receipt of sick money, being drunk on the Sabbath, striking another, "calling one another bye-names," coming into the clubroom in liquor, taking God's name in

vain. The Brotherhood of Maltsters added fines for drunkenness at *any* time, or for failure to attend the funerals of brothers or of their wives. The Glass-Makers (founded as early as 1755) added fines for failure in attending meetings, or for those who refused to take their turn in the rota of officers; for failing to keep silence when ordered, speaking together, answering back the steward, betting in the club, or (a common rule) disclosing secrets outside the society. Further,

> Persons that are infamous, of ill character, quarrelsome, or disorderly, shall not be admitted into this society. No Pitman, Collier, Sinker, or Waterman to be admitted . . .

The Watermen, not to be outdone, added a rule excluding from benefits any brother sick through "any illness got by lying with an unclean woman, or is clap't or pox'd." Brothers were to be fined for ridiculing or provoking each other to passion.

The Unanimous Society was to cut off benefits if any member in receipt of sick money was found "in ale-houses, gaming, or drunk." To maintain its unanimity there were fines for members proposing "discourse or dispute upon political or ecclesiastical matters, or government and governors." The Friendly Society of All Trades had a rule similar to "huffing" in draughts; there was a fine "if any member has an opportunity of fining his brother, and does not." The Cordwainers added fines for calling for drink or tobacco without leave of the stewards. The House-Carpenters and Joiners added a prohibition of "disloyal sentiments" or "political songs."

It is possible that some of these rules, such as the prohibition of political discourse and songs, should be taken with a pinch of salt. While some of these societies were select sick-clubs of as few as twenty or thirty artisans, meeting at an inn, others were probably covers for trade union activity; while at Newcastle, as at Sheffield, it is possible that after the Two Acts the formation of friendly societies was used as a cover for Jacobin organisation. (A "company" friendly society, in 1816, bore testimony to "the loyal, patriotic, and peaceable regulations" of many Newcastle societies, but complained that these regulations were often insufficient to prevent "warm debate and violent language.") The authorities were deeply suspicious of the societies during the war years, and one of the purposes of the rules was to secure registration with the local magistrates. But anyone familiar with procedure and etiquette in some trade unions and working-men's clubs today will recognise the origin of still-extant practices in several of the rules. Taken together, they indicate an attainment of self-discipline and a diffusion of experience of a truly impressive order.

Estimates of friendly society membership suggest 648,000 in 1793, 704,350 in 1803, 925,429 in 1815. Although registration with the magistrates, under the first Friendly Society Act of 1793, made possible the protection of funds at law in the event of defaulting officers, a large but unknown number of clubs failed to register, either through hostility to the authorities, parochial inertia, or through a deep secretiveness which, Dr. Holland found, was still strong enough to baffle his enquiries in Sheffield in the early 1840s. Nearly all societies before 1815 bore a strictly local and self-governing character, and they combined the functions of sick insurance with convivial club nights and annual "outings" or feasts. An observer in 1805 witnessed near Matlock –

> about fifty women preceded by a solitary fiddler playing a merry tune. This was a female benefit society, who had been to hear a sermon at Eyam, and were going to dine together, a luxury which our female benefit society at Sheffield does not indulge in, having tea only, and generally singing, dancing, smoking, and negus.

Few of the members of friendly societies had a higher social status than that of clerks or small tradesmen; most were artisans. The fact that each brother had funds deposited in the society made for stability in membership and watchful participation in self-government. They had almost no middle-class membership and, while some employers looked upon them favourably, their actual conduct left little room for paternalist control. Failures owing to actuarial inexperience were common; defaulting officers not infrequent. Diffused through every part of the country, they were (often heart-breaking) schools of experience.

In the very secretiveness of the friendly society, and in its opaqueness under upper-class scrutiny, we have authentic evidence of the growth of independent working-class culture and institutions. This was the sub-culture out of which the less stable trade unions grew, and in which trade union officers were trained.[2] Union rules, in many cases, were more elaborate versions of the same code of conduct as the sick club. Sometimes, as in the case of the Woolcombers, this was supplemented by the procedures of secret masonic orders:

> Strangers, the design of all our Lodges is love and unity,
> With self-protection founded on the laws of equity,
> And when you have our mystic rights gone through,
> Our secrets all will be disclosed to you.

After the 1790s, under the impact of the Jacobin agitation, the preambles to friendly society rules assume a new resonance; one of the strangest consequences of the language of "social man" of the philosophical Enlightenment is its reproduction in the rules of obscure clubs meeting in the taverns or "hush-shops" of industrial England. On Tyneside "Social" and "Philanthropic" societies expressed their aspirations in terms which ranged from throw-away phrases – "a sure, lasting, and loving society," "to promote friendship and true Christian charity," "man was not born for himself alone" – to more thundering philosophical affirmations:

> Man, by the construction of his body, and the disposition of his mind, is a creature formed for society . . .

> We, the members of this society, taking it into our serious consideration, that man is formed a social being in continual need of mutual assistance and support; and having interwoven in our Constitutions those humane and sympathetic affections which we always feel at the distress of any of our fellow creatures . . .

The friendly societies, found in so many diverse communities, were a unifying cultural influence. Although for financial and legal reasons they were slow to federate themselves, they facilitated regional and national trade union federation. Their language of "social man" also made towards the growth in working-class consciousness. It joined the language of Christian charity and the slumbering imagery of "brotherhood" in the Methodist (and Moravian) tradition with the social affirmations of Owenite socialism. Many early Owenite societies and stores prefaced their rules with the line from Isaiah (XLI, 6): "They helped every one his neighbour; and every one said to his brother, be of good courage." By the 1830s there were in circulation a score of friendly society or trade union hymns and songs which elaborated this theme.

Mr. Raymond Williams has suggested that "the crucial distinguishing element in English life since the Industrial Revolution is . . . between alternative ideas of the nature of social relationship." As contrasted with middle-class ideas of individualism or (at their best) of service, "what is properly meant by 'working-class culture' . . . is the basic collective idea, and the institutions, manners, habits of thought, and intentions which proceed from this." Friendly societies did not "proceed from" an idea; both the ideas and the institutions arose in response to certain common experiences. But the distinction is important. In the simple cellular structure of the friendly society, with its workaday ethos of mutual aid, we can see many features which were reproduced in more sophisticated and complex forms in trade unions, co-operatives, Hampden Clubs, Political Unions, and Chartist lodges. At the same time the societies can be seen as crystallising an ethos of mutuality very much more widely diffused in the "dense" and "concrete" particulars of the personal relations of working people, at home and at work. Every kind of witness in the first half of the nineteenth century – clergymen, factory inspectors, Radical publicists – remarked upon the extent of mutual aid in the poorest districts. In times of emergency, unemployment, strikes, sickness, childbirth, then it was the poor who "helped every one his neighbour." Twenty years after Place's comment on the change in Lancashire manners, Cooke Taylor was astounded at the way in which Lancashire working men bore "the extreme of wretchedness,"

> with a high tone of moral dignity, a marked sense of propriety, a decency, cleanliness, and order . . . which do not merit the intense suffering I have witnessed. I was beholding the gradual immolation of the noblest and most valuable population that ever existed in this country or in any other under heaven.

"Nearly all the distressed operatives whom I met north of Manchester . . . had a thorough horror of being forced to receive parish relief."

It is an error to see this as the *only* effective "working-class" ethic. The "aristocratic" aspirations of artisans and mechanics, the values of "self-help", or criminality and demoralisation, were equally widely dispersed. The conflict between alternative ways of life was fought out, not just between the middle and working classes, but within working-class communities themselves. But by the early years of the nineteenth century it is possible to say that collectivist values are dominant in many industrial communities; there is a definite moral code, with sanctions against the blackleg, the "tools" of the employer or the unneighbourly, and with an intolerance towards the eccentric or individualist. Collectivist values are consciously held and are propagated in political theory, trade union ceremonial, moral rhetoric. It is, indeed, this collective self-consciousness, with its corresponding theory, institutions, discipline, and community values which distinguishes the nineteenth-century *working class* from the eighteenth-century *mob.*

Political Radicalism and Owenism both drew upon and enriched this "basic collectivist idea." Francis Place may well have been right when he attributed the changed behaviour of Lancashire crowds in 1819 to the advance of political consciousness "spreading over the face of the country ever since the Constitutional and Corresponding Societies became active in 1792":

> Now 100,000 people may be collected together and no riot ensue, and why? . . . The people have an object, the pursuit of which gives them importance in their own eyes, elevates them in their own opinion, and thus it is that the very individuals who would have been the leaders of the riot are the keepers of the peace.

Another observer attributed the changes in Lancashire to the influence both of Cobbett and of the Sunday schools and noted a "general and radical change" in the character of the labouring classes:

> The poor, when suffering and dissatisfied, no longer make a riot, but hold a meeting – instead of attacking their neighbours, they arraign the Ministry.

This growth in self-respect and political consciousness was one real gain of the Industrial Revolution. It dispelled some forms of superstition and of deference, and made certain kinds of oppression no longer tolerable. We can find abundant testimony as to the steady growth of the ethos of mutuality in the strength and ceremonial pride of the unions and trades clubs which emerged from quasi-legality when the Combination Acts were repealed. During the Bradford woolcombers' strike of 1825 we find that in Newcastle, where the friendly society was so well rooted, the unions contributing to the Bradford funds included smiths, millwrights, joiners, shoemakers, morocco leather dressers, cabinetmakers, shipwrights, sawyers, tailors, woolcombers, hatters, tanners, weavers, potters and miners. Moreover, there is a sense in which the friendly society helped to pick up and carry into the trade union movement the love of ceremony and the high sense of status of the craftsmen's guild. These traditions, indeed, still had a remarkable vigour in the early nineteenth century, in some of the old Chartered Companies or Guilds of the masters and of master-craftsmen, whose periodical ceremonies expressed the pride of both the masters, and of their journeymen in "the Trade." In 1802, for example, there was a great jubilee celebration of the Preston "Guilds." In a week of processions and exhibitions, in which the nobility, gentry, merchants, shopkeepers, and manufacturers all took part, the journeymen were given a prominent place:

The Wool-Combers and Cotton Workers . . . were preceded by twenty-four young blooming handsome women, each bearing a branch of the cotton tree, then followed a spinning machine borne on men's shoulders, and afterwards a loom drawn on a sledge, each with work-people busily employed at them . . .

At Bradford, on the eve of the great strike of 1825, the woolcombers' feast of Bishop Blaize was celebrated with extraordinary splendour:

Herald, bearing a flag.
Twenty-four Woolstaplers on horseback, each horse caparisoned with a fleece.
Thirty-eight Worsted-Spinners and Manufacturers on horseback, in white stuff

waiscoats, with each a sliver of wool over his shoulder and a white stuff sash: the horses' necks covered with nets made of thick yarn.

And so on until we reach:

BISHOP BLAIZE
Shepherd and Shepherdess.
Shepherd-Swains.
One hundred and sixty Woolsorters on horseback, with ornamented
caps and various coloured slivers.
Thirty Comb-makers.
Charcoal Burners.
Combers' Colours.
Band.
Four hundred and seventy Wool-combers, with wool wigs, &c.
Band.
Forty Dyers, with red cockades, blue aprons, and crossed slivers of red and blue.

After the great strike such a ceremony could not be repeated.

This passage from the old outlook of "the Trade" to the duality of the masters' organisations, on the one hand, and the trade unions on the other, takes us into the central experience of the Industrial Revolution.[2] But the friendly society and trade union, no less than the organisations of the masters, sought to maintain the ceremonial and the pride of the older tradition; indeed, since the artisans (or, as they still are called, *tradesmen*) felt themselves to be the *producers* upon whose skill the masters were parasitic, they emphasised the tradition the more. With the repeal of the Combination Acts their banners moved openly through the streets. In London, in 1825, the Thames Ship Caulkers Union (founded in 1787) displayed its mottoes: "Main et Coeur," "Vigeur, Verité, Concorde, Dépêche," which reveal the pride of the medieval craft. The Ropemakers Union proceeded with a white banner on which was portrayed a swarm of bees around a hive: "Sons of Industry! Union gives Strength." (At the houses of masters who had granted them an increase, they stopped and gave a salute.) John Gast's Thames Shipwrights Provident Union, the pacemaker of the London "trades," outdid all with a blue silk banner: "Hearts of Oak Protect the Aged," a handsome ship drawn by six bay horses, three postillions in blue jackets, a band, the Committee, the members with more banners and flags, and delegations representing the trade from Shields, Sunderland, and Newcastle. The members wore blue rosettes and sprigs of oak, and in the ship were old shipwrights who lived in the union's almshouses at Stepney. At Nantwich in 1832 the Shoemakers maintained all the sense of status of the artisan's craft union, with their banner, "full set of secret order regalia, surplices, trimmed aprons . . . and a crown and robes for King Crispin." In 1833 the King rode on horseback through the town attended by train-bearers, officers with the "Dispensation, the Bible, a large pair of gloves, and also beautiful specimens of ladies' and gents' boots and shoes":

> Nearly 500 joined in the procession, each one wearing a white apron neatly trimmed. The rear was brought up by a shopmate in full tramping order, his kit packed on his back, and walking-stick in hand.

No single explanation will suffice to account for the evident alteration in manner of the working people. Nor should we exaggerate the degree of change. Drunkenness and uproar still often surged through the streets. But it is true that working men often appear most sober and disciplined, in the twenty years after the Wars, when most in earnest to assert their rights. Thus we cannot accept the thesis that sobriety was the consequence only, or even mainly, of the Evangelical propaganda. And we may see this, also, if we turn the coin over and look at the reverse. By 1830 not only the Established Church but also the Methodist revival was meeting sharp opposition in most working-class centres from free-thinkers, Owenites, and non-denominational Christians. In London, Birmingham, south-east Lancashire, Newcastle, Leeds and other cities the Deist adherents of Carlile or Owen had an enormous following. The Methodists had consolidated their position, but they tended increasingly to represent tradesmen and privileged groups of workers, and to be morally isolated from working-class community life. Some old centres of revivalism had relapsed into "heathenism." In Newcastle's Sandgate, once "as noted for praying as for tippling, for psalm-singing as for swearing," the Methodists had lost any following among the poor by the

1840s. In parts of Lancashire weaving communities as well as factory operatives became largely detached from the chapels and were swept up in the current of Owenism and free-thought:

> If it had not been for Sunday schools, society would have been in a horrible state before this time . . . Infidelity is growing amazingly . . . The writings of Garlic and Taylor and other infidels are more read than the Bible or any other book I have seen weeks after weeks the weavers assembled in a room, that would contain 400 people, to applaud the people who asserted, and argued that there was no God I have gone into the cottages around the chapel where I worship, I have found women assembled reading infidel publications . . .

Owenite and secular movements often took fire "like whins on the common", as revivalism had done before.

Engels, writing from his Lancashire experience in 1844, claimed that "workers are not religious, and do not attend church, with the exception of the Irish, a few elderly people, and the half-bourgeois, the overlookers, foremen, and the like." "Among the masses there prevails almost universally a total indifference to religion, or at the utmost, some trace of Deism . . ." Engels weakened his case by overstating it; but Dodd quoted a Stockport factory where nine out of ten did not attend any church, while Cooke Taylor, in 1842, was astonished at the vigour and knowledge of the Scriptures shown by Lancashire working men who contested Christian orthodoxies. "If I thought that the Lord was the cause of all the misery I see around me," one such man told a Methodist preacher, "I would quit his service, and say he was not the Lord I took him for." Similarly, in Newcastle in the Chartist years thousands of artisans and engineers were convinced free-thinkers. In one works employing 200 "there are not more than six or seven who attend a place of worship." "The working classes," said one working-man,

> are gathering knowledge, and the more they gather, the wider becomes the breach between them and the different sects. It is not because they are ignorant of the Bible. I revere the Bible myself . . . and when I look into it . . . I

find that the prophets stood between the oppressor and the oppressed, and denounced the wrong doer, however rich and powerful . . . When the preachers go back to the old book, I for one will go back to hear them, but not till then . . .

The Sunday schools were bringing an unexpected harvest.

The weakening hold of the churches by no means indicated any erosion of the self-respect and discipline of class. On the contrary, Manchester and Newcastle, with their long tradition of industrial and political organisation, were notable in the Chartist years for the discipline of their massive demonstrations. Where the citizens and shop-keepers had once been thrown into alarm when the "terrible and savage pitmen" entered Newcastle in any force, it now became necessary for the coal owners to scour the slums of the city for "candy-men" or rag-collectors to evict the striking miners. In 1838 and 1839 tens of thousands of artisans, miners and labourers marched week after week in good order through the streets, often passing within a few feet of the military, and avoiding all provocation. "Our people had been well taught," one of their leaders recalled, "that it was not riot we wanted, but revolution."

iv. Myriads of eternity

If we can now see more clearly many of the elements which made up the working-class communities of the early nineteenth century, a definitive answer to the "standard-of-living" controversy must still evade us. For beneath the word "standard" we must always find judgements of value as well as questions of fact. Values, we hope to have shown, are not "imponderables" which the historian may safely dismiss with the reflection that, since they are not amenable to measurement, anyone's opinion is as good as anyone else's. They are, on the contrary, those questions of human satisfaction, and of the direction of social change, which the historian ought to ponder if history is to claim a position among the significant humanities.

The historian, or the historical sociologist, must in fact be concerned with judgements of value in two forms. In the first instance, he is concerned with the values *actually held* by those who lived through the Industrial Revolution. The old and newer modes of production each supported distinct kinds of community with characteristic ways of life. Alternative conventions and notions of human satisfaction were in conflict with each other, and there is no shortage of evidence if we wish to study the ensuing tensions.

In the second instance, he is concerned with making some judgement of value upon the whole process entailed in the Industrial Revolution, of which we ourselves are an end-product. It is our own involvement which makes judgement difficult. And yet we are helped towards a certain detachment, both by the "romantic" critique of industrialism which stems from one part of the experience, and by the record of tenacious resistance by which hand-loom weaver, artisan or village craftsman confronted this experience and held fast to an alternative culture. As we see them change, so we see how we became what we are. We understand more clearly what was lost, what was driven "underground," what is still unresolved.

Any evaluation of the quality of life must entail an assessment of the total life-experience, the manifold satisfactions or deprivations, cultural as well as material, of the people concerned. From such a standpoint, the older "cataclysmic" view of the Industrial Revolution must still be accepted. During the years between 1780 and 1840 the people of Britain suffered an experience of immiseration, even if it is possible to show a small statistical improvement in material conditions. When Sir Charles Snow tells us that "with singular unanimity . . . the poor have walked off the land into the factories as fast as the 9th factories could take them", we must reply, with Dr. Leavis, that the "actual history" of the "full human problem [was] incomparably and poignantly more complex than that." Some were lured from the countryside by the glitter and promise of wages of the industrial town; but the old village economy was crumbling at their backs. They moved less by their own will than at the dictate of external compulsions which they could not question: the enclosures, the wars, the Poor Laws, the decline of rural industries, the counter-revolutionary stance of their rulers.

The process of industrialisation is necessarily painful. It must involve the erosion of traditional patterns of life. But it was carried through with exceptional violence in Britain. It was unrelieved by any sense of national participation in communal effort, such as is found in countries undergoing a national revolution. Its ideology was that of the masters alone. Its messianic prophet was Dr. Andrew Ure, who saw the factory system as "the great minister of civilization to the terraqueous globe," diffusing "the life-blood of science and religion to myriads . . . still lying 'in the region and shadow of death'." But those who served it did not *feel* this to be so, any more than those "myriads" who were served. The experience of immiseration came upon them in a hundred different forms; for the field labourer, the loss of his common rights and the vestiges of village democracy; for the artisan, the loss of his craftsman's status; for the weaver, the loss of livelihood and of independence; for the child, the loss of work and play in the home; for many groups of workers whose real earnings improved, the loss of security, leisure and the deterioration of the urban environment. R.M. Martin, who gave evidence before the Hand-Loom Weavers' Committee of 1834, and who had returned to England after an absence from Europe of ten years, was struck by the evidence of physical and spiritual deterioration:

> I have observed it not only in the manufacturing but also in agricultural communities in this country; they seem to have lost their animation, their vivacity, their field games and their village sports; they have become a sordid, discontented, miserable, anxious, struggling people, without health, or gaiety, or happiness.

It is misleading to search for explanations in what Professor Ashton has rightly described as "tedious" phrases – man's "divorce" from "nature" or "the soil." After the "Last Labourers' Revolt", the Wiltshire field labourers – who were close enough to "nature" – were far worse degraded than the Lancashire mill girls. This violence was done to *human* nature. From one standpoint, it may be seen as the outcome of the pursuit of profit, when the cupidity of the owners of the means of production was freed from old sanctions and had not yet been subjected to new means of social control. In this sense we may still read it, as Marx did, as the violence of the capitalist class. From another standpoint, it may be seen as a violent technological differentiation between work and life.

It is neither poverty nor disease but work itself which casts the blackest shadow over the years of the Industrial Revolution. It is Blake, himself a craftsman by training, who gives us the experience:

> Then left the sons of Urizen the plow & harrow, the loom,
> The hammer & the chisel & the rule & compasses . . .
> And all the arts of life they chang'd into the arts of death.
> The hour glass contemn'd because its simple workmanship
> Was as the workmanship of the plowman & the water wheel
> That raises water into Cisterns, broken & burn'd in fire
> Because its workmanship was like the workmanship of the shepherds
> And in their stead intricate wheels invented, Wheel without wheel,
> To perplex youth in their outgoings & to bind to labours
> Of day & night the myriads of Eternity, that they might file
> And polish brass & iron hour after hour, laborious workmanship,
> Kept ignorant of the use that they might spend the days of wisdom
> In sorrowful drudgery to obtain a scanty pittance of bread,
> In ignorance to view a small portion & think that All,
> And call it demonstration, blind to all the simple rules of life.

These "myriads of eternity" seem at times to have been sealed in their work like a tomb. Their best efforts, over a lifetime, and supported by their own friendly societies, could scarcely ensure them that to which so high a popular value was attached – a "Decent Funeral." New skills were arising, old satisfactions persisted, but over all we feel the general pressure of long hours of unsatisfying labour under severe discipline for alien purposes. This was at the source of that "ugliness" which, D.H. Lawrence wrote, "betrayed the spirit of man in the nineteenth century." After all other impressions fade, this one remains; together with that of the loss of any felt cohesion in the community, save that which the working people, in antagonism to their labour and to their masters, built for themselves.

"Thick Description: Toward an Interpretive Theory of Culture"

from *The Interpretation of Cultures: Selected Essays* (1973)

Clifford Geertz

Editors' introduction

It should not be surprising, after reading Raymond Williams's etymology of culture (see p. 15) in which many of the overlapping confusions and ambiguities of the term are laid bare – that defining the concept has remained a frustrating task for scholars in disciplines such as anthropology and geography. For Clifford Geertz, culture was by the 1960s stuck in a "conceptual morass" in which the term was being stretched to explain an eclectic array of human phenomena. "Theoretical diffusion" was, he argued, undermining the analytical power of culture and weakening the field of anthropology. Geertz's response to this situation is most succinctly laid out in his famous essay, "Thick Description," from which the following selection is excerpted.

By calling for a *semiotic* approach to culture, Geertz sought to distinguish culture from social structures and institutions which were often thought to regulate people's behaviors and practices. Culture was not, he argued, simply a function of people's material lives, and could not be reduced to a set of "laws" that linked economic, political, and social conditions to behaviors, beliefs, and practices. Rather, culture was that realm in which people interpreted and made meaning out of their lives. This meant that cultural analysis involved "sorting out the structures of signification . . . and determining their social ground and import." Geertz was essentially arguing that culture most fundamentally could not be viewed as a set of behaviors, practices, and beliefs, but rather was an ongoing construction of meaning as people continually reflected upon the significance of their lives. In this sense, culture was similar to language. It was a way of sharing meaning communicated through signs and symbols, "winks," "twitches," and "non-twitches," as Geertz puts it here. In the language of metaphysics, Geertz was shifting the question about culture from the realm of ontology (what *is* culture?) to that of epistemology (how do we *know* culture?). This shifted the goal from realizing a "complete" understanding of culture to one of studying the ongoing social contexts in which cultural meanings are being produced and how the production of culture matters in those contexts.

This shift had significant methodological and theoretical implications. The following selection focuses on Geertz's discussion of culture itself, rather than his discussion of ethnography as a method and cultural theory more broadly. However, a brief summary of his views of these topics will be helpful in grasping the overall significance of the essay. First and most important was the fact that Geertz's semiotic approach to culture made the ethnographic method an *interpretive* project. Such an approach challenged the pretensions of "scientific objectivity" that legitimized the ethnographic method as social science. Geertz was adamant that such a challenge did not foretell the doom of ethnography but rather provided a much needed clarification of exactly what ethnography was capable of doing. Rather than capturing "primitive facts in faraway places" and carrying them home "like a mask or carving," ethnography should be evaluated on its ability to clarify the ways other people understand their world: "whether it sorts winks from twitches and real winks from mimicked ones." There remains significant debate, of course, regarding both the interpretive authority of the ethnographer (how can the ethnographer's account be verified?) and the distinction between the ethnographer's interpretation of culture and that of the people about whom the ethnographer is writing (is this the author's understanding of these people's culture or is it the people's understanding?). One of the most difficult – and attractive – features of the semiotic approach to culture, then, is its blurring of the boundary between the world of the scholar and that of the informant, since both are always engaged in their projects of interpretation.

Second, because ethnography was necessarily place-based and focused on people's daily lives, its ability to provide generalization at broader scales was limited. Geertz argued on many occasions against the assumption that culture offered a gateway to understanding universal essences of whole nations or civilizations. Culture was not, in other words, a reservoir of meanings to which all people of a particular religion, ethnicity, or nation had access, but was rather an ongoing process of interpretation resulting from people negotiating the pathways of their lives in their particular corners of the world.

Third, this meant that cultural theory was necessarily grounded. A semiotic approach to culture would not allow abstraction away from the immediate contexts of cultural production. "Theoretical formulations," he wrote, "hover so low over the interpretations they govern that they don't make much sense or hold much interests apart from them." It follows of course that there is not much predictive capacity to cultural theory. This conclusion was of course cause for disappointment among his detractors, for Geertz was convinced that social science attempted grand theories across time and space at its peril.

Clifford Geertz (1926–2006) served on the faculty of the Institute for Advanced Study in Princeton, New Jersey. A prolific writer, he studied and published a great variety of work on religion, economic development, trade, village and family life, traditional political structures, and the nature of anthropological inquiry. Most of his fieldwork was carried out in Indonesia and Morocco. Aside from *The Interpretation of Cultures*, which was selected as one of the hundred most important books since World War II by *The Times Literary Supplement*, he is well known for *Negara: The Theater State in Nineteenth Century Bali* (1980), *Works and Lives: The Anthropologist as Author* (1988 – a winner of the National Book Critics Circle Award), *The Religion of Java* (1960), *Islam Observed: Religious Development in Morocco and Indonesia* (1968), and *The Politics of Culture: Asian Identities in a Splintered World* (2002).

The influence of Geertz's work has extended far beyond anthropology to include cultural geography, ecology, political science, and history. It would be hard to overstate the influence his work had on the debates within cultural geography in the late 1970s and early 1980s. James Duncan's critique of cultural geography in "The Superorganic in American Cultural Geography" (*Annals of the Association of American Geographers* 79, 2, 1980) relied heavily on Geertz's semiotic approach to culture. Indeed, Geertz's approach represented a considerable departure from the way most geographers conceptualized culture in their work, which tended to emphasize cultural ecology, landscape, and material culture. More to the point, however, would be the claim that cultural geography perhaps suffered the same "conceptual morass" that Geertz saw in anthropology. While Geertz's work was instrumental in efforts to redefine culture in geography, his approach has not had the same galvanizing effect in geography that it had in anthropology, and a lively debate has continued within cultural geography concerning how to define culture. It is doubtful that Geertz would have agreed with Duncan, who in 1994 (as also discussed in greater detail in the introduction to Part Two of the *Reader*) advocated viewing the field as a *heterotopia* – that is, a collection of incompatible approaches that, taken together, nevertheless make up some kind of whole.

■ ■ ■ ■ ■ ■

In her book, *Philosophy in a New Key*, Susanne Langer remarks that certain ideas burst upon the intellectual landscape with a tremendous force. They resolve so many fundamental problems at once that they seem also to promise that they will resolve all fundamental problems, clarify all obscure issues. Everyone snaps them up as the open sesame of some new positive science, the conceptual center-point around which a comprehensive system of analysis can be built. The sudden vogue of such a *grande idée*, crowding out almost everything else for a while, is due, she says, "to the fact that all sensitive and active minds turn at once to exploiting it. We try it in every connection, for every purpose, experiment with possible stretches of its strict meaning, with generalizations and derivatives."

After we have become familiar with the new idea, however, after it has become part of our general stock of theoretical concepts, our expectations are brought more into balance with its actual uses, and its excessive popularity is ended. A few zealots persist in the old key-to-the-universe view of it; but less driven thinkers settle down after a while to the problems the idea has really generated. They try to apply it and extend it where it applies and where it is capable of extension; and they desist where it does not apply or cannot be extended. It becomes, if it was, in truth, a seminal idea in the first place, a permanent and enduring part of our intellectual armory. But it no longer has the grandiose, all-promising scope, the infinite versatility of apparent application, it once had. The second law of thermodynamics, or the principle of natural selection, or the notion of unconscious motivation, or the organization of the means of production does not explain everything, not even everything human, but it still explains something; and our attention shifts to isolating just what that something is, to disentangling ourselves from a lot of pseudoscience to which, in the first flush of its celebrity, it has also given rise.

Whether or not this is, in fact, the way all centrally important scientific concepts develop, I don't know. But certainly this pattern fits the concept of culture around which the whole discipline of anthropology arose, and whose domination that discipline has been increasingly concerned to limit, specify, focus, and contain. It is to this cutting of the culture concept down to size, therefore actually insuring its continued importance rather than undermining it, that the essays below are all, in their several ways and from their several directions, dedicated. They all argue, sometimes explicitly, more often merely through the particular analysis they develop, for a narrowed, specialized, and, so I imagine, theoretically more powerful concept of culture to replace E.B. Tylor's famous "most complex whole," which, its originative power not denied, seems to me to have reached the point where it obscures a good deal more than it reveals.

The conceptual morass into which the Tylorean kind of *pot-au-feu* theorizing about culture can lead is evident in what is still one of the better general introductions to anthropology, Clyde Kluckhohn's *Mirror for Man*. In some twenty-seven pages of his chapter on the concept, Kluckhohn managed to define culture in turn as (1) "the total way of life of a people"; (2) "the social legacy the individual acquires from his group"; (3) "a way of thinking, feeling, and believing"; (4) "an abstraction from behavior"; (5) a theory on the part of the anthropologist about the way in which a group of people in fact behave; (6) a "storehouse of pooled learning"; (7) "a set of standardized orientations to recurrent problems"; (8) "learned behavior"; (9) a mechanism for the normative regulation of behavior; (10) "a set of techniques for adjusting both to the external environment and to other men"; (11) "a precipitate of history"; and turning, perhaps in desperation, to similes, as a map, as a sieve, and as a matrix. In the face of this sort of theoretical diffusion, even a somewhat constricted and not entirely standard concept of culture, which is at least internally coherent and, more important, which has a definable argument to make is (as, to be fair, Kluckhohn himself keenly realized) an improvement. Eclecticism is self-defeating not because there is only one direction in which it is useful to move, but because there are so many: it is necessary to choose.

The concept of culture I espouse, and whose utility the essays below attempt to demonstrate, is essentially a semiotic one. Believing, with Max Weber, that man is an animal suspended in webs of significance he himself has spun, I take culture to be those webs, and the analysis of it to be therefore not an experimental science in search of law but an interpretive one in search of meaning. It is explication I am after, construing social expressions on their surface enigmatical. But this pronouncement, a doctrine in a clause, demands itself some explication.

II

Operationalism as a methodological dogma never made much sense so far as the social sciences are concerned, and except for a few rather too well-swept corners – Skinnerian behaviorism, intelligence testing, and so on – it is largely dead now.

But it had, for all that, an important point to make, which, however we may feel about trying to define charisma or alienation in terms of operations, retains a certain force: if you want to understand what a science is, you should look in the first instance not at its theories or its findings, and certainly not at what its apologists say about it; you should look at what the practitioners of it do.

In anthropology, or anyway social anthropology, what the practioners do is ethnography. And it is in understanding what ethnography is, or more exactly *what doing ethnography is*, that a start can be made toward grasping what anthropological analysis amounts to as a form of knowledge. This, it must immediately be said, is not a matter of methods. From one point of view, that of the textbook, doing ethnography is establishing rapport, selecting informants, transcribing texts, taking genealogies, mapping fields, keeping a diary, and so on. But it is not these things, techniques and received procedures, that define the enterprise. What defines it is the kind of intellectual effort it *is*: an elaborate venture in, to borrow a notion from Gilbert Ryle, "thick description."

Ryle's discussion of "thick description" appears in two recent essays of his (now reprinted in the second volume of his *Collected Papers*) addressed to the general question of what, as he puts it, *"Le Penseur"* is doing: "Thinking and Reflecting" and "The Thinking of Thoughts." Consider, he says, two boys rapidly contracting the eyelids of their right eyes. In one, this is an involuntary twitch; in the other, a conspiratorial signal to a friend. The two movements are, as movements, identical; from an I-am-a-camera, "phenomenalistic" observation of them alone one could not tell which was twitch and which was wink, or indeed whether both or either was twitch or wink. Yet the difference, however unphotographable, between a twitch and a wink is vast; as anyone unfortunate enough to have had the first taken for the second knows. The winker is communicating, and indeed communicating in a quite precise and special way: (1) deliberately, (2) to someone in particular, (3) to impart a particular message, (4) according to a socially established code, and (5) without cognizance of the rest of the company. As Ryle points out, the winker has done two things, contracted his eyelids and winked, while the twitcher has done only one, contracted his eyelids. Contracting your eyelids on purpose when there

exists a public code in which so doing counts as a conspiratorial signal *is* winking. That's all there is to it: a speck of behavior, a fleck of culture, and – *voila!* – a gesture.

That, however, is just the beginning. Suppose, he continues, there is a third boy, who, "to give malicious amusement to his cronies," parodies the first boy's wink, as amateurish, clumsy, obvious, and so on. He, of course, does this in the same way the second boy winked and the first twitched: by contracting his right eyelids. Only this boy is neither winking nor twitching, he is parodying someone else's, as he takes it, laughable, attempt at winking. Here, too, a socially established code exists (he will "wink" laboriously, overobviously, perhaps adding a grimace – the usual artifices of the clown); and so also does a message. Only now it is not conspiracy but ridicule that is in the air. If the others think he is actually winking, his whole project misfires as completely, though with somewhat different results, as if they think he is twitching. One can go further: uncertain of his mimicking abilities, the would-be satirist may practice at home before the mirror, in which case he is not twitching, winking, or parodying, but rehearsing; though so far as what a camera, a radical behaviorist, or a believer in protocol sentences would record he is just rapidly contracting his right eyelids like all the others. Complexities are possible, if not practically without end, at least logically so. The original winker might, for example, actually have been fake-winking, say, to mislead outsiders into imagining there was a conspiracy afoot when there in fact was not, in which case our descriptions of what the parodist is parodying and the rehearser rehearsing of course shift accordingly. But the point is that between what Ryle calls the "thin description" of what the rehearser (parodist, winker, twitcher . . .) is doing ("rapidly contracting his right eyelids") and the "thick description" of what he is doing ("practicing a burlesque of a friend faking a wink to deceive an innocent into thinking a conspiracy is in motion") lies the object of ethnography: a stratified hierarchy of meaningful structures in terms of which twitchers, winks, fake-winks, parodies, rehearsals of parodies are produced, perceived, and interpreted and without which they would not (not even the zero-form twitches, which, *as a cultural category*, are as much nonwinks as winks are nontwitches) in fact exist, no matter what anyone did or didn't do with his eyelids.

Like so many of the little stories Oxford philosophers like to make up for themselves, all this winking, fake-winking, burlesque-fake-winking, rehearsed-burlesque-fake-winking, may seem a bit artificial. In way of adding a more empirical note, let me give, deliberately unpreceded by any prior explanatory comment at all, a not untypical excerpt from my own field journal to demonstrate that, however evened off for didactic purposes, Ryle's example presents an image only too exact of the sort of piled-up structures of inference and implication through which an ethnographer is continually trying to pick his way:

The French [the informant said] had only just arrived. They set up twenty or so small forts between here, the town, and the Marmusha area up in the middle of the mountains, placing them on promontories so they could survey the countryside. But for all this they couldn't guarantee safety, especially at night, so although the *mezrag*, trade-pact, system was supposed to have been legally abolished it in fact continued as before.

One night, when Cohen (who speaks fluent Berber), was up there, at Marmusha, two other Jews who were traders to a neighboring tribe came by to purchase some goods from him. Some Berbers, from yet another neighboring tribe, tried to break into Cohen's place, but he fired his rifle in the air. (Traditionally, Jews were not allowed to carry weapons; but at this period things were so unsettled many did so anyway.) This attracted the attention of the French and the marauders fled.

The next night, however, they came back, one of them disguised as a woman, who knocked on the door with some sort of a story. Cohen was suspicious and didn't want to let "her" in, but the other Jews said, "Oh, it's all right, it's only a woman." So they opened the door and the whole lot came pouring in. They killed the two visiting Jews, but Cohen managed to barricade himself in an adjoining room. He heard the robbers planning to burn him alive in the shop after they removed his goods, and so he opened the door and, laying about him wildly with a club, managed to escape through a window.

He went up to the fort, then, to have his wounds dressed, and complained to the local commandant, one Captain Dumari, saying he wanted his '*ar* – i.e., four or five times the value of the merchandise stolen from him. The robbers were from a tribe which had not yet submitted to French authority and were in open rebellion against it, and he wanted authorization to go with his *mezrag*-holder, the Marmusha tribal *sheikh*, to collect the indemnity that, under traditional rules, he had coming to him. Captain Dumari couldn't officially give him permission to do this, because of the French prohibition of the *mezrag* relationship, but he gave him verbal authorization, saying, "If you get killed, it's your problem."

So the *sheikh*, the Jew, and a small company of armed Marmushans went off ten or fifteen kilometers up into the rebellious area, where there were of course no French, and, sneaking up, captured the thief-tribe's shepherd and stole its herds. The other tribe soon came riding out on horses after them, armed with rifles and ready to attack. But when they saw who the "sheep thieves" were, they thought better of it and said, "All right, we'll talk." They couldn't really deny what had happened – that some of their men had robbed Cohen and killed the two visitors – and they weren't prepared to start the serious feud with the Marmusha a scuffle with the invading party would bring on. So the two groups talked, and talked, and talked, there on the plain amid the thousands of sheep, and decided finally on five hundred sheep damages. The two armed Berber groups then lined up on their horses at opposite ends of the plain, with the sheep herded between them, and Cohen, in his black gown, pillbox hat, and flapping slippers, went out alone among the sheep, picking out, one by one and at his own good speed, the best ones for his payment.

So Cohen got his sheep and drove them back to Marmusha. The French, up in their fort, heard them coming from some distance ("Ba, ba, ba," said Cohen, happily, recalling the image) and said, "What the hell is that?" And Cohen said, "That is my '*ar*." The French couldn't believe he had actually done what he said he had done, and accused him of being a spy for the rebellious Berbers, put him in prison, and took his sheep. In the town, his family, not having heard from him in so long a time, thought he was dead.

But after a while the French released him and he came back home, but without his sheep. He then went to the Colonel in the town, the Frenchman in charge of the whole region, to complain. But the Colonel said, "I can't do anything about the matter. It's not my problem."

Quoted raw, a note in a bottle, this passage conveys, as any similar one similarly presented would do, a fair sense of how much goes into ethnographic description of even the most elemental sort – how extraordinarily "thick" it is. In finished anthropological writings, including those collected here, this fact – that what we call our data are really our own constructions of other people's constructions of what they and their compatriots are up to – is obscured because most of what we need to comprehend a particular event, ritual, custom, idea, or whatever is insinuated as background information before the thing itself is directly examined. (Even to reveal that this little drama took place in the highlands of central Morocco in 1912 – and was recounted there in 1968 – is to determine much of our understanding of it.) There is nothing particularly wrong with this, and it is in any case inevitable. But it does lead to a view of anthropological research as rather more of an observational and rather less of an interpretive activity than it really is. Right down at the factual base, the hard rock, insofar as there is any, of the whole enterprise, we are already explicating: and worse, explicating explications. Winks upon winks upon winks.

Analysis, then, is sorting out the structures of signification – what Ryle called established codes, a somewhat misleading expression, for it makes the enterprise sound too much like that of the cipher clerk when it is much more like that of the literary critic – and determining their social ground and impact. Here, in our text, such sorting would begin with distinguishing the three unlike frames of interpretation ingredient in the situation, Jewish, Berber, and French, and would then move on to show how (and why) at that time, in that place, their copresence produced a situation in which systematic misunderstanding reduced traditional form to social farce. What tripped Cohen up, and with him the whole ancient pattern of social and economic relationships within which he functioned, was a confusion of tongues.

I shall come back to this too-compacted aphorism later, as well as to the details of the text itself. The point for now is only that ethnography is thick description. What the ethnographer is in fact faced with – except when (as, of course, he must do) he is pursuing the more automatized routines of data collection – is a multiplicity of complex conceptual structures, many of them superimposed upon or knotted into one another, which are at once strange, irregular, and inexplicit, and which he must contrive somehow to first grasp and then to render. And this is true at the most down-to-earth, jungle fieldwork levels of his activity: interviewing informants, observing rituals, eliciting kin terms, tracing property lines, censusing households . . . writing his journal. Doing ethnography is like trying to read (in the sense of "construct a reading of") a manuscript – foreign, faded, full of ellipses, incoherences, suspicious emendations, and tendentious commentaries, but written not in conventionalized graphs of sound but in transient examples of shaped behavior.

III

Culture, this acted document, thus is public, like a burlesqued wink or a mock sheep raid. Though ideational, it does not exist in someone's head; though unphysical, it is not an occult entity. The interminable, because unterminable, debate within anthropology as to whether culture is "subjective" or "objective," together with the mutual exchange of intellectual insults ("idealist!"–"materialist!"; "mentalist!"–"behaviorist!"; "impressionist!"–"positivist!") which accompanies it, is wholly misconceived. Once human behavior is seen as (most of the time; there *are* true twitches) symbolic action – action which, like phonation in speech, pigment in painting, line in writing, or sonance in music, signifies – the question as to whether culture is patterned conduct or a frame of mind, or even the two somehow mixed together, loses sense. The thing to ask about a burlesqued wink or a mock sheep raid is not what their ontological status is. It is the same as that of rocks on the one hand and dreams on the other – they are things of this world. The thing to ask is what their import is: what it is, ridicule or challenge, irony or anger, snobbery or pride, that, in their occurrence and through their agency, is getting said.

This may seem like an obvious truth, but there are a number of ways to obscure it. One is to imagine that culture is a self-contained "superorganic" reality with forces and purposes of its own; that is, to reify it. Another is to claim that it consists in the brute pattern of behavioral events we observe in fact to occur in some identifiable community or other; that is, to reduce it. But though both these confusions still exist, and doubtless will be always with us, the main source of theoretical muddlement in contemporary anthropology is a view which developed in reaction to them and is right now very widely held – namely, that, to quote Ward Goodenough, perhaps its leading proponent, "culture [is located] in the minds and hearts of men."

Variously called ethnoscience, componential analysis, or cognitive anthropology (a terminological wavering which reflects a deeper uncertainty), this school of thought holds that culture is composed of psychological structures by means of which individuals or groups of individuals guide their behavior. "A society's culture," to quote Goodenough again, this time in a passage which has become the *locus classicus* of the whole movement, "consists of whatever it is one has to know or believe in order to operate in a manner acceptable to its members." And from this view of what culture is follows a view, equally assured, of what describing it is – the writing out of systematic rules, an ethnographic algorithm, which, if followed, would make it possible so to operate, to pass (physical appearance aside) for a native. In such a way, extreme subjectivism is married to extreme formalism, with the expected result: an explosion of debate as to whether particular analyses (which come in the form of taxonomies, paradigms, tables, trees, and other ingenuities) reflect what the natives "really" think or are merely clever simulations, logically equivalent but substantively different, of what they think.

As, on first glance, this approach may look close enough to the one being developed here to be mistaken for it, it is useful to be explicit as to what divides them. If, leaving our winks and sheep behind for the moment, we take, say, a Beethoven quartet as an, admittedly rather special but, for these purposes, nicely illustrative, sample of culture, no one would, I think, identify it with its score, with the skills and knowledge needed to play it, with the understanding of it possessed by its performers or auditors, nor, to take care, *en passant*, of the reductionists and reifiers, with a particular performance of it or with some mysterious entity transcending material existence. The "no one" is perhaps too strong here, for there are always incorrigibles. But that a Beethoven quartet is a temporally developed tonal structure, a coherent sequence of modeled sound – in a word, music – and not anybody's knowledge of or belief about anything, including how to play it, is a proposition to which most people are, upon reflection, likely to assent.

To play the violin it is necessary to possess certain habits, skills, knowledge, and talents, to be in the mood to play, and (as the old joke goes) to have a violin. But violin playing is neither the habits, skills, knowledge, and so on, nor the mood, nor (the notion believers in "material culture" apparently embrace) the violin. To make a trade pact in Morocco, you have to do certain things in certain ways (among others, cut, while chanting Quranic Arabic, the throat of a lamb before the assembled, undeformed, adult male members of your tribe) and to be possessed of certain psychological characteristics (among others, a desire for distant things). But a trade pact is neither the throat cutting nor the desire, though it is real enough, as seven kinsmen of our Marmusha sheikh discovered when, on an earlier occasion, they were executed by him following the theft of one mangy, essentially valueless sheepskin from Cohen.

Culture is public because meaning is. You can't wink (or burlesque one) without knowing what counts as winking or how, physically, to contract your eyelids, and you can't conduct a sheep raid (or mimic one) without knowing what it is to steal a sheep and how practically to go about it. But to draw from such truths the conclusion that knowing how to wink is winking and knowing how to steal a sheep is sheep raiding is to betray as deep a confusion as, taking thin descriptions for thick, to identify winking with eyelid contractions or sheep raiding with chasing wooly animals out of pastures. The cognitivist fallacy – that culture consists (to quote another spokesman for the movement, Stephen Tyler) of "mental phenomena which can [he means 'should'] be analyzed by formal methods similar to those of mathematics and logic" – is as destructive of an effective use of the concept as are the behaviorist and idealist fallacies to which it is a misdrawn correction. Perhaps, as its errors are more sophisticated and its distortions subtler, it is even more so.

The generalized attack on privacy theories of meaning is, since early Husserl and late Wittgenstein, so much a part of modern thought that it need not be developed once more here. What is necessary is to see to it that the news of it reaches anthropology; and in particular that it is made clear that to say that culture consists of socially established structures of meaning in terms of which people do such things as signal conspiracies and join them or perceive insults and answer them, is no more than to say that it is a psychological phenomenon, a characteristic of someone's mind, personality, cognitive structure, or whatever, than to say that Tantrism, genetics, the progressive form of the verb, the classification of wines, the Common Law, or the notion of "a conditional curse" (as Westermarck defined the concept of *'ar* in terms of which Cohen pressed his claim to damages) is. What, in a place like Morocco, most prevents those of us who grew up winking other winks or attending other sheep from grasping what people are up to is not ignorance as to how cognition works (though, especially as, one assumes, it works the same among them as it does among us, it would greatly help to have less of that too) as a lack of familiarity with the imaginative universe within which their acts are signs. As Wittgenstein has been invoked, he may as well be quoted:

> We . . . say of some people that they are transparent to us. It is, however, important as regards this observation that one human being can be a complete enigma to another. We learn this when we come into a strange country with entirely strange traditions; and, what is more, even given a mastery of the country's language. We do not *understand* the people. (And not because of not knowing what they are saying to themselves.) We cannot find our feet with them.

IV

Finding our feet, an unnerving business which never more than distantly succeeds, is what ethnographic research consists of as a personal experience; trying to formulate the basis on which one imagines, always excessively, one has found them is what anthropological writing consists of as a scientific endeavor. We are not, or at least I am not, seeking to become natives (a compromised word in any case) or to mimic them. Only romantics or spies would seem to find point in that. We are seeking, in the widened sense of the term in which it encompasses very much more than talk, to converse with them, a matter a great deal more difficult, and not only with strangers, than is commonly recognized. "If speaking *for* someone else seems to be a mysterious process," Stanley Cavell has remarked, "that may be because speaking *to* someone does not seem mysterious enough."

Looked at in this way, the aim of anthropology is the enlargement of the universe of human discourse. That is not, of course, its only aim – instruction, amusement, practical counsel, moral advance, and the discovery of natural order in human behavior are others; nor is anthropology the only discipline which pursues it. But it is an aim to which a semiotic concept of culture is peculiarly well adapted. As interworked systems of construable signs (what, ignoring provincial usages, I would call symbols), culture is not a power something to which social events, behaviors, institutions, or processes can be causally attributed; it is a context, something within which they can be intelligibly – that is, thickly – described.

The famous anthropological absorption with the (to us) exotic – Berber horsemen, Jewish peddlers, French Legionnaires – is, thus, essentially a device for displacing the dulling sense of familiarity with which the mysteriousness of our own ability to relate perceptively to one another is concealed from us. Looking at the ordinary in places where it takes unaccustomed forms brings out not, as has so often been claimed, the arbitrariness of human behavior (there is nothing especially arbitrary about taking sheep theft for insolence in Morocco), but the degree to which its meaning varies according to the pattern of life by which it is informed. Understanding a people's culture exposes their normalness without reducing their particularity. (The more I manage to follow what the Moroccans are up to, the more logical, and the more singular, they seem.) It renders them accessible: setting them in the frame of their own banalities, it dissolves their opacity.

It is this maneuver, usually too casually referred to as "seeing things from the actor's point of view," too bookishly as "the *verstehen* approach," or too technically as "emic analysis," that so often leads to the notion that anthropology is a variety of either long-distance mind reading or cannibal-isle fantasizing, and which, for someone anxious to navigate past the wrecks of a dozen sunken philosophies, must therefore be executed with a great deal of care. Nothing is more necessary to comprehending anthropological interpretation, and the degree to which it *is* interpretation, than an exact understanding of what it means – and what it does not mean – to say that our formulations of other people's symbol systems must be actor-oriented.

What it means is that descriptions of Berber, Jewish, or French culture must be cast in terms of the constructions we imagine Berbers, Jews, or Frenchmen to place upon what they live through, the formulae they use to define what happens to them. What it does not mean is that such descriptions are themselves Berber, Jewish, or French – that is, part of the reality they are ostensibly describing; they are anthropological – that is, part of a developing system of scientific analysis. They must be cast in terms of the interpretations to which persons of a particular denomination subject their experience, because that is what they profess to be descriptions of; they are anthropological because it is, in fact, anthropologists who profess them. Normally, it is not necessary to point out quite so laboriously that the object of study is one thing and the study of it another. It is clear enough that the physical world is not physics and *A Skeleton Key to Finnegan's Wake* not *Finnegan's Wake*. But, as, in the study of culture, analysis penetrates into the very body of the object – that is, *we begin with our own interpretations of what our informants are up to, or think they are up to, and then systematize those* – the line between (Moroccan) culture as a natural fact and (Moroccan) culture as a theoretical entity tends to get blurred. All the more so, as the latter is presented in the form of an actor's-eye description of (Moroccan) conceptions of everything from violence, honor, divinity, and justice, to tribe, property, patronage, and chiefship.

In short, anthropological writings are themselves interpretations, and second and third order ones to boot. (By definition, only a "native" makes first order ones: it's *his* culture.) They are, thus, fictions; fictions, in the sense that they are "something made," "something fashioned" – the original meaning of *fictio* – *not* that they are false, unfactual, or merely "as if" thought experiments. To construct actor-oriented descriptions of the involvements of a Berber chieftain, a Jewish merchant, and a French soldier with one another in 1912 Morocco is clearly an imaginative act, not all that different from constructing similar descriptions of, say, the involvements with one another of a provincial French doctor, his silly, adulterous wife, and her feckless lover in nineteenth century France. In the latter case, the actors are represented as not having existed and the events as not having happened, while in the former they are represented as actual, or as having been so. This is a difference of no mean importance; indeed, precisely the one Madame Bovary had difficulty grasping. But the importance does not lie in the fact that her story was created while Cohen's was only noted. The conditions of their creation, and the point of it (to say nothing of the manner and the quality) differ. But the one is as much a *fictio* – "a making" – as the other.

Anthropologists have not always been as aware as they might be of this fact: that although culture exists in the trading post, the hill fort, or the sheep run, anthropology exists in the book, the article, the lecture, the museum display, or, sometimes nowadays, the film. To become aware of it is to realize that the line between mode of representation and substantive content is as undrawable in cultural analysis as it is in painting; and that fact in turn seems to threaten the objective status of anthropological knowledge by suggesting that its source is not social reality, but scholarly artifice.

It does threaten it, but the threat is hollow. The claim to attention of an ethnographic account does not rest on its author's ability to capture primitive facts in faraway places and carry them home like a mask or a carving, but on the degree to which he is able to clarify what goes on in such places, to reduce the puzzlement – what manner of men are these? – to which unfamiliar acts emerging out of unknown backgrounds naturally give rise. This raises some serious problems of verification, all right – or, if "verification" is too strong a word for so soft a science (I, myself, would prefer "appraisal"), of how you can tell a better account from a worse one. But that is precisely the virtue of it. If ethnography is thick description and ethnographers those who are doing the describing, then the determining question for any given example of it, whether a field journal squib or a Malinowski-sized monograph, is whether it sorts winks from twitches and real winks from mimicked ones. It is not against a body of uninterpreted data, radically thinned descriptions, that we must measure the cogency of our explications, but against the power of the scientific imagination to bring us into touch with the lives of strangers. It is not worth it, as Thoreau said, to go round the world to count the cats in Zanzibar.

[. . .]

VIII

There is an Indian story – at least I heard it as an Indian story – about an Englishman who, having been told that the world rested on a platform which rested on the back of an elephant which rested in turn on the back of a turtle, asked (perhaps he was an ethnographer; it is the way they behave), what did the turtle rest on? Another turtle. And that turtle? "Ah, Sahib, after that it is turtles all the way down."

Such, indeed, is the condition of things. I do not know how long it would be profitable to meditate on the encounter of Cohen, the sheikh, and "Dumari" (the period has perhaps already been exceeded); but I do know that however long I did so I would not get anywhere near to the bottom of it. Nor have I ever gotten anywhere near to the bottom of anything I have ever written about, either in the essays below or elsewhere. Cultural analysis is intrinsically incomplete. And, worse than that, the more deeply it goes the less complete it is. It is a strange science whose most telling assertions are its most tremulously based, in which to get somewhere with the matter at hand is to intensify the suspicion, both your own and that of others, that you are not quite getting it right. But that, along with plaguing subtle people with obtuse questions, is what being an ethnographer is like.

There are a number of ways to escape this – turning culture into art folklore and collecting it, turning it into traits and counting it, turning it into institutions and classifying it, turning it into structures and toying with it. But they *are* escapes.

The fact is that to commit oneself to a semiotic concept of culture and an interpretive approach to the study of it is to commit oneself to a view of ethnographic assertion as, to borrow W.B. Gallie's by now famous phrase, "essentially contestable." Anthropology, or at least interpretive anthropology, is a science whose progress is marked less by a perfection of consensus than by a refinement of debate. What gets better is the precision with which we vex each other.

This is very difficult to see when one's attention is being monopolized by a single party to the argument. Monologues are of little value here, because there are no conclusions to be reported; there is merely a discussion to be sustained. Insofar as the essays here collected have any importance, it is less in what they say than what they are witness to: an enormous increase in interest, not only in anthropology, but in social studies generally, in the role of symbolic forms in human life. Meaning, that elusive and ill-defined pseudoentity we were once more than content to leave philosophers and literary critics to fumble with, has now come back into the heart of our discipline. Even Marxists are quoting Cassirer; even positivists, Kenneth Burke.

My own position in the midst of all this has been to try to resist subjectivism on the one hand and cabbalism on the other, to try to keep the analysis of symbolic forms as closely tied as I could to concrete social events and occasions, the public world of common life, and to organize it in such a way that the connections between theoretical formulations and descriptive interpretations were unobscured by appeals to dark sciences. I have never been impressed by the argument that, as complete objectivity is impossible in these matters (as, of course, it is), one might as well let one's sentiments run loose. As Robert Solow has remarked, that is like saying that, as a perfectly aseptic environment is impossible, one might as well conduct surgery in a sewer. Nor, on the other hand, have I been impressed with claims that structural linguistics, computer engineering, or some other advanced form of thought is going to enable us to understand men without knowing them. Nothing will discredit a semiotic approach to culture more quickly than allowing it to drift into a combination of intuitionism and alchemy, no matter how elegantly the intuitions are expressed or how modern the alchemy is made to look.

The danger that cultural analysis, in search of all-too-deep-lying turtles, will lose touch with the hard surfaces of life – with the political, economic, stratificatory realities within which men are everywhere contained – and with the biological and physical necessities on which those surfaces rest, is an ever-present one. The only defense against it, and against, thus, turning cultural analysis into a kind of sociological aestheticism, is to train such analysis on such realities and such necessities in the first place. It is thus that I have written about nationalism, about violence, about identity, about human nature, about legitimacy, about revolution, about ethnicity, about urbanization, about status, about death, about time, and most of all about particular attempts by particular peoples to place these things in some sort of comprehensible, meaningful frame.

To look at the symbolic dimensions of social action – art, religion, ideology, science, law, morality, common sense – is not to turn away from the existential dilemmas of life for some empyrean realm of de-emotionalized forms; it is to plunge into the midst of them. The essential vocation of interpretive anthropology is not to answer our deepest questions, but to make available to us answers that others, guarding other sheep in other valleys, have given, and thus to include them in the consultable record of what man has said.

"The Concept(s) of Culture"

from *Beyond the Cultural Turn: New Directions in the Study of Society and Culture* (1999)

William Sewell, Jr.

Editors' introduction

William Sewell's analysis of the multiple uses of the concept of culture begins with the premise that the situation observed by Clifford Geertz in the late 1960s – that of a "a conceptual morass" resulting from culture's "theoretical diffusion" – has only grown in academia, rather than abated. Sewell argues that when Geertz wrote *The Interpretation of Cultures*, the concept of culture more or less "belonged" to anthropology. Since then, however, a great variety of disciplines have made culture central to their lines of inquiry: literary studies, sociology, political science, and of course geography. Indeed, while culture has always been, at least implicitly, a central feature of geography, geographers didn't begin to interrogate the concept of culture itself until the 1980s (but see Zelinsky, p. 113), and now debates over the idea of culture have become central to the work of cultural geography. Similarly, Sewell notes that that academic work on culture has in fact shifted from Geertz's call for an *interpretive* approach to culture to Abu-Lughod's call for writing *against* culture (see p. 50). Thus, scholarship has, Sewell argues, become ambivalent about the usefulness of the concept in academic inquiry, and in some cases has advocated "undoing" earlier knowledge built around the concept (this being Abu-Lughod's project). Like Geertz before him, then, Sewell sets out to inject some clarity into the situation. He argues that ambivalence about culture is unwarranted as long as we are clear about distinguishing the different uses and approaches to the concept.

Sewell boils the concepts of culture down to two very general approaches: First, culture as an abstract category of social life – that is, a category derived by scholars to help make sense and analyze a certain part of social life; and second, culture as distinct worlds of meaning denoting "a concrete and bounded world of beliefs and practices." That is, culture as the distinct way of life of a particular group of people. The former approach is essentially epistemological, focusing on culture as a way of knowing the social world, while the latter is essentially ontological, focusing on culture as an actually existing part of the world. In the latter approach, culture can be pluralized (there are many cultures around the world) while the former approach can only remain singular. Sewell goes on to present several categories of the epistemological approach, and finishes by presenting some helpful reminders about the nature of the ontological approach. Ultimately he finds that the epistemological approach is still useful in academic inquiry, while the ontological approach is questionable unless one heeds his cautions. Still, he doesn't reject an ontological approach to culture altogether because there remains something to be said for recognizing the distinct cultural differences around which social groups continue to organize and identify with.

It is also worth noting that Sewell devotes some attention to more recent approaches to culture as practice and/or performance (see selections by Latham, Yúdice, and McDowell and Court, pp. 68, 422, and 457 respectively), and he argues that such an approach is not incompatible with the Geertzian notion of culture as a system of meanings. Sewell points out some shortcomings of the Geertzian approach in order to explain the rise of alternative conceptions which sought to couch culture more in the realm of the non-cognitive or habitual practices of our daily lives. Sewell insists there is nothing mutually exclusive about these approaches. This argument is one that readers should keep in mind when considering the later readings by both Abu-Lughod and Latham in Part One.

William Sewell received his PhD in history in 1971 from the University of California at Berkeley. He is the Frank P. Hixon Distinguished Service Professor of History and Political Science at the University of Chicago, where he specializes in modern French social and cultural history, labor history, and social theory. Among many books and articles, he is the author of *Logics of History: Social Theory and Social Transformation* (2005) and the widely cited essay "A theory of structure: duality, agency and transformation" (*American Journal of Sociology* 98, 1, 1992: 1–29).

The two broad approaches to culture identified here by Sewell resonate with cultural geography in its earlier as well as more contemporary manifestations. While cultural geography has long professed a concern for the material expression of culture (that is, culture as distinct worlds of meaning) it has increasingly concerned itself with interrogating culture as an abstract category of social life. In this latter project, cultural geographers have considered the gamut of Sewell's different types, from "culture as learned behavior" (e.g. Zelinsky, p. 113) to "culture as creativity or agency" (e.g. Gibson's "Cultures at work," *Social and Cultural Geography* 4, 2, 2003) to "culture as practice" (e.g. Thrift's "Afterwords," *Environment and Planning D: Society and Space* 18, 2000).

[. . .]

During the 1980s and 1990s, the intellectual ecology of the study of culture has been transformed by a vast expansion of work on culture – indeed, a kind of academic culture mania has set in. The new interest in culture has swept over a wide range of academic disciplines and specialties. The history of this advance differs in timing and content in each field, but the cumulative effects are undeniable. In literary studies, which were already being transformed by French theory in the 1970s, the 1980s marked a turn to a vastly wider range of texts, quasi-texts, paratexts, and text analogs. If, as Derrida declared, nothing is extratextual ("il n'y a pas de hors-texte"), literary critics could direct their theory-driven gaze upon semiotic products of all kinds – legal documents, political tracts, soap operas, histories, talk shows, popular romances – and seek out their intertextualities. Consequently, as such "new historicist" critics as Stephen Greenblatt and Louis Montrose recognize, literary study is increasingly becoming the study of cultures. In history the early and rather self-conscious borrowing from anthropology has been followed by a theoretically heterogeneous rush to the study of culture, one modeled as much on literary studies or the work of Michel Foucault as on anthropology. As a conse-

quence, the self-confident "new social history" of the 1960s and 1970s was succeeded by an equally self-confident "new cultural history" in the 1980s.

In the late 1970s, an emerging "sociology of culture" began by applying standard sociological methods to studies of the production and marketing of cultural artifacts – music, art, drama, and literature. By the late 1980s, the work of cultural sociologists had broken out of the study of culture-producing institutions and moved toward studying the place of meaning in social life more generally. Feminism, which in the 1970s was concerned above all to document women's experiences, has increasingly turned to analyzing the discursive production of gender difference. Since the mid-1980s the new quasi-discipline of cultural studies has grown explosively in a variety of different academic niches – for example, in programs or departments of film studies, literature, performance studies, or communications. In political science, which is well known for its propensity to chase headlines, interest in cultural questions has been revived by the recent prominence of religious fundamentalism, nationalism, and ethnicity, which look like the most potent sources of political conflict in the contemporary world. This frenetic rush to the study of culture has everywhere been bathed, to a greater or lesser extent, in the pervasive transdisciplinary influence of the French poststructuralist trinity of Lacan, Derrida, and Foucault.

It is paradoxical that as discourse about culture becomes ever more pervasive and multifarious, anthropology, the discipline that invented the concept – or at least shaped it into something like its present form – is somewhat ambivalently backing away from its long-standing identification with culture as its keyword and central symbol. For the past decade and a half, anthropology has been rent by a particularly severe identity crisis, which has been manifested in anxiety about the discipline's epistemology, rhetoric, methodological procedures, and political implications. The reasons for the crisis are many – liberal and radical guilt about anthropology's association with Euro-American colonialism, the disappearance of the supposedly "untouched" or "primitive" peoples who were the favored subjects for classic ethnographies, the rise of "native" ethnographers who contest the right of European and American scholars to tell the "truth" about their people, and the general loss of confidence in the possibility of objectivity that has attended poststructuralism and postmodernism. As anthropology's most central and distinctive concept, "culture" has become a suspect term among critical anthropologists – who claim that both in academia and in public discourse, talk about culture tends to essentialize, exoticize, and stereotype those whose ways of life are being described and to naturalize their differences from white middle-class Euro-Americans. If Geertz's phrase "The Interpretation of Cultures" was the watchword of anthropology in the 1970s, Lila Abu-Lughod's "Writing against Culture" more nearly sums up the mood of the late 1980s and the 1990s.

[. . .]

WHAT DO WE MEAN BY CULTURE?

Writing in 1983, Raymond Williams declared that "culture is one of the two or three most complicated words in the English language." Its complexity has surely not decreased since then. I have neither the competence nor the inclination to trace out the full range of meanings of "culture" in contemporary academic discourse. But some attempt to sort out the different usages of the word seems essential, and it must begin by distinguishing two fundamentally different meanings of the term.

In one meaning, culture is a theoretically defined category or aspect of social life that must be abstracted out from the complex reality of human existence. Culture in this sense is always contrasted to some other equally abstract aspect or category of social life that is not culture, such as economy, politics, or biology. To designate something as culture or as cultural is to claim it for a particular academic discipline or subdiscipline – for example, anthropology or cultural sociology – or for a particular style or styles of analysis – for example, structuralism, ethno-science, componential analysis, deconstruction, or hermeneutics. Culture in this sense – as an abstract analytical category – only takes the singular. Whenever we speak of "cultures," we have moved to the second fundamental meaning.

In that second meaning, culture stands for a concrete and bounded world of beliefs and practices. Culture in this sense is commonly assumed to belong to or to be isomorphic with a "society" or with some clearly identifiable subsocietal group. We may speak of "American culture" or "Samoan culture," or of "middle-class culture" or "ghetto culture." The contrast in this usage is not between culture and not-culture but between one culture and another – between American, Samoan, French, and Bororo cultures, or between middle-class and upper-class cultures, or between ghetto and mainstream cultures.

This distinction between culture as theoretical category and culture as concrete and bounded body of beliefs and practices is, as far as I can discern, seldom made. Yet it seems to me crucial for thinking clearly about cultural theory. It should be clear, for example, that Ruth Benedict's concept of cultures as sharply distinct and highly integrated refers to culture in the second sense, while Claude Lévi-Strauss's notion that cultural meaning is structured by systems of oppositions is a claim about culture in the first sense. Hence their theories *of* "culture" are, strictly speaking, incommensurate: they refer to different conceptual universes. Failure to recognize this distinction between two fundamentally different meanings of the term has real consequences for contemporary cultural theory; some of the impasses of theoretical discourse in contemporary anthropology are attributable precisely to an unrecognized elision of the two. Thus, a dissatisfaction with "Benedictine" ethnographies that present cultures as uniformly well-bounded and coherent has led to what seem to me rather confused attacks on "the culture concept" in general – attacks that fail to distinguish Benedictine claims about the tight integration of cultures from Lévi-Straussian claims about the semiotic coherence of culture as a system of meanings. Conversely, anthropologists who defend the culture concept also tend to conflate the two meanings, regarding claims that cultures are rent with fissures or that their boundaries are porous as implying an abandonment of the concept of culture altogether.

Here, I will be concerned primarily with culture in the first sense – culture as a category of social life. One must have a clear conception of culture at this abstract level in order to deal with the more concrete theoretical question of how cultural differences are patterned and bounded in space and time. Once I have sketched out my own ideas about what an adequate abstract theory of culture might look like, I will return to the question of culture as a bounded universe of beliefs and practices – to the question of cultures in the Benedictine sense.

CULTURE AS A CATEGORY OF SOCIAL LIFE

Culture as a category of social life has itself been conceptualized in a number of different ways. Let me begin by specifying some of these different conceptualizations, moving from those I do not find especially useful to those I find more adequate.

Culture as learned behavior. Culture in this sense is the whole body of practices, beliefs, institutions, customs, habits, myths, and so on built up by humans and passed on from generation to generation. In this usage, culture is contrasted to nature: its possession is what distinguishes us from other animals. When anthropologists were struggling to establish that differences between societies were not based on biological differences between their populations – that is, on race – a definition of culture as learned behavior made sense. But now that racial arguments have virtually disappeared from anthropological discourse, a concept of culture so broad as this seems impossibly vague; it provides no

particular angle or analytical purchase on the study of social life.

A narrower and consequently more useful conceptualization of culture emerged in anthropology during the second quarter of the twentieth century and has been dominant in the social sciences generally since World War II. It defines culture not as all learned behavior but as that category or aspect of learned behavior that is concerned with meaning. But the concept of culture-as-meaning is in fact a family of related concepts; *meaning* may be used to specify a cultural realm or sphere in at least four distinct ways, each of which is defined in contrast to somewhat differently conceptualized noncultural realms or spheres.

Culture as an institutional sphere devoted to the making of meaning. This conception of culture is based on the assumption that social formations are composed of clusters of institutions devoted to specialized activities. These clusters can be assigned to variously defined institutional spheres – most conventionally, spheres of politics, economy, society, and culture. Culture is the sphere devoted specifically to the production, circulation, and use of meanings. The cultural sphere may in turn be broken down into the subspheres of which it is composed: say, of art, music, theater, fashion, literature, religion, media, and education. The study of culture, if culture is defined in this way, is the study of the activities that take place within these institutionally defined spheres and of the meanings produced in them.

This conception of culture is particularly prominent in the discourses of sociology and cultural studies, but it is rarely used in anthropology. Its roots probably reach back to the strongly evaluative conception of culture as a sphere of "high" or "uplifting" artistic and intellectual activity, a meaning that Raymond Williams tells us came into prominence in the nineteenth century. But in contemporary academic discourse, this usage normally lacks such evaluative and hierarchizing implications. The dominant style of work in American sociology of culture has been demystifying: its typical approach has been to uncover the largely self-aggrandizing, class-interested, manipulative, or professionalizing institutional dynamics that undergird prestigious museums, artistic styles, symphony orchestras, or philosophical schools. And cultural studies, which has taken as its particular mission the appreciation of cultural forms disdained by the spokesmen of high culture – rock music, street fashion, cross-dressing, shopping malls, Disneyland, soap operas – employs this same basic definition of culture. It merely trains its analytical attention on spheres of meaning production ignored by previous analysts and regarded as debased by elite tastemakers.

The problem with such a concept of culture is that it focuses only on a certain range of meanings, produced in a certain range of institutional locations – on self-consciously "cultural" institutions and on expressive, artistic, and literary systems of meanings. This use of the concept is to some extent complicit with the widespread notion that meanings are of minimal importance in the other "noncultural" institutional spheres: that in political or economic spheres, meanings are merely superstructural excrescences. And since institutions in political and economic spheres control the great bulk of society's resources, viewing culture as a distinct sphere of activity may in the end simply confirm the widespread presupposition in the "harder" social sciences that culture is merely froth on the tides of society. The rise of a cultural sociology that limited itself to studying "cultural" institutions effected a partition of subject matter that was very unfavorable to the cultural sociologists. Indeed, only the supersession of this restrictive concept of culture has made possible the explosive growth of the subfield of cultural sociology in the past decade.

Culture as creativity or agency. This usage of culture has grown up particularly in traditions that posit a powerful "material" determinism – most notably Marxism and American sociology. Over the past three decades or so, scholars working within these traditions have carved out a conception of culture as a realm of creativity that escapes from the otherwise pervasive determination of social action by economic or social structures. In the Marxist tradition, it was probably E.P. Thompson's *Making of the English Working Class* that first conceptualized culture as a realm of agency, and it is particularly English Marxists – for example, Paul Willis in *Learning to Labor* – who have elaborated this conception. But the defining opposition on which this concept of culture rests – culture versus structure – has also become pervasive in the vernacular of

American sociology. One clear sign that American anthropologists and sociologists have different conceptions of culture is that the opposition between culture and structure – an unquestioned commonplace in contemporary sociological discourse – is nonsensical in anthropology.

In my opinion, identifying culture with agency and contrasting it with structure merely perpetuates the same determinist materialism that "culturalist" Marxists were reacting against in the first place. It exaggerates both the implacability of socioeconomic determinations and the free play of symbolic action. Both socioeconomic and cultural processes are blends of structure and agency. Cultural action – say, performing practical jokes or writing poems – is necessarily constrained by cultural structures, such as existing linguistic, visual, or ludic conventions. And economic action – such as the manufacture or repair of automobiles – is impossible without the exercise of creativity and agency. The particulars of the relationship between structure and agency may differ in cultural and economic processes, but assigning either the economic or the cultural exclusively to structure or to agency is a serious category error.

This brings us to the two concepts of culture that I regard as most fruitful and that I see as currently struggling for dominance: the concept of culture as a system of symbols and meanings, which was hegemonic in the 1960s and 1970s, and the concept of culture as practice, which has become increasingly prominent in the 1980s and 1990s.

Culture as a system of symbols and meanings. This has been the dominant concept of culture in American anthropology since the 1960s. It was made famous above all by Clifford Geertz, who used the term "cultural system" in the titles of some of his most notable essays. The notion was also elaborated by David Schneider, whose writings had a considerable influence within anthropology but lacked Geertz's interdisciplinary appeal. Geertz and Schneider derived the term from Talcott Parsons's usage, according to which the cultural system, a system of symbols and meanings, was a particular "level of abstraction" of social relations. It was contrasted to the "social system," which was a system of norms and institutions, and to the "personality system," which was a system of motivations. Geertz and Schneider especially wished to distinguish the cultural system from the social system. To engage in cultural analysis, for them, was to abstract the meaningful aspect of human action out from the flow of concrete interactions. The point of conceptualizing culture as a system of symbols and meanings is to disentangle, for the purpose of analysis, the semiotic influences on action from the other sorts of influences – demographic, geographical, biological, technological, economic, and so on – that they are necessarily mixed with in any concrete sequence of behavior.

Geertz's and Schneider's post-Parsonian theorizations of cultural systems were by no means the only available models for symbolic anthropology in the 1960s and 1970s. The works of Victor Turner, whose theoretical origins were in the largely Durkheimian British school of social anthropology, were also immensely influential. Claude Lévi-Strauss and his many followers provided an entire alternative model of culture as a system of symbols and meanings – conceptualized, following Saussure, as signifiers and signifieds. Moreover, all these anthropological schools were in a sense manifestations of a much broader "linguistic turn" in the human sciences – a diverse but sweeping attempt to specify the structures of human symbol systems and to indicate their profound influence on human behavior. One thinks above all of such French "structuralist" thinkers as Roland Barthes, Jacques Lacan, or the early Michel Foucault. What all of these approaches had in common was an insistence on the systematic nature of cultural meaning and the autonomy of symbol systems – their distinctness from and irreducibility to other features of social life. They all abstracted a realm of pure signification out from the complex messiness of social life and specified its internal coherence and deep logic. Their practice of cultural analysis consequently tended to be more or less synchronic and formalist.

Culture as practice. The past decade and a half has witnessed a pervasive reaction against the concept of culture as a system of symbols and meanings, which has taken place in various disciplinary locations and intellectual traditions and under many different slogans – for example, "practice," "resistance," "history," "politics," or "culture as tool kit." Analysts working under all these banners object to a portrayal of culture as logical, coherent, shared,

uniform, and static. Instead they insist that culture is a sphere of practical activity shot through by willful action, power relations, struggle, contradiction, and change.

In anthropology, Sherry Ortner in 1984 remarked on the turn to politics, history, and agency, suggesting Pierre Bourdieu's key term "practice" as an appropriate label for this emerging sensibility. Two years later the publication of James Clifford and George Marcus's collection *Writing Culture* announced to the public the crisis of anthropology's culture concept. Since then, criticisms of the concept of culture as a system of symbols and meanings have flowed thick and fast. The most notable work in anthropology has argued for the contradictory, politically charged, changeable, and fragmented character of meanings – both meanings produced in the societies being studied and meanings rendered in anthropological texts. Recent work in anthropology has in effect recast culture as a performative term.

Not surprisingly, this emphasis on the performative aspect of culture is compatible with the work of most cultural historians. Historians are generally uncomfortable with synchronic concepts. As they took up the study of culture, they subtly – but usually without comment – altered the concept by stressing the contradictoriness and malleability of cultural meanings and by seeking out the mechanisms by which meanings were transformed. The battles in history have been over a different issue, pitting those who claim that historical change should be understood as a purely cultural or discursive process against those who argue for the significance of economic and social determinations or for the centrality of concrete "experience" in understanding it.

Sociologists, for rather different reasons, have also favored a more performative conception of culture. Given the hegemony of a strongly causalist methodology and philosophy of science in contemporary sociology, cultural sociologists have felt a need to demonstrate that culture has causal efficacy in order to gain recognition for their fledgling subfield. This has led many of them to construct culture as a collection of variables whose influence on behavior can be rigorously compared to that of such standard sociological variables as class, ethnicity, gender, level of education, economic interest, and the like. As a result, they have moved away from earlier Weberian, Durkheimian, or Parsonian conceptions of culture as rather vague and global value orientations to what Ann Swidler has termed a "tool kit" composed of a "repertoire" of "strategies of action." For many cultural sociologists, then, culture is not a coherent system of symbols and meanings but a diverse collection of "tools" that, as the metaphor indicates, are to be understood as means for the performance of action. Because these tools are discrete, local, and intended for specific purposes, they can be deployed as explanatory variables in a way that culture conceived as a translocal, generalized system of meanings cannot.

CULTURE AS SYSTEM AND PRACTICE

Much of the theoretical writing on culture during the past ten years has assumed that a concept of culture as a system of symbols and meanings is at odds with a concept of culture as practice. System and practice approaches have seemed incompatible, I think, because the most prominent practitioners of the culture-as-system-of-meanings approach effectively marginalized consideration of culture-as-practice – if they didn't preclude it altogether.

This can be seen in the work of both Clifford Geertz and David Schneider. Geertz's analyses usually begin auspiciously enough, in that he frequently explicates cultural systems in order to resolve a puzzle arising from concrete practices – a state funeral, trances, a royal procession, cockfights. But it usually turns out that the issues of practice are principally a means of moving the essay to the goal of specifying in a synchronic form the coherence that underlies the exotic cultural practices in question. And while Geertz marginalized questions of practice, Schneider, in a kind of *reductio ad absurdum*, explicitly excluded them, arguing that the particular task of anthropology in the academic division of labor was to study "culture as a system of symbols and meanings in its own right and with reference to its own structure" and leaving to others – sociologists, historians, political scientists, or economists – the question of how social action was structured. A "cultural account," for Schneider, should be limited to specifying the relations among symbols in a given domain of meaning – which he tended to render unproblematically

as known and accepted by all members of the society and as possessing a highly determinate formal logic.

Nor is the work of Geertz and Schneider unusual in its marginalization of practice. As critics such as James Clifford have argued, conventional modes of writing in cultural anthropology typically smuggle highly debatable assumptions into ethnographic accounts – for example, that cultural meanings are normally shared, fixed, bounded, and deeply felt. To Clifford's critique of ethnographic rhetoric, I would add a critique of ethnographic method. Anthropologists working with a conception of culture-as-system have tended to focus on clusters of symbols and meanings that can be shown to have a high degree of coherence or systematicity – those of American kinship or Balinese cockfighting, for instance – and to present their accounts of these clusters as examples of what the interpretation of culture in general entails. This practice results in what sociologists would call sampling on the dependent variable. That is, anthropologists who belong to this school tend to select symbols and meanings that cluster neatly into coherent systems and pass over those that are relatively fragmented or incoherent, thus confirming the hypothesis that symbols and meanings indeed form tightly coherent systems.

Given some of these problems in the work of the culture-as-system school, the recent turn to a concept of culture-as-practice has been both understandable and fruitful – it has effectively highlighted many of the earlier school's shortcomings and made up some of its most glaring analytic deficits. Yet the presumption that a concept of culture as a system of symbols and meanings is at odds with a concept of culture as practice seems to me perverse. System and practice are complementary concepts: each presupposes the other. To engage in cultural practice means to utilize existing cultural symbols to accomplish some end. The employment of a symbol can be expected to accomplish a particular goal only because the symbols have more or less determinate meanings – meanings specified by their systematically structured relations to other symbols. Hence practice implies system. But it is equally true that the system has no existence apart from the succession of practices that instantiate, reproduce, or – most interestingly – transform it. Hence system implies practice. System and practice constitute an indissoluble duality or dialectic: the important theoretical question is thus not whether culture should be conceptualized as practice or as a system of symbols and meanings, but how to conceptualize the articulation of system and practice.

[. . .]

CULTURES AS DISTINCT WORLDS OF MEANING

Up to now, I have been considering culture only in its singular and abstract sense – as a realm of social life defined in contrast to some other non-cultural realm or realms. My main points may be summarized as follows: culture, I have argued, should be understood as a dialectic of system and practice, as a dimension of social life autonomous from other such dimensions both in its logic and in its spatial configuration, and as a system of symbols possessing a real but thin coherence that is continually put at risk in practice and therefore subject to transformation. Such a theorization, I maintain, makes it possible to accept the cogency of recent critiques yet retain a workable and powerful concept of culture that incorporates the achievements of the cultural anthropology of the 1960s and 1970s.

But it is probably fair to say that most recent theoretical work on culture, particularly in anthropology, is actually concerned primarily with culture in its pluralizable and more concrete sense – that is, with cultures as distinct worlds of meaning. Contemporary anthropological critics' objections to the concept of culture as system and their insistence on the primacy of practice are not, in my opinion, really aimed at the concept of system as outlined above – the notion that the meaning of symbols is determined by their network of relations with other symbols. Rather, the critics' true target is the idea that cultures (in the second, pluralizable sense) form neatly coherent wholes: that they are logically consistent, highly integrated, consensual, extremely resistant to change, and clearly bounded. This is how cultures tended to be represented in the classic ethnographies – Mead on Samoa, Benedict on the Zuni, Malinowski on the Trobriands, Evans-Pritchard on the Nuer, or, for that matter, Geertz on the Balinese. But recent research and thinking about cultural practices, even in relatively "simple" societies, has turned this classic model on its head. It now appears that we should think of worlds of meaning as normally being contradictory, loosely integrated, contested, mutable, and highly permeable. Consequently the very concept of cultures as coherent and distinct entities is widely disputed.

Cultures are contradictory. Some authors of classic ethnographies were quite aware of the presence of contradictions in the cultures they studied. Victor Turner, for example, demonstrated that red symbolism in certain Ndembu rituals simultaneously signified the contradictory principles of matrilineal fertility and male bloodletting. But he emphasized how these potentially contradictory meanings were brought together and harmonized in ritual performances. A current anthropological sensibility would probably emphasize the fundamental character of the contradictions rather than their situational resolution in the ritual. It is common for potent cultural symbols to express contradictions as much as they express coherence. One need look no farther than the central Christian symbol of the Trinity, which attempts to unify in one symbolic figure three sharply distinct and largely incompatible possibilities of Christian religious experience: authoritative and hierarchical orthodoxy (the Father), loving egalitarianism and grace (the Son), and ecstatic spontaneity (the Holy Ghost). Cultural worlds are commonly beset with internal contradictions.

Cultures are loosely integrated. Classic ethnographies recognized that societies were composed of different spheres of activity – for example, kinship, agriculture, hunting, warfare, and religion – and that each of these component parts had its own specific cultural forms. But the classic ethnographers typically saw it as their task to show how these culturally varied components fit into a well-integrated cultural whole. Most contemporary students of culture would question this emphasis. They are more inclined to stress the centrifugal cultural tendencies that arise from these disparate spheres of activity, to stress the inequalities between those relegated to different activities, and to see whatever "integration" occurs as based on power or domination rather than on a common ethos. That most anthropologists now work on complex, stratified, and highly differentiated societies, rather than on the "simple" societies that were the focus of most classic ethnographies, probably enhances this tendency.

Cultures are contested. Classic ethnographies commonly assumed, at least implicitly, that a culture's most important beliefs were consensual, agreed on by virtually all of a society's members. Contemporary scholars, with their enhanced awareness of race, class, and gender, would insist that people who occupy different positions in a given social order will typically have quite different cultural beliefs or will have quite different understandings of what might seem on the surface to be identical beliefs. Consequently, current scholarship is replete with depictions of "resistance" by subordinated groups and individuals. Thus James Scott detects "hidden transcripts" that form the underside of peasants' deference in contemporary Malaysia and Marshall Sahlins points out that it was Hawaiian women who most readily violated tabus when Captain Cook's ships arrived – because the tabu system, which classified them as profane (*noa*) as against the sacred (*tabu*) men, "did not sit upon Hawaiian women with the force it had for men." Cultural consensus, far from being the normal state of things, is a difficult achievement; and when it does occur it is bound to hide suppressed conflicts and disagreements.

Cultures are subject to constant change. Cultural historians, who work on complex and dynamic societies, have generally assumed that cultures are quite changeable. But recent anthropological work on relatively "simple" societies also finds them to be remarkably mutable. For example, Renato Rosaldo's study of remote Ilongot headhunters in the highlands of northern Luzon demonstrates that each generation of Ilongots constructed its own logic of settlement patterns, kinship alliance, and feuding – logics that gave successive generations of Ilongots experiences that were probably as varied as those of successive generations of Americans or Europeans between the late nineteenth and late twentieth centuries.

Cultures are weakly bounded. It is extremely unusual for societies or their cultural systems to be anything like isolated or sharply bounded. Even the supposedly simplest societies have had relations of trade, warfare, conquest, and borrowing of all sorts of cultural items – technology, religious ideas, political and artistic forms, and so on. But in addition to mutual influences of these sorts, there have long been important social and cultural processes that transcend societal boundaries – colonialism, missionary religions, inter-regional trading associations and economic interdependencies, migratory diasporas, and, in the current era, multinational corporations and transnational nongovernmental organizations. Although these transsocietal processes are certainly more prominent in more recent history than previously, they are hardly entirely new. Think of the spread of such "world religions" as Islam, Christianity, Hinduism, or Buddhism across entire regions of the globe or the development of extensive territorial empires in the ancient world. I would argue that social science's once virtually unquestioned model of societies as clearly bounded identities undergoing endogenous development is as perverse for the study of culture as for the study of economic history or political sociology. Systems of meaning do not correspond in any neat way with national or societal boundaries – which themselves are not nearly as neat as we sometimes imagine. Anything we might designate as a "society" or a "nation" will contain, or fail to contain, a multitude of overlapping and interpenetrating cultural systems, most of them subsocietal, transsocietal, or both.

Thus all of the assumptions of the classic ethnographic model of cultures – that cultures are logically consistent, highly integrated, consensual, resistant to change, and clearly bounded – seem to be untenable. This could lead to the conclusion that the notion of coherent cultures is purely illusory; that cultural practice in a given society is diffuse and decentered; that the local systems of meaning found in a given population do not themselves form a higher-level, societywide system of meanings. But such a conclusion would, in my opinion, be hasty. Although I think it is an error simply to assume that cultures possess an overall coherence or integration, neither can such coherences be ruled out a priori.

[. . .]

It is no longer possible to assume that the world is divided up into discrete "societies," each with its corresponding and well-integrated "culture." I would argue forcefully for the value of the concept of culture in its nonpluralizable sense, while the utility of the term as pluralizable appears to me more open to legitimate question. Yet I think that the latter concept of culture also gets at something we need to retain: a sense of the particular shapes and consistencies of worlds of meaning in different places and times and a sense that in spite of conflicts and resistance, these worlds of meaning somehow hang together. Whether we call these partially coherent landscapes of meaning "cultures" or something else – worlds of meaning, or ethnoscapes, or hegemonies – seems to me relatively unimportant so long as we know that their boundedness is only relative and constantly shifting. Our job as cultural analysts is to discern what the shapes and consistencies of local meanings actually are and to determine how, why, and to what extent they hang together.

"Writing against Culture"

from Richard G. Fox (ed.) *Recapturing Anthropology: Working in the Present* (1991)

Lila Abu-Lughod

Editors' introduction

When Clifford Geertz wrote that a semiotic approach to culture challenged the assumption of a clear divide between the ethnographer and informant and therefore raised some important questions about the "objectivity" of cultural interpretation, a line of critical inquiry was launched that perhaps found its final apogee in the influential volume *Writing Culture* (1986). Edited by James Clifford and George Marcus, this book was the outcome of a seminar that sought to explore the making of ethnographic texts. *Writing Culture* considered the "politics and poetics" of ethnographies as "partial truths," "situated knowledges," and even "fictions." Influenced by poststructuralist theory, the volume's authors scrutinized the ways knowledge about culture was produced, how language itself structured such knowledge, and how scholarly interpretations should not be viewed as transparent media through which one might gain a complete understanding of other people. The authors of the volume raised such questions about the production of knowledge and the questionable objectivity of ethnographic accounts by subjecting them to a *textual* critique. If culture, in other words, was similar to language (that is, a system of signs and meanings), as Geertz had initially suggested, then it was time to challenge it with the same poststructural theories of language that were being applied throughout the humanities during the late 1970s and 1980s.

Lila Abu-Lughod begins her essay with the claim that the arguments made in *Writing Culture* need to be extended to a more radical conclusion. Rather than settle for new textual strategies in ethnography that acknowledge the "partial" and "situated" qualities of ethnographic texts, Abu-Lughod argues for strategies of writing *against* culture altogether. Culture, she argues, remains too laden with the assumptions of a divide between the knowledgeable scholar (that is, the "subject," the "self") and the person whose culture is under investigation (the "object", the "other"). *Writing Culture* did not go far enough to challenge this basic divide, Abu-Lughod argues, because it did not directly address the situations of feminist scholars and what she calls *halfies* (people of mixed national or cultural identity). Had feminist perspectives been considered, for instance, a more basic challenge to the *self–other divide* upon which ethnographic inquiry is based would have been revealed. The feminist argument Abu-Lughod references here is that the "self" is created by being contrasted to some "other." That self–other binary lies at the heart of our sense of identity and is expressed in many different ways (e.g. man/woman; straight/gay; local citizen/outside alien). But the most important part of recognizing this binary is to understand that it always entails some kind of *uneven or hierarchical* relationship. Because "culture" is the tool for creating this self–other binary in disciplines focusing on culture, such as anthropology or geography, it carries with it the baggage of hierarchy.

Anthropologists and cultural geographers have long recognized the colonial and imperialist contexts within which their forebears worked. These contexts serve as a focal point for much of the discussion in Part Two of the *Reader*. Clearly, nineteenth and early twentieth century scholars from Europe and North America who worked in Latin America, Africa, and much of Asia and Oceana carried out their work under the colonial flag and with the kind of impunity that their connections to imperial power afforded them. Yet Abu-Lughod finds that contemporary scholars have failed to really come to terms with the fact that the idea of culture as a "whole way of life" came about because European and North American scholars were able to study "others" in a colonial situation in which those scholars also held considerable *power* over those "others". Work on culture today needs to not simply acknowledge this history, Abu-Lughod argues, but actively work *against* it by developing critical challenges to the idea of culture as we know it today.

Lila Abu-Lughod received her PhD at Harvard University and teaches anthropology and Women's and Gender Studies at Columbia University. She has also taught at Williams College, Princeton University, and New York University. Among many books and articles, she is the author of *Dramas of Nationhood: The Politics of Television in Egypt* (2005) and *Writing Women's Worlds: Bedouin Stories* (1993).

Abu-Lughod's work is important in cultural geography not only because of her interrogation of the concept of culture itself, but because of her work bridging feminist theory, national identity, and popular culture. Her work on the Egyptian media, for instance, helps trace the linkages across scale between local practices of viewing television soap operas and larger practices of nation building. As with much contemporary cultural geography, culture here is viewed as "ordinary" and "situated" in a local context and place, and yet this does not mean it is not also part of the apparatus that builds larger-scale processes, such as the construction of a national identity. In addition, the feminist critique in Abu-Lughod's work has played a significant role in shaping debates and new directions in contemporary cultural geography. Work, for example, by Gillian Rose (*Feminism and Geography*, 1993), Geraldine Pratt (*Working Feminism*, 2004), and Nicky Gregson (*Second Hand Cultures*, 2003), demonstrates many of the approaches to writing "against culture" advocated here by Abu-Lughod, including a focus on everyday practice, situating the researcher in connection with her research subjects, and focusing on "ethnographies of the particular."

Writing Culture, the collection that marked a major new form of critique of cultural anthropology's premises, more or less excluded *two* critical groups whose situations neatly expose and challenge the most basic of those premises: feminists and "halfies" – (people whose national or cultural identity is mixed by virtue of migration, overseas education, or parentage). In his introduction, Clifford apologizes for the feminist absence; no one mentions halfies or the indigenous anthropologists to whom they are related. Perhaps they are not yet numerous enough or sufficiently self-defined as a group. The importance of these two groups lies not in any superior moral claim or advantage they might have in doing anthropology, but in the special dilemmas they face, dilemmas that reveal starkly the problems with cultural anthropology's assumption of a fundamental distinction between self and other.

In this essay I explore how feminists and halfies, by the way their anthropological practice unsettles the boundary between self and other, enable us to reflect on the conventional nature and political effects of this distinction and ultimately to reconsider the value of the concept of culture on which it depends. I will argue that culture operates in anthropological discourse to enforce separations that inevitably carry a sense of hierarchy. Therefore anthropologists should now pursue, without exaggerated hopes for the power of their texts to change the world, a variety of strategies for writing *against* culture. For those interested in textual strategies, I explore the advantages of what I call "ethnographies of the particular" as instruments of a tactical humanism.

SELVES AND OTHERS

The notion of culture (especially as it functions to distinguish "cultures"), despite a long usefulness, may now have become something anthropologists would want to work against in their theories, their ethnographic practice, and their ethnographic writing. A helpful way to begin to grasp why is to consider what the shared elements of feminist and halfie anthropology clarify about the self/other distinction central to the paradigm of anthropology. Marilyn Strathern raises some of the issues regarding feminism in essays ["Dislodging a worldview" in *Australian Feminist Studies* 1, 1985, and "An awkward relationship" in *Signs* 12, 1987] that both Clifford and Rabinow cited in *Writing Culture*. Her thesis is that the relationship between anthropology and feminism is awkward. This thesis leads her to try to understand why Feminist scholarship, in spite of its rhetoric of radicalism, has failed to fundamentally alter anthropology and why feminism has gained even less from anthropology than vice versa.

The awkwardness, she argues, arises from the fact that despite a common interest in differences, the scholarly practices of feminists and anthropologists are differently structured in the way they organize knowledge and draw boundaries, and especially in the nature of the investigators' *relationship to* their subject matter. Feminist scholars, united by their common opposition to men or to patriarchy, produce a discourse composed of many voices; they "discover the self by becoming conscious of oppression from the Other." Anthropologists, whose goal is "to make sense of differences," also constitute their "selves" in relation to an other, but do not view this other as "under attack."

In highlighting the self/other relationship, Strathern takes us to the heart of the problem. Yet she retreats from the problematic of power (granted as formative in feminism) in her strangely uncritical depiction of anthropology. When she defines anthropology as "a discipline that continues to know itself as the study of social behavior or society in terms of systems and collective representations," she underplays the self/other distinction. In characterizing the relationship between anthropological self and other as nonadversarial, she ignores its most fundamental aspect. Anthropology's avowed goal may be "the study of man [*sic*]," but it is a discipline built on the historically constructed divide between the West and the non-West. It has been and continues to be primarily the study of the non-Western other by the Western self, even if in its new guise it seeks explicitly to give

voice to the Other or to present a dialogue between the self and other, either textually or through an explication of the fieldwork encounter. And the relationship between the West and the non-West, at least since the birth of anthropology, has been constituted by Western domination. This suggests that the awkwardness Strathern senses in the relationship between feminism and anthropology might better be understood as the result of diametrically opposed processes of self-construction through opposition to others – processes that begin from different sides of a power divide.

[. . .]

If anthropology continues to be practiced as the study by an unproblematic and unmarked Western self of found "others" out there, feminist theory, an academic practice that also traffics in selves and others, has in its relatively short history come to realize the danger of treating selves and others as givens. It is instructive for the development of a critique of anthropology to consider the trajectory that has led, within two decades, to what some might call a crisis in feminist theory, and others, the development of postfeminism.

From Simone de Beauvoir on, it has been accepted that, at least in the modern West, women have been the other to men's self. Feminism has been a movement devoted to helping women become selves and subjects rather than objects and men's others. The crisis in feminist theory (related to a crisis in the women's movement) that followed on the heels of feminist attempts to turn those who had been constituted as other into selves – or, to use the popular metaphor, to let women speak – was the problem of "difference." For whom did feminists speak? Within the women's movement, the objections of lesbians, African-American women, and other "women of color" that their experiences as women were different from those of white, middle-class, heterosexual women problematized the identity of women as selves. Cross-cultural work on women also made it clear that masculine and feminine did not have, as we say, the same meanings in other cultures, nor did Third World women's lives resemble Western women's lives. As Harding puts it, the problem is that "once 'woman' is deconstructed into 'women' and 'gender' is recognized to have no fixed referents, feminism itself dissolves as a theory that can reflect the voice of a naturalized or essentialized speaker."

From its experience with this crisis of selfhood or subjecthood, feminist theory can offer anthropology two useful reminders. First, the self is always a construction, never a natural or found entity, even if it has that appearance. Second, the process of creating a self through opposition to an other always entails the violence of repressing or ignoring other forms of difference. Feminist theorists have been forced to explore the implications for the formation of identity and the possibilities for political action of the ways in which gender as a system of difference is intersected by other systems of difference, including, in the modern capitalist world, race and class.

Where does this leave the feminist anthropologist? Strathern characterizes her as experiencing a tension – "caught between structures faced with two different ways of relating to her or his subject matter." The more interesting aspect of the feminist's situation, though, is what she shares with the halfie: a blocked ability to comfortably assume the self of anthropology. For both, although in different ways, the self is split, caught at the intersection of systems of difference. I am less concerned with the existential consequences of this split . . . than with the awareness such splits generate about three crucial issues: positionality, audience, and the power inherent in distinctions of self and other. What happens when the "other" that the anthropologist is studying is simultaneously constructed as, at least partially, a self?

Feminists and halfie anthropologists cannot easily avoid the issue of positionality. Standing on shifting ground makes it clear that every view is a view from somewhere and every act of speaking a speaking from somewhere. Cultural anthropologists have never been fully convinced of the ideology of science and have long questioned the value, possibility, and definition of objectivity. But they still seem reluctant to examine the implications of the actual situatedness of their knowledge.

Two common, intertwined objections to the work of feminist or native or semi-native anthropologists, both related to partiality, betray the persistence of ideals of objectivity. The first has to do with the partiality (as bias or position) of the observer. The second has to do with the partial (incomplete) nature of the picture presented. Halfies are more associated with the first problem, feminists the second. The problem with studying one's own society is alleged to be the problem of gaining enough distance. Since, for halfies, the Other is in certain ways the self, there is said to be the danger shared with indigenous anthropologists of identification and the easy slide into subjectivity. These worries suggest that the anthropologist is still defined as a being who must stand apart from the Other, even when he or she seeks explicitly to bridge the gap. Even Bourdieu, who perceptively analyzed the effects this outsider stance has on the anthropologist's (mis)understanding of social life, fails to break with this doxa. The obvious point he misses is that the outsider self never simply stands outside. He or she stands in a definite relation with the Other of the study, not just as a Westerner, but as a Frenchman in Algeria during the war of independence, an American in Morocco during the 1967 Arab–Israeli war, or an Englishwoman in postcolonial India. What we call the outside is a position within a larger political-historical complex. No less than the halfie, the "wholie" is in a specific position vis à vis the community being studied.

The debates about feminist anthropologists suggest a second source of uneasiness about positionality. Even when they present themselves as studying gender, feminist anthropologists are dismissed as presenting only a partial picture of the societies they study because they are assumed to be studying only women. Anthropologists study society, the unmarked form. The study of women is the marked form, too readily sectioned off, as Strathern notes. Yet it could easily be argued that most studies of society have been equally partial. As restudies like Weiner's of Malinowski's Trobriand Islanders or Bell's of the well-studied Australian aborigines indicate, they have been the study of men. This does not make such studies any less valuable; it merely reminds us that we must constantly attend to the positionality of the anthropological self and its representations of others. James Clifford, among others, has convincingly argued that ethnographic representations are always "partial truths." What is needed is a recognition that they are also positioned truths.

Split selfhood creates for the two groups being discussed a second problem that is illuminating for anthropology generally: multiple audiences. Although all anthropologists are beginning to feel what might be called the Rushdie effect – the effects of living in a global age when the subjects of their studies begin to read their works and the governments of the countries they work in ban books and deny visas – feminist and halfie anthropologists struggle in poignant ways with multiple accountability. Rather than having one primary audience, that of other anthropologists, feminist anthropologists write for anthropologists and for feminists, two groups whose relationship to their subject matter is at odds and who hold ethnographers accountable in different ways. Furthermore, feminist circles include non-Western feminists, often from the societies feminist anthropologists have studied, who call them to account in new ways.

Halfies' dilemmas are even more extreme. As anthropologists, they write for other anthropologists, mostly Western. Identified also with communities outside the West, or subcultures within it, they are called to account by educated members of those communities. More importantly, not just because they position themselves with reference to two communities but because when they present the Other they are presenting themselves, they speak with a complex awareness of and investment in reception. Both halfie and feminist anthropologists are forced to confront squarely the politics and ethics of their representations. There are no easy solutions to their dilemmas.

The third issue that feminist and halfie anthropologists, unlike anthropologists who work in Western societies (another group for whom self and other are somewhat tangled), force us to confront is the dubiousness of maintaining that relationships between self and other are innocent of power. Because of sexism and racial or ethnic discrimination, they may have experienced – as women, as individuals of mixed parentage, or as foreigners – being other to a dominant self, whether in everyday life in the U.S., Britain, or France, or in the Western academy. This is not simply an experience of difference, but of inequality. My argument, however, is structural, not experiential. Women, blacks, and people of most of the non-West have been historically constituted as others in the major political systems of difference on which the unequal world of modern capitalism has depended. Feminist studies and black studies have made sufficient progress within the academy to have exposed the way that being studied by "white men" (to use a shorthand for a complex and historically constituted subject-position) turns into being spoken for by

them. It becomes a sign and instrument of their power.

Within anthropology, despite a long history of self-conscious opposition to racism, a fast-growing, self-critical literature on anthropology's links to colonialism, and experimentation with techniques of ethnography to relieve a discomfort with the power of anthropologist over anthropological subject, the fundamental issues of domination keep being skirted. Even attempts to refigure informants as consultants and to "let the other speak" in dialogic or polyvocal texts – decolonizations on the level of the text – leave intact the basic configuration of global power on which anthropology, as linked to other institutions of the world, is based. To see the strangeness of this enterprise, all that is needed is to consider an analogous case. What would our reaction be if male scholars stated their desire to "let women speak" in their texts while they continued to dominate all knowledge about them by controlling writing and other academic practices, supported in their positions by a particular organization of economic, social, and political life?

Because of their split selves, feminist and halfie anthropologists travel uneasily between speaking "for" and speaking "from." Their situation enables us to see more clearly that dividing practices, whether they naturalize differences, as in gender or race, or simply elaborate them, as I will argue the concept of culture does, are fundamental methods of enforcing inequality.

CULTURE AND DIFFERENCE

The concept of culture is the hidden term in all that has just been said about anthropology. Most American anthropologists believe or act as if "culture," notoriously resistant to definition and ambiguous of referent, is nevertheless the true object of anthropological inquiry. Yet it could also be argued that culture is important to anthropology because the anthropological distinction between self and other rests on it. Culture is the essential tool for making other. As a professional discourse that elaborates on the meaning of culture in order to account for, explain, and understand cultural difference, anthropology also helps construct, produce, and maintain it. Anthropological discourse

gives cultural difference (and the separation between groups of people it implies) the air of the
self-evident.

In this regard, the concept of culture operates
much like its predecessor – race – even though in
its twentieth-century form it has some important
political advantages. Unlike race, and unlike even
the nineteenth-century sense of culture as a synonym for civilization (contrasted to barbarism),
the current concept allows for multiple rather than
binary differences. This immediately checks the
easy move to hierarchizing, the shift to "culture" . . .
has a relativizing effect. The most important of
culture's advantages, however, is that it removes
difference from the realm of the natural and the
innate. Whether conceived of as a set of behaviors,
customs, traditions, rules, plans, recipes, instructions,
or programs . . . culture is learned and can change.

Despite its anti-essentialist intent, however, the
culture concept retains some of the tendencies to
freeze difference possessed by concepts like race.
This is easier to see if we consider a field in which
there has been a shift from one to the other.
Orientalism as a scholarly discourse (among other
things) is, according to Said, "a style of thought
based upon an ontological and epistemological
distinction made between the Orient and (most of
the time) 'the Occident'." What he shows is that
in mapping geography, race, and culture on to one
another, Orientalism fixes differences between
people of "the West" and people of "the East" in
ways so rigid that they might as well be considered
innate. In the twentieth century, cultural difference,
not race, has been the basic subject of Orientalist
scholarship devoted now to interpreting the "culture" phenomena (primarily religion and language)
to which basic differences in development, economic
performance, government, character, and so forth
are attributed.

Some anticolonial movements and present-day
struggles have worked by what could be labelled
reverse Orientalism, where attempts to reverse the
power relationship proceed by seeking to valorize
for the self what in the former system had been
devalued as other. A Gandhian appeal to the greater
spirituality of a Hindu India, compared with the
materialism and violence of the West, and an
Islamicist appeal to a greater faith in God, compared
with the immorality and corruption of the West,
both accept the essentialist terms of Orientalist

constructions. While turning them on their heads,
they preserve the rigid sense of difference based
on culture.

A parallel can be drawn with feminism. It is a
basic tenet of feminism that "women are made, not
born." It has been important for most feminists
to locate sex differences in culture, not biology
or nature. While this has inspired some feminist
theorists to attend to the social and personal
effects of gender as a system of difference, for many
others it has led to explorations of and strategies
built on the notion of a women's culture. Cultural
feminism takes many forms, but it has many of the
qualities of reverse Orientalism just discussed.
For French feminists like Irigaray, Cixous, and
Kristeva, masculine and feminine, if not actually male
and female, represent essentially different modes
of being. Anglo-American feminists take a different
tack. Some attempt to "describe" the cultural differences between men and women . . . Others try
to "explain" the differences . . . Much feminist theorizing and practice seeks to build or reform social
life in line with this "women's culture." There have
been proposals for a woman-centered university,
a feminist science, a feminist methodology in the
sciences and social sciences, and even a feminist
spirituality and ecology. These proposals nearly
always build on values traditionally associated in
the West with women – a sense of care and con-
nectedness, maternal nurturing, immediacy of
experience, involvement in the bodily (versus the
abstract), and so forth.

This valorization by cultural feminists, like reverse
Orientalists, of the previously devalued qualities
attributed to them may he provisionally useful in
forging a sense of unity and in waging struggles of
empowerment. Yet because it leaves in place the
divide that structured the experiences of selfhood
and oppression on which it builds, it perpetuates
some dangerous tendencies. First, cultural feminists
overlook the connections between those on each
side of the divide, and the ways in which they define
each other. Second, they overlook differences within
each category constructed by the dividing practices,
differences like those of class, race, and sexuality
(to repeat the feminist litany of problematically
abstract categories), but also ethnic origin, personal experience, age, mode of livelihood, health,
living situation (rural or urban), and historical
experience. Third, and perhaps most important, they

ignore the ways in which experiences have been constructed historically and have changed over time. Both cultural feminism and revivalist movements tend to rely on notions of authenticity and the return to positive values not represented by the dominant other. As becomes obvious in the most extreme cases, these moves erase history. Invocations of Cretan goddesses in some cultural-feminist circles and, in a more complex and serious way, the powerful invocation of the seventh-century community of the Prophet in some Islamic movements are good examples.

The point is that the notion of culture which both types of movements use does not seem to guarantee an escape from the tendency toward essentialism. It could be argued that anthropologists use "culture" in more sophisticated and consistent ways and that their commitment to it as an analytical tool is firmer. Yet even many of them are now concerned about the ways it tends to freeze differences. Appadurai, for example, in his compelling argument that "natives" are a figment of the anthropological imagination, shows the complicity of the anthropological concept of culture in a continuing "incarceration" of non-Western peoples in time and place. Denied the same capacity for movement, travel, and geographical interaction that Westerners take for granted, the cultures studied by anthropologists have tended to be denied history as well.

Others, including myself, have argued that cultural theories also tend to overemphasize coherence. Clifford notes both that the discipline of fieldwork-based anthropology, in constituting its authority, constructs and reconstructs coherent cultural others and interpreting "selves" and that ethnography is a form of culture collecting (like art collecting) in which "diverse experiences and facts are selected, gathered, detached from their original temporal occasions, and given enduring value in a new arrangement." Organic metaphors of wholeness and the methodology of holism that characterizes anthropology both favor coherence, which in turn contributes to the perception of communities as bounded and discrete.

Certainly discreteness does not have to imply value; the hallmark of twentieth-century anthropology has been its promotion of cultural relativism over evaluation and judgment. If anthropology has always to some extent been a form of cultural (self-) critique, that too was an aspect of a refusal

to hierarchize difference. Yet neither position would be possible without difference. It would be worth thinking about the implications of the high stakes anthropology has in sustaining and perpetuating a belief in the existence of cultures that are identifiable as discrete, different, and separate from our own. Does difference always smuggle in hierarchy?

In *Orientalism*, Said argues for the elimination of "the Orient" and "the Occident" altogether. By this he means not the erasure of all differences but the recognition of more of them and of the complex ways in which they crosscut. More important, his analysis of one field seeks to show how and when certain differences, in this case of places and the people attached to them, become implicated in the domination of one by the other. Should anthropologists treat with similar suspicion "culture" and "cultures" as the key terms in a discourse in which otherness and difference have come to have, as Said points out, "talismanic qualities"?

THREE MODES OF WRITING AGAINST CULTURE

If "culture," shadowed by coherence, timelessness, and discreteness, is the prime anthropological tool for making "other," and difference, as feminists and halfies reveal, tends to be a relationship of power, then perhaps anthropologists should consider strategies for writing against culture. I will discuss three that I find promising. Although they by no means exhaust the possibilities, the sorts of projects I will describe – theoretical, substantive, and textual – make sense for anthropologists sensitive to issues of positionality and accountability and interested in making anthropological practice something that does not simply shore up global inequalities. . . .

Discourse and practice

Theoretical discussion, because it is one of the modes in which anthropologists engage each other, provides an important site for contesting culture. It seems to me that current discussions and deployments of two increasingly popular terms – practice and discourse – do signal a shift away from culture. Although there is always the danger that these terms will come to be used simply as synonyms for culture, they were intended to enable us to analyze social life without presuming the degree of coherence that the culture concept has come to carry.

Practice is associated, in anthropology, with Bourdieu, whose theoretical approach is built around problems of contradiction, misunderstanding, and misrecognition, and favors strategies, interests, and improvisations over the more static and homogenizing cultural tropes of rules, models, and texts. Discourse has more diverse sources and meanings in anthropology. In its Foucauldian derivation, as it relates to notions of discursive formations, apparatuses, and technologies, it is meant to refuse the distinction between ideas and practices or text and world that the culture concept too readily encourages. In its more sociolinguistic sense, it draws attention to the social uses by individuals of verbal resources. In either case, it allows for the possibility of recognizing within a social group the play of multiple, shifting, and competing statements with practical effects. Both practice and discourse are useful because they work against the assumption of boundedness, not to mention the idealism, of the culture concept.

Connections

Another strategy of writing against culture is to reorient the problems or subject matter anthropologists address. An important focus should be the various connections and interconnections, historical and contemporary, between a community and the anthropologist working there and writing about it, not to mention the world to which he or she belongs and which enables him or her to be in that particular place studying that group. This is more of a political project than an existential one, although the reflexive anthropologists who have taught us to focus on the fieldwork encounter as a site for the construction of the ethnographic facts have alerted us to one important dimension of the connection. Other significant sorts of connections have received less attention. Pratt notes a regular mystification in ethnographic writing of "the larger agenda of European expansion in which the ethnographer, regardless of his or her own attitudes to it, is caught up, and that determines the ethnographer's own material relationship to the group under study." We need to ask questions about the historical processes by which it came to pass that people like ourselves could be engaged in anthropological studies of people like those, about the current world situation that enables us to engage in this sort of work in this particular place, and about who has preceded us and is even now there with us (tourists, travelers, missionaries, AID consultants, Peace Corps workers). We need to ask what this "will to knowledge" about the Other is connected to in the world.

These questions cannot be asked in general; they should be asked about and answered by tracing through specific situations, configurations, and histories. Even though they do not address directly the place of the ethnographer, and even though they engage in an oversystemization that threatens to erase local interactions, studies like those of Wolf [*Europe and the People without History*] on the long history of interaction between particular Western societies and communities in what is now called the Third World represent important means of answering such questions. So do studies like Mintz's [*Sweetness and Power*] that trace the complex processes of transformation and exploitation in which, in Europe and other parts of the world, sugar was involved. The anthropological turn to history, tracing connections between the present and the past of particular communities, is also an important development.

Not all projects about connections need be historical. Anthropologists are increasingly concerned with national and transnational connections of people, cultural forms, media, techniques, and commodities. They study the articulation of world capitalism and international politics with the situations of people living in particular communities. All these projects, which involve a shift in gaze to include phenomena of connection, expose the inadequacies of the concept of culture and the elusiveness of the entities designated by the term *cultures*. Although there may be a tendency in the new work merely to widen the object, shifting from culture to nation as locus, ideally there would be attention to the shifting groupings, identities, and interactions within and across such borders as well, If there was ever a time when anthropologists could consider without too much violence at least some communities as isolated units, certainly the nature of global interactions in the present makes that now impossible.

Ethnographies of the particular

The third strategy for writing against culture depends on accepting the one insight of Geertz's about anthropology that has been built upon by everyone in this "experimental moment" who takes textuality seriously. Geertz has argued that one of the main things anthropologists do is write, and what they write are fictions (which does not mean they are fictitious). Certainly the practice of ethnographic writing has received an inordinate amount of attention from those involved in *Writing Culture* and an increasing number of others who were not involved. Much of the hostility toward their project arises from the suspicion that in their literary leanings they have too readily collapsed the politics of ethnography into its poetics. And yet they have raised an issue that cannot be ignored. Insofar as anthropologists are in the business of representing others through their ethnographic writing, then surely the degree to which people in the communities they study appear "other" must also be partly a function of how anthropologists write about them. Are there ways to write about lives so as to constitute others as less other?

I would argue that one powerful tool for unsettling the culture concept and subverting the process of "othering" it entails is to write "ethnographies of the particular." Generalization, the characteristic mode of operation and style of writing of the social sciences, can no longer be regarded as neutral description. . . .

There are two reasons for anthropologists to be wary of generalization. The first is that, as part of a professional discourse of "objectivity" and expertise, it is inevitably a language of power. On the one hand, it is the language of those who seem to stand apart from and outside of what they are describing. . . . On the other hand, even if we withhold judgment on how closely the social sciences can be associated with the apparatuses of management, we have to recognize how all professionalized discourses by nature assert hierarchy. The very gap between the professional and authoritative discourses of generalization and the languages of everyday life (our own and others') establishes a fundamental separation between the anthropologist and the people being written about that facilitates the construction of anthropological objects as simultaneously different and inferior.

Thus, to the degree that anthropologists can bring closer the language of everyday life and the language of the text, this mode of making other is reversed. . . .

The second problem with generalization derives not from its participation in the authoritative discourses of professionalism but from the effects of homogeneity, coherence, and timelessness it tends to produce. When one generalizes from experiences and conversations with a number of specific people in a community, one tends to flatten out differences among them and to homogenize them. The appearance of an absence of internal differentiation makes it easier to conceive of a group of people as a discrete, bounded entity, like the "the Nuer," "the Balinese," and "the Awlad "Ali Bedouin" who do this or that and believe such-and-such. The effort to produce general ethnographic descriptions of people's beliefs or actions tends to smooth over contradictions, conflicts of interest, and doubts and arguments, not to mention changing motivations and circumstances. The erasure of time and conflict make what is inside the boundary set up by homogenization something essential and fixed. These effects are of special moment to anthropologists because they contribute to the fiction of essentially different and discrete others who can be separated from some sort of equally essential self. Insofar as difference is, as I have argued, hierarchical, and assertions of separation a way of denying responsibility, generalization itself must be treated with suspicion.

For these reasons I propose that we experiment with narrative ethnographies of the particular in a continuing tradition of fieldwork-based writing. In telling stories about particular individuals in time and place, such ethnographies would share elements with the alternative women's tradition discussed above. I would expect them to complement rather than replace a range of other types of anthropological projects, from theoretical discussions to the exploration of new topics within anthropology . . .

By focusing closely on particular individuals and their changing relationships, one would necessarily subvert the most problematic connotations of culture: homogeneity, coherence, and timelessness. Individuals are confronted with choices, struggle with others, make conflicting statements, argue about points of view on the same events, undergo ups and downs in various relationships and changes in their circumstances and desires, face new pressures, and fail to predict what will happen to them or those around them. So, for example, it becomes difficult to think that the term "Bedouin culture" makes sense when one tries to piece together and convey what life is like for one old Bedouin matriarch.

When you ask her to tell the story of her life, she responds that one should only think about God. Yet she tells vivid stories, fixed in memory in particular ways, about her resistances to arranged marriages, her deliveries of children, her worries about sick daughters. She also tells about weddings she has attended, dirty songs sung by certain young men as they sheared the elders' sheep herds, and trips in crowded taxis where she pinched a man's bottom to get him off her lap.

The most regular aspect of her daily life is her wait for prayer times. Is it noon yet? Not yet. Is it afternoon yet? Not yet. Is it sunset yet? Grandmother, you haven't prayed yet? It's already past sunset. She spreads her prayer rug in front of her and prays out loud. At the end, as she folds up her prayer rug, she beseeches God to protect all Muslims. She recites God's names as she goes through her string of prayer beads. The only decoration in her room is a photograph on the wall of herself and her son as pilgrims in Mecca.

Her back so hunched she can hardly stand, she spends her days sitting or lying down on her mattress. She is practically blind and she complains about her many pains. People come and go, her sons, her nephews, her daughter, her nieces, her granddaughters, her great-grandson. They chat, they confer with her about connections between people, marriages, kinship. She gives advice; she scolds them for not doing things properly. And she plays with her great grandson, who is three, by teasing, "Hey, I've run out of snuff. Come here so I can sniff your little tuber."

Being pious and fiercely preserving protocol in the hosting of guests and the exchanging of visits and greetings does not seem to stop her from relishing the outrageous story and the immoral tale. A new favorite when I saw her in 1987 was one she had just picked up from her daughter, herself a married mother of five living near Alamein. It was a tale about an old husband and wife who decide to go visit their daughters, and it was funny for the upside-down world it evoked.

This tale depicted a world where people did the unthinkable. Instead of the usual candy and biscuits, the couple brought their daughters sacks of dung for gifts. When the first daughter they stayed with went off to draw water from the well, they started dumping out all the large containers of honey and oil in her merchant husband's house. She returned to find them spilling everything and threw them out. So they headed off to visit the second daughter. When she left them minding her baby for a while, the old man killed it just to stop it from crying. She came back, discovered this and threw them out. Next they came across a house with a slaughtered sheep in it. They made belts out of the intestines and caps out of the stomachs and tried them on, admiring each other in their new finery. But when the old woman asked her husband if she didn't look pretty in her new belt he answered, "You'd be really pretty, except for that fly sitting on your nose." With that he smacked the fly, killing his wife. As he wailed in grief he began to fart. Furious at his anus for farting over his dead wife, he heated up a stake and shoved it in, killing himself.

The old woman chuckles as she tells this story, just as she laughs hard over stories about the excessive sexuality of old women. How does this sense of humor, this appreciation of the bawdy, go with devotion to prayer and protocols of honor? How does her nostalgia for the past – when the area was empty and she could see for miles around when she used to play as a little girl digging up the occasional potsherd or glass bottle in the area now fenced and guarded by the government Antiquities Organization; when her family migrated with the sheep herds and milked and made butter in desert pastures – go with her fierce defense of her favorite grandson, whose father was furious with him because the young man was rumored to have drunk liquor at a local wedding? People do not drink in the community, and drinking is, of course, religiously proscribed. What can "culture" mean given this old woman's complex responses?

"Beyond 'Culture': Space, Identity, and the Politics of Difference"

from *Culture, Power, Place: Explorations in Critical Anthropology* (1997)

Akhil Gupta and James Ferguson

Editors' introduction

In what ways has culture developed as a spatial concept? While cultural geographers have long been interested in cultural landscapes and the spatial expression of cultural practices and artifacts, the concept of culture itself was not the subject of much spatial theorizing among cultural geographers until interest in the work of Clifford Geertz, Raymond Williams, and other cultural theorists began to emerge in the early 1980s. And yet, culture is nevertheless a concept that has always had certain underlying spatial implications or assumptions. These spatial assumptions have in some ways been an important, if unexamined, part of cultural geography's traditional approach to its subject matter: cultures were assumed to have distinct landscapes and spatial territories or regions that could be mapped with boundaries that reflected these distinctions. While work by Geertz, Williams, and others raised questions that made it increasingly difficult to view culture as a "mappable object," a surge of new theoretical interest in the concept of space in the social sciences began to raise similar questions about the relationship between space and culture. These questions asked if space itself was not a neutral container "out there" but in fact a product of social relations, and thus shaped by the characteristics of those relations (see also the introduction to Part Five).

Akhil Gupta and James Ferguson's article occupies this intersection of interests – between critical interrogations of the concept of culture and those of the concept of space. They begin with the premise that most of the thinking about culture since the 1970s (see the essays by Geertz, Abu-Lughod, and Sewell, pp. 29, 50, and 40 respectively) has had little to say about the spatial assumptions underlying culture. They ask: What are the implications of the new thinking about space for the concept of culture and for our ideas about cultural difference? They are particularly interested in theories about space that came about as a result of feminist, poststructural, and postmodern theories. For an introduction to this thinking about space in relation to geography specifically, Edward Soja's *Postmodern Geographies* (1989) is a good place to start. Caren Kaplan's *Questions of Travel* (1996) also offers a feminist reading of the spatial theorizing that has influenced Gupta and Ferguson's work.

Noting that unprecedented human mobility has forced us to rethink our association of culture with a fixed place or territory, Gupta and Ferguson are interested in the ways that cultural practices and identities have become increasingly mixed. Spatial concepts such as "deterritorialization" and "borderland" (including related ideas of "hybridity" and "marginality") provide a framework, then, upon which to move beyond our assumptions about culture as spatially localized or fixed with clear boundaries and territory. Their interest in cultural mixing recalls Abu-Lughod's (p. 50) concern with "halfies." Gupta and Ferguson come to the similar conclusion that a focus on people who live in the borders between dominant societies or nations (and here borders is also a metaphor for people who identify, culturally, with more than one group) makes clear the fact that differences between cultures come about not because of their isolation from each other, but because of their *connections* with each other. Such a conclusion also suggests that along with difference comes the hierarchies of power. Culture is not only a concept that expresses difference between peoples, but also a concept that masks the uneven power relations between peoples, and these uneven power relations can only exist through connection, rather than isolation.

Akhil Gupta and James Ferguson both teach anthropology at Stanford University. Their work focuses on bringing ethnographic inquiry to bear on topics – for example, postcolonial state formation and the political economy of development – that have typically been the focus of less culturally-focused methodologies. Gupta is the author of *Postcolonial Developments: Agriculture in the Making of Modern India* (1998) and co-editor of *The Anthropology of the State* (2006). Ferguson's work includes *Expectations of Modernity: Myths and Meanings of Urban Life on the Zambian Copperbelt* (1999) and *Global Shadows: Africa in the Neoliberal World Order* (2006).

Gupta and Ferguson's work has much in common, has been influenced by, and in turn has influenced important developments in cultural geography. In particular, their work on the relationship between place and culture draws on conceptions in geography that view place in terms of connections rather than bounded isolation. Doreen Massey's idea of a *progressive sense of place* (p. 257), for instance, outlines a concept of place as a node of spatial relations; her approach offers an explicit basis for theorizing Gupta and Ferguson's "difference through connections" in spatial terms. More generally, much cultural geography has been inspired by an interest in hybridity, marginality, and borderlands that unsettle our traditional ideas about cultural difference. See, for instance, Tim Creswell's *In Place/Out of Place* (1996) and Rob Shields's *Places on the Margin* (1001).

[...]

Representations of space in the social sciences are remarkably dependent on images of break, rupture, and disjunction. The distinctiveness of societies, nations, and cultures is predicated on a seemingly unproblematic division of space, on the fact that they occupy "naturally" discontinuous spaces. The premise of discontinuity forms the starting point from which to theorize contact, conflict, and contradiction between cultures and societies. For example, the representation of the world as a collection of "countries," as on most world maps, sees it as an inherently fragmented space, divided by different colors into diverse national societies, each "rooted" in its proper place. It is so taken for granted that each country embodies its own distinctive culture and society that the terms "society" and "culture" are routinely simply appended to the names of nation-states, as when a tourist visits India to understand "Indian culture" and "Indian society" or Thailand to experience "Thai culture" or the United States to get a whiff of "American culture."

Of course, the geographical territories that cultures and societies are believed to map on to do not have to be nations. We do, for example, have ideas about culture areas that overlap several nation-states, or of multicultural nations. On a smaller scale perhaps are our disciplinary assumptions about the association of culturally unitary groups (tribes or peoples) with "their" territories: thus "the Nuer" live in "Nuerland" and so forth. The clearest illustration of this kind of thinking are the classic "ethnographic maps" that purported to display the spatial distribution of peoples, tribes, and cultures. But in all these cases, space itself becomes a kind of neutral grid on which cultural difference, historical memory, and societal organization [are] inscribed. It is in this way that space functions as a central organizing principle in the social sciences at the same time that it disappears from analytical purview.

This assumed isomorphism of space, place, and culture results in some significant problems. First, there is the issue of those who inhabit the border, what Gloria Anzaldúa calls the "narrow strip along steep edges" of national boundaries. The fiction of cultures as discrete, objectlike phenomena occupying discrete spaces becomes implausible for those who inhabit the borderlands. Related to border inhabitants are those who live a life of border crossings – migrant workers, nomads, and members of the transnational business and professional elite. What is "the culture" of farm workers who spend half a year in Mexico and half in the United States? Finally, there are those who cross borders more or less permanently – immigrants, refugees, exiles, and expatriates. In their case, the disjuncture of place and culture is especially clear: Khmer refugees in the United States take "Khmer culture" with them in the same complicated way that Indian immigrants in England transport "Indian culture" to their new homeland.

A second set of problems raised by the implicit mapping of cultures on to places is to account for cultural differences *within* a locality. "Multiculturalism" is both a feeble recognition of the fact that cultures have lost their moorings in definite places and an attempt to subsume this plurality of cultures within the framework of a national identity. Similarly, the idea of "subcultures" attempts to preserve the idea of distinct "cultures" while acknowledging the relation of different cultures to a dominant culture within the same geographical and territorial space. Conventional accounts of ethnicity, even when used to describe cultural differences in settings where people from different regions live side by side, rely on an unproblematic link between identity and place. While such concepts are suggestive because they endeavor to stretch the naturalized association of culture with place, they fail to interrogate this assumption in a truly fundamental manner. We need to ask how to deal with cultural difference, while abandoning received ideas of (localized) culture.

[...]

Last and most important, challenging the ruptured landscape of independent nations and autonomous cultures raises the question of understanding social change and cultural transformation as situated within interconnected spaces. The presumption that spaces are autonomous has enabled the power of topography successfully to conceal the topography of power. The inherently fragmented space assumed in the definition of anthropology as the study of cultures (in the plural) may have been one of the reasons behind the long-standing failure to write anthropology's history as the biography of imperialism. For if one begins with the premise that spaces have *always* been hierarchically interconnected, instead of naturally disconnected, then cultural and social change becomes not a matter of cultural contact and articulation but one of rethinking difference *through* connection.

[...]

It is for this reason that what Fredric Jameson has dubbed "postmodern hyperspace" has so fundamentally challenged the convenient fiction that mapped cultures on to places and peoples. In the capitalist West, a Fordist regime of accumulation, emphasizing extremely large production facilities, a relatively stable work force, and the welfare state combined to create urban "communities" whose outlines were most clearly visible in company towns. The counterpart of this in the international arena was that multinational corporations, under the leadership of the United States, steadily exploited the raw materials, primary goods, and cheap labor of the independent nation-states of the postcolonial "Third World." Multilateral agencies and powerful Western states preached and, where necessary, militarily enforced the "laws" of the market to encourage the international flow of capital, whereas national immigration policies ensured that there would be no free (that is, anarchic, disruptive) flow of labor to the high-wage islands in the capitalist core. Fordist patterns of accumulation have now been replaced by a regime of flexible accumulation – characterized by small-batch production, rapid shifts in product lines, extremely fast movements of capital to exploit the smallest differentials in labor and raw material costs – built on a more sophisticated communications and information network and better means of transporting goods and people. At the same time, the industrial production of culture, entertainment, and leisure that first achieved something approaching global distribution during the Fordist era led, paradoxically, to the invention of new forms of cultural difference and new forms of imagining community. Something like a transnational public sphere has certainly rendered any strictly bounded sense of community or locality obsolete. At the same time, it has enabled the creation of forms of solidarity and identity that do not rest on an appropriation of space where contiguity and face-to-face contact are paramount. In the pulverized space of postmodernity, space has not become irrelevant: it has been *re*territorialized in a way that does not conform to the experience of space that characterized the era of high modernity. It is this reterritorialization of space that forces us to reconceptualize fundamentally the politics of community, solidarity, identity, and cultural difference.

IMAGINED COMMUNITIES, IMAGINED PLACES

People have undoubtedly always been more mobile and identities less fixed than the static and typologizing approaches of classical anthropology would suggest. But today, the rapidly expanding and quickening mobility of people combines with the refusal of cultural products and practices to "stay put" to give a profound sense of a loss of territorial roots, of an erosion of the cultural distinctiveness of places, and of ferment in anthropological theory. The apparent deterritorialization of identity that accompanies such processes has made James Clifford's question [in *The Predicament of Culture*, 1988, p. 275] a key one for recent anthropological inquiry: "What does it mean, at the end of the twentieth century, to speak . . . of a 'native land'? What processes rather than essences are involved in present experiences of cultural identity?"

Such questions are, of course, not completely new, but issues of collective identity do seem to take on a special character today, when more and more of us live in what Edward Said has called "a generalized condition of homelessness," a world where identities are increasingly coming to be, if not wholly deterritorialized, at least differently territorialized. Refugees, migrants, displaced and stateless peoples – these are perhaps the first to live out these realities in their most complete form, but the problem is more general. In a world of diaspora, transnational culture flows, and mass movements of populations, old-fashioned attempts to map the globe as a set of culture regions or homelands are bewildered by a dazzling array of postcolonial simulacra, doublings and redoublings as India and Pakistan seem to reappear in postcolonial simulation in London, prerevolution Teheran rises from the ashes in Los Angeles, and a thousand similar cultural dramas are played out in urban and rural settings all across the globe. In this culture-play of diaspora, familiar lines between "here" and "there," center and periphery, colony and metropole become blurred.

Where "here" and "there" become blurred in this way, the cultural certainties and fixities of the metropole are upset as surely, if not in the same way, as are those of the colonized periphery. In this sense, it is not only the displaced who experience a displacement. For even people remaining in familiar and ancestral places find the nature of their relation to place ineluctably changed and the illusion of a natural and essential connection between the place and the culture broken. "Englishness," for instance, in contemporary, internationalized England is just as complicated and nearly as deterritorialized a notion as Palestinian-ness or Armenian-ness, for "England" ("the real England") refers less to a bounded place than to an imagined state of being or a moral location. Consider, for instance, the following quote from a young white reggae fan in the ethnically chaotic neighborhood of Balsall Heath in Birmingham [from Hebdige's *Cut 'n' Mix: Culture, Identity, and Caribbean Music*, 1987, pp. 158–159]:

> There's no such thing as "England" anymore . . . welcome to India, brothers! This is the Caribbean! . . . Nigeria! . . . There is no England, man. This is what is coming. Balsall Heath is the centre of the melting pot, 'cos all I ever see when I go out is half-Arab, half-Pakistani, half-Jamaican, half-Scottish, half-Irish. I know 'cos I am [half-Scottish/half-Irish] . . . who am I? . . . Tell me who I belong to? They criticize me, the good old England. Alright, where do I belong? You know, I was brought up with blacks, Pakistanis, Africans, Asians, everything, you name it . . . who do I belong to? . . . I'm just a broad person. The earth is mine . . . , you know we was not born in Jamaica . . . we was not born in "England." We were born here, man. It's our right. That's the way I see it. That's the way I deal with it.

The broadminded acceptance of cosmopolitanism that seems to be implied here is perhaps more the exception than the rule, but there can be little doubt that the explosion of a culturally stable and unitary "England" into the cut-and-mix "here" of contemporary Balsall Heath is an example of a phenomenon that is real and spreading. It is clear that the erosion of such supposedly natural connections between peoples and places has not led to the modernist specter of global cultural homogenization. But "cultures" and "peoples," however persistent they may be, cease to be plausibly identifiable as spots on the map.

But the irony of these times is that as actual places and localities become ever more blurred and indeterminate, *ideas* of culturally and ethnically distinct places become perhaps even more salient. It is here that it becomes most visible how imagined communities come to be attached to imagined places, as displaced peoples cluster around remembered or imagined homelands, places, or communities in a world that seems increasingly to deny such firm territorialized anchors in their actuality. In such a world, it becomes ever more important to train an anthropological eye on processes of construction of place and homeland by mobile and displaced people.

Remembered places have, of course, often served as symbolic anchors of community for dispersed people. This has long been true of immigrants, who use memory of place to construct their new lived world imaginatively. "Homeland" in this way remains one of the most powerful unifying symbols for mobile and displaced peoples, though the relation to homeland may be very differently constructed in different settings. Moreover, even in more completely deterritorialized times and settings – settings not only where "home" is distant but also where the very notion of "home" as a durably fixed place is in doubt – aspects of our lives remain highly "localized" in a social sense. We need to give up naïve ideas of communities as literal entities but remain sensitive to the profound "bifocality" that characterizes locally lived existences in a globally interconnected world and to the powerful role of place in the "near view" of lived experience.

The partial erosion of spatially bounded social worlds and the growing role of the imagination of places from a distance, however, themselves must be situated within the highly spatialized terms of a global capitalist economy. The special challenge here is to use a focus on the way space is imagined (but not *imaginary*) as a way to explore the mechanisms through which such conceptual processes of place making meet the changing global economic and political conditions of lived spaces – the relation, we could say, between place and space. For important tensions may arise when places that have been imagined at a distance must become lived spaces. Places, after all, are always imagined in the context of political-economic determinations that have a logic of their own. Territoriality is thus reinscribed at just the point it threatens to be erased.

[. . .]

As Malkki [see p. 275] shows, two naturalisms must be challenged here. The first is what we will call the ethnological habit of taking the association of a culturally unitary group (the "tribe" or "people") and "its" territory as natural, which we discussed in the previous section. A second and closely related naturalism is what we will call the national habit of taking the association of citizens of states and their territories as natural. Here the exemplary image is of the conventional world map of nation-states, through which schoolchildren are taught such deceptively simple-sounding beliefs as that France is where the French live, America is where the Americans live, and so on. Even a casual observer knows that not only Americans live in America, and it is clear that the very question of what is a "real American" is largely up for grabs . . . Both the ethnological and the national naturalisms present associations of people and place as solid, commonsensical, and agreed on, when they are in fact contested, uncertain, and in flux.

Much more-recent work in anthropology and related fields has focused on the process through which such reified and naturalized national representations are constructed and maintained by states and national elites. Such analyses of nationalism leave no doubt that states play a crucial role in the popular politics of place making and in the creation of naturalized links between places and peoples. But it is important to note that state ideologies are far from being the only point at which the imagination of place is politicized. Oppositional images of place have, of course, been extremely important in anticolonial nationalist movements, as well as in campaigns for self-determination and sovereignty on the part of contested nations such as the Hum, the Eritreans, the Armenians, or the Palestinians. Such instances may serve as a useful reminder, in the light of nationalism's often reactionary connotations in the Western world, of how often notions of home and "own place" have been empowering in anti-imperial contexts.

[. . .]

SPACE, POLITICS, AND ANTHROPOLOGICAL REPRESENTATION

[. . .]

Marjorie Shostak's *Nisa: The Life and Words of a !Kung Woman* (1981) has been very widely admired for its innovative use of life history and has been hailed as a noteworthy example of polyphonic experimentation in ethnographic writing. But with respect to the issues we have discussed here, *Nisa* is a very conventional and deeply flawed work. The individual, Nisa, is granted a degree of singularity, but she is used principally as the token of a type: "the !Kung." The San-speaking !Kung of Botswana (the "Bushmen" of old) are presented as a distinct, "other," and apparently primordial "people." Shostak treats the Dobe !Kung as essentially survivals of a prior evolutionary age: they are "one of the last remaining traditional gatherer-hunter societies," racially distinct, traditional, and isolated (p. 4). Their experience of "culture change" is "still quite recent and subtle" and their traditional value system "mostly intact" (p. 6). "Contact" with "other groups" of agricultural and pastoral peoples has occurred, according to Shostak, only since the 1920s, and only since the 1960s has the isolation of the !Kung really broken down, raising for the first time the issue of "change," "adaptation," and "culture contact" (p. 346).

The space the !Kung inhabit, the Kalahari Desert, is clearly radically different and separate from our own. Again and again the narrative returns to the theme of isolation: in a harsh ecological setting, a way of life thousands of years old has been preserved only through its extraordinary spatial separateness. The anthropological task, as Shostak conceives it, is to cross this spatial divide, to enter into this land that time forgot, a land with antiquity but no history, to listen to the voices of women which might reveal "what their lives had been like for generations, possibly even for thousands of years" (p. 6).

The exoticization implicit in this portrait, in which the !Kung appear almost as living on another planet, has drawn surprisingly little criticism from theorists of ethnography. Mary Louise Pratt has rightly pointed out the "blazing contradiction" between the portrait of primal beings untouched by history and the genocidal history of the white "Bushman conquest". As she says, "What picture of the !Kung would one draw if instead of defining them as survivors of the stone age and a delicate and complex adaptation to the Kalahari desert, one looked at them as survivors of capitalist expansion, and a delicate and complex adaptation to three centuries of violence and intimidation?" But even Pratt retains the notion of "the !Kung" as a preexisting ontological entity – "survivors," not products (still less, producers) of history. "They" are victims, having suffered the deadly process of "contact" with "us."

A very different and much more illuminating way of conceptualizing cultural difference in the region may be found in Wilmsen's devastating critique of the anthropological cult of the "Bushman" [in *Land Filled With Flies*, 1989]. Wilmsen shows how, in constant interaction with a wider network of social relations, the difference that Shostak takes as a starting point came to be produced in the first place – how, one might say, "the Bushmen" came to be Bushmen. He demonstrates that San-speaking people have been in continuous interaction with other groups for as long as we have evidence for; that political and economic relations linked the supposedly isolated Kalahari with a regional political economy both in the colonial and precolonial eras; that San-speaking people have often held cattle and that no strict separation of pastoralists and foragers can be maintained. He argues powerfully that the Zhu (!Kung) have never been a classless society and that if they give such an impression "it is because they are incorporated as an underclass in a wider social formation that includes Batswana, Ovaherero, and others" (p. 270). Moreover, he shows that the "Bushman/San" label has been in existence for barely half a century, the category having been produced through the "retribalization" of the colonial period, and that "the cultural conservatism uniformly attributed to these people by almost all anthropologists who have worked with them until recently, is a consequence – not a cause – of the way they have been integrated into the modern capitalist economies of Botswana and Namibia" (p. 12).

With respect to space, Wilmsen is unequivocal: "It is not possible to speak of the Kalahari's isolation, protected by its own vast distances. To those inside, the outside – whatever 'outside' there may have been at any moment – was always present. The appearance of isolation and its reality of dispossessed poverty are recent products of a process that unfolded over two centuries and culminated in the last moments of the colonial era" (p. 157). The process of the production of cultural difference, Wilmsen demonstrates, occurs in continuous, connected space, traversed by economic and political relations of inequality. Where Shostak takes difference as given and concentrates on listening "across cultures," Wilmsen performs the more radical operation of interrogating the "otherness" of the other, situating the production of cultural difference within the historical processes of a socially and spatially interconnected world.

What is needed, then, is more than a ready ear and a deft editorial hand to capture and orchestrate the voices of "others"; what is needed is a willingness to interrogate, politically and historically, the apparent "given" of a world in the first place divided into "ourselves" and "others." A first step on this road is to move beyond naturalized conceptions of spatialized "cultures" and to explore instead the production of difference within common, shared, and connected spaces – "the San," for instance, not as "a people," "native" to the desert, but as a historically constituted and depropertied category systematically relegated to the desert.

The move we are calling for, most generally, is away from seeing cultural difference as the correlate of a world of "peoples" whose separate histories wait to be bridged by the anthropologist and toward seeing it as a product of a shared historical process that differentiates the world as it connects it. For the proponents of "cultural critique," difference is taken as starting point, not as end product. Given a world of "different societies," they ask, how can we use experience in one to comment on another? But if we question a pregiven world of separate and discrete "peoples and cultures" and see instead a difference-producing set of relations, we turn from a project of juxtaposing preexisting differences to one of exploring the construction of differences in historical process.

[. . .]

In suggesting the requestioning of the spatial assumptions implicit in the most fundamental and seemingly innocuous concepts in the social sciences such as "culture," "society," "community," and "nation," we do not presume to lay out a detailed blueprint for an alternative conceptual apparatus. We do, however, wish to point out some promising directions for the future.

One extremely rich vein has been tapped by those attempting to theorize interstitiality and hybridity: in the postcolonial situation; for people living on cultural and national borders; for refugees and displaced peoples; and in the case of migrants and workers. The "syncretic, adaptive politics and culture" of hybridity, Homi K. Bhabha [in an interview in the journal *Emergences* 1, 1, 1989] points out, raises questions about "the imperialist and colonialist notions of purity as much as it question[s] the nationalist notions." It remains to be seen what kinds of politics are enabled by such a theorization of hybridity and to what extent it can do away with all claims to authenticity, to all forms of essentialism, strategic or otherwise. Bhabha points to the troublesome connection between claims to purity and utopian teleology in describing how he came to the realization that "the only place in the world to speak from was at a point whereby contradiction, antagonism, the hybridities of cultural influence, the boundaries of nations, were not sublated into some utopian sense of liberation or return. The place to speak from was through those incommensurable contradictions within which people survive, are politically active, and change." The borderlands make up just such a place of incommensurable contradictions. The term does not indicate a fixed topographical site between two other fixed locales (nations, societies, cultures) but an interstitial zone of displacement and deterritorialization that shapes the identity of the hybridized subject. Rather than dismissing them as insignificant, as marginal zones, thin slivers of land between stable places, we want to contend that the notion of borderlands is a more adequate conceptualization of the "normal" locale of the postmodern subject.

Another promising direction that takes us beyond culture as a spatially localized phenomenon is provided by the analysis of what is variously called "mass media," "public culture," and the "culture industry." (Especially influential here has been the journal *Public Culture*.) Existing symbiotically with the commodity form, profoundly influencing even the remotest people that anthropologists have made such a fetish of studying, mass media pose the clearest challenge to orthodox notions of culture. National, regional, and village boundaries have, of course, never contained culture in the way that the anthropological representations have often implied. But the existence of a transnational public sphere means that the fiction that such boundaries enclose cultures and regulate cultural exchange can no longer be sustained.

[...]

The reconceptualization of space implicit in theories of interstitiality and public culture has led to efforts to conceptualize cultural difference without invoking the orthodox idea of "culture." This is as yet a largely unexplored and underdeveloped area. We do, clearly, find the clustering of cultural practices that do not "belong" to a particular people" or to a definite place. [In *Postmodernism, or, The Cultural Logic of Late-Capitalism*, 1991], Jameson has attempted to capture the distinctiveness of these practices in the notion of a "cultural dominant," whereas Ferguson [in *Expectations of Modernity*, 1999] proposes an idea of "cultural style" that searches for a logic of surface practices without necessarily mapping such practices on to a "total way of life" encompassing values, beliefs, attitudes, and so on, as in the usual concept of culture. We need to explore what Bhabha calls "the uncanny of cultural difference": "Cultural difference becomes a problem not when you can point to the Hottentot Venus, or to the punk whose hair is six feet up in the air; it does not have that kind of fixable visibility. It is as the strangeness of the familiar that it becomes more problematic, both politically and conceptually ... when the problem of cultural difference is ourselves-as-others, others-as-ourselves, that borderline."

Why focus on that borderline? We have argued that deterritorialization has destabilized the fixity of "ourselves" and "others." But it has not thereby created subjects who are free-floating nomads, despite what is sometimes implied by those eager to celebrate the freedom and playfulness of the postmodern condition. As Martin and Mohanty point out [in an essay on feminist politics in *Feminist Studies / Critical Studies*, 1986], indeterminacy too has its political limits, which follow from the denial of the critic's own location in multiple fields of power. Instead of stopping with the notion of deterritorialization, the pulverization of the space of high modernity, we need to theorize how space is being reterritorialized in the contemporary world. We need to account sociologically for the fact that the "distance" between the rich in Bombay and those in London may be much shorter than that between different classes in "the same" city. Physical location and physical territory, for so long the *only* grid on which cultural difference could be mapped, need to be replaced by multiple grids that enable us to see that connection and contiguity – more general, the representation of territory – vary considerably by factors such as class, gender, race, and sexuality and are differentially available to those in different locations in the field of power.

"Research, Performance, and doing Human Geography: Some Reflections on the Diary-Photograph, Diary-Interview Method"

from *Environment and Planning A* 35 (2003): 1993–2007

Alan Latham

Editors' introduction

When Clifford Geertz proposed an interpretive, semiotic, approach to culture (p. 29), he likened "writing culture" to writing "fictions." By this he meant that scholars themselves are in the business of "fashioning" culture in their work, just as their informants do in their daily lives. This blurring of the boundary between scholar and informant led to a critical inquiry among many cultural scholars of the rhetorical and narrative strategies used in ethnographic writing, an inquiry best exemplified in the volume *Writing Culture* (1986). The focus on the way culture is "written" resulted in an increased awareness of culture as a kind of *representation*. That is, if culture was "fashioned" as a way of interpreting and making meaningful our world, then it was also a way of *representing* the world. In other words, a great deal of work on culture began to explore the ways people make their world meaningful by "*re*-presenting" it as an object of interpretation, reflection, contemplation. The focus on representation enabled scholars to see the ways particularly dominant or powerful ideas (e.g. those promoted by ruling elites, powerful states, or socially privileged groups) shaped the ways people made their world meaningful, and thus helped theorize the relationship between culture and society in important new ways. In urging us to write "against" or "beyond" culture, for example, Abu-Lughod (p. 50) and Gupta and Ferguson (p. 60) illustrate how understanding culture as representation helps us see unequal (or "hierarchical") social relations hidden within the concept of culture.

But this way of approaching culture has also been unsatisfactory for many scholars. One problem was the nagging difficulty of sorting out the scholar's "fashioning" from that of his or her informants. This was, for instance, the basis of Vincent Crapanzano's critique of Geertz, articulated in his chapter in *Writing Culture*. Crapanzano argued that it is seldom clear in Geertz's work whether his informants really share his interpretation of cultural practices. Similarly, some scholars argued that informants are generally not particularly reflective or thoughtful about the meaning of their lives, and that they seldom engage in interpretive cognition (unless of course asked to do so by scholars!). While culture as representation tells us much about the ways scholars have made their world meaningful, and about culture as an epistemological category of knowledge, there is suspicion that for many people meaning comes about in ways that don't necessarily involve conscious reflection. People may derive meaning less from creating and interpreting symbols around them than from their emotions, from their movements from one place to another, or from their embodied senses. In short, meaning could be derived from a broader range of senses and activities than the "cognitive" activity of interpretation.

Alan Latham's article explores both the significance of approaching culture as *practice* rather than representation, and the methodological implications of taking this kind of "nonrepresentational" approach to cultural research. Not surprisingly, his exploration focuses less on new ways to *write* about culture (that is, to represent it) than on new ways of *doing fieldwork* which enable us to understand culture as something people *do* in their daily lives, rather than something they think about. In terms of theory, his article builds on a great deal of work by cultural geographers that emerged in the late 1990s, particularly that of Nigel Thrift. Thrift's project has been to outline a non-representational theory for cultural geography that understands a broad spectrum of non-cognitive ways in which people make their world meaningful. In particular, Thrift is interested in the metaphor of performance as an alternative way of conceptualizing culture as a kind of practice, rather than a "web of meaning." Beginning with the claim that viewing culture as representation has had the unfortunate effect of making cultural geography less empirically focused (that is, too focused on culture as "text"), Latham turns to the metaphor of performance to suggest ways in which culture should be reconceptualized and research reframed. Such a reframing will, he argues, enable cultural geographers to better understand the ways people make places and cultures out of the "performances" of their everyday lives.

There is a considerable body of scholarship on performance and practice that Latham references in this article. Among the most influential of these beyond geography have been Erving Goffman's *The Presentation of Self in Everyday Life* (1959) and *Behavior in Public Places* (1963), Howard Garfinkel's *Studies in Ethnomethodology* (1967), Michel de Certeau's *The Practice of Everyday Life* (1984), Pierre Bourdieu's *Outline of a Theory of Practice* (1972/1977), and Judith Butler's *Gender Trouble* (1990). Within geography, key explorations of these ideas can be found in two special issues of *Environment and Planning D: Society and Space* 18: 4–5 (2000). Nigel Thrift has written numerous pieces on the subject, such as "The still point" in *Geographies of Resistance*, edited by Pile and Keith (1997) and "Afterwords" in *Environment and Planning D: Society and Space* 10, 2 (2000). Additional perspectives can be found in Gillian Rose's "Performing Space," in *Human Geography Today*, edited by Massey, Allen, and Sarre (1999), Catherine Nash's "Performativity in practice: some recent work in cultural geography" in *Progress in Human Geography* 24 (2000), and Jon May's "A little taste of something exotic: the imaginative geographies of everyday life geography" in *Geography* 81 (1996).

Alan Latham teaches geography at University College London. He is the author of numerous journal articles and book chapters on sociality and urban life, globalization and the cultural economy of cities, and corporeal mobility. Particularly related to this selection is his guest-edited issue of *Environment and Planning A* 35: 11 (2003), "Making place: performance, practice, and space."

1 INTRODUCTION

... Over the last couple of decades we have seen something of a revolution in ways we frame what it is that geography is concerned with. We have seen that it is as much about discourses as about 'actual' events; that things that seem small and everyday can be as interesting and complex as phenomena that appear much larger and more general; that our own ways of writing the world are bound up with that world's constitution. But we do not seem to have made much progress in rethinking what this should mean to us as researchers ... The result has been that, rather than simply freeing us from the burdens of an earlier physical-science-based paradigm of social scientific investigation and opening up new research possibilities, the cultural critiques of the 1990s have in certain respects enfeebled human geography as *an empirical discipline*. ...

The aim of this paper is to contribute to this opening of methodological horizons within human geography. Specifically, I want to contribute to the emerging discussion on the uses and limits of the metaphor of performance as a way to frame the research process. What I want to show is how reframing research as creative, performative practice allows the researcher to address some novel questions about the cultures of everyday urban experience that more conventional, representationally oriented, methods fail to address adequately. I want also to demonstrate how such a reframing involves a reappraisal of our relationship to our research subjects and the narratives they offer. Thus, I am interested in the ideas of performance and practice on two discrete levels. First, I seek to articulate an understanding of everyday urban public culture as embodied practice – a practice that is creative, pregnant with possibilities, but nonetheless located within particular networks of power/knowledge. Second, drawing on this conceptualisation of everyday life (or 'ordinary culture'), I attempt to outline how the processes of 'fieldwork' and interpretation can embody, enact and thus respect the creativity of social practice whilst still offering useful (and critical) accounts of that practice.

2 JOSEPH'S PONSONBY ROAD

Started work, Star Graphics, 8.00 a.m. 208 Ponsonby Rd, opposite Franklin Rd. 10.30 a.m., Morning Tea. Left work to get coffee at "Duo," walked – just across Rd, opposite "Tuatara." Talked with Scottie (who works there and has become somewhat of a friend), asked how Weekend was etc. Ordered Single Flat White, which Scott added his artistic touch to by drawing a pattern in the froth with a spoon. (He always does this!) Sat outside and flicked through "Herald," while drinking and having a smoke. Scottie came out and joined Me, as there was no one else in the cafe. Talked some more. Joined a few minutes later by Gail (fellow patron, and friend of Scott). A nice unplanned encounter. Went back to work, at approx 10.45 a.m. Coffee was great as usual. (Research diary entry, Joseph, 27, actor, copy-shop assistant, coffee drinker)

Every weekday morning, almost without fail, Joseph Ryman wanders across the road from his work in a local copy shop and drinks a mid-morning coffee at Duo Café. Duo is part of Ponsonby Road – a sprawling, charming mess of a street skirting the western margins of downtown Auckland. Originally a retail and service centre for the Victorian and Edwardian villas on the slopes either side of it, Ponsonby Road has evolved over the past couple of decades into a prosperous hospitality strip, home to over sixty restaurants, cafes, and bars. The road is a curiosity. Its architecture is almost uniformly shabby, notable only for the hard-nosed veracity with which it narrates the

uninspired, sometimes bizarre, tastes of Auckland's property owners over the past century. Single storey nineteenth-century weatherboard buildings – little more than sheds – share the road with freshly constructed neotraditionalist terraces, bland 1970s and 1980s concrete and glass boxes, and a huge white Mississippi riverboat of a building that thrusts out of the ground in a chaos of balustrades, bargeboards, and corrugated roofing. Only the few relatively intact turn-of-the-twentieth-century buildings offer any sense of coherence or quality. Yet it is one of the most fashionable parts of Auckland. Its ambience is worldly, confident, cosmopolitan. As one walks along it, past the studied yet casual stylishness of establishments such as Atlas Power Café, Tuatara, Masala, One Red Dog, Atomic Café, Dizengoff, it is hard not to be impressed by a similarity to places such as Melbourne's Brunswick Street, Oxford Street in Sydney, even London's Stoke Newington Church Street or parts of Amsterdam's Spuistraat.

Joseph, with his beautifully coifed short black hair, confident casual dress – red New Balance Classic trainers, designer jeans, loose-fitting short-sleeved shirt – and easy style, leaning back sipping his flat white at one of the aluminium footpath tables outside Duo, underlines this impression of cosmopolitan knowingness. This picture of Joseph – aged 27, actor, copy-shop assistant, coffee drinker, dandy – encapsulates much of what is interesting about Ponsonby Road. Traditionally, New Zealand has been defined by a limited, intensely masculine, Calvinistic public culture. This culture was and remains intensely antiurban, seeing the city as corrupt and emasculating. Over the past 25 years, and most strikingly in the 1990s, however, the country's larger cities have seen the development of a strong, self-consciously urban, public culture. The evolution of this new urban public culture – for want of a more felicitious phrase – marks a shift in the way a significant proportion of New Zealanders make sense of their world. This shift is evident in a whole number of areas: in accepted notions of masculinity and femininity, in an openness (indeed, obsession) with difference, whether it be sexual, ethnic, or simply lifestyle based; in an increased confidence that New Zealand (or New Zealand's larger cities at least) is part of a wider cosmopolitan community. This is the cultural milieu in which Joseph makes sense, from which he gains his confidence. And it is a culture that has been built in significant ways through places such as the cafés, restaurants, and bars along Ponsonby Road.

The question confronting me when I began to study Ponsonby Road (and two other similar places) was how to interpret and understand this new urban culture. How to make sense of Joseph? Intellectually, Ponsonby Road is engaging precisely because it seems to embody a multitude of processes transforming Western cities. Given the similarity of Ponsonby Road to other, globally oriented, Pacific Rim cities it is hard to resist, for example, reading Joseph and his Ponsonby Road coffee drinking as a cipher for some kind of overarching process – globalization, time–space compression, McDonaldization. The shift in the way men and women relate to each other on a day-to-day basis within public space, and the unfamiliar and often ambiguous gender performances which are a part of this, also resonate with trends analysed elsewhere. And the more general questions that have shaped political arguments about the road – arguments about deviancy, difference, and mainstream norms of social behaviour – flow directly into ongoing debates about what (and for whom) the public and quasi-public spaces of the city should be for. And yet, if we return to Joseph, to what he is doing in Duo, a limit on these generalisations is apparent. If we can see elements of the above trends in Joseph's actions, what is also apparent is how he is engaged in an (often subtle) dialogue with the people and objects in the cafés, bars, and other places he uses. One can begin to see a little of what I mean by this in Joseph's diary entry at the start of this section. The timing of Joseph's near-daily 10.30 a.m. coffee visit to Duo is structured by the demands of his work obligations. However, the actual feel and content of the visit is generated through how Joseph works the possibilities of being in Duo. His conversation with Scottie the barista is a careful improvisation involving a subtle mix of interest and nonchalance. The "somewhat of a friend[ness]" relationship Joseph has with Scottie is something that has been nurtured and sustained with dexterity. Similarly the casual encounter with Gail ("fellow patron, and friend of Scott," and later we discover a friend of Joseph, too) is part of the fragile texture of friendship and community which is essential to the webs of sociality which make up Ponsonby Road. My point is not that the interpretative work of Joseph negates the aim of attempting to delineate general trends, or tendencies.

Rather, it nudges at a need to recognise the centrality of everyday social practice in the articulation of these tendencies. And it demands methodological and interpretative strategies that build this recognition into their very core . . .

3 THEORISING EVERYDAY LIFE

. . . Everyday life and everyday culture are two of the great frontiers of contemporary human geography. . . . [T]he pages of geography journals now teem with an expanding array of articles on topics as diverse as men's lifestyle magazines; gentrification and the art of dining in ethnic restaurants; the sexual politics of lipstick lesbians and gay skinheads; popular photography and the touristic gaze; women hobos and urban graffiti artists; car-boot sales; shopping malls and the politics of hanging out; popular music; and the skills of supermarket shopping. Even that arduous weekly trip to the gym has been opened up to the inquiring cultural geographer. These articles – diverse though they undoubtedly are – are united by a conviction that everyday life is a key realm where social power is exercised and maintained, and the everyday simultaneously opens-up new realms of resistance to mainstream networks of power/knowledge. . . .

. . . [Nigel Thrift's work on the practices of the everyday suggests that there are] at least three crucial elements that any accounts of everyday life must contain if they are to be plausible and interesting. First, they must be respectful of the social practices through which the everyday unfolds. They must recognise that much social practice is different (but certainly not inferior) to more contemplative academic modes of being in the world – embedded as they are in the noncognitive, preintentional and commonsensical. Second, they must contain a sense that practices (and thus the subjectivities and agencies of which they are a part) are shot through with creativity and possibility (even though these are "constrained" and limited by existing networks of association). Third, the everyday should not be viewed as a world apart from more rationally grounded realms of social action such as "the state," "the economic," "the political," or whatever. Rather, what needs to be recognised is how all elements of social life, all institutions, all forms of practice are in fact tied together with the work of getting on from day-to-day.

[These criteria suggest three observations about cultural geography:]

1. Cultural geography's revival was largely built upon a commitment to a particular politics of representation, and it remains obsessively focused on representation. This obsession not only implicitly downgrades the importance of practice, stressing as it does the symbolic over the expressive, "responsive and rhetorical" [to quote Thrift] dimensions of language. It also has an alarming tendency [as Thrift argues] to slip into simplistic (and often exaggerated) narratives "based on highly romantic stereotypes of both politics and persons." Thus, to take an example close to the concerns of this paper, white professionals living in an ethnically diverse area of North London, and eating out at its ethnic restaurants, are not reaching out towards some kind of engagement with the existing community (ambiguous, limited, and inadequate though that may be). No! They are [as Jon May has argued] "eating the Other," and are implicated, despite their protestations, in a process of cultural imperialism intricately bound within a complex historical geography of racisms!

2. This example leads neatly to the second limitation. In too much culturally inflected work the everyday is reified as a pure, pristine realm, heroically unbowed by the grubby domination of the powerful. Not only does this unnecessarily romanticize the everyday as a mystical counterweight to domination – a romanticism embodied in the much-quoted claim of Michel Foucault [in *Power/Knowledge*, 1980, p. 142] that "there are no relations of power without resistance" (a romanticising of resistance that is all too evident in Pile's assertion [in *Geographies of Resistance*, 1997] that if "power seems to be everywhere [. . . it is also] open to gaps, tears, inconsistencies, ambivalences, possibilities for inversion, mimicry, [and] parody"). It also drifts towards a view of everyday practices as escaping completely the grasp of the social researcher, whilst simultaneously disavowing the constitutive role of these practices to networks of domination.

3. Lastly, in large part because of its obsession with issues of representation, the cultural turn has not equipped human geography to study anything but a relatively narrow range of social theoretical questions. We simply do not have the methodological resources and skills to undertake research that takes the sensuous, embodied, creativeness of social practice seriously. Indeed, counterintuitive though it may sound [to quote Thrift], "cultural geography is not empirical enough." This is a problem that runs deep. In part the difficulty derives from an unwillingness to experiment with techniques that go beyond the now canonical cultural methods: in-depth interviews, focus groups, participant observation of some form or other. This is a conservatism that is reflected in the methodological content and focus of a number of recent (and generally very good) geography textbooks aimed at introducing undergraduates to qualitative research. But even where attempts are made to reach beyond the limitations of these methods – as is thankfully becoming a little more common – the accounts produced are uncomfortably similar to those that preceded them . . .

But how then can we approach studying the ordinary, the everyday, in ways that actively engage embodiments of social practice as Thrift urges us to? What kinds of methodologies should we employ if we are to be more sensitive to the creativity of practice? . . . I want to suggest that, rather than ditching the methodological skills that human geography has so painfully accumulated, we should work through how we can imbue traditional research methodologies with a sense of the creative, the practical, and being with practiceness that Thrift is seeking. Pushed in the appropriate direction there is no reason why these methods cannot be made to dance a little.

4 PERFORMING RESEARCH: PART ONE – FEELING TOWARDS A METHOD

[. . .]

Let us return to Joseph, and the questions posed about researching him and Ponsonby Road at the end of section 2. Joseph is – as we already know – a subtle and socially sophisticated inhabitant of Ponsonby Road. He knows the casual but intricate etiquette of café usage, how to carry through a drifting conversation with Scottie as he attends to his barista work, how to work in Gail when she arrives, and he possesses a keen sense of the significance of self-presentation. He is also thoughtful and articulate. Yet, when asked about why he likes Duo, how he would describe his relations to Scottie or indeed Gail, how he learnt to be so adept at doing coffee, he feels put on the spot.

Questions such as these were important for me as I groped to understand something of [what Raymond Williams called] the "structure of feeling," the tissue of relationships and events, within which the communities of sociability woven through Ponsonby Road were enacted. And, as I will try to demonstrate, making sense of and respecting the reasons why Joseph had difficulty in answering questions about his time spent on Ponsonby Road is centrally important in conceiving methodologies that take the flow of practice and its complex embodied intersubjectivities seriously.

So why did Joseph have difficulty answering? There were, I think, three reasons. The first reason was simply that a good number of these questions simply are not those that Joseph would have much reason to think about in any depth in the usual course of events. The relationships that form the context through which his life is lived are not always under scrutiny or the object of constant deliberation. Indeed, this kind of self-reflection seems somehow out of tune with the ethos of Joseph's (and, as I was to come to appreciate, with many other of my respondents') friendships and social relations on the road (and indeed elsewhere).

The second reason, one closely related to the first, was how I framed my questions. My questions were those of the social scientist, and as such they demanded a style and logic that was not necessarily aligned with the way Joseph thought about his day-to-day life. He does not, for example, need a reason why he likes Duo and it is almost (but not entirely) unreasonable to demand that he has one. Acknowledging this difference not only requires recognition of the need to gain a sense of the frame of reference through which an individual encounters and negotiates his or her world. It also means acknowledging and accepting that accounts offered by people may appear by their very nature "indistinct," "self-contradictory," or "incomplete" .
. .

This brings us to the third reason for Joseph's inarticulateness . . . For Joseph, a great deal of what he knows and does on Ponsonby Road has accumulated through straightforward usage. Joseph knows what to do, and has an intuitive knowledge of what Ponsonby Road is about, that, if not exactly subconscious, is in certain respects nonconscious, noncognitively oriented, or, as Anthony Giddens [in *The Constitution of Society*, 1984, p. 7] puts it, is profoundly "practical." This

knowledge is by no means itself inarticulate – the expressiveness of Joseph's (and others') use of Ponsonby Road is witness to that. But its logic and sense is not ordered through the discursive and, if we are to find ways of properly accounting for these, we too must think beyond the discursive.

In approaching Ponsonby Road and thinking about methodology, it was initially the problem of how to "get at" these practical, routine, knowledges that most concerned me. This was for two reasons. I am interested in the ways in which urban places, particularly urban public places, become through the sensuous interweaving of the lives and daily projects of the thousands of individuals who daily dwell within them. And, as I have suggested with the example of Joseph, a great deal of this "making place" becomes through the work of embodied routine, routines of occupation, and use. Second, it also seemed that one of the most problematic dimensions for the researcher studying the sociality of public spaces (that is, places where people are routinely subject to interaction with strangers) are precisely these routine, noncognitive, embodied aspects and the solidarities that they form: if they are noncognitive, and in large part nonverbal, how can they be included within research? Assuming that they are not entirely of a knowledge that Michel de Certeau [in *The Practice of Everyday Life*, 1984, p. 93] evocatively characterised as being "as blind as two lovers in each other's arms," one answer is to try to construct a sensitively structured technique through which research subjects can find a space for reflecting upon these practices.

[. . .]

Slowly it dawned on me that, if the world could productively be viewed in terms of sets of practical performances and enactments, the research process itself could, too, be framed as a kind of performance. . . .

5 PERFORMING RESEARCH: PART TWO – THE DIARY–PHOTOGRAPH, DIARY–INTERVIEW METHOD

Diary continued . . .

6.30 p.m.
Now Wendy's is a whole new experience. A fast food, fast package and container meal. You notice everything is wrapped, cartoned,

sacheted or shrink-wrapped, plastic, cardboard and free . . . on the serviettes. And trayed. The trays always remind me of BOARDING SCHOOL. I guess it's similar to an airline meal that never gets off the ground. You don't get a "FASTEN YOUR SEAT BELTS" sign or self-control air conditioning above your head. And everyone is facing different directions. However, Isaac enjoyed it. He looks very trendy in peak cap with Red jersey and wearing SOUTH PARK t-shirt underneath "THE MANY DEATHS OF KENNY." He's lining up pure cane sugar on the table, cutting it up and using the coffee straws to SUCK it up his nose. "I've seen them do it on the movies and we are talking about DRUGS at school. They are BAD for you and it wrecks your MIND and you forget everything. CANNABIS, MARIJUNA, CRACK, POT, COCAINE, CAFFEINE, CABBAGE WEED." ISAAC. B. Well that came out of nowhere. This diary is working wonders already. My name is Paul, Paul Rennie Brown and Wendy's is my middle name. (Research diary entry Paul, 42, estate agent, father, paraglider)

The metaphor of performance – surprisingly, given its current popularity – has a well-established lineage of usage within the social sciences and humanities. Within sociology, ethnomethodologists and symbolic interactionists such as Erving Goffman and Howard Garfinkel drew heavily on dramatological metaphors in their research into everyday interactions. More recently, Judith Butler has used the term "performativity" to theorise how gender is reproduced through everyday social practices. Equally, the more radical appropriation of performance advocated by Thrift draws on work from theatre studies and performance art rooted in a heterodox tradition which arguably reaches back to Dada, and includes the agitprop theatre of the 1960s and 1970s, Situationist International with their *dérivés* and *détournements*, community theatre, and body art.

As Nicky Gregson and Gillian Rose have argued, it is the work of Goffman and Garfinkel that has most influenced work in human geography. However, one of the primary inspirations of my turn to the performed was more prosaic. The idea of day-to-day life as involving an element of performance is pervasive in contemporary popular culture (this is as true in New Zealand as it is throughout much

of the Western world). This can be seen in the heightened attention to self-presentation and self-fashioning evident in the evolution of many postwar urban subcultures. It is also evident in the popularity and success of 'reality'-based and diary-based programmes on television and radio. In reflecting on this popular culture, it also occurred to me that rather than just using writing (the diary) and talk (the diary–interview) it also made sense to try and draw more directly on people's visual imaginations. Hence, I provided each of my diarists with a disposable camera with which they were asked to take photographs of interesting and/or significant places and events of their week.

[. . .]

What does this mean in practical terms? I want to highlight two areas that define my own engagement with the performative, practice-oriented nature of social life:

5.1 The partialness of accounts

I have in the preceding argument repeatedly stressed the importance of recognising the degree to which the world is made through the work of practical, sensual, social action. If we leave aside for the moment the not-insubstantial question of the solidity and enduring nature of the institutions reproduced through this practical, sensual, social action, such an ontology demands that we preserve a sense of openness and possibility within our accounts of the world even when these accounts are about the ways in which certain institutions, certain facts, certain ways of thinking and acting appear utterly natural and immutable. If this ontological stance is fundamentally optimistic in tone, it nonetheless has some important implications for how we understand the reach and certainty of the knowledge we as social scientists produce. First, it suggests we need, in interpreting interviews and related empirical material, to be more sensitive than we have been in the past to the partial-ness and moment-ness of the accounts offered. An interview, even a series of interviews or diaries and diary–interviews, does not provide a definitive account of an event, place, or individual. . . .

. . . Just as Joseph negotiates Ponsonby Road anew each time he uses it, the interview, too, is a negotiation of a relationship to the events outlined in his diary. The more he and I talk about it, the more detail and perspectives I get on Joseph's relationship to Ponsonby Road. But this is not leading to a single unified truth about either Joseph or Ponsonby Road. At the same time, attention to the rhetorical content helps to make apparent through the gaps and ambiguities of this account interesting aspects of his relationships to others and the world. This is worth reiterating. The notion of the interview as a kind of performance helps us to avoid thinking of the self as funda-mentally an issue of depth. As David Silverman [in *Qualitative Research*, 1997] has argued, the very idea of the interview is bound up with a hermeneutics of the soul that is similarly closely related to the technologies of the confessional, and those of the mass media. All too often this works towards a "reconstruction of a common and uni-tary construction of the self" (p. 248). The notion of performance helps to deflect us away from looking for depth (in the sense of a single unified truth) and directs us towards detail (in the sense of a fuller and more variegated picture of the interviewee).

[. . .]

6 CONCLUSION

. . . [T]he argument of this paper is rooted in a con-viction that the metaphor of performance offers more than yet another new way of doing human geography. Although the arguments of writers such as Thrift and others exploring ideas of per-formance can be read as an effort to establish something like a new paradigm within human geography, they do not have to be. Rather, the tone of their writing can be seen more in terms of an attempt to alter the style in which human geo-graphy is done. Approached from the appropriate angle, the movement towards a framing of the social world based around terms such as enactment, performance, and practice offers a possibility for a

range of creative dialogues between already-established forms of human geographic writing and, more obviously, novel approaches to doing human geography. The sense of playfulness, as-if-ness, plurality, combined with a genuine curiosity about the ways that social life is ordered and carried through, does not only encourage us to explore new realms of social action. That is to say, it not only encourages us to think about a wide range of social phenomena such as the body, emotions, nonhuman objects, the everyday, in ways that take us beyond an obsession with a politics of rep-resentation. It also presents an opportunity to reinterpret and reappropriate established method-ologies and ways of writing human geography that transcend the anxious culture of critique which has marked so much of the turn towards the cul-tural. Indeed, in place of this anxious culture it is possible to see the emergence of an energetic methodological pluralism that is both reinvigorat-ing and transforming the ways in which we think about human geography.

Clearly, to realise the opportunities of this con-temporary interest in performance requires more than simply trying to reframe our theoretical talk in terms of practice and performance. It requires a broadminded openness to methodological experi-mentation and pluralism within human geography, and the allowance of a certain amount of method-ological naivete. . . . [T]his experimentation can be relatively modest. After all, the purpose behind the diaries, photographs, and interviews produced with Joseph, Miranda, Paul, and others was to try and build up an account of Auckland's public life: (a) that was respectful to the people and commu-nities involved in its making; and (b) that had a cer-tain truthfulness (a truthfulness consisting both of an intellectual rigour as well as a certain emotional resonance). Such an approach, in dialogue with the more radical methodological accounts being developed by people such as Pratt and Thrift, can help make for a more dynamic and more empiric-ally engaging style of human geography.

PART TWO

Cultural Geography: A Transatlantic Genealogy

Somewhere a Man's Shoes are Wet', by Brian Taylor. Courtesy of Modernbook Gallery, Palo Alto, California

INTRODUCTION TO PART TWO

It is tempting to imagine cultural geography as a coherent sub-field of geography with its own distinct intellectual roots and a clear lineage from founding to present. Such an approach is appealing because we like to think of our present scholarship as building upon a firm foundation laid by our forebears, and extending that foundation into important new directions that would not have been possible without the work that had come before. But this kind of history would be a vast over-simplification meant to serve our present needs more than our understanding of the past. There has been a need, it seems, to view cultural geography as a single field of study, one that can, for instance, be summarized neatly in readers such as this one. Indeed, cultural geography has been reinvented many times, particularly in the United States, beginning with Wagner and Mikesell's *Readings in Cultural Geography* (1962). Recently there has been a spate of readers and handbooks, each putting its own particular stamp on the roots and contemporary developments in the field (for instance, Foote *et al.*'s *Re-reading Cultural Geography*, 1994, Anderson *et al.*'s *Handbook of Cultural Geography*, 2003, Duncan *et al.*'s *A Companion to Cultural Geography*, 2004, and Thrift and Whatmore's *Cultural Geography*, 2004). This is *not* to suggest that these readers display a "presentist" history of cultural geography, one that represents all past scholarship in the field as building the foundation upon which our current work as cultural geographers rests.

Far from it. More than anything, a perusal of the various compilations, collections, and companions to cultural geography reveals it to be an eclectic field with a broad range of scholarly topics, many of which seemingly have little in common with each other. In an article in one of these collections (Foote *et al.*'s *Re-reading Cultural Geography*), James Duncan went so far as to call cultural geography a "heterotopia," by which he meant that it did not so much share "a common intellectual project" as an "institutional site." Duncan's use of the term *heterotopia* comes from French philosopher and social critic Michel Foucault, and describes a space that contains within it, or juxtaposes, several incompatible sites. It is an "other" space, a space of difference.

But a heterotopia is not simply an abstract space of difference. Foucault meant it to identify actual sites or places in which seemingly incompatible differences were – however awkwardly – brought together. Foucault drew on the medical term "heterotopia" (which means the displacement of an organ from its normal position) to suggest those spaces in society which served as "counter-utopias." Thus, whereas a utopia was a kind of pure space that did not really exist – literally a "non-place" – but which expressed the social norms that dominated our ideas of what kinds of spaces *ought* to exist, a heterotopia was an actual space within which various incompatible sites in fact did exist. A theater stage could be viewed as a heterotopia. And while Foucault referred to psychiatric hospitals, prisons and other spaces of deviance and crisis as heterotopias, his primary example of such a space was the garden: "The traditional garden of the Persians was a sacred space that was supposed to bring together inside its rectangle four parts representing the four parts of the world, with a space still more sacred than the others that was like an umbilicus, the navel of the world at its center (the basin and water fountain were there); and all the vegetation of the garden was supposed to come together in this space, in this sort of microcosm."[1]

In thinking about cultural geography as a heterotopia, then, we are reminded that Foucault viewed such spaces as constituted by *discipline* and *power*. Heterotopias were, perhaps paradoxically, the outcome of particular social *orderings* of the world, but orderings that resulted in contradictory spaces of difference, rather than perfect spaces of utopia. How then might the broader disciplinary ordering of geography have resulted in Duncan's heterotopia of cultural geography? How might cultural geography today represent a space of difference that has been constituted by, as Foucault would have put it, particular "regimes of truth" that have defined the norms of geographical knowledge? This question is important for at least two reasons: First, it is important to recognize a scholarly field like cultural geography as itself constituted through the social relations within which scholars are situated. In this sense, ideas and knowledge do not exist apart from the social situations in which they are produced. It is incumbent on any student of cultural geography, then, to understand the social contexts within which cultural geography has been produced as a field of knowledge about the world. Second, understanding the production of knowledge as socially situated requires that we also view our current scholarship in the same way, recognizing that what are sometimes easily viewed today as the moral or ethical failings of our forebears cannot be safely stashed away in the dustbin of history. The production of knowledge, both in the past and in the present, must be recognized as infused with social relations of power.

In Part Two, we focus on past works of cultural geography with an eye toward understanding some of the social contexts within which cultural geography has been produced. The section features original articles by Friedrich Ratzel, Paul Vidal de la Blache, Carl Sauer, W.G. Hoskins, and Wilbur Zelinsky. Each of these is introduced with some brief discussion that situates the work of these authors in a broader social context. Additionally, we have included three articles by contemporary scholars – Karl Ditt, Brian Graham, and Pyrs Gruffudd – examining the work of Franz Petri, Estyn Evans, and H.J. Fleure respectively, as examples of current scholarship that makes productive use of viewing past geographers as situated within particular social contexts. Taken as a whole, the section seeks to demonstrate that cultural geography – like all academic disciplines – has always been subject to the social orderings that constitute knowledge at any given point in history.

In subtitling this section "A Transatlantic Genealogy" we reference another term, *genealogy*, used by Foucault to describe the historical study of the disciplining practices that bring a person's subjectivity into being. Such practices emerge from the historical contexts in which particular sets of ideas achieve a kind of "common sense," and these ideas are then given stability and power within particular institutions (for example, educational, governmental, medical). Foucault used the term to describe a means of analyzing the ways people's perceptions, experiences, and interpretations of the world are disciplined by discourse; that is, by the socially accepted ways of saying things, of commonsense ideas. Because discourses reflect the particular social power relations that hold sway in a given historical period, genealogy was a study of how particular discourses have emerged historically and how they have shaped subjectivity during particular historical periods. A genealogy, then, is an exploration of the determinate historical conditions under which statements are combined and regulated to form and define a distinct field of knowledge, forming a particular "regime of truth." A genealogy seeks to address questions like these: Under what historical conditions do particular truths emerge and achieve power? How do particular discursive formations come about?

In referencing Foucault's genealogy here, we suggest that cultural geography reflects not simply the social contexts in which it has been produced, but the fact that a history of cultural geography cannot assume that our current knowledge has been created through a straightforward process of building upon and improving the ideas of the past. Foucault argued, instead, that discourses are historically discontinuous, and that different historical eras are marked by different *epistemes*. In *The Order of Things* (1966/1970), Foucault used the term *episteme* to describe the commonsense assumptions that provided the basis for the kinds of knowledge and discourses that were possible during a particular historical period. He outlined three distinct epistemes, each with its own dominant system of reproducing knowledge: the Renaissance, in which knowledge was reproduced primarily by resemblance; the classic period, in which representation dominated; and the modern period, in which structuralism was the primary framework in which knowledge about the world was reproduced. Thus, knowledge that made sense during one era might be viewed in another as complete nonsense.

This is not to suggest that the work of early cultural geographers should be viewed as nonsense. Rather, it is to suggest that past cultural geographies inform our present work in complex and often unexpected ways. It is to further suggest that terms like "culture", "landscape", "civilization", "nation", or "nature" don't necessarily maintain consistent meanings, and instead must be understood as carrying within them some of the commonsense notions of the particular historical periods in which they're being used. Culture has been, for example, viewed at times as that which separates humans from nature, yet it has also been viewed as something that reflects nature's influence on humans. Culture has been viewed as producing sub-national regions of distinct life-ways, yet has also been mobilized to describe the unity of nation-states. These couplings of various ideas have produced powerful explanations in geography. The idea that nature is an influential force on human behavior, for example, was easily viewed as common sense within geography during the early decades of the twentieth century. Yet today cultural geographers are likely to think of *environmental determinism* as an embarrassing idea implicating academic geography in the reproduction of racist ideas that explained Euro-American power in terms of climatic advantages afforded to the white-skinned peoples of northwestern Europe. On the other hand, there have been criticisms of cultural geography today for its timidity in exploring environmental influences on culture.[2] We find such criticism unwarranted, however, and have thus chosen to devote a whole section of the *Reader* to the theme of nature (see Part Four). In the present part, the selections by Ratzel, Vidal, Sauer, and Hoskins each display a particular perspective on the relationship between nature and culture that reflect the common sense of particular social and historical contexts.

The nation as a scale of analysis is another socially and historically situated idea explored in this part. Today, it is often noted that geographers have been supportive of projects of nation-building and this has sometimes meant their complicity in projects of colonialism, imperialism, and fascism. And while it is certainly important to acknowledge "past sins," our goal here is not to tell a story of past shame and current redemption. The past certainly has no monopoly on shameful scholarship. We are more interested, however, in situating past scholars so as to view them as shaped – in Foucault's terms – by particular epistemes, to view them from the perspective of Foucault's genealogy, rather than with the conceit of political progressiveness. Obviously, recognizing social and historical contexts does not mean we should not see racism or fascism for what they are. But it is important to remind ourselves that ideas viewed as shameful today cannot be relegated safely to a past beyond which we have now progressed. If scholars as brilliant as Friederich Ratzel and Paul Vidal de la Blache may now be taken to task for their imperialist attitudes or racist assumptions (as indeed they should be), then humility demands that we accept the probability of our own moral failings too.

It is for this reason that we seek to avoid constructing a "traditional" cultural geography as a foil with which to establish some inherent progressiveness in a "new" cultural geography. Dividing cultural geography into traditional and new halves – which happened after a volley of incisive critiques were launched in the 1980s – risks the assumption that we're in fact talking about past and present versions of the same beast. Instead, a genealogical view of cultural geography insists that *all* scholarship be situated within the contexts that lend power and legitimacy to our ideas.

Finally, something must be added here about the "transatlantic" part of this part's title. Although we have included selections from Continental scholars like Ratzel and Vidal de la Blache, as well as Ditt's piece on Petri, the bulk of the material in this section – and in the *Reader* overall – is by scholars working in either the United States or Great Britain and Ireland. Another heterotopic quality of cultural geography, then, might be the locating of both American and Anglo-Irish scholarship in the same disciplinary space. And indeed, many scholars would say that the cultural geographies practiced on either side of the Atlantic often appear so different as to question the merits of their sharing the same label. In the United States, cultural geography developed largely out of the influence of Carl Sauer (see p. 96), with a focus on landscape and rich descriptions of historical change as manifest in the changing material artifacts of human settlement and work on the land. While Sauer himself drew much inspiration from the German *Landschaft* school, as well as from the Vidalian tradition of rich, descriptive regional monographs, American cultural geography developed within an early and mid-twentieth century disciplinary context that was increasingly hostile toward the "unscientific" qualities of landscape description. Yet, cultural geography maintained a significant position in American geography, largely due to its popularity as an introduction for students new to the discipline. Even today, many introductory human geography courses taught in the United States are essentially descriptive surveys of the world's cultural landscapes.

In Britain, cultural geography emerged from a strong tradition not in landscape description but in social history. While W.G. Hoskins (see p. 105) practiced a kind of cultural geography that would have been very recognizable to Carl Sauer and his students in the United States, his work was not recognized in Britain as cultural geography *per se*, but rather "landscape history". Cultural geography, instead, drew more from work on the relationship between social and cultural change (as evidenced, for example, in the selection by E.P. Thompson in the *Reader*, see p. 20). And while material manifestations of such change in the landscape might be relevant for study, it was not the central concern of cultural geography in Britain.

Several decades have now passed since British and American cultural geographers began to engage each other's work significantly, and as a result there has been a tremendous amount of fertilization across the Atlantic. But it is important to note that any genealogy of cultural geography must take into account this spatial divide and the intellectual continuities and discontinuities that continue to define it.

NOTES

1 Foucault, M., "Of other spaces" [*Des espaces autres*], trans. J. Miskowiec, *Diacritics* 16, 1 (1986): 25–26.
2 See, for instance, critiques by Noel Castree, "Differential geographies: place, indigenous rights and 'local' resources," *Political Geography* 23 (2004): 133–67; and Arturo Escobar, p. 287 of this *Reader*.

"Culture"

from *Völkerkunde* (1885–1888), translated as *The History of Mankind* by A.J. Butler (1896)

Friedrich Ratzel

Editors' introduction

Friedrich Ratzel (1844–1904) is perhaps best known as a political geographer, owing to his development of the concept of *Lebensraum* – literally "living space" or, the geographical area within which living organisms develop. *Lebensraum* was something that territorial states – like Germany – needed if they were to grow and mature, just as any vibrant organism needs "growing room". Without *Lebensraum*, the state would – like a houseplant – wither and die in the struggle for survival with other organism-states. One could thus read Ratzel's work as a precursor to the Social Darwinism that influenced a great deal of early twentieth century social science. His *Politische Geographie* (1897), indeed, explained Germany's expansionist ambitions in terms of a Darwinian struggle for survival.

Yet while Ratzel is well known for his political geography, his two-volume *Anthropogeographie* (1881 and 1891) established him as a major figure in the study of culture and its environmental influences. Linking the evolutionary ecology of Charles Darwin and Ernst Haeckel with patterns of human settlement and cultural development, Ratzel's work has typically been viewed as a manifesto on environmental causes of human behavior, an interpretation credited with inspiring what has come to be known as *environmental determinism* in the work of American geographers such as Ellen Semple and Ellsworth Huntington.

Ratzel made a clear distinction between the concepts of nature and civilization, and although his legacy is associated with environmental determinism, his actual claims are somewhat more subtle and complicated. Ratzel viewed nature and culture as opposing forces struggling for dominance over the course of human progress. The volume *Völkerkunde* (1885–1888), from which this selection is taken, was Ratzel's contribution to ethnographic theory generally, and the study of the relationship between culture and nature more specifically. Echoing the general sentiments of late nineteenth century cultural theorists such as E.B. Tylor and Lewis Henry Morgan, Ratzel argued that the study of culture allowed students to appreciate the deep roots of humankind in the natural world as well as humankind's ability to free itself from nature through culture.

Völkerkunde divides the human world into "natural races" and "cultured races." The latter are those races that have been liberated from the soil, whereas the former are still bound to the natural world. Ratzel assumed that such a divide marked a trajectory of historical progress; "cultured races" were more advanced. He noted, for example, that while many "natural races" were disappearing as a result of colonization and industrialism, there was "consolation" in the knowledge "that a great part of them is being slowly raised by the process of intermixture" with "cultured races." It is easy to see in Ratzel's views here a racist claim of Europe's inherent superiority over its colonized subjects. His work was indeed later invoked as a scientific justification for the rise of Nazism in the twentieth century. Yet while Ratzel's views could certainly be condemned today as ignorant and racist, he viewed culture in the more relativist terms of an anthropologist than a white racial supremacist.

Ratzel insisted, for example, that *all* races, whether "natural" or "cultured," have culture. All humans, he argued, are born with the same basic faculties of reason and intelligence; but some are hindered by internal social and external environmental conditions more than others. "Every people has intellectual gifts," Ratzel argued. "Each can claim a certain sum of knowledge and power which represents *its* civilization. But the difference between the various 'sums of acquirement of the intelligence' resides not only in their magnitude but in their power of growth. To use an image, a civilized race is like a mighty tree ... There are plants which die off every year ... The distinction lies in the power of retaining, piling up and securing the results of each individual year's growth ... Civilization is the product of many generations of men. ... The development of civilization is a process of hoarding."

Like a tree, then, civilization was "rooted" in the soil. While human civilizations could thus be compared to natural organisms, culture was something that ultimately made us different from the other flora and fauna of the natural world. Ratzel was fond of pointing out that culture also denoted the tillage of the ground. It revealed both our connection to the soil and our domination over nature by our intellect. And it was through the tillage of the soil, with its associated divisions of labor, that the most advanced civilizations emerged.

Ratzel observes, however, that throughout history there runs a struggle between the settled civilizations of the tiller and the empires of the nomadic herdsman. In the selection below we see, for instance, Ratzel's argument that when considering the origins of ancient Egyptian culture, one must look for a broader context of connections across the Afro-Eurasian landmass, and that Egypt acquired its culture through immigrants from Asia. He also notes that Chinese culture – long regarded as fostered in rooted seclusion – is as much a product of connections across Asia as of isolation. Civilization may be like a tree, then, but culture grows and spreads. Ratzel's claims here could again be viewed merely as a thinly veiled celebration of the spread of European culture to the "natural races," justifying with science the colonial ambitions of Germany. But it is also worth pointing out that his views of the importance of cultural intermixing and connections across space suggest a more complex understanding of cultural change.

There is a wealth of scholarship on Ratzel's political geography and on his legacy in the discipline more generally. A good introduction can be found in W.D. Smith's "Friedrich Ratzel and the Origins of *Lebensraum*" (*German Studies Review* 3, 1980). In reading Ratzel from the perspective of contemporary cultural geography, we learn that his work was firmly situated within a social context of European colonialism which understood European civilization as the apogee of cultural development. We also learn that descriptions of cultural difference could be undertaken only in a way that accounted for such difference in terms of an assumed continuum of progress and advancement toward a particular understanding of what it meant to be "civilized". This represents of course a quite different approach to understanding difference than typically taken in scholarship today, in which structures of unequal social power might be taken into account, instead of groups occupying different positions along a single timeline of historical development.

* * * * * *

In regard to the growth and existence of culture, the condition holds good that culture is promoted by whatever fixes the movable human being, and the thing that most obviously has this effect is fertility of soil combined with a tolerable climate. The fixed man applies to nature a measure quite other than that applied by the man of fleeting abode; he asks, "Where have we the guarantee of a permanent stay?" Speaking of the Chaco, Dobrizhoffer says: "The Spaniards look upon it as the rendezvous of all wretchedness, but the savages, as their promised land and their Elysium." The Europeans who made their way to America did not begin by setting up tents and making pasture grounds on the virgin soil; they built houses and cities of stone. Cortes conquered Mexico in 1521, and in that year was laid the foundation of the stone cathedral; which looks as if they meant to stay. At that date mankind had long learnt on what soil culture would successfully take root. Mexico alone, with its plateau growing wheat like Castile, received the honourable name of New Spain. In the warm but temperate climate, and on good agricultural soil, it was hoped that a scion of the old Spanish culture would most speedily take root. Thus with a deep, almost instinctive knowledge of the necessity for a soil favourable to tillage, culture spread over the New World.

The material life of the peoples freed itself earlier than the spiritual from the bonds in which it had been held by indolence, insecurity, lack of necessaries, and of intercourse. A great list of inventions form the basis of what we call semi-culture. Weapons and tools of compound construction, like crossbows, removable armour, harpoons, ploughs, harrows, carts, drills, potters' wheels, rudders, sailing and outrigged boats, are found far down in the lower stages. They all involve increased labour, and labour gives them their value. Jacquemont prophesied that Spanish America within the tropics would relapse to its condition before 1492. "It will become a land without population, without wealth, because it can do without labour." Culture has ever retrograded where labour has slackened. The saying "labour ennobles" is universally true; labour has created the nobility of mankind. The most laborious of the semi-cultured races, the Chinese, stands in respect highest among the peoples of Asia. After labour itself, division of labour is unquestionably the most important condition of progress in culture; a it resides primarily in the organisation of the uniform crowd according to social functions.

Early in our first volume we referred to the intimate alliance between culture and agriculture; its significance for the cultured races remains to be spoken of. From Japan to Egypt it affords the basis of the food-supply, and is in such esteem that the plough was not deemed unmeet for the hand of the emperor. The salvation of titled land from the influx of nomads is the aim of endless fights between tillers and herdsmen. The efforts of civilised states are directed to the gaining of an independent food-supply for their people, and being indebted to no one for it. In China the highest praise given to an emperor is that he fed his people in peace. Everywhere the better tillage of the ground is what most marks the agriculture of the cultured races. Thus we get rotation of crops manuring, terrace-cultivation, irrigation, the plough, the harrow. These implements obviously indicate a boundary line in culture. The plough especially denotes a different economical system: the large farm with slaves and draught cattle becomes necessary as soon as large areas are brought under tillage. In Eastern Europe the steppe-country still possesses heavier ploughs and knows the use of them better than the forest-country. But among all races which have the plough, spade-husbandry, gardening, is also found. The choice of plants also is different. Grain of all kinds, good for storing, predominates rice in Eastern Asia, millet in India, wheat in Western Asia; also pulse everywhere. The banana, of which it may be said, as of the manna of the Israelites, "it tempered itself to every man's liking," and generally the whole family of fruits and roots yielding easily and abundantly, but not highly nutritious, shows a marked decline. The varieties of grain come from the natural grass-lands of Asia; and the turf from which they spring was trodden by the progenitors of the ox and the horse. The most important domestic animals and plants have been gained from the steppe. Generally the conditions of the Old World were the most favourable for the selection of cultivable plants and domesticable animals, and Asia could offer the more important kinds in largest number.

Compared with nomadism, agriculture is endowed with a share of the power of waiting which belongs in the greatest measure to the higher, the sedentary culture. The greater the capital of labour which is put into the ground which bears the crops, or the more toilsomely built huts and houses, temples and fortifications, the more firmly does the man cleave to it, first physically then mentally. Gunnar in the Njáls Saga refuses to leave his home now that "the cornfields are white to harvest, and the home mead is mown," and stays to meet his death. The nomad, even when he roams within narrow limits, has a new home at least in every season of the year; the farmer holds tight to his as the centuries go round. When the nomad puts two miles behind him between winter and summer, the tiller of the ground at most lays a new field to the old. Fixed frontiers come with a fixed station. How closely is the delimitation landmark bound up with agriculture! When Horace praises a country life, he does not forget the gods of the boundaries.

Agriculture serves the most immediate need, and leaves the creation of exchange-values and objects of luxury to cattle-breeding, hunting, fishing. It is cattle-breeding that first forms a capital; the herd is a travelling treasury. If agriculture produces the most important components of food, it does not provide each day for the day's consumption. The barn no less than the plough belong to agriculture, whether it take the form of the store-hut on poles, as found from the Niger to the Amos, or the earthenware urn of the Kaffirs, or the baked underground vault of Arabia and Tibet. Field-crops ought not, like the millet of the negroes, to perish so soon that beer has to be brewed in order to utilize them. A peculiarity of all tropical cereals is that you cannot bake what we should call bread from them; only the *kissere* of the Arabs, leathery tough dampers that have to be toasted on an iron plate, can be made of the leavened dough. Bread in the European sense is indeed unknown to any Asiatic race. In place of it rice, in wet or at least moist preparations, appears as the staple of food in Eastern and Southern Asia. Yet however this may preponderate, there is no cultured race that eats rice and rice only. Meat and fish with other nitrogenous foods, for example beans, take their place beside it. Indeed among all cultured races the variety of foods is great, and the sense of taste appeals at a very early stage. A liking for insects and worms is no sign of low culture. It is not only among Arabised negro tribes that locusts, water-beetles maggots, form much-prized dainties; the like is found in India and China. The Arab proverb says, "a locust in the hand is worth six in the air." Indeed the caprices of taste in ancient Rome and modern Europe have been known to go further.

The silently creative activity of culture is not measured by increased mileage, but by the growth of the number which can live permanently in a narrow area. On rich soil and with vigorous labour populations grow dense, and this is what culture needs. The great facts of the spread of mankind over the earth, in greater and less density, stand in cause and effect in the closest connection with the development of culture. Where the population is thinly scattered over wide regions, there culture is low. In the Old World the steppe-zone is everywhere thinly peopled, while the countries round the Mediterranean – Egypt, Southern Arabia, India, China, Japan – are thickly so. Six-sevenths of the population of the Earth belong today to the lands of culture. China and India number 700 millions; a corresponding area of the Central Asiatic nomad region in Mongolia, Thibet, and Eastern Turkestan, scarcely a sixtieth of that. To the stage of culture corresponds the manner of its diffusion. When it becomes conscious of this, it also strives to

disseminate itself. Europeans were allowed not only by their superiority in everything to do with culture, but also by the rapid increase in their numbers, to diffuse themselves rapidly over the earth; but it was by them too that the wish to leave no gaps in the land was raised to a principle of policy. Obstructive natives were simply shoved aside. Even a cruel "natural" race was never able to depopulate a country like Cuba in a few generations and furnish it with a new population; but civilization managed it.

Agriculture occupies its territories otherwise than by warlike conquest. The former covers tract after tract gradually but with permanent success the latter stakes out a wide frontier. The former travels step by step, the latter flies swiftly over wide spaces. Hence the former is certain in its consequences, if only time be allowed it, while the latter is transitory, or at least incalculable. The average rapidity with which white men moved westward, until they made the mighty leap from the Missouri to the Pacific, was twenty miles a year. In three centuries China has won for culture her territory outside the Great Wall, once the nursery of the most dangerous nomad hordes; and in the same time Russia has carried a band of culture all across Northern Asia to the Pacific. Before this slow but sure progress not only the "natural" races, but at last the nomads too, have to give way. The best land is withdrawn from them by agricultural colonies, the indispensable water comes into the possession of the settlers who therewith fertilise the sand and bind it together, the nomad is cast out of the grass-land into the scrub and thence into the desert. There he becomes poor and perishes. How and where he has accommodated himself to a settled life we shall have to show.

It is a law in the development of culture that the higher the point it has attained the more obscure are its beginnings. For it is always turning over its own soil, and the new life destroys the remains of the old upon which it has come into bloom. In the soil of the Old World civilizations, stone implements alone testify of earlier conditions. But as we know not the age of the stone tools and weapons found in the earth, so we do not know the circumstances of those who used them. They give no clear answer to questions as to the age of culture. Living traces of a Stone Age at least make us acknowledge that the length of the interval and the height of the stage which divide the possession of iron from the use of stone must not be overestimated. Even now, the Nubian Arabs find a stone knife specially suitable for circumcision, also for shaving the head. Pliny says that in Syria the balsam was obtained from the trees with knives of stone bone, or glass, since the use of iron tools caused the stem to wither. Schweinfurth's view, that the small, hardly-used stone weapons found by Lenz and others in the Sahara, were only made in later times for religious or superstitious purposes, looks convincing. Discoveries of stone articles in India and Japan show that there the use of stone weapons and implements has not very long been extinct. Excellent stone implements in great numbers also lie in the soil of Egypt, so that we may safely assume a Stone Age for that country. The bridge from it to the epoch of culture passes through the dearth of iron which characterized ancient Egypt.

[. . .]

May not the origins of this culture have lain elsewhere? The further we go into the inner nature of Egyptian culture, the more clearly it is manifest that it must not be regarded as an isolated phenomenon. Special as may be the stamp of it, its fundamental ideas agree with what meets us further eastward. Writing, religious conceptions, astronomical and mathematical science, and technical capacity, the theocratic government, the organisation in castes, the forms underlying architecture and sculpture; all equally underlie the culture of Mesopotamia, of Eastern and Southern Asia.

Three groups of facts combine to prove an extra-African origin for the Egyptians. Physiological characteristics point to a connection with the races of Western Asia and Southern Europe. In their paintings the Egyptians distinguished themselves from all other Africans by the colour – black for the southern men, grey for the older Libyan, white and reddish for the younger. Again, neither in the oldest monuments, nor in the post-Christian Coptic manuscripts, does the language show any trace of African affinities; nay, it is almost impossible, says Brugsch, "to mistake the close relations which formerly prevailed between the Egyptians and the so-called Indo-Germanic and Semitic races." Lastly, the oldest abodes of culture lie in the Nile delta, in the outward parts, or Lower Egypt which looks towards Arabia, Phoenicia, Palestine – that is, towards Western Asia and the Mediterranean, and in the transition-country between Asia and Africa. The further we

proceed up the Nile, while the stamp of antiquity disappears upon the monuments, the more apparent is the decline in style, beauty, and skill. And when we finally advance to Ethiopia, where, according to the old notion, the cradle of the Egyptian race was to be sought, we find, to quote Brugsch again "as the culmination of intellectual faculty and artistic development in Ethiopia, a helpless imitation of Egyptian knowledge in all that concerns science and art." Asia alone, in various favoured spots, can point to early developments of culture; while Africa, even to the most zealously-enquiring observation, can show only beginnings, and even of these the originality is still doubtful.

The difficulty of the question lies in the fact that at the moment when the Egyptians step into history they are already so decisively linked with their soil as practically to justify their own tradition that they are aboriginal. No trace is found of the instability of immigrants. "Immigration," no doubt, is not applicable to whole races, only to fragments, who find people at home there before them, and impress their stamp on these in proportion to their own number and force. This is colonisation. The conclusion is not remote: that a race already settled, extending over a great part of North and East Africa, received the germs of its culture through immigration from without. The question of descent may, therefore be solved thus: that a foreign origin is not provable for the major part of the people of Egypt. But the connection with other cultures presupposes partial immigration from Asia, and permanent intercourse with it. Since, in ancient times, so copious elements of culture only entered in company with men, an admixture of Asiatic blood became also certain.

The voyages of the Egyptians to Punt, the land of balsam, whence they themselves traced their descent, preceded by centuries Solomon's voyage to Ophir. Egyptian culture was not always a thing apart. To the northward it had the most expansive race of the world at that time – the Phoenicians – and Phoenician settlements to the north and west. As for Southern Arabia, there is no doubt that the herdsmen of the Arabian plains did not always exercise the influence that has made the land lie idle. The fertility of the soil, the favourable position for trade and seafaring, the denser population, could once have freer effect. The people of Katanieh, in South Arabia, bore, perhaps, the greatest resemblance to their nearest neighbours in

Mesopotamia. They had a complicated system of worship, religious monuments, written and pictorial, political institutions, flourishing cities, an elaborate social organisation. On the coast of South Arabia once lay marts for Indian and East African goods.

But the history of the interaction between Egypt and the neighbouring people is obscure just in those departments that are of most importance for our insight into the course of the world's history. It was only in comparatively recent times that Egypt came into contact with the states of Mesopotamia, which we must regard as connected of old by access to a common store of culture. But the origin of its culture and of its people leads us to Asia. Not only does one endmost link in the chain of Old World civilizations allow itself to be joined on to the rest; an explanation of its existence is possible only upon this supposition. At the other end, similarly apart, we find a region of similar, perhaps even older, culture in China, and its daughter-states Corea and Japan. Some have seen in Buddha a fugitive priest of Isis, and through that close bonds must have united Egypt and China; while others have assumed for China a wholly independent development. The former notion, though fabulous in form, has a germ of truth; the latter, expressed in Peschel's commendation of the Chinese as self-taught, in contrast to the European "pupils of nations historicaly buried," is not only unhistorical, but most of all ungeographical.

Curiously like the country of Egypt is that which lies between the Euphrates and the Tigris – a great oasis, surrounded by a mostly desert region, rising in the north and east to heights which form its limit; lying, too, in a kindred climate, and a gift of the waters in both senses, namely as an alluvial land, and as a land whose fertility must be called into life by inundations and artificial irrigation. The resemblance is so great that the idea of kinship forces itself on us. Here, too, culture has travelled up the river, after both mythically and literally rising out of the water. In the oldest times, which lie even further back than those of Egypt, it had its seat in Babylonia, not reaching Assyria till later. In the very oldest traces we meet with hieroglyphic writing, like that of Egypt the result of allegory evolved in the single form of cuneiform writing, and with it the same delight in recording, the same care of tradition, even monumental

tradition, which builds pyramids to put temples on – less durable, however, than that of Egypt, for Mesopotamian culture works only in clay. Examining the inner life we find a numerous priesthood no less powerful, to whom in a sense the thing belongs, whose verbose reports of victories and triumphal butcheries remind us in their very style of the historical tablets of the pharaohs. Religion – dispersed among the powers and phenomena of nature with the sun as supreme – astronomy, surveying, were the priests' affair nor could science here, any more than in Egypt, set itself free from their astrology and magic, even though in observation it made progress.

We have less information about ancient Babylonian art than about Egyptian; but we know that here, too, the best work in art is the most recent. In artistic endowments the Babylonians and Assyrians are far behind the Egyptians, but their enormous luxury favoured the lesser arts, The question of Accadians and Sumerians, the alleged Turanian forerunners and creators of Babylonian and Assyrian culture, must be left to historical enquirers. For the Hyksos, too, a Central Asian origin is held probable. For the present we have to do only with Semites, either settled as in Babylonia and Assyria, or as nomad invaders like the Chaldeans, who conquer, and build on with the copious materials amassed by their creative predecessors.

In the south and east, Asia has ripened yet other civilizations – the Indian and the Chinese – the former borne by Aryans, the latter by races of Mongol stock; nor are these dead. Chinese culture stands next in age to those of the Hamites and Semites; and in its deeper layers much remains, in vestiges hidden under the guise of a certain originality, to recall Babylon and Memphis. It is misleading to seek the chief characteristic in the history of Chinese politics and culture, as in Egypt, in their seclusion; nor must we too rashly emphasise the contrast between the Chinese and the inhabitants of the borderlands on the west and south of the continent. It is said that beyond the Belur Dagh everything, conquest and commerce alike, pushes westwards, as the Phoenicians, Nebuchadnezzar, Cyrus; on the hither side people are content with themselves, and here, therefore, culture, furthered by nature, develops far earlier, more abundantly and completely, but remains stationary for lack of rivals or dangers. At any rate, on the eastern side of Asia, there is no question of the separation and reunion of Aryan, Chaldean, Egyptian culture, of a fertilising exchange, such as has woven the most abundant threads in the web of our civilization. The Chinese saw no race near them which they could recognise as their equal, or to which they did not feel themselves far superior by what they had achieved. Japan and Corea were only outliers of Chinese culture. Something of the same kind occurred temporarily in the west – in Egypt; but Egypt could not remain so long aloof. The Chinese, Japanese, and Coreans are the only peoples whose exclusiveness has lasted almost till to-day. Undoubtedly it has had a profound influence not only on what the Chinese have done, but in a degree on what they are.

They did not, however, shut themselves up from the first, and with conscious purpose. There was a period of active intercourse with the west and the east, which is not wholly prehistoric. Great powers in Chinese life have made their entry from without, if not with pomp and sound of trumpets. All the same, they came in. We see Buddhism and Mohammedanism become powerful in the secluded land; Christianity, yet more powerful, in the Nestorian time; and, again, at the beginning of the Manchu dynasty, in the victorious missions of the Jesuits. When we look at the facts we see that what is important in Chinese culture is not isolation but connection. The Chinese of the last thousand years or so have lived in tranquil seclusion, but ideas which in common underlie the old culture have become great in combination and union. They belong to an age so remote that the history of the cultured races does not reach back to it. But their recurrence among the poor stunted possessions of the "natural" races indicates the old combination. Not only in this case, but in the study of every sphere of culture, even the Egyptian, the highest place among the great problems is always taken by the enquiry into its connections and relations, its give and take in the ebb and flow of the current of culture and intellect. Here the interest of the special history passes into that of the history of mankind. All other questions are for us of only preparatory significance.

Among the instruments of culture, of which the acquisition is, by Chinese tradition, ascribed to the Emperor Hwang-Ti, many point to Western Asia.

Like Nakhunte, the god of Susiana, this mythical sovereign founded a cycle of twelve years, and settled the year at 360 days, divided into twelve months, with an intercalary month. The names of the months have the same meaning as in Babylonia. His observatory recalls similar works in that region. With those astronomers of Western Asia, ancient China shares not only the pre-eminence of star-gazing among the sciences, but also the intimate way in which, as astrology, it is interwoven with all affairs of life. The Chinese are the only nation of the present day among whom may be seen the preponderance with which this science of superstition was invested in Mesopotamia of old. They also know five planets, four of which have names of equivalent meaning to those assigned to them in Babylonia; and about them was entwined a web of prognostics and prophecies which again recalls Western Asia. In considering the common store of culture, great weight has always been rightly attached to the remarkable agreement of astronomical notions which connects East, South, and West Asia. In the common subdivision of the ecliptic zone into twenty-seven or twenty-eight parts, designated, with reference to the intricate path of the moon, as lunar "stations" or houses, lies a strong proof of an exchange of ideas. The stars of this zone leave wide room for caprice in the selection of constellations; yet the subdivision is so alike among the three races as to exclude the assumption of an original difference. The Arabic lunar circle, which varies from the other in very few cases, is mentioned in the Koran as known to everyone. Among the Indians, whose lunar circle shows the most peculiarities, there is no mention of it before 1150 B.C. In all the old Chinese literature, a general knowledge of it is presumed; and it was certainly known by 2300 B.C. May we, with Richthofen, assume that these "stations" had a common origin in the ancestral abodes of Central Asia? For the moment let us only call attention to the fact, that this authority does not look for the first beginnings of Chinese culture on Chinese soil, except as concerns an imperfect tillage of the ground and the silk industry. But the question of "whence?" can look for an answer only in the west; and this pushes the origin of this so-called peculiar civilization near to the roots of that in Western Asia . . .

"The Physiogamy of France"

from *Tableau de la géographie de la France*
(1903), translated as *The Personality of France*
by H.C. Brentnall (1928)

Paul Vidal de la Blache

Editors' introduction

President Charles de Gaulle once famously said that it was impossible to govern a country with 365 different kinds of cheese. Had he been alive at the time, Paul Vidal de la Blache (1845–1918) might have responded that such variety is *precisely what unifies* France. And while de Gaulle was being explicit about the political implication's of France's inherent cultural diversity, Vidal's work describing that diversity could also be interpreted as carrying a clear political message. For Vidal's work – which claimed that environmental variability created conditions in which distinct local ways of life were assimilated into an overarching French culture – offered a strong argument for France's governability. Such an argument was, in fact quite useful as an instrument of political nationalism, and in this we see another inflection of that issue raised in the Introduction to this *Reader*, that the political is never far from the cultural.

A contemporary of Friedrich Ratzel's (see p. 83), Vidal's work is often compared with his German counterpart, and while both scholars were equally devoted to celebrating the glories of their respective nations through their scholarship, they differed somewhat in how they understood the relationship between humans and their environment. Vidal's work reflected less of an intellectual debt to Darwinian evolution, and whereas Ratzel's legacy is typically associated with *environmental determinism*, Vidal's work has been linked to the somewhat vague concept of *possibilism*. This term is meant to convey Vidal's belief that the natural environment presented a range of possibilities for societies to make use of. The human geographer's task, for Vidal, was to account for distinct *genres de vie* ("life-styles" or translated in the selection below as "modes of existence") in terms of understanding how societies transformed their environments in response to the constraints of those environments. Ultimately, there is in fact less distance separating Vidal and Ratzel than one would assume, given that "possibilism" and "environmental determinism" are typically contrasted with each other. Both scholars approached geography as a dynamic relationship between humans and their environment. And both understood culture as the key to humankind's ability to transcend its environmental constraints while remaining "rooted" to a particular physiogamy. Vidal's belief in the dynamism of culture allowed him, most significantly, to argue for the national unity of France despite the great diversity of its physical environments. Culture's ability to overcome such diversity and assimilate toward larger scales of expression offered a basis for France's development as a unified nation, rather than resulting in a collection of distinct societies living under the constraints of their regional environments.

Vidal's most enduring contribution to geography was his *Tableau de la géographie de la France* (1903), from which the following selection is taken. The *Tableau* was his fifth book, and quickly became a model for

the regional geography monograph; it was studied in universities throughout the world. Indeed, Vidal is credited with making the "regional monograph" one of the most significant cultural geography texts, and this model
influenced the approach taken by Carl Sauer in the United States. By the 1960s, of course, such an approach
was being roundly criticized as too descriptive, atheoretical and apolitical. In this context it is worth noting,
however, that the *Tableau* was written as the introductory volume to a much larger project: Ernest Lavisse's
twenty-seven-volume *Histoire de France*, which covered the period leading up to the 1789 revolution. Thus,
the *Tableau* focused on France's natural history and long-term social and cultural development, rather than
on issues – such as industrialization and urbanization – of Vidal's own time.

While the *Tableau*, then, does focus on rural and "traditional" themes in its descriptions of France's distinct *genres de vie*, it is apolitical only on the surface. There is an unmistakably political dimension to the
Tableau, illustrated in the selection offered here. Vidal was a strident nationalist, and sought to demonstrate
in the *Tableau* the unity that made France a distinct nation with a heritage that was deeply rooted in the very
soil of the land. Conceptualized as *genres de vie*, "culture" for Vidal was what people *did* with the resources
offered to them by their environments. Culture was the outcome of people eking out a "good life" from what
was offered up by a particular slice of land. Situating Vidal in a particular socio-historical context in which
environmental constraints remained significant in determining transport, communications, and other forms of
connection across space, it is important to note how a description of the heterogeneity of France's physical
features is central to Vidal's construction of a *unified* France.

Vidal taught for twenty-two years, beginning in 1877, at the prestigious Ecole Normale Supérieure in Paris,
and then earned the position of Chair of Geography at the Sorbonne, in 1898. He trained a whole generation of French geographers, including Jean Brunhes and Emmanuel de Martonne (who edited the posthumously published volume *Principes de géographie humaine*, 1921). Vidal's work, and that of his students,
cemented regional studies, or chorology, as one of the cornerstones of methodology in human geography, a
legacy most significantly continued in the United States by Carl Sauer and his students at the University of
California at Berkeley (see p. 96).

English language examples of "Vidalian" regional geography can be found in the works of Paul Claval (*An
Introduction to Regional Geography*, 1998) and Jean Brunhes (*Human Geography*, 1952). The classic English
language study of Vidal and his legacy in geography is Anne Buttimer's *Society and Milieu in the French
Geographic Tradition* (1971), while the broader intellectual context of Vidal's work is laid out in Paul
Rabinow's *French Modern* (1989). An insightful study of Vidal's links to the "Lamarckian" paradigm in nineteenth century evolutionary social science is offered by Kevin Archer ("Regions as social organisms: the Lamarckian
characteristics of Vidal de la Blache's regional geography," *Annals of the Association of American Geographers* 83, 3, 1993). And an insightful study of Vidal's early geographical thinking can be found in Howard
Andrews's "The early life of Paul Vidal de la Blache and the makings of modern geography" (*Transactions of
the Institute of British Geographers* n.s. 11, 1986). Finally, an excellent French language biography of Vidal
can be found in Sanguin's *Vidal de la Blache 1845–1918: un génie de la géographie* (1993).

To meet the diversity of influences which beset and
pass her borders, France has recourse to her powers
of assimilation. She transforms what she receives.
Disparities lose their sharpness, invasions their
violence. There must be something in her nature
that smoothes away angularities and softens contours. Wherein does her secret lie?

Varieties of Soil and Climate. – The keynote of
France is variety. The causes of that variety are

complex. They are due in great measure to the soil;
and so derive from the long series of geological
experiences that country has passed through.
France carries the marks of upheavals of every age.
She belongs to one of those regions of the globe
– and they are not so common as is generally
believed – which have been remoulded again and
again, and with many later readjustments, by the
subterranean forces. Even those parts which
entered on a state of quiescence long ago have not

lost the traces of the convulsions they formerly endured. Secular erosion may soften outlines and reduce elevations, but it is less successful in annihilating the essential properties of soils. There is a district in Brittany, round Tréguier, which owes its peculiar fertility to the material ejected from a volcano that has been extinct since the Primeval period. Yet no vestige of its former existence has been visible for ages past in the form of the relief. Actually, the phases of France's highly complicated geological history are still quite commonly recorded in her soil.

[. . .]

Variety in Southern France. – We should first of all differentiate the South-East, the Mediterranean South, from the South-West, or Atlantic South. When we speak of the South, the Midi, it is the former that presents itself primarily to the mind's eye – the more distinct, or, to use Madame do Sévigné's expression, the more *excessive*, of the two. Yet we have only to travel thirty miles west of Narbonne, and the olive, that inseparable companion of the Mediterranean, disappears. A little farther on, and the vineyards that nowadays carpet the plains also cease. Fields of wheat and maize, first clumps, then little woods of the British oak, build up little by little a landscape of a wholly different appearance. As we get farther from the Mediterranean and nearer to Toulouse, we pass by degrees from a region where rains are light and, what is more, unevenly apportioned to one where rains are more abundant and better distributed, reaching, in Upper Languedoc, Quercy, Agenais and Armagnac, their maximum in spring. The transition is a gradual one: the increase in the summer rains, which fall so rarely on the shores of the Mediterranean, is perceptible by the time we reach Carcassonne, and becomes clearly marked between that town and Toulouse. Gradually, too, but this is farther inland, the winds, whose wild descant rises so clamorously round the Mediterranean, breathe in a less violent strain. Softened by the rain and swept by milder airs, the soil resolves itself into a loam of a brown or light yellow colour. Maize, which needs the spring rains, disputes the ground with wheat.

There are therefore at least two Souths in the South of France. By the Mediterranean, in Roussillon and Lower Languedoc, and on the limestones of Provence, we have the more clearly marked variety, due, in the main, to the impress that summer

leaves upon the landscape. When the countryside has endured several weeks of drought – perhaps a hundred days on end with a temperature of more than 68°F, and everything is covered by a cloak of dust, the mind is haunted at moments by that image of death which is associated with summer in some mythologies of the ancient world and Mexico. The moisture has sought refuge in the subsoil, where the long roots of the trees and shrubs burrow in search of it. The rivers hide their waters under a bed of pebbles. On the rocky hillsides no trace remains of the wealth and variety of flowers that bloomed in spring. But the cyclonic rains which the latter half of September usually brings with it put an end to this crisis of the year. In the Mediterranean region October and November are preeminently the rainy months. With the passing of summer the sharp contrasts of temperature appear again, whose effect, though sometimes treacherous, is tonic and bracing on the whole, and one of the characteristics of the Provençal climate.

[. . .]

Variety in Northern France. – The variety in the North is equally great, but different in kind. It is made up of subtle shades rather than of sharp contrasts, and blends in a quieter colour-scheme.

In the North the relief is more uniform. However short their acquaintance with the contours of the southern landscape, few travellers fail to experience a sensation of regret, a tinge of sadness, as their eyes meet the unbroken lines and languishing horizons that confront them once the Central Highlands are crossed.

[. . .]

Varieties due to Different Soils and Aspects. – Now imagine within the picture-frame of Northern France every shade of difference that a changeable climate and a great variety of soils can produce. For here, more than elsewhere, the change in life-forms proceeds by successive additions and subtractions by touches added one moment to be erased the next. Spring makes its appearance sooner in the valley of the Rhine than in the rest of Germany, and sooner in the Ile-de-France than in the valley of the Rhine. Lorraine has several features still in common with Central Europe: summer rains are pronounced, and the rugged table-lands of Lorraine and Burgundy are indebted to them for the preservation of their forests which, once destroyed, are so difficult to re-establish. Another advantage which the East owes to its more continental position is a longer duration of the bright autumn weather, which helps the vine to ripen. Lying near to the limits where continental and maritime influences meet yet still open to those of the South, the country between the Rhine and Paris derives from its unstable climatic equilibrium a more delicate response to the slightest variations in altitude, aspect and soil.

Thence come in endless variety little changes in the scenery. We note, for instance, the differences between the slopes up which the rainy west winds climb and those across the watersheds. The limestone escarpments of Mâconnais, with their bright tones and crumbling heaps of loose stone draped in a delicately chiselled vegetation of creepers and convolvulus, reminded Lamartine of pictures of Greece. Indeed, between watery Bresse and the dismal table-lands of Auxois the lines of eastward spreading hills have a luminous quality which we shall not find again in our northward journey. Taking advantage of slopes facing continuously in the same direction, the chestnut and even the almond extend into the folds of the valleys of Alsace. The eastern flanks of the ridges of Lorraine are hollowed into combes, in which the reflected light and warmth bring vines to maturity. Near Metz they shelter veritable orchards. The rich crops that love the sun, vines, fruit-trees and walnuts, extend to the foot of the Ardennes, which protect them from the north wind; and with them a vegetation which, in the wealth and elegance of its forms, heralds the approach of the South or reminds us of it still.

Botanical geographers inform us that, of the principal factors governing vegetation – water, heat and soil – soil acquires its greatest importance in the transition climates. The observation applies with particular force to the North of France. Anyone who crosses the country from east to west, say, from Metz to Rheims, or from Nancy to Paris, soon sees a new type of landscape replacing, in Porcien, Argonne, Perthois and Vallage, the table-lands and limestone ridges. For the moment the vine disappears. The increasing number of trees, sometimes massed in forests, sometimes scattered along the hedgerows or in the fields and pastures; the association of broom, birch and heather in the waste places; the ponds and soggy ground whose vicinity is proclaimed by muddy foot-paths that never

dry out – everything would seem to indicate a change of climate. Yet none has occurred. The sole cause of the alteration is the appearance of a narrow but lengthy line of clays extending from the Oise to the Loire, from Thiérache to Puisaye, over which we can still trace one of the greatest forest belts of the France of an earlier day.

We know that in Northern France a series of different strata are arranged concentrically about the Ile-de-France. Thus as one comes towards Paris from the east, the nature of the soil changes at almost every step. This arrangement lends itself to landscapes suggestive at one time of the north, at another of the south. The eye misses and recovers by turns characteristics which it is wont to associate with each, and the alternations will only cease as the proximity of the English Channel and the North Sea becomes more apparent. Then the greater frequency of cloudy skies and rainy days and a marked decrease in summer temperatures, combined with the earlier arrival of the autumn rains, produce in their turn a noticeable effect upon the face of Nature. The vine, prematurely overtaken by the rains of September, leaves us finally west of Paris, and the apple tree takes its place. The beech, which, in the east, preferred the mountains and the hills, comes nearer to the plains. Still a little sickly-looking at Fontainebleau, more vigorous at St. Gobain, it becomes the dominant tree on the slopes of the Normandy valleys. It flourishes there, as on the shores of the Danish gulfs, or *foehrden*, in the misty atmosphere through which Ruysdael loves to show its white trunk gleaming. But Picardy and part of Normandy consist of table-lands of loam, testing on a permeable subsoil which drains their surface effectively. The soil mitigates in some sort by its dryness the effects of the climate. Pastures and meadow-lands are the rule on the clays of the Auge district of Normandy, but they are the exception on these table-lands, where wheat, whose deep roots save it from the need of constant moistening, finds itself in a Promised Land.

Between the two types represented by the North of France, the Ile-de-France plays the intermediary part it assumes in almost every relation. Nature languishes on the rolling plains of Berry and Champagne, but revives again in the Ile-de-France The flinty sands of Fontainebleau shelter in their setting of running water a warm-climate flora and a fauna which includes a few wholly southern forms that have found refuge in this oasis. The recesses of the deeply etched valleys enclose orchards of figs. In such features as these the Ile-de-France might remind us of the South. But it possesses also its damp forest, and, above all its great arable table-lands extending from Paris towards Picardy and Vexin.

[. . .]

What strikes one first of all in the general physiognomy of the country is its wide-reaching differences. On a surface representing only one-eighteenth part of Europe we see regions like Flanders or Normandy on the one hand, Béarn, Roussillon or Provence on the other – regions whose affinities are with Lower Germany and England, or with the Asturias and Greece. No other country of similar extent includes such diversities. How, then, does it happen that these disparities have not operated to produce centrifugal movements? Immigrants have not been lacking on the shores of France, Saxons, Scandinavian or others; yet we never find that these groups have succeeded, even if they tried, in forming isolated populations, turning their backs on the interior, as certain maritime tribes of Lower Germany, like the Frisians or Batavians, have done.

Bonds between North and South. – The reason lies in the fact that between those opposite poles of France Nature exhibits a wealth of tones that cannot be found elsewhere. If North and South stand out in sharp relief, between them there lies a whole series of intermediate shades. Climatic, geological and topographical causes are continually interfering to weld South and North together till their identities are lost; yet anon they reappear. France is so placed with regard to the continental and oceanic influences which wage an ever indecisive war within her borders, that from one side or the other plants and crops find scope to spread and take advantage of the thousand and one opportunities afforded by the varying relief and soils. The blending of North and South is more clearly shown in certain transitional regions like Burgundy and Touraine, which represent, to extend the phrase of Michelet, "the *bonding* element in France." But in truth this blend may be called the very France. The general impression suggests a mean in which all discordant tints melt into a series of graduated shades.

Modes of Existence. – Hence the great variety of products to which the soil of France lends itself, a variety which acts as a safeguard for the inhabitants, whom it enables to counteract the failure of one crop by the success of another in the same year. "The great advantage," wrote an English consul recently, "that the small tenant-farmer or small proprietor has in France, lies in the differences of climate, which favour the growth of various articles and small products that do not succeed in our country." It is these small products which render possible the ideal long cherished by the inhabitants of old France, and still firmly rooted here and there, of having all the necessities and conveniences of life at command and obtaining them all at one's own door. Such a desire must assuredly have been evoked by those "blessed lands," to be found on every side, in which it is not extravagant to dream of a life of abundance, sufficing in great measure to itself. Apply the notion more widely, and it will be found to correspond pretty closely to what the average Frenchman thinks of France. It is the abundance of the "good things of the earth," to adopt the phrase so dear to the old folks, which they identify with the name. Germany to the German is first and foremost a racial conception. What the Frenchman chiefly values in France, as is proved by his regrets when he quits her shores, is the goodness of the soil and the delight of living there. For him she is the country of countries, something, that is to say, closely bound up with his instinctive ideal of life.

Nevertheless there are bad districts in France as well as good. There are some which man adorned with flattering epithets and which were contrasted, formerly, at any rate, in the popular mind and speech with less favoured lands, forced to replace the rows of subsistence, wheat, wine and the rest, by sorry expedients. The farmer in the good districts despises the land that will not feed its man. A note of compassion tempered with mockery would welcome dwellers on unfertile soils devoted to buckwheat or the chestnut, or in districts incapable of supplying their own needs and forced to procure them from their neighbours. The poor inhabitants of Vôge used to excite this sentiment when they visited their rich neighbours in the Comté in quest of potash to fertilize their beaten sandstone soils, where trees grow more freely than wheat. When Rabelais somewhere wishes to describe the destitution of Panurge, he finds no expression more to his purpose than to display him to us "in such bad equipage that he looked like an apple-gatherer of the country of Perche."

In all districts, favoured and unfavoured alike, abundance and prosperity awaken the same desires and ideas. The principal sign of luxury is abundance of linen, a feature much less evident in neighbouring countries. Among the greet majority of the rural districts of France there is little difference in the food consumed, or even in the cooking of it, despite a few ingredients which are matters of controversy between North and South. The peasant of Champagne whom Talus depicts eating his soup at the door of his house might be found in a similar attitude similarly employed anywhere in France. When we see in the pictures of those rare painters, like Lenain, who have not disdained to paint the peasant, the attitude and physiognomy of the rustics of the seventeenth century, we recognize them again as their descendants of the present day. They have just the slow gestures of those men whose food is bread, sitting heavily on their wooden stools round a frugal loaf, and sipping ever and anon their wine like men who know its worth.

Bread, vegetables of various kinds, meat of which poultry and pip contribute the larger share – such is the food we should expect on a soil devoted mainly to cereals and the kind of stock dependent on them. Wheat is the staff of life in Southern Europe, and it so happens that France's principal wheat lands lie in the North. The uniformity in food between the North and the South of France is as marked as the difference in this respect between the French and the English or even the Germans. The French peasant's appreciation for white bread, his love of vegetables and his ingenuity in growing them, arouse the interest and curiosity of the neighbouring Teutonic peoples. In his account of the French campaign Goethe notes the antagonism between the two peoples on the question of bread: "White bread and black bread are the shibboleths, the rallying-cries that tell the French from the Germans." The Breton fishermen, all gardeners, more or less, on their mild, moist seaboard, astonish the English crews in Newfoundland when they contrive to grow a few ingredients for a salad on that barren coast. In the seventeenth century French refugees transformed the dreary *Moabit* in the sandy suburbs of Berlin with their vegetable and garden plots.

An all-pervading atmosphere, instilling ways of feeling, methods of expression, tricks of speech and a particular kind of sociability, has enveloped the various populations whom fate has brought together on the soil of France. Nothing has done more to draw the different elements into one. There is always a certain bitterness in the contact of men of different races. The Celt has never forgiven the Anglo-Saxon, nor the German the Slav. Born of pride, these antagonisms are excited and exacerbated by contiguity. But in France there is nothing of this sort. How can men withstand a power of which they are unaware, that takes possession of them without their suspecting it – a power that emanates from their deepest-rooted habits and brings them into closer and ever closer association? A little sooner or a little later, all in turn have signed the covenant.

There is, then, a beneficent power, a *genius loci*, which has rendered a national existence possible for France, and which imparts to it an element of wholesomness – something indefinable that rises superior to territorial divergences. It balances them and combines them into a single whole; yet the variations persist; they have still to be reckoned with, and the study of them is the necessary counterpart to that study of more universal relations on which we have been engaged.

"The Morphology of Landscape"

from *University of California Publications
in Geography* 2, 2 (1925): 19–54

Carl Sauer

Editors' introduction

For a while, Carl Sauer (1889–1975) was something of a lightning rod in the "culture wars" within geography, and his name is still sometimes invoked as the paradigmatic example of the kind of cultural geography that many geographers since the 1980s have seen themselves moving beyond. As head of the Geography Department at the University of California at Berkeley from 1923 to 1954, Sauer – more than anyone else – shaped the intellectual content of American cultural geography in the first half of the twentieth century. By the 1980s, however, his legacy was undergoing a significant re-evaluation by a new generation of cultural geographers influenced by recent developments in American anthropology (see Geertz, p. 29) and British cultural studies (see Williams, Thompson, pp. 15 and 20). Peter Jackson (*Maps of Meaning*, 1989), for example, referred to Sauer's "excessive focus on the material elements of culture and their representation in the landscape," and Don Mitchell (*Cultural Geography: A Critical Introduction*, 2000) has noted that the kind of cultural geography launched by Sauer was increasingly "irrelevant" to the social worlds that most geographers live in today: "As American (and British) cities burned in the wake of race riots, the collapse of the manufacturing economy, and fiscal crisis upon fiscal crisis, American cultural geographers were content to fiddle with the geography of fenceposts and log cabins..." (p. 35). And a textbook on methodology in human geography began by contrasting Sauer's approach to fieldwork with that of British feminist geographer Linda McDowell (see p. 457) in order to highlight the ways geographers have only recently begun to interrogate critically some of the assumptions (or lack thereof) underlying their approaches to research (see Paul Cloke *et al., Practising Human Geography*, 2004). In this way Sauer's name has been repeatedly invoked to represent a "traditional" kind of cultural geography that was essentially descriptive and atheoretical.

It is perhaps ironic, then, that Carl Sauer's best known work – "The Morphology of Landscape" – is not a descriptive regional monograph at all, but a sustained and systematic treatise on methodology in geography. It is also one of his only writings to present something approaching an explicit conceptualization of culture. "Morphology" was Sauer's attempt to bring into American geography insights from both the German *Landschaft* school and the regional monographs of Vidal de la Blache (see p. 90) and his students in France.

Sauer prefaces his essay by noting the need to re-examine the "common ground" upon which the discipline of geography is established. Such a need comes about, Sauer claims, due to developments in Europe in which Vidal and his students in France, and Hettner, Passarge, and Krebs in Germany, were "reasserting more and more the classical tradition of geography as chorologic relation." This meant the European field was moving well beyond the American focus – inspired by Ratzel but exemplified by Ellen Semple – on environmental causes of human geographic patterns. Sauer sought to recover for American geography, in other words, a regionalist tradition that described the dynamic interaction between humans and their environment from a long-term historical perspective. As a graduate student at the University of Chicago, Sauer had attended Semple's lectures, but he came to view the idea of environmental determinants of human behavior as unfounded scientifically and even questionable morally. Sauer felt that the evidence supported a view that did not hold nature constant, but regarded "the scene" of human action (that is, the human–environment relationship) as constantly changing.

"Morphology" sought to systematize such a view by proposing *landscape* as the organic unit upon which the ever-changing human–environment relationship could be observed, measured, and recorded (see also introduction to Part Three). Culture played the key role as the agent of change emanating from the human side of that relationship. Thus, for Sauer, culture, rather than environment, was the dynamic, causal agent of change. Sauer was not concerned to articulate the precise workings of culture itself – that being the task, in his view, of anthropology – but he did offer a conceptual and philosophical basis for his argument by outlining the methodology of morphology – that is, the study of structural change. Morphology – as originally conceived by the German philosopher Goethe – did not concern itself with explaining the general causes of change, but rather sought to merely describe the changing "architecture of organisms." Thus, an analogy could be drawn if landscape could be viewed as a kind of organism. As such, landscape morphology could be studied in the same way that Goethe proposed for biological organisms. Geography's task was to systematically describe the form of landscape by isolating its constitutive elements and the changes those elements experienced.

Culture was the most significant of these constitutive elements, but as already noted, Sauer was not himself concerned with "inner workings" of culture itself, but rather with the *outcomes* of culture, its *imprints* on the landscape. Such an approach has been criticized for ignoring individuals and the relations among them and focusing instead on their material artifacts in the landscape.

But it is also important to recognize that Sauer's work, taken as a whole, was framed by a deep concern over the ways that industrialization and modernization were not simply changing the landscape, but more importantly transforming our attitudes toward and understandings of the land and our relationship to it. Sauer may have ignored individuals as such, but he believed that there was need for an appreciation of distinct cultural groups and their unique ways of shaping the land. It is thus useful to situate Sauer's work in a more general early twentieth century intellectual climate of concern about the impact that rapid industrialization and urbanization was having on the local *genres de vie* so celebrated by Vidal (see p. 90).

The most complete collection of Sauer's writings can be found in John Leighly's edited volume *Land and Life* (1963). There have been numerous books and articles examining Sauer's scholarship and life, including essays by Michael Williams ("'The apple of my eye': Carl Sauer and historical geography," *Journal of Historical Geography* 9, 1, 1983) and Martin Kenzer ("Milieu and the 'intellectual landscape': Carl O. Sauer's undergraduate heritage," *Annals of the Association of American Geographers* 75, 2, 1985). Criticism of Sauer's legacy began with the publication of James Duncan's "The superorganic in American cultural geography" (*Annals of the Association of American Geographers* 79, 2, 1980), while Sauer's legacy was defended by Marie Price and Martin Lewis's "The reinvention of cultural geography," *Annals of the Association of American Geographers* 83, 1, 1993).

Summary of the objective of geography. – The task of geography is conceived as the establishment of a critical system which embraces the phenomenology of landscape, in order to grasp in all of its meaning and color the varied terrestrial scene. Indirectly Vidal de la Blache has stated this position by cautioning against considering "the earth as 'the scene on which the activity of man unfolds itself,' without reflecting that this scene is itself living" [Vidal, *Principles of Human Geography*, 1922]. It includes the works of man as an integral expression of the scene. This position is derived from Herodotus rather than from Thales. Modern geography is the modern expression of the most ancient geography.

The objects which exist together in the landscape exist in interrelation. We assert that they constitute a reality as a whole that is not expressed by a consideration of the constituent parts separately, that area has form, structure, and function, and hence position in a system, and that it is subject to development, change, and completion. Without this view of areal reality and relation, there exist only special disciplines, not geography as generally understood. The situation is analogous to that of history, which may be divided among economies, government, sociology, and so on; but when this is done the result is not history.

THE CONTENT OF LANDSCAPE

Definition of landscape. – The term "landscape" is proposed to denote the unit concept of geography, to characterize the peculiarly geographic association of facts. Equivalent terms in a sense are "area" and "region." Area is of course a general term, not distinctively geographic. Region has come to imply, to some geographers at least, an order of magnitude. Landscape is the English equivalent of the term German geographers are using largely, and strictly has the same meaning: a land shape, in which the process of shaping is by no means thought of as simply physical. It may be defined, therefore, as an area made up of a distinct association of forms, both physical and cultural.

The facts of geography are place facts; their association gives rise to the concept of landscape. Similarly, the facts of history are time facts; their association gives rise to the concept of period. By definition the landscape has identity that is based on recognizable constitution, limits, and generic relation to other landscapes, which constitute a general system. Its structure and function are determined by integrant, dependent forms. The landscape is considered, therefore, in a sense as having an organic quality. We may follow Bluntschli in saying that one has not fully understood the nature of an area until one "has learned to see it as an organic unit, to comprehend land and life in terms of each other." It has seemed desirable to introduce this point prior to its elaboration because it is very different from the unit concept of physical process of the physiographer or of environmental influence of the anthropogeographer of the school of Ratzel. The mechanics of glacial erosion, the climatic correlation of energy, and the form content of an areal habitat are three different things.

Landscape has generic meaning. – In the sense here used, landscape is not simply an actual scene viewed by an observer. The geographic landscape is a generalization derived from the observation of individual scenes. Croce's remark that "the geographer who is describing a landscape has the same task as a landscape painter" has therefore only limited validity. The geographer may describe the individual landscape as a type or possibly as a variant from type, but always he has in mind the generic, and proceeds by comparison.

An ordered presentation of the landscapes of the earth is a formidable undertaking. Beginning with infinite diversity, salient and related features are selected in order to establish the character of the landscape and to place it in a system. Yet generic quality is nonexistent in the sense of the biologic world. Every landscape has individuality as well as relation to other landscapes, and the same is true of the forms that make it up. No valley is quite like any other valley; no city the exact replica of some other city. In so far as these qualities remain completely unrelated they are beyond the reach of systematic treatment, beyond that organized knowledge that we call science. "No science can rest at the level of mere perception . . . The so-called descriptive natural sciences, zoology and botany, do not remain content to regard the singular, they raise themselves to concepts of species, genus, family, order, class, type" [Croce]. "There is no idiographic science, that is, one that described the individual merely as such. Geography was formerly idiographic; it has long since attempted to become nomothetic, and no geographer would hold it at its previous level" [Croce]. Whatever opinion one may hold about natural law, or nomothetic, general, or causal relation, a definition of landscape as singular, unorganized, or unrelated has no scientific value.

Element of personal judgment in the selection of content. – It is true that in the selection of the generic characteristics of landscape the geographer is guided only by his own judgment that they are characteristic, that is, repeating; that they are arranged into a pattern, or have structural quality, and that the landscape accurately belongs to a specific group in the general series of landscapes. Croce objects to a science of history on the ground that history is without logical criteria: "The criterion is the choice itself, conditioned, like every economic art, by knowledge of the actual situation. This selection is certainly conducted with intelligence, but not with the application of a philosophic criterion, and is justified only in and by itself. For this reason we speak of the fine tact, or scent, or instinct of the learned man" [Croce]. A similar objection is sometimes urged against the scientific competence of geography, because it is unable to establish complete, rigid, logical control and perforce relies upon the option of the student. The geographer is in fact continually exercising freedom of choice as to the materials he includes in his observations, but he is also continually drawing inferences as to their relation. His method, imperfect as it may be, is based on induction; he deals with sequences, though he may not regard these as a simple causal relation.

If we consider a given type of landscape, for example a North European heath, we may put down notes such as the following:

The sky is dull, ordinarily partly overcast, the horizon is indistinct and rarely more than a half-dozen miles distant, though seen from a height. The upland is gently and irregularly rolling and descends to broad, flat basins. There are no long slopes and no symmetrical patterns of surface form. Water-courses are short, with clear brownish water, and perennial. The brooks end in irregular swamps, with indistinct borders. Coarse grasses and rushes form marginal strips along the water bodies. The upland is covered with heather, furze, and bracken. Clumps of juniper abound, especially on the steeper, drier slopes. Cart traces lie along the longer ridges, exposing loose sand in the wheel tracks, and here and there a rusty, cemented base shows beneath the sand. Small flocks of sheep are scattered widely over the land. The almost complete absence of the works of man is notable. There are no fields or other enclosed tracts. The only buildings are sheep sheds, situated usually at a distance of several miles from one another, at convenient intersections of cart traces.

The account is not that of an individual scene, but a summation of general characteristics. References to other types of landscape are introduced by implication. Relations of form elements within the landscape are also noted. The items selected are based upon "knowledge of the actual situation," and there is an attempt at a synthesis of the form elements. Their significance is a matter of personal judgment. Objective standards may be substituted for them only in part, as by quantitative representation in the form of a map. Even thus the personal element is brought only under limited control, since it still operates in choosing the qualities to be represented. All that can be expected is the reduction of the personal element by agreement on a "predetermined mode of inquiry," which shall be logical.

Extensiveness of areal features. – The content of landscape is something less than the whole of its visible constituents. The identity of the landscape is determined first of all by conspicuousness of form, as implied in the following statement [by Passarge, 1919]: "A correct representation of the surface form, of soil, and of surficially conspicuous masses of rock, of plant cover and water bodies, of the coasts and the sea, of areally conspicuous animal life and of the expression of human culture is the goal of geographic inquiry." The items specified are chosen because the experience of the author has shown their significance as to mass and relation. The chorologic position necessarily recognizes the importance of areal extensiveness of phenomena, this quality being inherent in the position: Herein lies an important contrast between geography and physiography. The character of the heath landscape described above is determined primarily by the dominance of sand, swamp, and heather. The most important geographic fact about Norway, aside from its location, probably is that four-fifths of its surface is barren highland, supporting neither forests nor flocks, a condition significant directly because of its extensiveness.

Habitat value as a basis for the determination of content. – Personal judgment of the content of landscape is determined further by interest. Geography is distinctly anthropocentric, in the sense of value or use of the earth to man. We are interested in that part of the areal scene that concerns us as human beings because we are part of it, live with it, are limited by it, and modify it. Thus we select those qualities of landscape in particular that are or may be of use to us. We relinquish those features of area that may be significant to the geologist in earth history but are of no concern in the relation of man to his area. The physical qualities of landscape are those that have habitat value, present or potential.

The natural and the cultural landscape. – "Human geography does not oppose itself to a geography from which the human element is excluded; such a one has not existed except in the minds of a *few* exclusive specialists" [Vidal, 1922]. It is a forcible abstraction, by every good geographic tradition *a tour de force*, to consider a landscape as though it were devoid of life. Because we are interested primarily in "cultures that grow with original vigor out of the lap of a maternal natural landscape, to which each is bound in the whole course of its existence" [Spengler, 1920] geography is based on the reality of the union of physical and cultural elements of the landscape. The content of landscape is found therefore in the physical qualities of area that are significant to man and in the forms of his use of the area, in facts of physical background and facts of human culture. A valuable discussion of this principle is given by Krebs under the title "Natur- und Kulturlandschaft."

For the first half of the content of landscape we may use the designation "site," which has become well established in plant ecology. A forest site is not simply the place where a forest stands; in its full connotation, the name is a qualitative expression of place in terms of forest growth, usually for the particular forest association that is in occupation of the site. In this sense the physical area is the sum of all natural resources that man has at his disposal in the area. It is beyond his power to add to them; he may "develop" them, ignore them in part, or subtract from them by exploitation.

The second half of landscape viewed as a bilateral unit is its cultural expression. There is a strictly geographic way of thinking of culture; namely, as the impress of the works of man upon the area. We may think of people as associated within and with an area, as we may think of them as groups associated in descent or tradition, in the first case we are thinking of culture as a geographic expression, composed of forms which are a part of geographic phenomenology. In this view there is no place for a dualism of landscape.

[. . .]

FORMS OF LANDSCAPE AND THEIR STRUCTURE

The division between natural and cultural landscapes. – We cannot form an idea of landscape except in terms of its time relations as well as of its space relations. It is in continuous process of development or of dissolution and replacement. It is in this sense a true appreciation of historical values that has caused the geomorphologists to tie the present physical landscape back into its geologic origins, and to derive it therefrom step by step. In the chorologic sense, however, the modification of the area by man and its appropriation to his uses are of dominant importance. The area before the introduction of man's activity is represented by one body of morphologic facts. The forms that man has introduced are another set. We may call the former, with reference to man, the original, natural landscape. In its entirety it no longer exists in many parts of the world, but its reconstruction and understanding are the first part of formal morphology. Is it perhaps too broad a generalization to say that geography dissociates itself from geology at the point of the introduction of man into the areal scene? Under this view the prior events belong strictly in the field of geology and their historical treatment in geography is only a descriptive device employed where necessary to make clear the relationship of physical forms that are significant in the habitat.

The works of man express themselves in the cultural landscape. There may be a succession of these landscapes with a succession of cultures. They are derived in each case from the natural landscape, man expressing his place in nature as a distinct agent of modification. Of especial significance is that climax of culture which we call civilization. The cultural landscape then is subject to change either by the development of a culture or by a replacement of cultures. The datum line from which change is measured is the natural condition of the landscape. The division of forms into natural and cultural is the necessary basis for determining the areal importance and character of man's activity. In the universal, but not necessarily cosmologic sense, geography then becomes that part of the latest or human chapter in earth history which is concerned with the differentiation of the areal scene by man.

The natural landscape: geognostic basis. – In the subsequent sections on the natural landscape a distinction is implied between the historical inquiry into origin of features and their strictly morphologic organization into a group of forms, fundamental to the cultural expression of the area. We are concerned alone with the latter in principle, with the former only as descriptive convenience.

The forms of the natural landscape involve first of all the materials of the earth's crust which have in some important measure determined the surface forms. The geographer borrows from the geologist knowledge of the substantial differences of the outer lithosphere as to composition, structure, and mass. Geology, being the study of the history of these materials, has devised its classification on the basis of succession of formations, grouped as to period. In formations *per se* the geographer has no interest. He is concerned, however, with that more primitive phase of geology, called geognosy, which regards kind and position of material but not historical succession. The name of a geologic formation may be meaningless geographically, if it lumps lithologic differences, structural differences, and differences in mass under one term. Geognostic condition provides a basis of conversion of geologic data into geographic values. The geographer is interested in knowing whether the base of a landscape is limestone or sandstone, whether the rocks are massive or intercalated, whether they are broken by joints or are affected by other structural conditions expressed in the surface. These matters may be significant to the understanding of topography, soil, drainage, and mineral distribution.

The application of geognostic data in geographic studies is usual in a sense, areal studies being hardly feasible without some regard for the underlying materials. Yet to find the most adequate analysis of the expression of the underlying materials in the surface it is probably necessary to go back to the work of the older American and British geologists, such as Powell, Dutton, Gilbert, Shaler, and Archibald Geikie. In the aggregate, of course, the geologic literature that touches upon such matters is enormous, but it is made up of rather incidental and informal items, because landscape is not in the central field of interest of the geologist. The formal analysis of critical geognostic qualities and their synthesis into areal generalizations has not

had a great deal of attention. Adequately comparable data are still insufficient from the viewpoint of geography. In briefest form Sapper has lately attempted a general consideration of the relation of geologic forms to the landscapes of varying climates, thereby illuminating the entire subject of regional geography.

Rigorous methodologist that he is, Passarge has not failed to scrutinize the geographic bearing of rock character and condition, and has applied in intensive areal study the following observations (somewhat adapted):

Physical resistance
    Soft, easily eroded formations
    Rocks of intermediate resistance
        much broken (*zerklüftet*)
        moderately broken
        little broken
    Rocks of high resistance
        as above
Chemical resistance and solubility
    Easily soluble
        highly permeable
        moderately permeable
        relatively impermeable
Moderately subject to solution and chemical alteration
    as above
Resistant

In a later study he added provision for rocks notably subject to creep (*Fluktionsfähig*). An interpretation of geologic conditions in terms of equivalence of resistance has never been undertaken for this country. It is probably possible only within the limits of a generally similar climatic condition. We have numerous classifications of so-called physiographic regions, poorly defined as to their criteria, but no truly geognostic classification of area, which, together with relief representation, and climatic areas, is alone competent to provide the base map of all geographic morphology.

The natural landscape: climatic basis. – The second and greater link that connects the forms of the natural landscape into a system is climate. We may say confidently that the resemblance or contrast between natural landscapes in the large is primarily a matter of climate. We may go further and assert that under a given climate a distinctive landscape will develop in time, the climate ultimately cancelling the geognostic factor in many cases.

Physiography, especially in texts, has, largely, either ignored this fact or has subordinated it to such an extent that it is to be read only between the lines. The failure to regard the climatic sum of physiographic processes as differing greatly from region to region may be due to insufficient experience in different climatic areas and to a predilection for the deductive approach. Most physiographic studies have been made in intermediate latitudes of abundant precipitation, and there has been a tendency to think of the agencies in terms of a standardized climatic milieu. The appreciation even of one set of phenomena, as for example drainage forms, is likely to be too much conventionalized by applying the schematism of standardized physiographic process and its results to New England and the Gulf states, to the Atlantic and the Pacific coasts, not to mention the deserts, the tropics, and the polar margins.

But, if we start from the areal diversity of climates, we consider at once differences in penetration of heat and cold diurnally and seasonally, the varying areal expression of precipitation as to amount, form, intensity, and seasonal distribution, the wind as a factor varying with area, and above all the numerous possibilities of combination of temperature, precipitation, dry weather, and wind. In short, we place major emphasis on the totality of weather conditions in the molding of soil, drainage, and surface features. It is geographically much more important to establish the synthesis of natural landscape forms in terms of the individual climatic area than to follow through the mechanics of a single process, rarely expressing itself individually in a land form of any great extent.

The harmony of climate and landscape, insufficiently developed by the schools of physiography, has become the keystone of geographic morphology in the physical sense. In this country the emergence of this concept is to be sought largely in the studies in the arid and semi-arid West, though they did not result at once in the realization of the implied existence of a distinct set of land forms for every climate. In the morphologic form category of soils, the climatic factor was fully discovered first at the hand of Russian students, and was used by them as the primary basis of soil classification in a more thoroughgoing manner than that which had been applied to topographic forms. Under the direction of Marbut the climatic system has become basal to the work of the United States Bureau of Soils. Thus the ground was prepared for the general synthesis of physical landscape in terms of climatic regions. Most recently, Passarge, using Koppen's climatic classification, has undertaken a comprehensive methodology on this basis.

The relation of climate to landscape is expressed in part through vegetation, which arrests or transforms the climatic forces. We therefore need to recognize not only the presence or absence of a cover of vegetation, but also the type of cover that is interposed between the exogenous forces of climate and the materials of the earth and that acts on the materials beneath.

Diagrammatic representation of the morphology of the natural landscape. – We may now attempt a diagram of the nature of physical morphology to express the relation of landscape, constituent forms, time, and connecting causal factors [Figure 1]. The thing to be known is the natural landscape. It becomes known through the totality of its forms. These forms are thought of not for and by themselves, as a soil specialist would regard soils, for example, but in their relation to one another and in their place in the landscape, each landscape being a definite combination of form values. Behind the forms lie time and cause. The primary genetic bonds are climatic and geognostic, the former being in general dominant, and operating directly as well as through vegetation. The "X" factor is the pragmatic "and," the always unequated remnant. These factors are justified as a device for the connection of the forms, not as the end of inquiry. They lead toward the concept of the natural landscape which in turn leads to the cultural landscape. The character of the landscape is determined also by its position on the

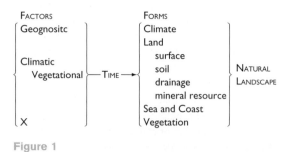

Figure 1

timeline. Whether this line is of determinate or infinite length does not concern us as geographers. In some measure, certainly, the idea of a climax landscape is useful, a landscape that, given a constancy of impinging factors, has exhausted the possibilities of autogenous development. Through the medium of time the application of factor to form as cause-and-effect relation is limited; time itself is a great factor. We are interested in function, not in a determination of cosmic unity. For all chorologic purposes the emphasis in the diagram lies at its right hand; time and factor have only an explanatory descriptive role.

This position with reference to the natural landscape involves a reaffirmation of the place of physical geography, certainly not as physiography nor geomorphology as ordinarily defined, but as physical morphology, which draws freely from geology and physiography certain results to be built into a view of physical landscape as a habitat complex. This physical geography is the proper introduction to the full chorologic inquiry that is our goal.

[. . .]

The extension of morphology to the cultural landscape. – The natural landscape is being subjected to transformation at the hands of man, the last and for us the most important morphologic factor. By his cultures he makes use of the natural forms, in many cases alters them, in some destroys them.

The study of the cultural landscape is, as yet, largely an untilled field. Recent results in the field of plant ecology will probably supply many useful leads for the human geographer, for cultural morphology might be called human ecology. In contrast to the position of Barrows in this matter, the present thesis would eliminate physiologic ecology or autecology and seek for parallels in synecology. It is better not to force into geography too much biological nomenclature. The name ecology is not needed: it is both morphology and physiology of the biotic association. Since we waive the claim for the measurement of environmental influences, we may use, in preference to ecology, the term morphology to apply to cultural study, since it describes perfectly the method.

Among geographers in America who have concerned themselves with systematic inquiry into cultural forms, Mark Jefferson, O.E. Baker, and M. Aurousseau have done outstanding pioneering.

Brunhes' "essential facts of geography" represent perhaps the most widely appreciated classification of cultural forms. Sten De Geer's population atlas of Sweden was the first major contribution of a student who has concentrated his attention strictly on cultural morphology. Vaughan Cornish introduced the concepts of "march," "storehouse," and "crossroads" in a most valuable contribution to urban problems. Most recently, Walter Geisler has undertaken a synthesis of the urban forms of Germany, with the deserved subtitle, "A contribution to the morphology of the cultural landscape." These pioneers have found productive ground; our periodical literature suggests that a rush of homesteaders may soon be under way.

Diagrammatic representation of the morphology of the cultural landscape. – The cultural landscape is the geographic area in the final meaning (*Chore*). Its forms are all the works of man that characterize the landscape. Under this definition we are not concerned in geography with the energy, customs, or beliefs of man but with man's record upon the landscape. Forms of population are the phenomena of mass or density in general and of recurrent displacement, as seasonal migration. Housing includes the types of structures man builds and their grouping, either dispersed as in many rural districts, or agglomerated into villages or cities in varying plans (*Städtebild*). Forms of production are the types of land utilization for primary products, farms, forests, mines, and those negative areas which he has ignored. [Figure 2]

The cultural landscape is fashioned from a natural landscape by a culture group. Culture is the agent, the natural area is the medium, the cultural landscape the result. Under the influence of a given culture, itself changing through time, the

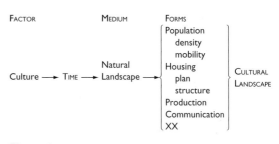

Figure 2

landscape undergoes development, passing through phases, and probably reaching ultimately the end of its cycle of development. With the introduction of a different – that is, an alien – culture, a rejuvenation of the cultural landscape sets in, or a new landscape is superimposed on remnants of an older one. The natural landscape is of course of fundamental importance, for it supplies the materials out of which the cultural landscape is formed. The shaping force, however, lies in the culture itself. Within the wide limits of the physical equipment of area lie many possible choices for man, as Vidal never grew weary of pointing out. This is the meaning of adaptation, through which, aided by those suggestions which man has derived from nature, perhaps by an imitative process, largely subconscious, we get the feeling of harmony between the human habitation and the landscape into which it so fittingly blends. But these, too, are derived from the mind of man, not imposed by nature, and hence are cultural expressions.

[. . .]

BEYOND SCIENCE

The morphologic discipline enables the organization of the fields of geography as positive science. A good deal of the meaning of area lies beyond scientific regimentation. The best geography has never disregarded the esthetic qualities of landscape, to which we know no approach other than the subjective. Humboldt's "physiognomy", Banse's "soul," Volz's "rhythm," Gradmann's "harmony" of landscape, all lie beyond science. These writers seem to have discovered a symphonic quality in the contemplation of the areal scene, proceeding from a full novitiate in scientific studies and yet apart therefrom. To some, whatever is mystical is an abomination. Yet it is significant that there are others, and among them some of the best, who believe that, having observed widely and charted diligently, there yet remains a quality of understanding at a higher plane that may not be reduced to formal process.

[. . .]

"The Industrial Revolution and the Landscape"

from *The Making of the English Landscape* (1955)

W.G. Hoskins

Editors' introduction

While there is a separate section in the *Reader* focusing on the concept of landscape itself, we include this selection from W.G. Hoskins's famous *The Making of the English Landscape* (1955) here in order to recognize the importance of landscape history, a field of study that was not central to cultural geography in Britain yet which deserves a place in the field's history nevertheless, particularly due to its affinity with the work of Carl Sauer and many of his students in the United States (see p. 96). Hoskins's work reminds us of Sauer's ideas of "landscape morphology," which suggested a deeply historical approach to the study of landscapes. And like Sauer, Hoskins's work betrayed an attitude that was generally conservative and concerned to describe the landscape changes associated with industrialization and urbanization. But Hoskins was perhaps somewhat more explicit than Sauer in accounting for these changes with an aesthetic eye. He wrote, for example, that "Since the last years of the nineteenth century ... and especially since the year 1914, every single change in the English landscape has either uglified it or destroyed its meaning, or both."

Yet there are other more important differences between Hoskins and Sauer. Whereas Sauer was concerned to cement landscape morphology as the foundation of a scientifically legitimate academic discipline, Hoskins saw landscape as a text from which to read the past. Trained in history, Hoskins wrote history by observing the landscape. Perhaps the closer American comparison with Hoskins would be J.B. Jackson's vernacular landscape essays (see pp. 53 and 220). But, again, our task here is less to analyze Hoskins's conception of landscape, and more to situate him within the broader social and historical context of late industrial development. In this regard, Hoskins's folk-cultural or vernacular proclivities found echoes in the work of other scholars whose work is included or referenced in this section, particularly H.J. Fleure, Estyn Evans, Patrick Geddes, and Paul Vidal de la Blache. And like these scholars, Hoskins's work was also engaged in a broader political project of nation building. *The Making of the English Landscape* could be read as a kind of guidebook for the lay geographer–traveler, a key to the clues, imbedded in the landscape, that together narrated the history of England as a nation. Near the beginning of the book, Hoskins wrote, "What I have done is to take the landscape of England as it appears today, and to explain as far as I am able how it came to assume its present form, how the details came to be inserted and when. At all points I have tried to relate my explanation to the things that can be seen today by any curious and intelligent traveller going around his native land." Thus, while his work reads as a fascinating history unveiled by rich landscape description, there are passages in the selection below where Hoskins sounds more like an art critic examining a series of paintings. He describes the landscape, in other words, with the scrutiny of a particular point of view.

W.G. Hoskins (1908–1992) was, beginning in 1931, Lecturer in Commerce and then Reader in English Local History at University College, Leicester. In 1952 he became Reader in Economic History at Oxford, and in 1965 was appointed Hatton Professor of English History at the University of Leicester. Beyond the academy, he was very active in local history and preservation work, serving on local history committees and county archives. In 1976 Hoskins wrote and presented the BBC television series *The Landscape of England*.

By far his most well known work is *The Making of the English Landscape*. It has long been a standard text in local history, and the book's introductory passage outlining a thousand years of English history encapsulated in the view of Steeple Barton from Hoskins's study window has become the definitive introduction to the field of landscape history. Yet Hoskins was a prolific writer and published dozens of books and essays on landscape history, including *The Midland Peasant* (1957), *Local History in England* (1959), *Two Thousand Years in Exeter* (1960). His work has remained influential among vernacular and landscape historians, and in the fields of landscape design and architecture. It remains worth pointing out, however, that the bulk of his influence has been felt outside of academic geography. This is perhaps unfortunate, given the obvious intellectual correlations between Hoskins and some of the key figures in cultural geography.

THE EARLY INDUSTRIAL LANDSCAPE

England was still a peaceful agricultural country at the beginning of the seventeenth century. Though she was passing through what has been called her first industrial revolution, there was as yet little to show for it in the landscape. Quarries and coal-pits were numerous in certain localities, salt-works and glassworks were flourishing, the cloth industry was growing; but so far as the visible signs upon the face of the country were concerned it was all a mere scratching on the surface. Neither Leland nor Camden has much to say about industry in England; and there was nothing that could be specifically called an industrial landscape. Perhaps the multitude of coal-pits near the Tyne were beginning to wear that look, and Camden observed in the 1580s that Sussex 'is full of iron mines, all over it; for the casting of which there are furnaces up and down the country, and abundance of wood is yearly spent; many streams are drawn into one channel, and a great deal of meadow ground is turned into ponds and pools for the driving of mills by the flashes, which, beating with hammers upon the iron, fill the neighbourhood round about it, night and day with continual noise'. The iron industry, centred in the Wealden woods, was steadily changing the face of the landscape in this region from the middle of the sixteenth century onwards, and a good deal remains to be seen by the historically minded traveller.

By the end of the seventeenth century the industrial landscape was much more evident. Yarranton in 1677 thought there were more people within a radius of ten miles of Dudley, and 'more money returned in a year', than in the whole of four Midland farming counties. This was pretty certainly an exaggeration, but it shows unmistakably that the Black Country (though this name had yet to be invented) was in process of creation.

The early industrial landscapes differed essentially from those that developed with steam-power. They showed a thick scattering of settlement, of cottages and small farmhouses dotted about all over the place, and a corresponding splitting up of fields into small crofts and paddocks. It was a 'busy' landscape, full of detail and movement, like one of Breughel's paintings, not a massive conglomeration of factories and slums. The Black Country in its early days was still country, 'a countryside in course of becoming industrialized; more and more a strung-out web of iron-working villages, market-towns next door to collieries, heaths and wastes gradually and very slowly being covered by the cottages of nailers and other persons carrying on industrial occupations in rural surroundings' [W.H.B. Court, *The Rise of the Midland Industries, 1600–1838*, 1938, p. 22]. The typical figure was that of the craftsman-farmer, combining, say, a smithy with a smallholding, living in his own small balanced economy; hence the minuteness of the detail in the picture. One still finds traces of this kind of landscape on the fringes of the Black Country, as for example in the hamlet of Lower Gornal, in the hills to the north-west of Dudley.

Defoe gives us a splendid picture of an industrial landscape in the time of Queen Anne or shortly after. It is the landscape of the cloth industry in the neighbourhood of Halifax before the revolutionary changes brought about by the invention of power-driven machinery:

> The nearer we came to Hallifax, we found the houses thicker, and the villages greater in every bottom; and not only so, but the sides of the hills, which were very steep every way, were spread with houses, and that very thick; for the land being divided into small enclosures, that is to say, from two acres to six or seven acres each, seldom more; every three or four pieces of land had a house belonging to it.
>
> . . . This division of the land into small pieces, and the scattering of the dwellings, was occasioned by, and done for the convenience of the business which the people were generally employ'd in . . .

This particular landscape had its origin in two sources – the outcropping of coal, and the presence of running water everywhere, even on the tops of the hills. Wherever Defoe passed a house he found a little rill of running water.

> If the house was above the road, it came from it, and cross'd the way to run to another; if the house was below us, it cross'd us from some other distant house above it, and at every considerable house was a manufactory or workhouse, and as they could not do their business without water, the little streams were so parted and guided by gutters and pipes, and by turning and dividing the streams, that none of those houses were without a river, if I may call it so, running into and through their work-houses.

The coal-pits near the tops of the hills were worked in preference to those lower down, for various reasons. The coal was easier to come at, water presented less of a drainage problem, and the pack-horses could go up light and come down laden. Every clothier kept a horse or two, to carry his coal from the pit, to fetch home his wool and his provisions from the market, to take his yarn to the weavers, his cloth to the fulling-mill and finally to the cloth market to be sold. He also kept two or three cows for the sustenance of the family, and so required two, three, or four pieces of enclosed land around his house.

> Having thus fire and water at every dwelling, there is no need to enquire why they dwell thus dispers'd upon the highest hills. . . . Among the manufacturers houses are likewise scattered an infinite number of cottages or small dwellings, in which dwell the workmen which are employed, the women and children of whom are always busy carding, spinning, & c. so that no hands being unemploy'd, all can gain their bread, even from the youngest to the ancient; hardly any thing above four years old, but its hands are sufficient to itself. . . . After we had mounted the third hill, we found the country one continued village, tho' mountainous every way, as before; hardly a house standing out of a speaking distance from another, and . . . we could see that almost at every house there was a tenter, and almost on every tenter a piece of cloth, or kersie, or shalloon, for they are three articles of that country's labour; from which the sun glancing, and, as I may say, shining (the white reflecting its rays) to us, I thought it was the most agreeable sight that I ever saw, for the hills, as I say, rising and falling so thick, and the valleys opening sometimes one way, sometimes another, so that sometimes we could see two or three miles this way, sometimes as far another; sometimes like the streets near St Giles's, called the Seven Dials; we could see through the glades almost every way round us, yet look which way we would, high to the tops, and low to the bottoms, it was all the same; innumerable houses and tenters, and a white piece upon every tenter.

[. . .]

WATER-POWER AND THE EARLY MILLS

Early inventions in most industries – except in those requiring large amounts of fixed capital, like the iron industry – benefited the small man, or at least kept him in business. Kay's flying shuttle (1733) and Hargreaves's spinning jenny (1767) multiplied the output of domestic workers in the textile industry without compelling them to enter mills or factories. Not until the application of water-power to machinery, and a consequent great increase in the size of machines, do we begin to see the large factory as an element in the landscape. Before that time the largest unit of production was what Defoe calls in Yorkshire the 'work-house'. But the great revolution was on its way.

The first true factory built in England was the silk mill built for John and Thomas Lombe at Derby in 1718–22. It was five or six storeys high, employed three hundred men, and was driven by the water-power of the river Derwent. It was, as Mantoux says, in every respect a modern factory, with automatic tools, continuous and unlimited production, and specialized functions for the operatives. Within fifty years there were several silk factories employing four hundred to eight hundred persons, but the silk industry was of secondary importance and did not initiate the factory system. It was when power reached the cotton, woollen, and iron industries that the face of the country really began to change on a large scale, and that was not until the 1770s.

Matthew Boulton opened his great Soho factory, in the still unravished country outside Birmingham, in 1765, and shortly afterwards began the manufacture of steam engines. Wedgwood's new large factory at Etruria in the Potteries was opened in 1769. Richard Arkwright, the greatest of the new industrial capitalists, erected his first spinning mill, worked by horses, at Nottingham in 1768, but his second factory, built on a much larger scale at Cromford on the Derwent in 1771, was driven by water power. In the 1760s, too, the Darbys enlarged their ironworks at Coalbrookdale in Shropshire to the largest works of any kind in the kingdom. With these four large-scale factories, the creation of the modern industrial landscape may be said to have begun.

The new mills, factories and works tended to be in more or less remote places, partly because of the need to be near a falling stream for the supply of power, and later to escape too close an inspection and regulation of their uninhibited activities. One finds these early mills therefore, often windowless and deserted today, in the upper reaches of the moorland valleys on either side of the Pennines. Coalbrookdale, then a romantically beautiful valley, was chosen by the Darbys for their ironworks

because here a rapid stream entered the broad navigable waterway of the Severn. Water was needed in the iron industry both for power and for the transport of heavy materials. It was not long before the ravishing of this scene attracted the lament of the poets. Anna Seward, 'The Swan of Lichfield', mourned over 'Coalbrook Dale' in a poem written about 1785:

> Scene of superfluous grace, and wasted bloom,
> O, violated Colebrook! in an hour,
> To beauty unpropitious and to song,
> The Genius of thy shades, by Plutus brib'd,
> Amid thy grassy lanes, thy wildwood glens,
> Thy knolls and bubbling wells, thy rocks, and streams,
> Slumbers! – while tribes fuliginous invade
> The soft, romantic, consecrated scenes…

Some ten years earlier, Arthur Young had already noted the discord between the natural beauty of the landscape and what man had done to it, but he saw, too – and painters also were on the verge of seeing it – that an unrestrained industrial landscape has a considerable element of sublimity about it. 'That variety of horrors art has spread at the bottom [of Coalbrookdale]; the noise of the forges, mills, etc., with all their vast machinery, the flames bursting from the furnaces with the burning of the coal and the smoak of the lime kilns, are altogether sublime.'

The scale of the new industries brought about a number of visual changes, some of them unexpected. The large sums of fixed capital sunk in the factory buildings and the machinery, and the fact that water-power, unlike human labour, needed no rest, demanded that the new buildings be used by night as well as by day. Shifts of labour were therefore organized, and these tall fortress-like structures were lit from top to bottom at night, and presented something new and dramatic to those who had the leisure to stay outside and contemplate it with detachment. So we get Joseph Wright of Derby as early as 1780 painting Arkwright's cotton mill by night – tiers of tiny yellow lights in the immemorial country darkness of the Derwent valley, the isolated forerunner of those tremendous galaxies of light that one now sees from the Pennine Moors after sundown.

In the eighth book of *The Excursion*, Wordsworth sees the other side of this romantic scene:

When soothing darkness spreads
O'er hill and vale, and the punctual stars,
While all things else are gathering to their homes,
Advance, and in the firmament of heaven
Glitter – but undisturbing, undisturbed;
As if their silent company were charged
With peaceful admonitions for the heart
Of all-beholding Man, earth's thoughtful lord;
Then, in full many a region, once like this
The assured domain of calm simplicity
And pensive quiet, an unnatural light
Prepared for never-resting labour's eyes
Breaks from a many-windowed fabric huge;
And at the appointed hour a bell is heard,
Of harsher import than the curfew-knoll
That spake the Norman Conqueror's stern behest –
A local summons to unceasing toil!
Disgorged are now the Ministers of day;
And, as they issue from the illumined pile,
A fresh band meets them, at the crowded door –
And in the courts – and where the rumbling stream,
That turns the multitude of dizzy wheels,
Glares, like a troubled spirit, in its bed,
Among the rock below. Men, maidens, youths,
Mother and little children, boys and girls,
Enter, and each the wonted task resumes
Within his temple, where is offered up
To Gain, the master idol of the realm,
Perpetual sacrifice.

[. . .]

In the textile districts the new industrial land-scape lay in the valley bottoms, which had been comparatively ignored in Defoe's day, when the thickest settlement was on the hillside. Now, down in the bottoms, arose the new many-storeyed mills, some of them handsome buildings not too unlike the plain country houses of the time. Around them grew up short streets of cottages for the workpeople, run up so quickly that they look as though they were planted flat on the surface, without any foundations; but still there was no congestion. The water-power age produced hamlets, at the most small villages, gathered around a new mill. Around Ashton-under-Lyne, for example, where it was reckoned there were nearly a hundred cotton mills within a ten-mile radius – all on the river Tame or its tributaries – we find hamlets in the 1790s with the significant names of Boston, Charlestown and Botany Bay.

[. . .]

The Derwent valley, which exemplifies along its bottom so much industrial history of the water-power age, attracted large mills from the beginning by reason of its fast-flowing river; but not every-one admired the result as Wright of Derby did. Uvedale Price in his *Essays on the Picturesque* (1810) observed:

> When I consider the striking natural beauties of such a river as that at Matlock, and the effect of the seven-storey buildings that have been raised there, and on other beautiful streams, for cotton manufactories, I am inclined to think that nothing can equal them for the purpose of disbeautifying an enchanting piece of scenery; and that economy had produced, what the greatest ingenuity, if a prize were given for ugli-ness, could not surpass.

Mills arose in the remote valleys below the moors, and hamlets and villages quickly clustered around them. But established towns too were advancing over the surrounding fields. Trees and hedges were torn up, red-brick or grit-stone streets, short and straight, multiplied every year, even before the age of steam: Sheffield, Birmingham, Liverpool, Manchester, all were on the move. According to Langford, 'The traveller who visits [Birmingham] once in six months supposes himself well acqu-ainted with her, but he may chance to find a street of houses in the autumn, where he saw his horse at grass in the spring.' The population of the town doubled in the last forty years of the eighteenth century (35,000 people in 1760; 73,000 in 1801), but it was as yet far from being the dark and horrible landscape that it eventually became. Even in the early years of the nineteenth century the middle-class streets had 'prospects' of the country and the older working-class houses at least still had gardens. The dirt and over-crowding came with the steam age in the nineteenth century.

Sheffield, on the other hand, was 'very populous and large' in Queen Anne's time when Defoe traversed it, and its houses were already 'dark and black' from the smoke of the forges. Two genera-tions later the population had trebled and the pall of industrial smoke had become permanent. As Anna Seward saw it:

Grim Wolverhampton lights her smouldering fires,
And Sheffield, smoke-involv'd; dim where she stands
Circled by lofty mountains, which condense
Her dark and spiral wreaths to drizzling rains
Frequent and sullied . . .

In Lancashire and the Potteries the worst had still to come. Chorley was, when Aikin wrote (1795), 'a small, neat market town' with its river flowing through a pleasant valley, turning 'several mills, engines and machines'. It possessed the first water-driven factory to be erected in Lancashire (1777). Preston was 'a handsome well-built town, with broad regular streets, and many good houses. The earl of Derby has a large modern mansion in it. The place is rendered gay by assemblies and other places of amusement, suited to the genteel style of the inhabitants.' Aikin notes that the cotton industry had just come to the town. In the south of the county what was to be the most appalling town of all – St Helens – was just beginning to defile its surroundings. The British Plate Glass Manufactory had been erected at Ravenhead, near the village in 1773, and other glassworks followed. And about the year 1780 'a most extensive copper-work' was erected to smelt and refine the ore from Paris mountain in Anglesey. The atmosphere was being poisoned, every green thing blighted, and every stream fouled with chemical fumes and waste. Here, and in the Potteries and the Black Country especially, the landscape of Hell was foreshadowed.

STEAM-POWER AND SLUMS

[. . .]

We are not concerned here with the general effects upon industry and the English economy of the use of steam-power, but with its visible effects upon the landscape, and these are now obvious enough. Steam-power meant a new and intense concentration of large-scale industry and of the labour-force to man it. It meant that manufacturers no longer needed to seek their power where there was fast-running water, especially in the higher reaches of lonely dales, but found it near the canals which brought coal to them cheaply, or directly upon the coalfields themselves. So emerged what Wordsworth called 'social Industry'. No longer need they go out into the wilderness and

create a village or a hamlet to house their labour. Manufacturers ran up their mills, factories and works on the edge of existing towns, and their workers were housed in streets of terrace-houses built rapidly on the vacant ground all around the factory.

Industry spread over the lower-lying parts of the towns, leaving the hills for the residences of the well-to-do, but this was not a conscious piece of 'zoning'. Large-scale industries in pit-railway days needed canal-side sites both for bringing in their coal and other raw materials and for taking away their heavy products. Thus they chose the flatter and lower ground where the canals lay. Moreover, it was the low-lying areas that were vacant when the industrialists appeared on the scene, for earlier generations had wisely avoided building on them wherever they could. The sites were there waiting. And again, it was easier and cheaper to build on a flat site than on a hillside. As a consequence most of the new streets of working-class houses were also built on land that presented difficult drainage problems (not that anyone except the victims gave much thought to this), and the sanitary conditions soon became appalling. The slums were born. The word *slum*, first used in the 1820s, has its origin in the old provincial word *slump*, meaning 'wet mire'. The word *slam* in Low German, Danish and Swedish, means 'mire': and that roughly described the dreadful state of the streets and courtyards on these undrained sites. It need hardly be said that the industrialist of the Steam Age did not build his own house near the works, as the country factory owners had done. He went to dwell on the 'residential heights' and walked down to the mill each day.

But there is more meaning in the word *slum* than simply a foul street or yard: it denotes also a certain quality of housing. In the early nineteenth century the quality of working-class houses, as structures, deteriorated rapidly. The industrialists of the water-power age, out in the open country, had put up houses for their workpeople – as at Cromford, Mellor and Styal, where many of them may still be seen – which were, in Professor Ashton's words, 'not wanting in amenity and comfort' [The *Industrial Revolution, 1760–1830*, 1948, p. 160] and even possessed a certain quality of design and proportion. These decent working-class houses were put up in the 1770s and 1780s, where land was cheap and when building materials were plentiful, wages in the building trades relatively low, and money relatively cheap.

With the outbreak of twenty years' war in 1793, the price of materials and wages in the building trades both began to rise steadily. Interest rates, too, increased and remained high for a generation. Since at least two-thirds of the rent of a house consists of interest charges, the rise in interest rates alone was sufficient to bring about a drastic reduction in the size and quality of working-class houses in order to preserve an 'economic rent'. Further, land inside the older towns was acquiring a scarcity value, above all in the towns that were surrounded by open fields, so that they could not grow outwards, and a steady rise in the price of land for building was added to the rise in the price of borrowed money. Possibly, too, the building trade was invaded by a new class of speculator who made conditions even worse than they need have been by extracting high profits out of the unprecedented demand for cheap houses. No one has studied this particular class of parasite, how he worked, or what fortunes he made. One often wonders in what opulence his descendants live today forgetful, or perhaps ignorant, of the origin of their wealth. Their forebears would make a fruitful study.

Bad materials and fewer of them, and bad workmanship, reduced the costs of building. Houses run up in the courts of Birmingham in the 1820s and 1830s cost £60 each to build. Birmingham specialized in close, dark and filthy courtyards: there were over two thousand of these in the town in the 1830s, and many of their houses were built back to back in order to get the maximum number on to each expensive acre. The local medical men did not object, but rather commended them for their cheapness. At first some of them had a deceptive brightness, but their abominable quality soon revealed itself and decay rapidly set in. Decent people moved out if they could, and the born-squalid moved in: the swamp of the slums spread a few years behind the speculative builder everywhere.

Open spaces inside the older towns vanished rapidly. The last remnant of Birmingham Heath was enclosed in 1799, and was built over forthwith with eight new streets. Precisely the same thing was happening around the Lancashire towns also, where the ancient commons were enclosed and grabbed by the private speculator for building, as at Oldham. Only Preston managed to save its commons from the vultures, and to transform some of them eventually into public parks.

Not only the commons but the large gardens of the eighteenth-century bourgeoisie disappeared under bricks and mortar. The house of Baskerville, the eminent Birmingham printer, was sold in 1788 and the seven acres of land that surrounded it were advertised as 'a very desirable spot to build upon'. In these older towns, too, the large houses of the middle class were divided into tenements to house the swarming population, and factories and warehouses went up on their gardens and orchards. Slowly the other features of the industrial towns were added: Anglican churches, Nonconformist chapels, schools and public houses. Public parks came in the 1840s, and public libraries a few years later; later still perhaps the grandiose Town Hall, by no means always to be despised as architecture.

Entirely new towns grew out of hamlets in the industrial north and Midlands. The germ of Middlesbrough was a single farmhouse near the banks of the unsullied Tees in 1830: by 1880 it was a town of more than fifty thousand people. Barrow-in-Furness, too, sprang from a single house, grew into a fishing village of about three hundred people by the 1840s, and by 1878 was a town of forty thousand. South Shields, St Helens and Birkenhead all shot up quickly during the first half of the nineteenth century. 'Meanwhile,' said Wordsworth in *The Excursion* (1814):

Meanwhile, at social Industry's command,
How quick, how vast an increase! From the germ
Of some poor hamlet, rapidly produced
Here a huge town, continuous and compact,
Hiding the face of earth for leagues – and there,
Where not a habitation stood before,
Abodes of men irregularly massed
Like trees in forests, – spread through spacious tracts,
O'er which the smoke of unremitting fires
Hangs permanent, and plentiful as wreaths
Of vapour glittering in the morning sun.
And, wheresoe'er the traveller turns his steps,
He sees the barren wilderness erased,
Or disappearing . . .

Nor was the industrial landscape represented solely in the great towns, for between them stretched miles of torn and poisoned country-side – the mountains of waste from mining and other industries; the sheets of sullen water, known as 'flashes', which had their origin in subsidence of the surface as a result of mining below; the disused pit-shafts; the derelict and stagnant canals. The train-journey between Leeds and Sheffield shows one this nineteenth-century landscape to perfection. In the Lancashire township of Ince there are today twenty-three pit-shafts covering 199 acres, one large industrial slag-heap covering six acres, nearly 250 acres of land under water or marsh due to mining subsidence, another 150 acres liable to flooding, and thirty-six disused pit-shafts. This is the landscape of coal-mining. As for the Black Country, one can hardly begin to describe it. Dickens has an horrific description of it in *The Old Curiosity Shop* (1841), when it had reached the rock bottom of filth and ugliness, and of human degradation. The early industrialists were not 'insensitive to the appeal of the country: the beauty of Cromford and Millers Dale suffered little by the enterprise of Arkwright, and stretches of the Goyt and the Bollin owe something to Oldknow and the Gregs' [Ashton, *Industrial Revolution*, p. 157]. But the later industrialists, the heirs of the steam age, were completely and grotesquely insensitive. No scruples weakened their lust for money; they made their money and left behind their muck.

The industrial landscape is not confined to the north of England and the west Midlands. In Cornwall for instance one finds two distinct landscapes of industry, one dead, the other still active. Over central Cornwall, particularly to the north-west of St Austell, are the spoil-heaps of the china-clay industry, an almost lunar landscape that one sees gleaming on the horizon from almost any hill-top in the county. And there is the equally striking landscape of the vanished tin-mining industry: the windowless engine-houses, the monolithic chimney stacks against the skyline, the ruined cottages of an old mining hamlet, and the stony spoil-heaps – a purely nineteenth-century landscape, and perhaps because of its setting, the most appealing of all the industrial landscapes of England, in no way ugly but indeed possessing a profound melancholy beauty. Just across the Devonshire border is the old mining landscape of Blanchdown, west of Tavistock, where, in the middle decades of the nineteenth century, the Devon Great Consols was the richest copper mine in the world: now its miles of spoil-heaps have created a silent and desolate beauty of their own, and foxes and snakes haunt the broken buildings and the glades between.

There is a point, as Arthur Young saw, when industrial ugliness becomes sublime. And indeed the new landscape produced some fine dramatic compositions such as the railway viaduct over the smoking town of Stockport; or the sight of Bradford at night from the moorland hills to the north; or the smoky silhouette of Nottingham on a winter evening as seen from the south-bound train on the Eastern Region line; or the city of Sheffield in full blast on a murky morning; even (one thinks sometimes) the sight of long gas-lit streets of red brick working-class houses in a Victorian town with not a tree or a bush in sight: only the lamps shining on pavements blanched by the autumn evening wind.

"Process"

from *The Cultural Geography of the United States* (1973)

Wilbur Zelinsky

Editors' introduction

The Cultural Geography of the United States (1973) is probably Wilbur Zelinsky's most influential work. At the time of its publication it represented one of the only sustained attempts in American cultural geography to blend a comprehensive regional monograph in the tradition of Carl Sauer (see p. 96) or Paul Vidal de la Blache (see p. 90) with a detailed accounting of culture as the fundamental agent of landscape change and place-based identity. As with a long line of geographers before him, going back to Ratzel (see p. 83), Zelinsky notes the importance of culture as *the* basic trait that separates humans from the rest of the "natural world." And he shares with Sauer – his PhD advisor – the idea of culture as the basic agent of landscape change and thus central to the concerns of the geographer. But Zelinsky goes beyond Sauer in theorizing the "inner workings" of culture that Sauer was content to leave up to anthropology.

For Zelinsky, the study of culture is at the very frontiers of science, calling for "sophisticated techniques of the highest order." This is because objectivity is so difficult to achieve in the study of culture, particular when the culture under study is one's own. And if objectivity was less of a concern to a landscape historian like W.G. Hoskins (see p. 105) or the great essayist of vernacular landscape, J.B. Jackson (see p. 153), it was viewed by Zelinsky as a central concern if cultural geography was to be called a science. In 1973, Zelinsky claimed that cultured had only recently emerged as an important variable in the explanation of individual and social behavior. The cultural geographer, therefore, had an obligation to subject culture to analytical scrutiny, just as Sauer had done with landscape half a century earlier.

Zelinsky's focus on culture thus marks an important point in a broader trajectory of culture in geography. In the broader context of the human sciences during the twentieth century, culture was becoming an increasingly important variable in accounting for patterns of human behavior. For geographers in the United States, this trajectory was marked first by the rejection of environmental determinism and the articulation of culture as an agent of change. But by the time of Zelinsky's writing, culture was widely accepted as an independent variable and, as such, demanded a degree of conceptual sophistication that was less necessary during the era of Sauer, Evans, Fleure, or Hoskins.

But Zelinsky's book was not just a treatise on the concept of culture in geography. It was more significantly a statement about the unity of the United States as a discrete culture region and, thus, a nation with a distinctive identity that could be accounted for scientifically. As such, *The Cultural Geography of the United States* drew directly from the Vidalian tradition that also inspired Sauer, in which a national identity was defined as something greater than the sum of its distinctive local parts. For Zelinsky, vernacular culture explains national identity, and he sees the nation as the spontaneous will of "the people" rather than something consciously fabricated by a "cabal of nation-builders." Thus, early in the book, Zelinsky makes three important claims regarding the cultural geography of the United States: "1. Useful nonstereotypic statements can be made about the cultural idiosyncrasies (that is, national character) of an ethnic group taken as a whole; 2. the population of the United States does indeed form a single large, discrete ethnic group; 3. statements about the character of the larger community cannot be, indeed should not be, transferred to individuals because of sharp discontinuities of scale."

In this way, Zelinsky established the culture of the United States as something that was *superorganic*, that is, greater than the sum of its parts, something that exists only as a larger-scale collective, but with its own agentive powers to shape individual behavior. Such an approach to culture required a particular methodology, and the first part of the selection below details Zelinsky's method of cultural study. This section also lays out the six distinct processes he identifies as having shaped American culture through time and space. The second part of the selection offers a discussion of the American house, illustrating the six distinct processes and Zelinsky's general methodological scheme for analyzing culture.

It is important to recognize that Zelinsky's work is perhaps less significant for the methodological treatise it laid out than for the vision of national cultural identity that it illustrates. It ought to thus be clearly situated within a context in which – during the late 1960s and early 1970s – many were questioning the supposedly unifying elements of American identity. *The Cultural Geography of the United States* was written as much in defense of that identity as it was to shore up the science of culture in geography.

As mentioned above, Wilbur Zelinsky was a student of Carl Sauer at Berkeley, and received his PhD in 1953. After several short-term appointments, he joined the faculty at Penn State in 1963 and taught there until his retirement. The author of many essays in the *Annals of the Association of American Geographers* and *The Geographical Review*, his research has focused primarily on American vernacular culture, ethnicity, and identity. His most recent book is *The Enigma of Ethnicity: Another American Dilemma* (2001). Zelinsky's articulation of a "superorganic" idea of culture has become a focus of critique among more recent generations of cultural geographers, most explicitly in James Duncan's "The superorganic in American cultural geography," *Annals of the Association of American Geographers* 70 (1980). While the concept has become nearly synonymous with the so-called "Berkeley school" of cultural geography initiated by Sauer, the key issue is perhaps less the extent to which early American geographers embraced the concept – which was originally proposed by the anthropologists Alfred Kroeber and Robert Lowie in the 1920s – but more the fact that it was being embraced and systematically articulated by American cultural geographers like Zelinsky long after it had been discredited and abandoned in anthropology (see Geertz, p. 29). Nor had the concept ever caught on in Europe. *The Cultural Geography of the United States* thus illustrates the extent to which American cultural geography had diverged from broader trends in the study of culture by the 1970s.

EXPLAINING THE SPATIAL ASPECTS OF CULTURAL CHANGE

What processes have been most influential, within the total cultural system, in shaping the geography of the country? And how have they operated? The question may also be rephrased to read: How and why has American culture changed through time and space? ... [S]everal distinct processes have been at work. These are:

1   The selective transfer of immigrants and cultural traits from the Old World;

2   The interaction of the newcomers among themselves and with new habitats in several early cultural hearths;

3   Differential participation by various groups in the advance of the settlement frontier from these cultural hearths;

4   Differential mobility of different groups of people during the post-pioneering period;

5   The spatial diffusion of a great range of specific innovations;

6   Deep structural change in society and culture that is expressed at different times and rates in different tracts.

All these mechanisms of change have been, or will be, noted (though in rather different order); but special attention should be accorded the one that may be the most important, yet the least amenable to direct observation: the deep structural changes experienced by a society and culture as a community evolves upward from a relatively primitive set of conditions toward an ever more complex civilized existence. The vast question of the degree to which the evolutionary paths of developing societies are followed in blind obedience to fundamental historical laws or, on the contrary, are coincidental in nature or the result of contacts among different societies, is one of the most difficult and controversial facing the cultural anthropologist . . .

In any case, if one were to amass all possible data on local inventions, the diffusion of innovations, interaction with the local habitat, the spatial movement of people and influences, and all the other discrete events that contribute to culture formation there would still be a large, unexplained residuum. Much of what has happened in the slow character-building process in the culture history of a locality would seem to be transacted at the unlit subterranean levels of consciousness as a series of extremely gradual, subtle shifts in modes of thinking, feeling, and impulse in response to basic alterations in socioeconomic structure and ecological patterns. There is no reason to believe that such has not also been the case with the United States and its various subregions, even within the relatively brief time this society has existed.

SOME BASIC CULTURAL PROPOSITIONS

Before we can begin exploring the how and why of spatial and temporal shifts in American culture we must take a hard definitional look at what is being studied: the concept of culture. Only within the past few decades have students of mankind begun to recognize the existence of an entity called "culture," something within, yet beyond the minds of individual human beings. It is a very large, complex assemblage of items, which, taken together, may be as important a variable as any in explaining the behavior of individuals or societies or the mappable patterns of activities and man-made objects upon the face of the earth. The late emergence of any semblance of a "science of culture" can be attributed in part to the extraordinary

difficulty of observing and objectively measuring the characteristics of so complex and elusive a phenomenon. The idea that the traditional ways of thinking and acting of one's group are not absolute and that there is some coherence in the seemingly chaotic kaleidoscope of beliefs and customs of alien societies required a bold leap of the anthropological imagination.

The history of scientific thought also helps account for our inability to offer a fully rigorous definition of culture or to suggest many firm ideas about the structure of cultural systems or the laws governing their behavior through space and time. The physical and chemical properties of inorganic matter are the most obvious, measurable items for the curious mind searching after some underlying order in the universe; and their study initiated formal modern science, as we know it. The methodical observation of plants and animals appeared soon after as another scientifically respectable endeavor, despite the greater difficulties involved. Very much later, the human mind began to be inquisitive about itself and the properties of our nervous system and personality, and the science of psychology was born. It is when the scientist approached groups of things in complex interaction that both observation and analysis posed the most formidable challenges. Sociology and political science are new, and their achievements relatively modest. So are the "scientific" approach to history and the study of ecology, that is, the "societal" aspects of plants and animals coexisting within specific habitats. The systematic understanding of the culture of human groups calls for sophisticated techniques of an even higher order. Not the least of the problems is achieving adequate objectivity, a relatively minor matter in the physical and biological disciplines. The student of culture must somehow strip himself of his native preconceptions. It is difficult enough to do so when looking at alien folk; it entails a near miracle when investigating one's own culture, as in the present work.

Much ink has been spilled in the effort to reach a satisfactory definition of culture. Perhaps the most successful to date is that offered by Kroeber and Kluckhohn in 1952 after an exhaustive critique of the literature:

Culture consists of patterns, explicit and implicit, of and for behavior acquired and transmitted by symbols, constituting the distinctive achievement of human groups, including their embodiment in artifacts; the essential core of culture consists of traditional (i.e., historically derived and selected) ideas and especially their attached values; culture systems may, on the one hand, be considered as products of action, on the other as conditioning elements of further action.

The full exegesis of this statement could, and did, require a full volume; and some cultural anthropologists would take exception to all or part of the definition. But all would agree that culture is an assemblage of learned behavior of a complexity and durability well beyond the capacities of nonhuman animals. Following the Kroeber–Kluckhohn formulation, culture can be regarded as the structured, traditional set of patterns for behavior, a code or template for ideas and acts. It is highly specific to each cultural and subcultural group, and survives by transfer not through biological means but rather through symbolic means, substantially but not wholly through language. In its ultimate, most essential sense, culture is an image of the world, of oneself and one's community.

It will be helpful to spell out several general attributes of culture implied in the definition above that are of importance to geographers and other students. First of all, culture is indeed an exclusively human achievement. In fact, it is the critical human attribute, the one exclusive possession that sets mankind far apart from all other organisms. With the appearance of cultural behavior at least one million years ago during the organic evolution of man-like creatures, true human beings can be identified. . . .

The power wielded over the minds of its participants by a cultural system is difficult to exaggerate. No denial of free will is implied, nor is the scope for individual achievement or resourcefulness belittled. It is simply that we are all players in a great profusion of games mid that each cultural arena the entire team, knowingly or not, follows the local set of rules, at most bending them only slightly. Only a half-wit or a fool would openly flout them. But as in chess, the possibilities for creativity and modulation are virtually infinite. It is enough to have experienced "cultural shock," the sudden immersion in another culture without special briefing, or the almost equally painful reentry into one's own community after such an episode,

to realize that many of the habits one regards as natural or logical are so only for one's own group. Most of the norms, limits, or possibilities of human action thus are set as much or more by the configuration of the culture as by biological endowment or the nature of the physical habitat. . . .

A cultural system is not simply a miscellaneous stockpile of traits. Quite to the contrary, its many components are ordered. Moreover, the totality of culture is much greater than the simple sum of the parts, so much so that it appears to be a superorganic entity living and changing according to a still obscure set of internal laws. Although individual minds are needed to sustain it, by some remarkable process culture also lives on its own, quite apart from the single person or his volition, as a sort of "macro-idea," a shared abstraction with a special mode of existence and set of rules. This point becomes clearer if one examines some specific cultural complexes that share this attribute of superorganic existence with the total culture. Thus an economic system – *vide* the famous "Invisible Hand" of Adam Smith – has been perceived to evolve and act according to its own private code, at least in pre-Keynesian times, without the effective intervention of its participants. Languages constantly change, though at variable speeds, in accordance with complex rules we are only slowly beginning to grasp, but utterly without calculation or effort by their speakers. Any sensitive linguistic observer who has watched the dizzy pace at which American English has altered during the past generation can testify to a sense of lying helpless in the path of large anonymous forces. Similarly, there is a distinctive personality to be recognized in almost any viable organization – church, college, army. corporation, or government bureau – that almost literally lives and breathes, persists and develops, quite independently of the personal sentiments of its members.

The nation-state idea is perhaps the neatest illustration of the transpersonal character of cultural systems; and the origin, growth, and perpetuation of the idea of a United States of America is a superb example. Whatever the genesis of the idea, it was certainly not the conscious fabrication of any identifiable cabal of nation-builders, but rather a spontaneous surge of feeling that quickly acquired a force and momentum of its own. This idea was so powerful that millions of men were ready to sacrifice their lives for it. Individuals who entertain the nation-state idea are born and die, and some may even have doubts or reservations; but the idea marches on, quite clearly beyond the control of anyone. Even if it were a matter of dire necessity, it seems impossible to devise any program, excluding mass annihilation, whereby the idea of a United States or a Russia, France, or Germany could be disinvented.

The structure of cultural systems is rather loose and open. If we regard culture as a system in the technical sense of the term, it is rather special by reason of both complexity and sheer size. Probably the only other system of greater magnitude is that including all interacting subsystems on and near the face of the earth that comprise total terrestrial reality and the subject matter of geography. There may be certain quintessential ideas and practices that cannot be tampered with without profoundly revising the larger cultural pattern; but, in the main, the total structure can absorb much change or contamination, including addition or subtraction of elements in specific departments of culture without great impact upon the whole. Thus two centuries of radical technological and economic change have not basically revised the structure of the American family – at least not yet. And the near-disappearance of men's straw hats, electric trolleys, or Spencerian handwriting seems to have had minimal effect upon basic American life patterns. But if one could imagine anything as unimaginable as a mass conversion to ascetic Buddhism, the substitution of the Arabic language for English, the abolition of the

achievement motive, or the adoption of a joint family system, the reverberations all through the cultural matrix would be fast and shattering.

Even among so-called "primitive folk," a single cultural system encompasses an enormous range of information. Each cultural group has a certain common fund of traits – a full count of the individual bits of information would probably run well into the millions – that is acquired, usually quite unconsciously, during the early months and years of childhood. But, in addition, there are any number of special groups or activities, ranging from a half-dozen or so in the simplest of societies to literally hundreds of thousands in the most complex, each with its own distinctive subculture . . .

The problem of how to take an inventory of all traits or complexes that make up the total system, or how to classify the full range of subcultures and other major dimensions present within a given community, has not been solved. But the situation can be indicated roughly in diagrammatic form (Figure [1]). Consider the full rectangular solid as representing the total culture, with one dimension equivalent to the range of traits and complexes that make up the totality of any culture or subculture. A second dimension (here the vertical one) represents an additive (and overlapping) set of subcultures; for example, males and females, farmers, ditchdiggers, Presbyterians, mountain climbers, convicts, Freemasons, bowlers, and drug addicts. (The number of strata shown in the diagram is, of course, highly schematic.) The third dimension indicates variability through space, that is, the set

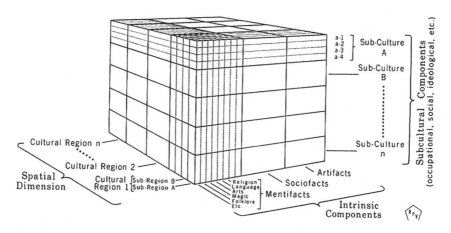

Figure [1] A schematic three-dimensional representation of cultural systems

of cultural regions and subregions that one can identify within the territorial range of the cultural system in question. Further dimensions – especially time – might be added to this scheme, but not without transcending the possibilities of graphic representation. Note that if the cube is sliced either horizontally or vertically (at an angle normal to the spatial face), the result is a regional or subcultural parcel of cultural phenomena that run the gamut of human ideas and practices: for example, courtship, facial expression, superstition, pronunciation pattern, motor skills, social etiquette, and burial customs.

For our purposes, it will suffice to adopt one of the simplest ways of categorizing the components of a culture, among the almost limitless array of possible schemes. This is a tripartite classification into artifacts, sociofacts, and mentifacts. Artifacts are those elements of culture that are directly concerned with matters of livelihood or, somewhat more broadly, the entire technology of supplying wanted goods and services. The variety of artifacts can barely be suggested: all tools, weapons, and other man-made objects; manufacturing in all its many aspects; the shelter system; the production of food and drink; the transportation system; medicine; property-holding and land-use systems; clothing; and many other phenomena. Sociofacts are those phases of the culture most directly concerned with interpersonal relations: kinship and family systems; political behavior; education; social etiquette; voluntary organizations; reproductive behavior; child rearing; and a host of others. Mentifacts are basically cerebral, psychological, or attitudinal in character, and include religion, along with other ideological baggage, magic and superstition, language, music, dance, and other arts, funerary customs, folklore, the basic value system, and abstract concepts of all sorts. In a sense, the mentifactual is "the innermost, least mutable, holiest" and most precious segment of the culture – the glue holding together the entire cultural mass and setting its tone and direction.

In practice, it is hard to find any single facet of culture that is purely artifactual, sociofactual, or mentifactual. These arbitrary categories are interdependent to a marked degree. For example, house design and construction, which might appear to be wholly technological or artifactual, is, in fact, closely associated with the nature of the family and social system and so also with religious or cosmological ideas. In any case, if the total culture can be seen as a loose, yet somehow structured, assemblage of an almost innumerable set of elements, viewed from another angle, it is also a package with many subcultural compartments, each with a decided amount of autonomy, and to many of which the individual may belong simultaneously. To make this thought more concrete, consider the single man, the microcosmic building block of a larger cultural universe, who carries his own unique collection of cultural attributes, and may also be a participant in many subcultural groups. Imagine someone who is, among other things, a Czech-American Lutheran plumber, a member of the VFW, an ardent Cleveland Indian fan, a radio ham, a regular patron of a particular bar, and a member of a car pool, the local draft board, the Book-of-the-Month Club, and the Republican party, and a parent whose son attends a particular college. Each of these subcultures will tend to have its own array of gear and physical arrangements, spectrum of economic and social beliefs and practices, cluster of abstract concepts, and, not least important for our purpose, distributional spread in physical space. The one man and his friends and associates move through many worlds. It is only by taking into account the relevant multiplicity of components and dimensions at the particular scale chosen for observation, and also the fact that they are changing through time, that a realistic understanding of a given culture can be reached.

[. . .]

The one attribute of cultural systems that most particularly interests us is the fact that they almost always have spatial dimensions, that is, they exist within certain localities Moreover, the general nature of the locality seems somehow to shape the culture and, in turn, to he influenced by it. In the strictest sense, the spatial component of culture is incidental. As a cerebral entity, a culture may flourish, move about, and propagate itself solely within the heads of a number of footloose individuals. Such extreme cases do occur, of course, but normally the facts of location and the processes of interaction with other localized or spatially structured phenomena do matter greatly. In fact, the territorial dimension is strong enough that it seems fitting to accord the regional aspect of culture an importance rivaling the technological, social, or ideological. This statement is especially valid when representatives of a culture have become deeply rooted in a specific place.

A particular culture, or combination of sub-cultures, helps impart to an area much of its special character and behavioral design – which, fundamentally, is what geography is all about. Conversely, the character of a given place may be a strong formative influence in the genesis of a particular culture, which is the ultimate concern of the cultural anthropologist. This is especially so, it must be stressed again, after a cultural group has become established in a certain tract, or when such a group is transferred en masse to another tract. We do not yet understand the nature of the process. The facts of proximity or remoteness with respect to other cultures and the ease of travel or communication certainly enter into the equation. Many of the elements of the inanimate and the biotic environment are emphatically relevant. The perception of local opportunity with respect to all manner of economic and social activity – a special place as glimpsed through the special lenses of the culture – would also appear to figure importantly. The regrettable fact that we cannot analyze all these many, still quite mysterious place-oriented interactions does not exorcize them. It is this zone of intersection between cultural process and the total character of places that is the special domain of the cultural geographer . . .

[. . .]

THE AMERICAN HOUSE

. . . We are concerned here with the dwellings of the great majority of the American population – what might be fairly characterized as "folk housing" – and buildings that never enjoyed the professional attention of architects. Although basically an artifact and one serving some urgent physical functions, the house is also the product of a complex set of societal and psychological factors, all filtered through the sediments of history. It is really as much sociofact or mentifact as artifact. When fully interpreted, the form and uses of the house tell us much, not only about the physical locale and the technology of the place and era, but also about the source and dates of the builder or renovator, the contacts and influences he experienced, his ethnic

affiliation, and possibly also class, occupation, and religion. In a very real sense, the house is the family's universe in microcosm, the distillation of past experience and a miniature model of how it perceives the outer world, as it is or perhaps even more as it should be.

In essence, the American house is a European import. Or, rather, it is a uniquely new object reconstituted from a number of earlier European fragments, which then evolved in a special way in accordance with the peculiarities of American life. Ideas, usually subconscious, as to the proper way to construct a dwelling varied from place to place along the colonial Atlantic Seaboard, depending upon time, sources, and conditions of colonization. Quite early, distinct regionalisms in house styles began to develop. But equally early, the designs, and often some of the materials, of the homes for the wealthy were imported intact from northwest Europe. It was only about the time of American Independence that an indigenous professionalism in architecture for homes and public structures began to develop. Yet, if the homes of the common people of Massachusetts, the Hudson Valley, and North Carolina were all distinct from the beginning, they shared nevertheless some unmistakably American traits. These, in turn, reflect the primordial notions about house morphology underlying folk architecture throughout a good part of Europe. These could surface only under the relatively primitive conditions of new North American society. There was little borrowing of aboriginal building techniques, and then only locally and temporarily. And no African heritage can be discerned among the structures built by or for the slaves, except perhaps in the still unstudied rural Negro churches.

Each of the three principal colonial culture hearths – southern New England, the Midland, and the Chesapeake Bay area – developed its own set of house types and other sorts of buildings at an early date, and so also did a number of minor subregions. We can trace the westward thrust of settlers and ideas into the continental interior from these seedbeds of American style quite precisely in the field, at least up until the mid-nineteenth century, by plotting the location of surviving examples of these regional types. And we can also observe the ways in which various strands of culture flowed together and sometimes produced new regional blends. In fact, the quasi-archaeological technique of studying older dwellings is one of the better, if more laborious, ways of charting the past or present extent of culture areas (or the microgeography of older cities) and of gaining insight into the historical geography of American ideas.

As in other departments of cultural practice, there is a striking depletion of individuality and inventiveness in building styles and ornamentation as one moves away from the early communities in the East to the relatively accessible regions of the West. In part this was presumably the result of relative isolation during formative years. In place of the truly riotous exuberance of form within a single long-lived New England village or even a single block in an eastern Pennsylvania borough – a variety, by the way, still somehow harmonious – there is the monotony and stunted imagination of the Middle Western or far Western residential neighborhood. This tendency is even more marked within business districts. An attenuation of style is observable even within so simple a category as techniques of log-house construction; and in a study of Georgia examples, a striking contrast "as seen between the earlier, richer repertory of log-house Forms in the (older) North and the later, stripped-down set found toward the (younger) South.

The structures built after about 1850 tell quite a different story from their predecessors. By that date there had been sufficient mingling and hybridization of the original colonial styles within the great central expanses of the country to produce a recognizably national group of building types; and the pervasive new modes of communication and manufacturing had started to iron out regional departures from mass norms. For the past century or so, domestic building styles are closely correlated with date, and are much more sensitive to rate of diffusion down a social or cultural hierarchy than to territorial location, Yet, however standardized American building practices may be becoming, the house, old or new, is still packed densely with information about the national ethos, the dealings between man and habitat, and the changing configurations of our cultural geography. Surely the geographer must concern himself with the form and meaning of the objects he studies if any real sense is to be made of their spatial array or of their processual linkages with other phenomena in space and time.

Several attributes of the American house, past and present, bespeak important peculiarities of the national character. Perhaps the most obvious is the lavish use of space, both in the sheer size of the house proper and in the largeness of the residential lot. Furthermore a disproportionately small fraction of the population live in apartment buildings or other multiunit structures; many of those who do are college students, convicts, the indigent or ailing elderly, and other institutionalized populations. Except in the most congested of urban settings, as in New York City, Boston, or San Francisco, the one-family dwelling is a freestanding unit, with at least a token patch of space between it and its neighbors. Row housing is a phenomenon restricted to urban neighborhoods in the Northeast that were built about two centuries ago, apparently in imitation of Northwest European models.

Although much of this expansiveness might be explained away, the residual cultural factor bulks large. Land was cheap and abundant, and still is relative to land in most parts of the world; but the urge toward very large, isolated, individual properties seems to go beyond any rational economic reckoning. This is most obvious in the isolated farmstead, for which no convincing argument can be made in terms of transportation systems or social utility. If the propensity of Americans toward larger lots is being more fully realized now with growing affluence and the lateral spread of cities, the cubic volume of the structure has been decreasing. This is largely because of rising costs of materials and labor, the scarcity of servants, and the shrinkage in average size of household. But American homes are still bulky by any universal criteria; and those of the nineteenth century middle class were often of incredible proportions, even after making all allowances for number of hired hands, children, and other kinfolk, or the provision of closets and storerooms for the accumulations of a super-productive economy. One cannot help but speculate that these dimensions reflect an optimistic, aggressively extroverted view of the world and the American's place in it.

The fact that Americans may be profligate with space but niggardly with time also appears in their building technology. Great store is set upon quickness of construction, and it is not by chance that the United States has originated or perfected most leading methods of building prefabrication, or that the technique of balloon frame construction, one that reduced costs and workdays for wooden edifices so markedly, won such instant, universal acceptance. The commercial hotel as we now know it and its up-to-date offspring, the motel, both of which were nurtured in, and are ubiquitous throughout, North America, also embody the themes of haste and transience. Note also the emphasis on the garage, often an integral, conspicuous part of the house, sometimes even threatening to dominate the nonautomotive segment of the structure. The stress on transience, clearly evident in the early shelters of pioneer settlers, has waxed rather than waned in recent decades. Few Americans build a house with the intention of occupying it for a lifetime or passing it on to their children; indeed our population is so mobile that every other family changes its abode every decade. These urges toward transience and mobility receive their ideal embodiment in that superlatively American invention, the house trailer, which is virtually nonexistent outside North America. Nomadism as a way of life, though not yet well documented by the social scientist, may have made marked progress dining the 1960s among some members of the so-called "counterculture," for whom the VW microbus or some other such vehicle in nearly constant motion may be the only true home. But even among the most respectable strata of American society, transience is revered. Witness the recent popularity of high-rise office and apartment buildings designed to obsolesce and be razed after a few years, but only after the maximum tax advantage has been squeezed out of them.

Partly because the ordinary American house is so transitory a phenomenon but basically because of a fundamental restlessness of character, we have witnessed a dizzying procession of building styles and fads, one following hard upon the heels of another. As already noted, the modern American house (or commercial or public building) is a much better indicator of date than of locality. So avid is the appetite for novelty that the architect has ransacked virtually every historic era and most regions of the world in search of inspiration.

A cheerful extroversion of personality is writ large in the American house and its surroundings, which do double service as status symbols as well as shelters. The penchant for large glass windows, a trend that shows no sign of abating, goes beyond a normal craving for natural illumination and creates some serious problems in heating and upkeep. With the advent of the picture window craze, it becomes especially clear that the house is designed to serve as a display case, to advertise to the world at large the opulence and amiability of the household. The same outward-going personality is visible in the extravagant development of the porch. Although the ultimate origins of the porch (or portico or piazza) are obscure – the British may have hit upon the idea in the West Indies or India – no other national group has seized upon the device with such enthusiasm. During its apogee around 1900, few self-respecting American houses were without one, and many houses were encased with a porch on two or three sides and possibly on the second as well as the ground level. These open-air extensions of the house – literally a perpetual "open house" – were the stages upon which much of the social life of the family was enacted during the warmer seasons. This obliteration of the distinction between inside and outside, totally at variance with Northwest European antecedents, is carried even further in much avant-garde architecture, especially in the Pacific Coast states.

The same impulses that favored the efflorescence of window and porch seem to lie behind the almost pathological fervor with which grass lawns are tended – and front fences or hedges are frowned upon. (Do we have here the democratization of the British baronial estate?) There is still much to be learned about the culture of a group, including the American, through the microgeographic analysis of their house gardens. However, one peculiarity of American landscaping leaps to the eye: contrary to general usage, gardens are not invariably private spaces behind walls or hedges, but are often aggressively public, placed on the street side of the house. And since the lawn itself is basically ornamental or symbolic, intended much more for show than for any sort of play or foot traffic, it must be considered along with shrubs and flowers as a badge of membership in a cheerful, outgoing democratic society, but one in which the privileges of a powerful individualism must also be made manifest. The lawn is also significant as a shorthand symbol for the edenic ideal that is so strong an undercurrent in American thought. It is also appropriate to indicate here a rather cavalier disregard of environmental conditions on the part of designer and homeowner, an attitude stemming from an overriding self-confidence and material abundance as much as from ignorance. Except for a tiny minority of interesting exceptions, ideas are *imposed* upon the land, however inappropriately (for example, lawns on the sands of Florida or picture windows in the subarctic). Climate, slope, drainage, geology, soil, and natural vegetation seem to matter little.

Paradoxically, despite its openness, the American house also attests to the supremacy of the private individual. We have already noted the aversion to inhabiting multifamily structures and the impulse to create token open spaces between neighbors. But it is in its internal arrangements, with the great stress upon isolation, the multiplicity of doors and closed spaces, and the segregation of specific functions, that the pervasive American privatism comes fully to the fore. In addition, taking the house and grounds as a single entity, there is the starkest kind of contrast between the American's attitude toward his private bubble of space and that toward all public spaces. All self-respecting householders spend an inordinate amount of time caring for yard and garden and on keeping the interior as antiseptic and spotless as human ingenuity can manage. But public spaces, including sidewalks, thoroughfares, roadsides, public vehicles, parks, and many public buildings reveal a studied neglect and frequently such downright squalor that it is difficult to believe one is encountering a civilized community.

Finally, the American house quite neatly illustrates an all-powerful mechanistic vision of the world, for it is a carefully manipulated machine, a working model of what Americans feel the cosmos fundamentally is – or could be induced to be. The internal physiology of the American house represents a truly awe-inspiring triumph of the mechanical arts. The list of inventions promoting domestic comfort and convenience attributable to American ingenuity is long and fascinating. Elaborate and ultimately effortless central heating (and cooling) systems have been devised; then, like the other wonders noted below, made available to the world at large. The water supply has been brought indoors, and thoroughly rationalized. American plumbing is the eighth wonder of the world, and the Great American Bathroom a veritable glittering cathedral of cleanliness. The kitchen is also a marvel of efficiency and clever design, incorporating a multitude of American "firsts." Similarly advanced are lighting, electrical wiring, laundry systems, and rubbish disposal. As already intimated, the layout of the house is thoroughly programmed, with a specific function, and usually no other, designated for each space. Thus we have carried to its logical extreme that spatial separation of place of work and place of residence that is so distinctive and important a feature of the larger landscape. This statement also applies to the farmer who resides at his workplace but works in different buildings on or near the farmstead from that in which he eats and sleeps. In almost every respect, then, the American house is a completely appropriate capsule world, fleshing out the main principles, myths, and values of the larger cultural system.

"The Idea of German Cultural Regions in the Third Reich: The Work of Franz Petri"

from *Journal of Historical Geography* 27: 2 (2001): 241–258

Karl Ditt

Editors' introduction

David Livingstone's historiography of disciplinary geography, *The Geographical Tradition* (1992), begins with the provocative question: "Should the history of Geography be X-rated?" Livingstone was referring to an article with a similar title that appeared in the journal *Science* in 1974 about whether the questionable behavior of some famous scientists should be hidden from students. Historians had been uncovering details of scientists' lives that questioned whether they made good models for students to follow in their own scientific aspirations. Livingstone's point is not that the "geographical tradition" should remain hidden, but that a great deal can be learned from its critical exposure.

One important chapter in the history of geography that deserves examination in this regard is the relationship between academic geography and the rise of fascism in early twentieth century Germany. To this end, Karl Ditt examines the relationship between German "cultural region" (*Kulturraumforscher*) academics and the rise of National Socialism and the Nazi Party in the 1920s and 1930s. He focuses on the work of Franz Petri (1903–1993) who wrote what Ditt refers to as "the most significant contribution to cultural region research work during the Third Reich." Ditt's essay serves both to address directly the cooperation of scholars with the National Socialists, and also to examine the development of the concept of "culture region."

The *culture region* has a particular legacy in cultural geography, one which links cultural geography to nationalist political movements of various kinds. Indeed, culture itself is a concept that, as Don Mitchell has argued in *Cultural Geography: A Critical Introduction* (2002), is "politics by another name." And while much cultural geography in America may have been largely irrelevant to political movements, in Europe geographers invoked culture in a much more explicitly political fashion. This was possible because by the beginning of the twentieth century geography – particularly in Germany – had been reformulated as a science of regions. And if geographers were to avoid a (by then discredited) environmental determinist approach to regions (that is, focusing on their physical characteristics and how these influenced human behavior), they would have to emphasize the *cultural distinctiveness* of regions as an outcome of human action on the land. This was, essentially, the approach that Carl Sauer (see p. 96) had also advocated – in "The morphology of landscape" – as the way forward for cultural geography in the United States as well.

The political character of this work derived from the fact that culture regions were equated with organisms. Drawing social science analogies from fields such as evolutionary biology was a common practice among nineteenth century intellectuals, as was demonstrated in Geography by Friedrich Ratzel (see p. 83) and Paul Vidal de la Blache (see p. 90). Ratzel's work, in particular, made clear the connection between the organic analogy and politics. His concept of *Lebensraum* ("living space") at once defined the characteristics of a national culture while establishing an argument justifying Germany's territorial expansion into neighboring lands – on the basis that the territorial state, like a living organism, needed "room to grow" as it developed and matured.

The *Kulturraumforscher* took a similarly organic approach to the nation as deeply "rooted" in the land, but emphasized less *Lebensraum*'s implications of territorial expansion and more the demarcating and securing of original boundaries that had, since ancient times, defined the land that nurtured a distinctive national culture (see also Maalki, p. 275).

As Ditt makes clear, Franz Petri sought to establish the boundaries of a German culture region based on a particular set of distinctively "German" characteristics. The organic analogy made it necessary to see culture regions as internally coherent "wholes" (rather than, say, assemblages of heterogeneous and unrelated components). Although Ditt claims that Petri himself had no explicit ideological or geopolitical objectives, the article makes clear how his work was nevertheless used precisely for ideological and geopolitical purposes. It was the organic analogy that made such uses possible, and this raises important questions regarding the extent to which scholars can believe themselves to be working above the fray of politics.

Karl Ditt is a specialist in Westphalian history, and the economic, social and cultural history of nineteenth and twentieth century Germany. Since 1989 he has held the position of Historian (*Wissenschaftlicher Referent*) in the Westfälisches Institut für Regionalgeschichte (Westphalian Institute of Regional History) in Münster. His publications include *Industrialisierung, Arbeiterschaft und Arbeiterbewegung in Bielefeld 1850–1914* (1982); the co-edited *Raum und Volkstum. Die Kulturpolitik des Provinzialverbandes Westfälen 1923–1945* (1988), and *Agrarmodernisierung und ökologische Folgen* (Westfälen vom späten 18. bis 20. Jahrhundert, 2001).

For further reading on the relationship between the culture region and German nationalism during the early twentieth century, see Boa and Palfreyman's *Heimat: A German Dream: Regional Loyalties and National Identity in German Culture 1890–1990* (2000).

[. . .]

The precursor to cultural region research work was the working approach adopted towards the German language atlas which had been organized by Walter Mitzka and Ferdinand Wrede since the end of the nineteenth century. They had distributed thousands of questionnaires, by means of which they were able to establish the local terms for specific objects and concepts in the German Reich. On the basis of these results they then drew up local dialect boundaries. Theodor Frings, Hermann Aubin and Franz Steinbach, the so-called Rhineland school, adopted this approach. They did not attempt to establish local dialect areas but the boundaries and distinguishing marks of "cultural regions" or "historical landscapes." By this they understood any historical regions which were remarkable for the concentration of primary social and cultural evidence and the attitudes and behaviour of their inhabitants as revealed in forms of settlements, dialects, manners and morals, laws, etc. The intersection of the various areas of dissemination and the clustering of their borders – which were regarded as sharply defined lines but not as transition zones or thresholds – delineated a "cultural region." Because of their internal correlation, and in spite of exchanges with other regions, cultural regions were regarded as being not only durable and individual but also variable "regional organisms." For the academics involved in cultural region research, the driving forces behind regional formations and transformations seemed to be settlement movements, economic and cultural processes of dispersal and exchange, communication networks, and political and confessional (territorial) decisions.

In the 1920s, research into *Volk* and cultural regions received a considerable impetus. Part of the reason for this can be found in its innovative academic approach, whose interdisciplinary procedures and broad-ranging methods yielded fresh results. It also corresponded to a political need following Germany's defeat in World War I. Academics felt compelled to enquire much more deeply into the two factors which they considered the natural and cultural bases, and also the strengths of the Germanic people: *Volk* and *Raum*. In the 1920s and 1930s, such an interdisciplinary approach to historical research was not only innovative, it was a modern form of historiography to set alongside the dominant politically orientated territorial and national account of history. In addition, the particular aims of cultural region research very quickly aroused political interest. Following the territorial losses after the treaty of Versailles, those politicians involved in the foreign policy of the German Reich were naturally in search of any support to strengthen their claims that the ceded areas were German. And even regional politicians within the German Empire sought out allies, since many *Länder* and provinces at the end of the 1920s feared losses or hoped for gains as a result of the debate about territorial reorganization in Germany. In both cases, scholarly arguments concerning the historicity of boundaries and regions, and the assertion of the existence of "*Volk* soil" and "cultural soil," "core regions," "regional constants," "regional communities" or "regional organisms" were a welcome legitimization and help in political confrontations.

Given the strength of such scholarly innovation and the political significance of this form of regional research, it is no surprise that in the Third Reich historical studies which concentrated on both *Volk* and cultural regions were quickly established and promoted financially, institutionally and politically. Franz Petri belonged to the young generation of historians who profited from such measures. He attempted to investigate the wellsprings of the German language beyond the present north-western boundaries of the German Reich. The starting point for his work had been established by his mentor, the cultural historian Franz Steinbach, who had cast doubt on the traditional view that the linguistic boundary between France and Germany mirrored the settlement boundary of the ancient Germanic (Frankish) tribes (see Figure 1).

Steinbach suspected that Germanic-Frankish settlements extended much further West and that the linguistic boundary represented a line of withdrawal. Petri adopted his supposition and began to look for evidence of Frankish land appropriation beyond the linguistic boundary, in France, Belgium and the Netherlands. His work concentrated particularly on the search for linguistic and archaeological evidence: place and field names and burial sites (see Figures 2 and 3).

On the basis of a comprehensive collection of material, he claimed to establish a Frankish settlement region extending to the border of Brittany in the West, to the Loire in the South and to the head streams of the Mosel, Maas and Marne. Looking eastwards he regarded the Frankish region as ending at the Teutoburg Forest. A map drawn up by the anthropologist Egon Freiherr von Eickstedt and based on the examination of skeletons confirmed Petri in his view that in the early Middle Ages this territory had been settled by tall, long-skulled Germans. Evidence from legal history and buildings also pointed in the same direction.

Petri summarized his evidence as such:

> Frankish land appropriation and the foundation of the Empire shifted the centrepoint of Frankish power and culture from the former original territories on the North Sea and in the lower Rhine valleys to the newly conquered areas in the Seine and middle Rhine.

Petri defined the areas between the Seine, the sources of the rivers Mosel, Maas and Marne and the Netherlands as the "core areas of Frankish culture." The Frankish conquest of Gaul had not simply led to a superficial military occupation but rather to an occupation by the Frankish people: "The character of the Frankish settlement in Walloon and Northern France [was] utterly Germanic," he claimed. Over the centuries the region had developed its own *Volkskultur*. Petri modified Steinbach's thesis somewhat by declaring that contemporary linguistic boundaries represented not a "line of withdrawal" but a "line of balance" which had developed as a result of the cultural confrontation between Romans and Germans around the year 1000 AD Petri's conclusions divided early Medieval Gaul into two halves: one in the North dominated by the Germans, the other in the south dominated by the Romans. For Petri, like Steinbach, the causes of the Germanic withdrawal from Gaul could be found in the "civilizing" superiority of the Roman world, in its urban culture through which Roman civilization and Christianity had diffused.

Petri regarded his researches into the western boundaries of the German *Volk* not only as scientific evidence but also – like his mentor Steinbach – as a contribution to the political debates around these boundaries. He stressed the "relevance to the

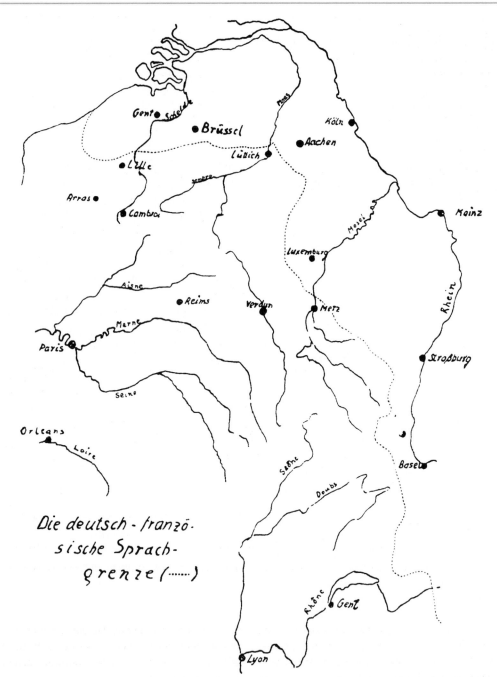

Die deutsch - franzö-
sische Sprach-
grenze (......)

Figure 1 Petri's sketch map of the Franco-German linguistic border. *Source* Franz Petri, "Deutsche Sprachgrenze im Westen," *Rheinische Heimblätter* 11 (1934): 445

present day" of his early Medieval research and closed his work by quoting a famous sentence by Goethe, the credo of many of his fellow *Volk* and cultural region researchers: "Take hold of what you inherit from your fathers in order to possess it. This is as valid for *Volk* history as it is in life." In other statements Petri tended to underline not the differences between the Germans and the

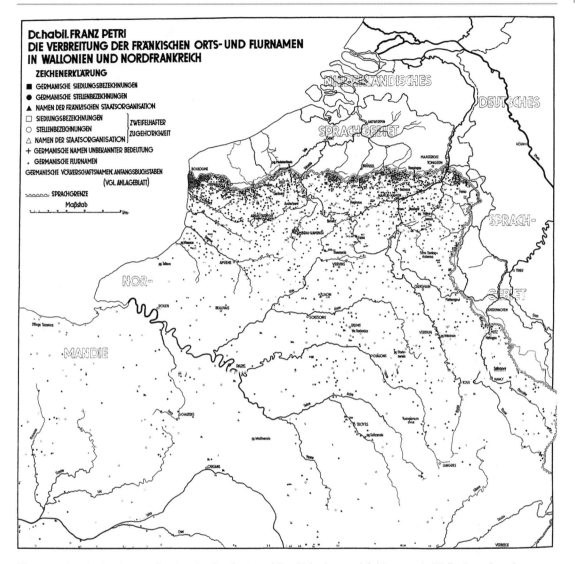

Figure 2 Petri's sketch map showing the distribution of Frankish place and field names in Wallonia and northern France. *Source* Franz Petri, *Volkserbe*, appendix

French but the common factors which bound them together. "The essential French character had received a constitutionally significant shot of German blood and German essence . . . Is it not time for an epoch which has become increasingly aware of the final *Volk* powers residing in it to reach back to and rise above any contradictory elements and concentrate on those *Volk* elements which bind people together, and which belong to an even older and more elementary layer of our common German–French past?" To sum up: In 1936 Petri offered an alternative interpretation of the fruits of

his research on the early Medieval period to legitimize claims for German expansion into the West but also to reinforce the common origins of Germany and France.

With his concept of "*Volkserbe*" (Volk inheritance), Petri had extended the work and the theses of his predecessor, Franz Steinbach. Steinbach in turn took up Petri's results, classified them in the historiography of the subject and in doing so sharpened them. He now regarded the era around 500 AD as a time in which "an overwhelmingly German-speaking population" stretched "almost

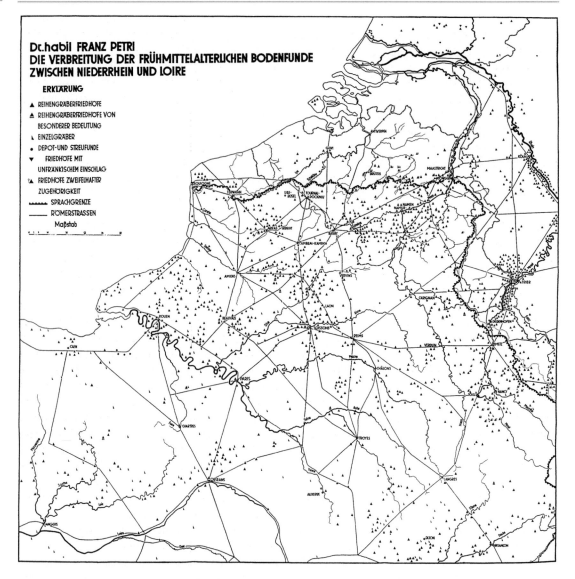

Figure 3 Petri's sketch map showing the distribution of early medieval burial sites between the lower Rhine and the Loire. *Source* Petri, *Volkserbe*, appendix

to the Loire". Petri's work soon entered the canon of literature dealing with *Volk* and cultural region: and quickly became a model of its type. Scholars in the fields of archaeology, early medieval history, linguistics, art and law adopted Petri's findings. Petri himself was able to publish his work on a wide scale and extended his influence with literary reviews and criticisms of other works on the theme. He was ultimately regarded as one of the leading German experts on cultural regions covering north-west Europe and in 1942 he was offered,

and accepted, a chair on the subject at the University of Cologne. The questions, methods and results which can be found in Petri's work were typical for the work of the majority of the younger academics involved in cultural and *Volk* region research in the 1930s. In the 1920s the defence of German borders and the establishment of cultural regions within Germany had been at the forefront of academic concerns. As Germany became politically and militarily stronger they began to concentrate more on examining "German *Volkstum*" and "German *Volksboden*" beyond the boundaries of the German Reich. Their researches stretched from ancient times to early modern history and, with regard to Eastern Europe, were often bound up with the opinion that the German *Volk*, i.e. the "Aryan race" was superior to the East European *Völker* i.e. the "Slavic race," and that German culture had always been fertile. Furthermore, the general opinion was that current political boundaries were not only unjust but arbitrary, and gave rise to conflict and further injustice. They were thus in need of an objective, scientifically researched basis which would allow boundaries to be redrawn around a self-enclosed German settlement area, for "kindred blood belonged to kindred blood." Since the German empire of the nineteenth and twentieth centuries was significantly smaller than the territories which had been invaded by the Germanic tribes or the area of the medieval German Empire, the results of this research confirmed the view that the current boundaries of the German empire were too constrained. The consequences of such a view went far beyond a revision of the Treaty of Versailles. On the other hand, when it came to studying the existence of alien *Volkstum* and alien "soil" within the boundaries of the German empire, such a question was considered to be outside German historians' areas of interest.

Despite the political opportunity afforded by Petri's results and Steinbach's more extensive conclusions, both academics had their critics during the Third Reich. First, Petri's philological work was criticized both for its methodology and its results. Critics disputed how much genuine worth could be attributed to place and field names in establishing the general state of linguistics and settlement in any particular era. They further cast doubt on the method whereby countless Romanic place and field names as well as word endings were traced back to Germanic predecessors. Critics pointed out that countless names which Petri had regarded as signs of the era of Frankish settlement stemmed from earlier or later times: when, for example, Germans worked for the Romans, either as auxiliary troops or as prisoners. Or during later waves of emigration after the Frankish era. Finally, they threw into doubt the interpolations which Petri had developed from a small and confined amount of linguistic evidence and which he had extrapolated to embrace much larger areas of territory. Philologists tended to speak of mixed areas and linguistic islands and rejected Petri's thesis of an all-embracing Frankish regional settlement.

Second, there were grave doubts as to the archaeological evidence, especially with regard to its amount and its assignation to Frankish types of graves. Third, linguistic and archaeological evidence only partly overlapped. Fourth, historians generally questioned where the Frankish tribes could have acquired their population potential, for only such potential could validate Petris and Steinbach's claims for a heavily populated settlement in the northern region of Gaul. Fifth, for this reason, doubt was cast on the claims of Petri and Steinbach that Frankish tribes had heavily populated the areas of northern Gaul, and an explanation was demanded as to why the alleged withdrawal from these areas had come to a halt along a line which happened to coincide precisely with contemporary boundaries, and why a linguistic boundary had also sprung up there. To sum up, Steinbach's and Petri's thesis of a massive and heavily populated Frankish immigration were criticized both empirically and theoretically. That is to say, critics thought that Germanic influences in Walloon and Northern Gaul had been vastly overrated. Their own work tended to support the traditional interpretation that Frankish emigration westward was halted by the Gallo-Romans in an area coinciding with contemporary linguistic boundaries. Or that there was a century-long process of seepage, with occasional intermingling of populations where both languages were used: in the latter case, however, there was a very clear Gallo-roman dominance. Petri's results and their sharpening by Steinbach were therefore a subject of great controversy in the Third Reich. But despite the massive criticism, his findings were adopted into *Volk* historiography.

"The Search for the Common Ground: Estyn Evans's Ireland"

from *Transactions of the Institute of British Geographers*, new series 19 (1994): 183–201

Brian J. Graham

Editors' introduction

Emyr Estyn Evans (1905–1989) was a student of H.J. Fleure's (see Gruffudd, p. 138) and became head of the Department of Geography and then Director of the Institute of Irish Studies at Queen's University in Belfast. He was one of the most influential geographers in Ireland. In this selection, Brian Graham traces Evans's intellectual lineage in the Vidalian regionalist tradition (see Vidal, p. 90) and explores the implications of Evans's geography for our understanding of Irish identity. Graham's account is particularly significant, given the partition of Northern Ireland. Evans's life and work, Graham notes, were in many ways devoted to demonstrating that the north of Ireland was a distinct culture region, and this obviously linked Evans firmly with unionist ideology. However, Graham argues that Evans's geography was in fact more nuanced and complex that this, and that his understanding of Ireland's distinct regional cultures led to a view of "Irishness" as constituted by diverse cultural identities, rather than a single homogeneous one.

As Franz Petri's links to the rise of German fascism make clear (see Ditt, p. 123), the intersection of culture and geography has political implications. In particular, geographers working on culture regions often conducted their studies with the objective of contributing to a larger project of nationalism and nation building. Thus, Vidal de la Blache's "Physiogamy of France" (p. 90) focused on how a distinctive French culture was built upon a diverse collection of *genres de vie*, and H.J. Fleure's regional studies of Wales contributed to a narrative of distinctive Welsh national identity (see Gruffudd, p. 138). Graham places Evans squarely within this context (noting, as well, the links between the work of Evans and that of Carl Sauer (p. 96)), but at the same time, he interrogates – via Evans's work – the assumption that a regional culture necessarily derived from internal coherence and homogeneity. Rather, a distinctive regional culture – and hence a national culture – could be constituted by a harmonious blend of local folk-culture traditions.

Graham also raises a number of criticisms of Evans's work, noting its material, or "artifactual," bias in treating culture as an assemblage of tangibles, rather than a symbolic realm of meaning. In this critique we see the clear disjuncture between more contemporary notions of culture and those of earlier cultural geographers. Graham also argues that Evans's approach was overly focused on rural peasant society at a time when Ireland was undergoing significant urbanization, industrialization, and working-class formation. Evans also got his history wrong in his view of Western Ireland's isolated folk cultures as the refugees of an ancient pre-Catholic and thus more "pure" Irish society. Similar criticisms have been leveled against the whole of early twentieth century cultural geography. But Graham also insists that despite these shortcomings, "Evans's work retains a significant relevance to contemporary Ireland through particular aspects of the vision of identity which it proffers." That vision is one of "common ground" amid diversity.

Brian Graham teaches geography at the University of Ulster. He is the co-author of *A Geography of Heritage: Power, Culture, Economy* (2000), and has edited numerous collections on Irish and European cultural identity and heritage, including *In Search of Ireland: A Cultural Geography* (1997) and *Senses of Place: Senses of Time* (2005). He continues to write about the politics of culture and heritage in Northern Ireland.

... [This] paper has two aims. In the first instance, it seeks to relate something of Evans and his relationship to geographical knowledge in general, and that of Ireland in particular. Evans's geography is seen as a cultural product, derived from the interaction of a particular geographical philosophy with the specific social and political circumstances of Ireland. It was a resource for himself and his own agenda as he worked in one part of a newly-partitioned island, both elements of which were struggling to establish their own identities, primarily through the adoption of mutually exclusive cultural and political discourses. However, throughout his work Evans was disinclined to be explicit about class, power, religion or politics; indeed, in *Irish Heritage* (1942, 2), he consciously eschews such controversial realms. Nevertheless, this discussion concludes that Evans did have a political agenda, signified particularly by a refusal to accept the assumptions of traditional Irish nationalism. The difficulty is that this dimension was never made explicit in his published work, creating an unresolved dissonance between its exploration of Irish identity and the political expressions thereof.

Secondly, therefore, the paper attempts an interrogation of Evans's geography in terms of the insights which it might offer into the political, economic and cultural conflict engendered by the existence within Ireland of contested bases for social understanding; his work remains a resource to be used in contemporary analyses of the nature and meaning of Irishness and its inevitable sub-text of partition. While issues of regional identity are much to the fore in these debates on Ulster's meaning and location within Ireland, there is little overt cognizance of Tuan's observations [in "Language and the making of place: a narrative-description approach," *Annals of the Association of American Geographers* 81, 1991] that regions may

have no existence outside the consciousness of the geographers who may persuade other people to accept these entities. Evans is often linked with John Hewitt in attempts to establish an Ulster identity which, although it may owe something to both Britain and Ireland, is primarily particular to the province itself. According to Hill [in "Regions: identity and power: a Northern Ireland perspective," in P. Drisceoil, ed., *Culture in Ireland*, 1993], for example Evans argued that archaeology and folk-life showed that the two communities in the north of Ireland, however deeply divided by religion, shared an outlook on life and a common heritage different from that prevailing in the south. Politically, the concept of a native Ulster tradition, which is broader than either Protestantism or Catholicism, can find expression either as Ulster nationalism or, less radically, as a form of regional identity within a greater United Kingdom, itself regionally disparate.

Consequently, Evans's views on regional distinctiveness can be depicted as no more than a convenient prop to unionist ideology which, intellectually, minimizes their importance in the wider Irish and even international contexts ... During a long academic career spent entirely in Belfast, Evans, a complex man in life and attitudes, had many disputes with southern Irish academics, in particular the archaeologist Ruadhri de Valera. Inevitably, political motives and interpretations could be, and were, attributed to their contrasting views on the primacy of particular cultural influences in early Irish society ... Although both academic and personal factors were involved here, such evidence can be used to depict Evans as a unionist who maintained only token links with the rest of Ireland. Conversely, however, I argue here that Evans's *oeuvre* also seems to address the intellectually more acceptable concept of a regionally diverse Ireland in which

Ulster is but *one* variant of a heterogeneous vision of Irishness. . . .

[. . .]

EVANS IN HIS GEOGRAPHICAL CONTEXT

. . . . [Evans's] *oeuvre* demonstrates a very high degree of internal consistency and coherence, the product of a sustained loyalty to a set of particular geographical principles. . . . [T]he most immediate influence upon Evans was the holistic philosophy of H.J. Fleure. In turn [according to David Livingstone, *The Geographical Tradition*, 1992], Fleure's ideas were based upon a reworked version of Patrick Geddes's Darwinian vision that the 'delineating of regional particularities' was crucial to the evolutionary 'promise of scientific synthesis' [see p. 138].

[. . .]

Fleure's direct influences upon Evans were heavily modified by their mediation, first through a Vidalian perspective and then through a long-lasting admiration for the *Annales* school of *géohistoire*. Paralleling Fleure's concern for peasant culture, Paul Vidal de la Blache emphasized the significance of ordinary people and their environment: to him, the region was not simply a convenient framework but rather a social reality. Although their approaches were very different, both Evans and T. Jones Hughes, one of his most influential contemporaries in the study of Irish geography, shared this idea that the landscape was a democratic text recording the history of the undocumented. Vidal and, somewhat later, Carl Sauer, another major influence upon Evans, believed that landscape was indicative of a harmony between human life and the *milieu* in which it was lived. . . . Vidalian ideas, such as *genre de vie*, *milieu* and *personnalité*, were crucial to the emergence of the *Annales* school and its concerted attempt to map and explain the complex reality of human life by reference to local and regional studies. It became a tenet of *géohistoire*, particularly as interpreted by Fernand Braudel, that any social reality must be referred to the space, place or region within which it existed. From the evidence of his written work at least, Evans was prepared to incorporate these Vidalian-derived ideas into his geographical philosophy.

Late in his career, he continued to profess admiration for Marc Bloch, Lucien Febvre and, perhaps above all, Braudel, applauding – like Sauer – the French genius for regional synthesis. From Vidal too, Evans took the idea of the *pays* as the geographical mediation of synthesis and continuity, the product of 'man's [*sic*] interaction with his physical environment over centuries' [A.R.H. Baker, "Reflections on the relations of historical geography and the *Annales* school of history," in Baker and Gregory, *Explorations in Historical Geography*, 1984] larger generalizations could emerge only gradually from a series of detailed and exact case studies of various *pays*.

. . . Clearly, his work is framed within a possibilist epistemology. Evans was to centre his work on that interaction between people and their environment, best summed up in Febvre's famous dictum:

> there are nowhere necessities, but everywhere possibilities and man as master [*sic*] of the possibilities, is the judge of their use.

Further, Evans maintained a strong relationship with Sauer and other North American geographers who had rejected determinism and shared a similar conception of geography as culture history in its regional articulation (Livingstone 1992, 297). These influences meshed with Fleure's theory of regions as places of lived experience, and his concept of contact zones was to inform Evans's geography of Ireland: regions were not just the 'product of a symbiotic union of people and places' but also the 'consequences of the shifting relationships between people and people' (Livingstone 1992, 285). To Evans, geography was 'the common ground between the natural world and cultural history' [R.E. Glasscock, "Obituary: E. Estyn Evans, 1905–1989," *Journal of Historical Geography* 17, 1991]. In his written work at least, Evans was to remain largely aloof from the political repercussions of these ideas. He was to retain also a certitude about this geographical philosophy and consequently his work itself constitutes a continuity, informed by these epistemological principles. This geographical heritage interweaves with, and underscores, his interpretation of Ulster and that province's relationship with the remainder of Ireland.

THE MOTIFS OF EVANS'S GEOGRAPHY OF IRELAND

My reading of Evans's work elicits five major motifs, all heavily dependent upon and intertwined with each other. Together, they constitute his contribution to defining Ireland's conceptual space. It is argued here that the motifs are directly derivative of the epistemological influences outlined above, mediated through a consistent attempt to define Ulster's cultural space within Ireland. This is an explicit rather than hidden agenda but, while it may have been stimulated by Evans's political aspirations as well as his geographical philosophy, the motifs are worked through in cultural terms alone. They are regionalism, human society and environment; the common ground – peasants and rurality; continuity; Ulster; and finally Irishness.
. . .

Regionalism, human society and environment

First, and basic to every other aspect of Evans's methodology, was a belief that the relationship between people and their environment is expressed within a regional dimension, itself a condition upon people's behaviour. Thus, we have the sustained emphasis on the possibilist perspective of human society being shaped by and, in turn, shaping the environment together with the concomitant connotation of place defined as a land and its people. Consequently, occupying the core of Evans's geographical cosmos was the holistic belief that people – with their shared paste cultural artefacts, values, beliefs and emotions – and land

> go together and have shaped each other, and you cannot understand one apart from the other.

He visualized the most genuine bonds as occurring in the *pays*, areas which were much smaller spatially than the four provinces of Ireland. The examples he most often quoted were the Kingdom of Mourne and West Cork, areas sufficiently small that interpretation might be checked against observation and local knowledge. His study of the former is now widely recognized as a classic and eloquent account, in the French style, of a small

but – physically and culturally – highly distinctive piece of County Down.

More generally, as noted earlier, Evans tended to believe that innovations were diffused from Scotland, through the east and south of Ireland towards its north and west. Consequently relicts, such as the open field agriculture in Gweedore, County Donegal, became part of a complex of cultural survivals, persisting in this far corner of the island. The west became the real Ireland for Evans, the *pays* where the peasant folk-culture of the common person remained, if not untouched, at least identifiable. Ironically, this was exactly the same Ireland which the Gaelic League, for example, defined – from very different premises – as the heartland of the island's cultural consciousness, the region of unspoilt beauty where the influences of modernity were at their weakest . . .

The argument contained in the pamphlet, *Ulster: the common ground* (1984), is perhaps the clearest account of Evans's culturally heterogeneous and regionalist view of Ulster and Ireland, albeit one characterized by distinct tinges of ethnic stereotyping. Evans saw the hidden closed-in drumlin lands of south Ulster as a Protestant landscape, occupied by a people of limited vision and imagination, marooned in their 'psychic stockade' to use Foster's graphic phrase [from *Colonial Consequences*, 1991, 159]. In contrast there is the other tradition of Ulster – the open, naked bogs and hills, the lands of the poetic and visionary in the Ulster soul – 'the spiritual hinterland of ancient memories of freedom and passion' (Foster 1991, 159). Evans believed that this diversity could be reconciled as a single theme with many variations, the *personnalité* of Ulster deriving from the fusion of many such small *pays*. He recognized that the landscape and material heritage was also a potent source of dissension, but argued that we must live with and exploit it as a total inheritance irrespective of formal creeds . . .

The common ground – peasants and rurality

Evans saw peasant culture as both the product of the mediation of the human–environment relationship and a repository of the vitality and continuity of lasting social values which urbanism tends to destroy. His geographical theme was very much the common people and the land itself, the land that they've helped to make: because the land is far older than us all, far older than all human cultures.

Perforce, this was almost entirely a rural world, full of resonances – transposed to Ireland – of Fleure's ideas about the furthest fringes of Wales being the 'ultimate refuge' of the true values and visions of Welshness. As Fleure himself wrote in a tribute to Evans:

> Ireland has been looked upon as an ultimate corner of western Europe, a treasury of the past, the last place to which a culture would spread and the last place in which an out-of-date culture would linger.

To some extent, it was seen in this way by Evans too, for – as has been argued here – the epicentre of his vision of Ireland was the west in general and the far north-west in particular. It is a powerful imagery but, in a telling point, Whelan [in "Beyond a paper landscape: J.H. Andrews and Irish historical geography," in Aelen and Whelan, *Dublin: City and Country*, 1992] observes that by the nineteenth century the west of Ireland was scarcely a far-flung periphery but, due to emigration, looked to and had intimate connections with North America.

Nevertheless, for Evans one of the common bonds that linked all the peoples of Ireland was this loyalty to local traditions and regions. Because of his belief in the symbiosis of human society and its physical environment, his particular definition of heritage included rural settlements, oral traditions, beliefs, languages, arts and crafts, a folk-culture embodied in environmental relationships . . .

Continuity

The concept of a continuity, dependent on 'habitat and heritage', constitutes the third motif of Evans's *oeuvre*, Further, it is perhaps the one most crucial to an understanding of his perspectives on Ulster and Ireland, a reason why it has become one of the more controversial aspects of his work. 'I have tried,' Evans wrote, 'to read the rural landscape and have come to see it as the key to the continuity of Irish history.' There is something here akin to the

idea – expressed amongst others by Seamus Heaney – that continuity and stability are to be found in the Irish land rather than in its people . . . Evans's concept of continuity invokes far more than the mere long-term survival of artefacts and customs through time, representing instead a belief that the very particularities of cultures are forged through 'a renewal of the old in contact with the new'. Thus, the centrality of continuity to Evans's representation of Ireland originated in Fleure's moral geography and ideas on regional conceptualization mediated through the notion of the *pays* as an embodiment of – and control on – peasant values.

Despite the evidence of urbanization and industrialization, Evans saw Ireland as having preserved 'to a remarkable degree, the customs and social habits of the pre-industrial phase of western civilization'. Although this could be held to imply cultural stasis, Evans did not see the island as a mere repository of archaic cultural artefacts. . . .

. . . I read the key element in Evans's perception of continuity as remaining the pluralistic idea of successive immigrant groups adapting to pre-existing societies which their arrival must also have changed, a concept illuminated by Fleure's concern for cultural contact zones. By continuity, therefore, Evans meant not simply the survival of townlands and place-names or customs through time, but the constant renewal of the old through its contact with new ideas and cultures. He saw Irish culture as diverse, the archaeological record demonstrating the enrichment of that cultural continuity, as well as its enduring characteristics. The Norse were absorbed and the Normans failed, their presence leading nevertheless to an enduring Anglo-Irish tradition. Evans did not use the word assimilation, referring instead to a process of absorption. It was precisely the clash of native and newcomer which struck the sparks in Irish culture and consequently Evans's interpretation is the very antithesis of the nationalist image which depicts Gaelic Ireland assimilating almost seamlessly the encroachments of various invaders. . . .

Ulster

. . . Evans was an Ulsterman by adoption and the nine-county province, which became the laboratory for his geographical ideas, constitutes the fourth motif of his *oeuvre*. He objected to the usurpation of the place-name, 'Ulster', by 'extreme Protestant spokesmen' to define the six counties. The boundary of Evans's Ulster was not that of partition but, characteristically, the difficult drumlin belt, stretching east–west from County Down to Donegal Bay, through Cavan, Monaghan and Fermanagh. His central concern was, I feel, to define the meaning of Ulster, to isolate its *personnalité*. It is my argument that he saw it as one manifestation of Irishness separate from, but part of, a larger entity which, itself, was far from homogeneous.

[. . .]

Irishness

The issue of Irishness constitutes the final motif of Evans's *oeuvre*, one of apparent contemporaneity, given the continuing debate referred to above. A logical extension of the conception of a geographical world, defined by the small scale of the *pays*, is the notion that although it may constitute a single theme, Ireland's character is defined by its numerous regional variations. Indeed, most European nations have evolved through such a fusion of regional loyalties. This relatively small island of 70 000 sq. km is characterized by regions which have long developed their own orientations and experiences, even if the compacting insular qualities of the place have meant that these experiences have had to be contained and shared within a narrow, often introverted, ground. It is within such a context that much of Evans's work can be read as a rejection of the homogenizing and sectarian certainties of orthodox Irish nationalism in favour of plurality or heterogeneity. . . .

A CRITIQUE OF EVANS'S IDEAS

[. . .]

. . . Clearly, [Evans's] visions of Ulster and Ireland are both flawed, largely through the inadequacy of his geographical philosophy and methodology in dealing with political discourse and the exigencies of an increasingly urbanized and industrialized world. The Vidalian notion of focusing on the creative power of human groups to adapt themselves to and, within limits, mould the natural environment

no longer has much relevance, even in Ireland. Although the country may be one of the least urbanized in Europe, 60 per cent of its population live in towns and cities, while employment in agriculture has dwindled commensurately. Over and above these various reservations, we can identify three specific grounds on which issue can be taken with Evans's view of Ireland.

First, it has been criticized for its artefactual rather than ideational or humanistic basis. As John Hewitt observed, Evans's emphasis upon folklife was largely affirmed through the study of material culture, particularly house-types and implements such as the spade, as opposed to things of the spirit – expressed in ballads, poetry and speech. As observed earlier, this is not to say that Evans was unappreciative of the oneness of culture or of the socio-cultural contexts of artefacts, for he regarded the rich heritage of Irish folklore as being a measure of the intimate association between people and their immediate surroundings. However, the balance of his published work tends to focus upon external rather than internal processes. This is a geography which personified places with, to appropriate Baker's comment on historical geography's 'false consciousness', a more limited reference to the peoples who inhabited them. This artefactual emphasis partly accounts for Evans's failure to engage power, class, religion or politics. His methodology, particularly in its bias against documentary sources, was ill-equipped to consider such issues, Only on the rarest of occasions did he allude to the political repercussions of his ideas for the island of Ireland: inevitably, their outcome would be 'a federal solution of some kind'. Nor did he address the issue that identity is manipulated through the exercise of power as one means of capturing the past in order to legitimate the present in which that past is seen to culminate. (It is within this context that the Ulster Folk and Transport Museum is so important. In creating this, Evans worked very closely with Terence O'Neill, later Prime Minister of Northern Ireland. Detailed investigation is required to establish the agendas to which both men were working, but I suspect that their motives were not entirely coincident.)

Secondly, serious questions must attach to Evans's views concerning continuity. As I have endeavoured to show, these originate from his epistemological orientation towards the timelessness of peasant cultures but also, and perhaps more seri-ously, from an attempt to refute the sequestration of Irishness by the Gaelic myth. Thus, Evans was forced to try to demonstrate the pre-Celtic antiqu-ity of the critical artefactual elements of peasant life, including settlement forms and field systems. Although it requires rather more research, justi-fication and qualification, Whelan's critique of this approach [in "Settlement and society in eighteenth-century Ireland," in Dawe and Foster, eds., *The Poet's Place: Ulster Literature and Society*, 1991] contains much of merit. Following Jones Hughes ["Society and settlement in nineteenth-century Ireland," *Irish Geography* V, 1965], he argues that peripheral western areas such as County Donegal were not refuges of some long-established folklife but actually experienced close and permanent set-tlement by farming peoples only in the eighteenth century in the wake of the Ulster Plantations and population increase. It was Evans's argument that a particularly Irish variant of openfield – rundale – and its associated nucleated settlement form – the clachan – which he held to be characteristic of such western regions, dated back to the Iron Age. Thus, the examples recorded by the first edi-tion Ordnance Survey six-inch maps of it 1840 were cultural survivals in an inaccessible region, dem-onstrative of long-term continuity. In contrast, Whelan argues cogently both for the recency and polygenetic origins of the complex of rural settle-ment recorded by these maps. Indeed, Evans did point to the lateness of much settlement in the Gweedore area of Donegal, one of his postulated refuge areas, but failed to address the implications for his earlier argument. Whelan's case does not invalidate Evans's claim for long-term continuity of material artefacts: permanent settlement along Donegal's wild and inaccessible Atlantic fringe dating back as early as the Early Bronze Age. It does serve, however, to demonstrate that Evans was incorrect both in positing a single characteristic Irish field and settlement system and in assuming a uniform origin for the elements of that complex.

. . . As I have suggested, Evans's ideas on con-tinuity were constructed from his wider *geograph-ical* philosophy, mediated through an aversion to traditional Irish nationalism; only the latter point makes them specific to Ireland as his generalist work on peasant societies amply demonstrates. Fur-ther, Evans believed in far more than the mere

preservation of folk-culture through time. To reiter-ate, although internally inconsistent, his view of continuity emphasized the renewal of the old in contact with the new. This distinguishes it from the perpetual survival of a set of fixed ethnic cer-tainties, one of the guiding tenets of traditional Irish nationalism. . . .

[As a final area of criticism] Evans considered cultural renewal in a highly selective fashion, a per-spective which, through its exclusion of influences not fitting into the peasant complex or amenable to his methodology, led to a most particular view of Ulster in Ireland. The rebirth took place within the framework of the *pays*, but many elements of the Irish past were given short shrift. Those included the influence of the Anglo-Normans, admittedly muted in Ulster but nonetheless significant, and the landlord-inspired improvements which trans-formed rural and urban landscapes in the eighteenth and early nineteenth centuries. But perhaps most significantly, there was almost no place in Evans's world for the transformations of Irish social space occasioned by the nineteenth-century penetration of industrial capitalism and urbanization, particu-larly into the north-east of the island. Ironically, even the remote 'rundale' areas of west Donegal were affected by this process: Evans's Gweedore 'peas-antry' developing very largely into a rural proletariat heavily involved in migrant labouring, particularly in Glasgow, instituting a cultural linkage that per-sists today. Evans, it might be concluded, created a representative Ulster and Ireland to which the working classes could not relate – except insofar as their ancestors came from the West – and the Irish state under de Valera was prepared to endorse that particular image in its cultural rep-resentations. His selective vision of continuity, derived from the belief in the primacy of peasant culture, meant that the *personnalité* of Ulster could never be fully developed.

CONCLUSIONS: EVANS'S GEOGRAPHY AS A RESOURCE FOR CONTEMPORARY IRELAND

Consequently, Evans's geographical philosophy produced a less than absorbing interest in certain elements of the diversity of his own landscape. Although his later work in particular does have something to say about the post-Plantation hetero-geneity of the Irish landscape, this topic is neither central to his arguments nor is it treated in a par-ticularly convincing fashion. Despite this serious reservation, I believe that it can be argued that Evans's work retains a significant relevance to contemporary Ireland through particular aspects of the vision of identity which it proffers. In the con-text of the contested bases of social understanding in the island, it is a text which addresses the vari-ety and diversity of Ireland – the traits submerged in the monolith of the Gaelic myth – while acknow-ledging it as a source of dissension, Evans be-lieved that it was those characteristics of our island which we had to learn to live with and exploit. Any essential unity emerges from this diversity and not through assumptions of a false homogeneity: to Evans, Sinn Féinism was the apotheosis of Irishness.

Evans's ideas must, therefore, be distinguished carefully from the agendas of those who seek to demonstrate Ulster's separateness from the remainder of Ireland. . . .

[. . .]

Evans's search for a common ground – his per-sonal agenda – which could emerge only by largely neglecting urbanization and almost entirely disregarding religion, politics and social conflict, is still with us through its incarnation as the Ulster Folk and Transport Museum. It remains to be estab-lished just how far this institution has shaped the perceptions which Ulster's peoples hold of their heritage. If they choose to believe that the con-temporary world is a divergence from a past communality then, no matter how unreal Evans's images of Ireland may now seem, they would possess a continuing relevance as a resource for con-temporary society and the efforts to effect some form of reconciliation between its conflicting ele-ments. In a formidable listing of negatives, *The Opsahl Report on Northern Ireland*, the most recent (unofficial) investigation into the perplexing and ambiguous questions surrounding the contested bases of social understanding in Northern Ireland, concluded that there was no realistic prospect of any form of Irish unity in the foreseeable future. One major difficulty identified was the loss of a sense of Irishness amongst Protestants, who believe that Irish history, culture and language have been expropriated by nationalists as political weapons. Despite all the qualifications voiced here, Evans's life work addressed this very dilemma, and that is why his geography still matters. As Tuan argues, geographers can 'create place by their eloquence', and few have written as poetically, or with such emotional attachment to place, as Evans did about Ulster and its *pays*. In so doing, he denied the exclu-sivity of Irish Ireland. Thus, his geography can be appropriated to support the contemporary idea that the explanation of the complexities of Irish cultural identity are to be found in 'a plurality of continuities, interlocking, full of complexity' [G. Ó Tuathaigh, "The Irish-Ireland idea: rationale and rel-evance," in Longley, ed., *Culture in Ireland: Division or Diversity*, 1991]. The heritage defined by it – in the sense of meanings attached to inanimate object – urges all Ulster people to accept their Irishness. Evans's Ireland encompasses Planter and Gael, his work arguing for the centrality of immigration and colonization in the sense that both Irish and Ulster identity have been forged through the continual renewal of the old in contact with the new. . . .

"Back to the Land: Historiography, Rurality and the Nation in Interwar Wales"

from *Transactions of the Institute of British Geographers* n.s. 19 (1994): 61–77

Pyrs Gruffudd

Editors' introduction

One of the fairly consistent themes that threads its way throughout this part of the *Reader* is the close association between cultural geography and nation building (see also Part Five). Geographers have long been enlisted in the project of "imagining" the nation. In their descriptions of landscape history (Hoskins, p. 105), or the regional ways of life based on the interactions between physical environment and culture (Vidal, p. 90), or by delimiting the boundaries of racial culture regions (Ditt, p. 123), geographers participated in an unacknowledged cultural politics that belied the presumption of geography as an "objective" and "value-neutral" science.

Much of this cultural politics was expressed in the presumed link between rural folk cultures and a "pure" national identity. It was often in the local or regional cultural traditions that geographers and other early twentieth century intellectuals found the scattered remnants of a "truer" national culture that had survived beyond the industrializing and urbanizing places. By the 1980s, of course, this kind of cultural geography of ethnic salvage was being subjected to criticism as irrelevant to the concerns of contemporary society. The rural, pre-industrial focus of Vidal's *Tableau de la géographie de la France*, of Evans's studies of northern Ireland, of Sauer's cultural landscapes, have all been criticized for an anti-urban and anti-industrial romanticism that saw in peasant folk culture a kind of purity in which the true identity of a people could be found.

In Pyrs Gruffudd's study of the "back to the land" movements of inter-war Wales, however, a more complicated interpretation of this romanticism is offered. While it can be demonstrated that scholars such as Carl Sauer and W.G. Hoskins viewed the cultural landscape with conservative eyes, it would be wrong to presume a necessary link between the regional monograph tradition and an anti-modernist intellectual attitude. Vidal himself celebrated the industrial development of France. And, as Gruffudd argues here, many inter-war Welsh intellectuals like H.J. Fleure (1877–1969) regarded the rural cultural landscape as a site of "utopian fusions of tradition and modernity which challenged the polarized notion of rural stagnation and urban modernization."

H.J. Fleure was raised in Guernsey and studied natural sciences at Aberystwyth in 1897 and, later, in Zurich. Fleure returned to Aberystwyth to lecture in zoology and geology but by 1908 was lecturing in geography, and in 1917 was appointed Chair of Geography and Anthropology. One of Fleure's more well known students was Estyn Evans (see Graham, p. 130). Like many human scientists of his time, Fleure was influenced by the neo-Lamarckian version of evolutionary theory, which held that organisms modify themselves in response to the changes in their environment. Yet his work emphasized more of the interplay between humans and their environment, along the lines of the possibilism associated with Vidal de la Blache and his students (see p. 90). Fleure was also influenced by the biologist and sociologist, Patrick Geddes, and the Regional Survey movement that Geddes initiated.

Pyrs Gruffudd is a Senior Lecturer in Geography at the University of Wales Swansea. His research focuses on the cultural geography of twentieth century Wales, with specific attention to questions of landscape, Welsh identity, conservation and planning. He is the author of numerous journal articles on the cultural geography of Wales, and is co-editor of *Cultural Geography in Practice* (2004).

INTRODUCTION

Several recent texts have argued that the nation and national identity are fluid, contextual and contested. They must, it is argued, be read in the context of discourses as diverse as textual narrative, sexuality, and patriotism. Nations are now as much imagined as they are material entities . . . It is this process of cultural and geographical imagining which is the theme of this paper. According to Anthony Smith [*The Ethnic Origin of Nations*, 1900], 'legends and landscapes' are integral features of the national imaginings of both 'ethnic' and 'civic' nations. Territory is nationalized through ethnic and historical associations and achieves significance in a symbolic sense: 'a land of dreams is far more significant than any actual terrain'. Within cultural geography, the landscape has been seen [by Stephen Daniels in *Fields of Vision*, 1993] as an arena for national symbolism: 'as exemplars of moral order and aesthetic harmony, particular landscapes achieve the status of national icons'. But whilst Daniels stresses the role of the arts in the articulation and negotiation of national identity, he notes that scholars and professionals – geographers included – have been enlisted too. National identity is, therefore, a complex zone of convergence of a number of discourses – political, artistic, academic – which are not merely reflections of some social reality but serve to constitute that reality.

My concern in this paper is not with aesthetic representations of the nation but with the imagined grounding of a nation in a particular environment and the presumed moral attributes of that environment. The paper is concerned with 'back to the land' tendencies in interwar Wales and with some of their political resonances. A yearning for the 'spiritual wholeness of the countryside is a common theme amongst nationalists and intellectuals and it

motivated the 'invention of tradition' in Wales during the Romantic period. But I would argue that the rural imagination presented here is not an uncomplicated romanticism. It is, rather, a dynamic engagement with a 'place on the margin', to borrow from Rob Shields. The rural becomes almost a liminal zone which is seen as occupying a ground between tradition and modernity and the societies they represent. The polarity between the two cultural outlooks is blurred and the possibility of a new social formation emerges. . . .

ESSENCES OF LOCALITY

[. . .]

[H.J. Fleure's] research supported what has been called the social-Lamarckism of Patrick Geddes, an extended understanding of human types, including mental, spiritual and social characteristics. It was in this way that Fleure contributed to debates on identity in Wales, an identity historically seen as being rooted in the rural and in the traditional. Between 1905 and 1916 2,500 individuals were surveyed in Wales by Fleure and his colleague T.C. James, with name, sex, location, family history and a total of 19 physical characteristics – including head shape and skin pigmentation – being recorded. Data relating to individuals, however, were mapped only if all four grandparents came from within a 12–15 mile radius; in this way, people were read as 'concentrated essences of that locality'. Mapping demonstrated that Wales was characterized by marked regional differentiation, understood as the result of interplay between heredity and environment. The latter had protected local distinction, Wales having experienced only limited effects of modem population movements. What Fleure called the simple folk of

Wales represented types of humankind whose distinctions dated from a remote past. These types were easily distinguishable; Welshmen could instinctively 'tell' someone was from a particular district, This subterranean geography of Welshness was further strengthened by Fleure's humanist insistence that local types be studied in relation to natural regions rather than administrative units. In this way, Welshness was constructed as an organic unity between humans and environment. The basic Welsh physical inheritance, Fleure claimed, was from the Palaeolithic era when north-west Europe after the Ice Age was in the hands of a remnant population – strongly built people with dark colouring, long heads and deep-set eyes. Fleure argued that such people survive on the Plynlymon [sic] moorlands and in the Black Mountain country of Carmarthenshire. The scattered farms on Pumlumon had in fact 'yielded more than seven adult male cases of unusually complete survival of physical characteristics we generally associate with the earliest type of modern man'. Upland Wales – its geography as rugged and inhospitable as in prehistory – was seen, therefore, as a refuge from what Fleure called 'the new world-life outside'. The fundamental type – the 'little dark people' – were a predominant element in both rural and industrial Wales. But on the coastal lands could be found a type which revealed something of the historical and cultural geography of Wales. These men of 'stalwart build' and dark colouring were found 'in nearly all the fishing harbours' of Cornwall and also in the Hebrides, Ireland, Brittany, Spain and southern Italy, proving Welsh links with seafaring European nations and the process of cultural diffusion along the western seaways.

It should be stressed that Fleure challenged notions of racial purity and that his analysis of Welsh physical types was based in large part on the effects of culture contact and mixing. He consistently attacked the idea of national types, arguing in 1922 that all humans were

> mosaics of inheritances and that a 'race-type' exists mainly in our own minds and should not be used without great reserve in scientific discussion.

He was active in anti-racist campaigns and was particularly critical of racial theories, like the Nazis' Nordic Myth, where science was used to veil sinister political propaganda. Whilst there is an ongoing debate about the historiography of eugenics, we can, nonetheless, locate Fleure within a progressive, reforming strand of a movement which was not, according to Searle [*Eugenics and Politics in Britain*, 1976], peripheral or crankish 'but an important challenge to politicians and academic theorists alike'. Whilst Fleure rejected the notion of a 'Celtic race', he still felt that 'Celtic types' were physically and socially disadvantaged within industrial society and his 'moral geography', discussed below, was a feature of 'positive' eugenics concerned with social and environmental reform.

[. . .]

A MORAL TOPOGRAPHY

[Fleure believed that] Wales in particular was 'a refuge of old ways and old types' where continuity and persistence were revealed by archaeology and anthropology. Geography caused this continuity for

> The physical features of the country, the framework of mountain-moorland that separates the Wye and Severn region from the valleys that radiate out to the sea, have broken the face of many waves of change ere they have reached the quiet western cwms.

In the remote western areas could be found racial remnants and a persisting folk way of life, a localism distilled by undisturbed centuries. In *The personality of Britain*, the archaeologist Cyril Fox (1932), Director of the National Museum of Wales, drew on Fleure's work to argue that Highland and Lowland Britain were influenced by different culture streams. The Lowlands were susceptible to rapid cultural change but in the Highland region new cultures were absorbed, thus generating cultural continuity. For Fox

> There is a wide range of evidence of this such as the survival of the Celtic language, the persistence in the west of very old racial stocks, and the persistence of tribal custom, the importance of kinship and clan.

More broadly, Fleure saw the Celtic fringe of Europe as

the ultimate refuge in the far west, wherein persist, among valleys that look towards the sunset, old thoughts and visions that had else been lost to the world.

In this broad and admittedly fragmentary sense the work of Fleure and his colleagues echoed a broader European conceptual tradition of [what Langton called] 'habitat, economy and society'. Elsewhere, the countryside, and in particular the hill country, was seen as keeping alive inheritances from the past. The German countryside, for instance, was [according to Farr, "'Tradition' and the peasantry: on the modern historiography of rural Germany," in Evans and Lee, eds., *The German Peasantry*, 1986] seen 'as a reservoir of traditionalism, and . . . the peasantry as an arsenal of pre-modern characteristics'. This was later corrupted into the Nazis' 'Blood and Soil' philosophy. In France, rural sociology and ethnology flourished after the Great War and in Scandinavia the study of agrarian history and the roots of folk life was perceived to be of great contemporary importance. Fleure praised Denmark's role as a laboratory of experiments in the modernisation of peasant life without setting the peasants adrift.

The stream of inspiration

But this European concern with the rural was not necessarily a nostalgic response to modernization. In many cases it represented an attempt to theorize the perceived spiritual importance of the remote rural areas and their peoples, seen as wellsprings of civilization. [For Fleure] . . . The Little Dark People, the basic Welsh type, contributed large numbers of church ministers, for the moorland people's idealism 'usually expresses itself in music, poetry, literature and religion rather than in architecture, painting and plastic arts generally'. Fairer, Nordic types and the darker, coastal types were more prominent in commerce and the former were also astute politicians. So the geography that allowed for recovery of 'survivals' on Pumlumon also allowed for the recovery of a storehouse of values protected by social continuity. In this sense, [what David Livingstone, *The Geographical Tradition*, 1992, called] the regionalizing ritual was also a moral one. Fleure was convinced that the peasantry cherished universal and abiding values, and that

peasant life retained a vital diversity. To urbanites and suburbanites it might, Fleure admitted, seem mere fond sentimentalism to admit to an interest in old countryside traditions. Times, they would stress, had changed but Fleure noted that

some of us doubt whether change is always progress and whether change, as it affects us, is not often a specialisation in certain directions that cuts off possibilities in others.

Fleure was not anti-urban; he saw the potential of the city as the social expression of the better elements of the human soul but, under the conditions of industrial capitalism, the effects on both people and the environment were devastating. Hence Fleure's support for various town planning movements, including Patrick Geddes's 'Civics'. According to Fleure, modem society threatened to neglect the spirit in a pursuit of materialism: 'It is a case of cheap goods and cheap food for cheap people in cheap houses and cheap towns'. Modernization was seen as having a detrimental effect on personality in contrast to the rich diversity of peasant life. Whilst the latter's lifelong sequences perhaps limited initiatives, it also protected people from becoming the flotsam and jetsam of slum and suburb. In the rural west, according to Fleure, the personality was fully developed and even simple working folk would eagerly discuss philosophy or religion.

[. . .]

In Fleure's opinion, the peasantry was of importance in combating the materialism of *laissez-faire*. In 1921 he argued that civilization's one hope of avoiding collapse was to have a stream of supply from the rural areas where the treasures of ancient inspiration survived. Wales was seen as

a fount whence may well up streams of inspiration refreshing to the aded and overstrained business life of our perplexed modem England.

Fleure argued that British life had been enriched by migrants from the Celtic west seeking employment in arenas like the army, the churches or politics where the social characteristics attributed to their physical types might best be utilized. The Celtic west had been that spring of an ancient cultural tradition with its vision and its dreams that has given its men a quality we need to keep us fresh. The miners of South Wales have been preserved from some of the worst evils of industrialism by these contacts and few who have known them well would *dispute* the statement that they are a specially valuable element in our British population.

[. . .]

NATIONALISM AND RURALISM

[. . .]

Iorwerth Peate (1901–82) was one of Fleure's first students of geography and anthropology at Aberystwyth, and he later pursued doctoral research under his supervision on the links between physical type and Welsh language dialect (Peate 1926a). A carpenter's son from rural Wales, Peate shared Fleure's concern about cultural insensitivity and particularly the tendency of modernization to eradicate local differences. He was a passionate defender of the *gwerin* – the common folk – and shared with Fleure and Stapledon a belief in their essential wisdom and spirituality. According to Peate, to discuss religion, literature or politics in Wales, one naturally went to the carpenter, shoemaker, truly Welsh miner or blacksmith. Peate saw rural society as emerging from a living tradition and, ultimately, a living language. Thus folk life assumed immense significance in the context of modern challenges to cultural continuity.

For Peate, Wales was a refuge from the waves of new cultures advancing from the east. In the west could be found

folk songs, superstitions, crafts, the gentle bearing of the poor, and a host of other things which are like the fragments of a dream lost in the uproar of industry's juggernaut.

Peate's particular interest was the craft industries of rural Wales. When he edited Fleure's *Festschrift* in 1930, he contributed a discussion of Welsh wood turners and their trade in which he outlined the geography of production and marketing and traced European influences on style (again tying Wales into a broader European network). But new influences were increasingly evident:

the introduction of German-made spoons into Pembrokeshire and of Woolworth spoons into the large towns is slowly destroying the remnants of the turner's trade.

As the English observer of rural life, George Sturt, had observed, aesthetic changes reflected cultural change. Peate was convinced that the spiritual basis of Welsh rural life – and indeed civilization as a whole – depended on the preservation of rural industrial organization based on a combination of agriculture, industry, and the crafts. The village had always been, he argued, a self-sufficient community

where work and leisure, individual enterprise and co-operation were combined to produce a rural polity which seems to be far nearer perfection than the unhealthy striving of those communities where poverty is extreme and wealth out of all proportion to the needs of those who enjoy it.

The community was an organic, self-sufficient and cooperative system which encouraged courtesy, artistry and kindness. However,

The shoddy furniture of the cities and the short-lived manufactures of the mass-production firms have found their way into the countryside, and the result is not only a deterioration of the common necessities of life, but a disintegration also of rural society.

'A call to nationhood'

Peate was amongst the first members of Plaid Cymru, the Welsh nationalist party, founded in 1925 and helped edit *Y ddraig goch*, its monthly paper. According to D.H. Davies, during its first twenty years Plaid Cymru was not a political party at all but a cultural and educational movement seeking to elicit a sense of common ethnic identity and to 'resist and reverse all those trends that were assimilating Wales into England'. Whilst this sense of identity was overwhelmingly focused on the language, the concept of cultural continuity was understood in geographical terms which echoed the work of Fleure and others. Key components in this reconstruction of identity were the appeal of the rural and of the *gwerin*. *Y ddraig goch* was one method through which this geographical and cultural message was diffused. In 1926, Peate argued in its pages that Wales was an 'immortal nucleus' containing the core of the western World's traditions and that it was incumbent on the Welsh to ensure its perpetuation by resisting English cultural encroachment. Saunders Lewis, the party's President, emphasized this view, arguing that Wales was a European nation rather than a British region. He claimed that in Britain the European Latin tradition was only represented by the Welsh, and, like Fleure, chose to regard the Welsh 'not as a people driven headlong to the West and the mountains before a swift and irreversible Anglo-Saxon onslaught' but as Britons that had identified themselves with Roman cultural and spiritual ideals including Christianity. Modern Celtic cultural continuity in western Britain was, therefore, a rebuttal of Anglo-Saxon culture and politics. Given a choice between the Empire and the League of Nations, the Welsh – claimed Lewis – would opt for the League and for a quasi federal Europe of small nations. He urged that the claims of Welsh nationhood be considered in the context of European history, with the Middle Ages before the rise of the modern state, as the ideal.

Thus Saunders Lewis and Iorwerth Peate tied Wales firmly into the historical and contemporary currents of European civilization. . . . *Y ddraig goch* contained European and world political analysis and, whilst some looked to Ireland, many drew inspiration from European culture and politics. . . .

BACK TO THE LAND

Amongst the currents of political and social thought strong in interwar Europe was, as we have seen, an idealization of the rural population and of the rural areas as sustaining 'national' characteristics. This idealization underpinned various movements aimed at shifting the orientation of society, both ideologically and physically, 'back to the land'. Many of these movements were on the Right, the extreme example being the Nazi 'Blood and Soil' ideologues. British Fascist and Conservative groups also promoted similar ideas. But this rural idealism was also central in many socialist and distributist movements. The Danish cooperative movements, for instance, influenced the work of many Plaid

Cymru members, and in Britain the Arts and Crafts and Garden City movements were founded on the romantic socialist opposition between industrial oppression of the proletariat and rural liberation. In the South Wales coalfield, the Quakers sought to alleviate deprivation by introducing crafts workshops and farm units. These fragmentary influences contributed to an anti-industrial and anti-urban sentiment within Plaid Cymru which argued that these influences were anathema to Welshness. Ambrose Bebb – echoing George Stapledon – blamed the education system for causing rural depopulation by failing to inculcate rural values:

> How sad it is . . . to see arising generation after generation of boys and girls, who swarm together to the schools of the plains, there to drink from a poison which visibly weakens them and makes them unable to perceive the majesty of the high pastures and the shepherd's life, the romance of farming the land, and of passionately smelling the fertile soil; but who rather set off in cowardly fashion, with neither valour nor heroism, for the lazy, inert abodes of the towns and cities.

In Bebb's opinion

> One of Wales's greatest needs today is not only to keep her Sons on the land, but to bring back from the city to the land the masses who flowed there during recent years.

The party's chief agricultural adviser, Moses Gruffudd, argued that

> Placing the people back on the land is not only appropriate, but is essential *if* the Welsh nation is to live. The Welsh nation is a nation with its roots in the country and the soil.

This back to the land sentiment clearly echoed the academic interpretation of rural virtue. But academics like Fleure and Stapledon also proposed plans for rural rejuvenation which also came to influence broader political discourses, in part through the agency of Iorwerth Peate. Fleure was committed to the active role of education in the life of the community, thus maintaining a tradition established at Aberystwyth when the college was founded in 1872 with over 100,000 donations of under half a crown from the *gwerin*. He believed that Britain's well-being depended on the return of vitality to provincial life and that the universities might be instrumental in this renewal as, he argued, had been the case at Aberystwyth. His work in adult education and with the Regional Survey movement and his belief in the university's role as provider of specialist advice, were part of this reciprocal relationship between institution and community. Survey should, he argued, foster social renewal by inculcating citizenship and a spiritual awareness of place:

> My first plea then is that in our work we should cultivate the master [sic] light of memories and traditions, the deep intuitions of life, and that we can do this very forcefully by encouraging direct observation and study of the surroundings in which we live.

This awareness and local patriotism would then facilitate a re-birth founded on tradition, in a report for an influential planning campaign group, Fleure addressed the social and economic future of Cardigan Bay and thus applied his philosophical understanding of rural society. He blamed *laissez faire* for the decay of the traditional industries and noted some of the consequences, most notably the increasing role of tourism in coastal villages. A population which came to depend entirely on the city tourist was socially degenerate and Fleure urged action for reasons as much moral and civic as financial. He advocated scientific support from the university and a centralized and cooperative authority to control local industries. Themes in Geddes's sociology, such as 'Civic sympathy' and 'social reintegration', were crucial elements in Fleure's vision of social evolution. Technology also played a part. Afforestation and hydroelectric power might evolve alongside fisheries – a pattern then characteristic of the Alpine countries and Scandinavia – and indeed hydroelectric power (HEP) was seen as crucial in transforming what Fleure called 'regions of difficulty'. This was part of a broader trend of evangelizing on behalf of the new technology which Bill Luckin [in *Questions of Power*, 1990] has termed 'techno-arcadianism' whereby the old, moral order is re-established on modern, technological foundations.

[. . .]

The rebirth of the nation

. . . Peate's vision was far from being a nostalgic retreat to the past. He again provides a link between academic and political discourses on the nature of Welsh rural society. Peate's proposed action to reinvigorate rural Wales was based on his academic studies of crafts and social organization. In a critique of the 1942 Scott Report on *Land utilization in rural areas*, Peate attacked the division between urban and rural, claiming it was sentimental and indeed immoral, having no application in Wales. Rural Wales, Peate argued, had always been characterized by the dual foundation of agriculture and industry. No-one, given care in planning and design, needs fear the destruction of beauty by industrialization. Industry could indeed beautify:

> We must face these facts rather than live in a sentimental mist and be content with the persistent feebleness of the countryside. There are dynamic foundations to true beauty.

Peate, like Fleure, called for the development of HEP and forestry and argued for the introduction into rural Wales of new 'mobile' industries like plastics, located in well-planned additions to existing settlements. But this was not merely a countryside planning argument; it was calculated to re-establish the moral geography of the organic community. Small factories could breathe new life into declining districts, stemming population flow and re-establishing the old social organization and its moral basis of cooperation on a new, technological foundation. As Luckin puts it, triumphalist enthusiasts for electricity

> came to stress 'natural' connections between farming, the revival of the 'organic' village community, and the new form of energy as a stimulant to rural crafts and industries.

This back to the land ideology was adopted by Plaid Cymru who derived widespread inspiration for their notion of a national plan. Nationalist economic analysts, drawing on the work of George Stapledon and on the experiences of other small European nations, argued in 1939 that Wales could afford self-government. Another inspiration was the integrated, democratic development scheme of the Tennessee Valley Authority. Saunders Lewis argued for a Welsh National Development Council to guide the de-industrialization under way in the depression years for the benefit of Wales. Former industrial workers should, he argued, he settled in farming colonies with the policy operating alongside slum clearance in the urban areas. But the policy was not solely a response to industrial decline. It had both economic and ideological coherence as illustrated by Lewis's *Ten points of policy*, published in 1938. He argued that agriculture should be the primary industry of Wales 'and the foundation of its civilization' and that South Wales must be de-industrialized 'for the moral health of Wales' and of the region's population. . . .

CONCLUSION

Throughout the debates on the primacy of rural values, Peate argued that the traditions of the past had to be acknowledged in the solution of present problems, one of the themes of the Regional Survey movement. In all replanning, the Welsh nation should, he argued, look to its own traditions. In *Y ddraig goch* he advocated the role of agriculture and market gardening as complements to mining and industry in the revival of South Wales. HEP and large-scale forestry could also form the basis of a new, Welsh rural culture:

> We cry for old methods in vain: we attempt to revive the dead in vain, but on the grave of the old methods, we can build new factories and keep alive, in the sound of the machines of this age, the spirit of the rich culture we have inherited from the old craftsmen of Wales.

In Peate's techno-arcadianism, the old order was revived on the foundation of new, modem industries. Such ideas, expressed also by Fleure and Stapledon, begin to make 'the rural' more complex in the interwar period. In essence, we cannot see the relationship between country and town as a simple polarization of tradition and modernity, or stagnation and progress. In his study of Weimar and Nazi Germany, Jeffrey Herf identifies what he calls 'reactionary modernism' – an assimilation of technological advance into anti-capitalist romanticism. The Right were particularly adept at straddling the tradition–modernity divide, harnessing historical idealism in conjunction with the promises of an ordered, scientific twentieth century, producing [what Cullen calls] 'a strange contrast . . . between modern and anti-modern themes'.

But the key idea here is, perhaps, the tension between materialism and idealism. Herf claims that reactionary modernists did not see materialism and technology as identical. Technology and idealism could, therefore, be reconciled and, whilst the Right reconciled them around National Socialism, elsewhere in interwar Europe – including Wales – they were reconciled around the notion of a 'moral geography'. Modernist notions of progress, utopianism and democracy represented for some by technology were allied to a rural idealism focused on morality and cultural continuity. The move back to the land was not necessarily a regressive or reactionary step, but one which challenged dominant ideas of 'progress': ideas based on industrial capitalism and urban life. It asserted that certain values of community and artistry, apparently denied by urban civilization, could be re-captured in the rural areas. In the case of Welsh nationalism, a move back to the land could also reunite a culture with its European influences. Whilst modernity is generally cast in opposition to notions of romanticism or nostalgia, this move back to the land advocated its own version of progress, founded on an utopian fusion of past and future. Iorwerth Peate summarized this new relationship between the traditional and the modern by drawing on Lewis Mumford's *Technics and civilization*. As Mumford himself put it

> [W]ith a change in ideals from material conquest, wealth, and power, to life, culture, and expression the machine like the menial with a new and more confident master, will fall back into its proper place: our servant, not our tyrant.
>
> [. . .]

PART THREE

Landscape

Courtesy of Alex Dorfsman

INTRODUCTION TO PART THREE

As we have seen in Part Two, cultural geography has changed and evolved over time, developing import-ant divergences in different countries' traditions. However, for most cultural geographers, *landscape* has provided an enduring core idea across time and place.

The evolution of the word 'landscape' points to the diverse meanings behind the term. In this sec-tion, both W.J.T. Mitchell and J.B. Jackson carefully unpack the etymology of the term. In its Old English and various Germanic usages, words such as *landscipe*, *landschaften*, and *landtschap* referred to a land under identifiable ownership by an individual or a group. It was a short step from this association to more formal administrative divisions of land, as well as legal and political representation based on identification with particular lands, which did occur throughout much of Northern Europe. Thus the spread of capitalism, with the private ownership of land as one of its legal-institutional pillars, provided the larger context for this particular development of the term in Northern Europe. In the Romance languages, the French *paysage* and Spanish *paisaje* invoked a sense of a cohesive region, smaller than today's nation-states, which possessed a distinctive local character. These terms, and their meanings, are still important in France and Spain today, where regional variation in dialect, cuisine, vegetation, and so forth can be striking (see also Vidal, p. 90).

In the early seventeenth century, Dutch *landschap* painters began to employ landscape in a pictorial manner closer to the way it is popularly understood today: as scenery. This understanding of landscape was not limited to painters, but used also by the theater and landscape architects. Landscape thus acquired a highly visual character. The ascendance of the visuality of landscape went hand-in-hand with changes in the scope and nature of power relationships. To represent something is to turn it into an object; as John Berger has famously commented: "Oil paintings often depict things. Things which in reality are buyable. To have a thing painted and put on a canvas is not unlike buying it and putting it in your house."[1] The enframing of sweeping vistas of horizon and ground, the swell of mountains, and the curve of shore so came to shape notions of landscape that the visual representations of landscape became at least as – if not even more – important than the literal land that was depicted. You can experience the import-ant influence of landscape representations yourself on your next holiday or vacation, by paying close attention to the 'scenic views' indicated on signs in national parks, historic landmarks, and the like. Did you already know, more or less, what to expect at these places? Do the 'scenic views' look like post-cards or paintings? If inclement weather, natural disasters such as fire, or the presence of other tourists alters or obstructs your expected view, are you disappointed?

Modern cultural geographers are somewhat divided about the meaning of landscape, what the term does and does not encompass, and how to best study it. The modern academic use of the term *land-scape*, at least in the United States, is associated with Carl O. Sauer's particular approach to appre-hending the world around him, as detailed in his essay, published in 1925, titled "The Morphology of Landscape" (see p. 96). For Sauer, man-made cultural processes worked to shape natural surroundings, the result of which was the visible world around us: the cultural landscape. It was the task of the geo-grapher to provide a detailed description of an area, and to then meticulously uncover the layers of human activity that had shaped the visible landscape in particular ways. Sauer's *morphological* approach quickly became engraved on the American cultural geography scene through the founding of the so-called Berkeley school (of Geography, at the University of California, Berkeley) in the early 1930s. With Sauer at the helm, hiring key colleagues and training a number of students who would carry his legacy forward, this particular approach to landscape dominated American cultural geography through the 1950s.

Yet Sauer himself was profoundly influenced by deeper, European roots (see the introduction to Part Two). Sauer openly acknowledged his debt to nineteenth century German geographers and their systematic approach to studying the visible elements of landscape. The notion of landscape as cohesive assemblage of natural and cultural features, small enough to be captured at a glance, harkens as well to a long tradition of French geographers and their interest in regional variation. As discussed in the previous section, British approaches to the landscape, by contrast, had long emphasized the history of place in landscape analysis. Even more so than in the German tradition, landscape expressed the culmination of layers of intense, deep, and often fraught engagement between human societies and the natural world around them. Indeed, as evidenced by J.B. Jackson's focus on cultural history in his approach to landscape, the Berkeley school itself had developed significant divergence from Sauer's original morphological approach by the mid-twentieth century. In the excerpt from Jackson's work included in this part, you will note that cultural history becomes a more important aspect of American landscape studies by the 1950s, bringing this thread of American cultural geography closer in line with the British tradition.

By the 1960s, such approaches were seen by many as too descriptive, subjective, and particularistic, and as such rather unscientific. In general, landscape interpretation as a field of interest among human geographers waned, such that by 1983 the *Dictionary of Concepts in Human Geography* would assert that landscape has declined in importance in recent years due to an increasing emphasis on scientific analysis, theory, and model building. Many human geographers instead turned their attention to what were thought to be more objective, quantitative, and law-seeking (*nomothetic*) approaches. These endeavors were increasingly assisted (some might say driven) by computer-based data analysis. Indeed, a case might be made that GIS (Geographic Information Systems) today facilitate a particular *sort* of approach to landscape, one that is deeply rooted in measurable data compiled and analyzed by computer.

Not all cultural geographers put their interpretive landscape approaches on mothballs, however. An enduring tradition of exploring the human need for connection to place, how humans dwell, and people's relationship with their surroundings was revitalized in the 1970s (see also the introduction to Part Five). This humanistic current in geography emphasized the affective, perceptual, and experiential dimensions of landscape. Yi-Fu Tuan's enduring characterization of geography as "the study of the Earth as the home of people"[2] invokes the notion of *home* that is at the heart of work by humanistic cultural geographers. Home invokes attachment, affection, and an existential assessment of human's place on Earth, literally and figuratively speaking. Thus nostalgic landscapes can exist in dreams and memories and landscapes can be acted upon (and act) at emotional levels involving love or hate. Ultimately, landscapes allow humans to *dwell* in the world, according to cultural geographers of a humanistic bent.

In another current within human geography, the advent of critical perspectives such as Marxism, feminism, and the general rise of social theory in the 1970s and 1980s brought a less particularistic, and at the same time more politicized, approach to landscape. It was understood by such scholars that landscapes reflect societal power relations, and could not simply be taken at face value as the sum of their material elements. In other words, there is more to landscape than meets the eye. In addition, it became increasingly accepted that landscape does not merely *reflect* power in society; it also *acts* to reproduce, naturalize, as well as to contest, power relations. Dominant actors in society shape landscapes to reflect their ideals, concerns, and priorities, while subordinate voices are literally written out of the landscape. In other words, landscape was far from the passive written record of human activities, as in Sauer's morphological approach. Rather, the landscape itself is an active player in human affairs. Much as with the term nature (discussed in Part Four, 'Nature', see p. 201), landscapes encode and naturalize relations of domination and subordination, particularly with regard to women, racialized minorities, and conquered peoples. Moreover, their elements facilitate and perpetuate unequal power relations. Examples of critical approaches to landscape included in this part are Gillian Rose's feminist understanding of landscape as facilitating the objectification of women as well as land, Don Mitchell's Marxist-inspired reading of California's agricultural landscape as one involving an exploitative relationship with agricultural laborers, and W.J.T. Mitchell's emphasis on the importance of the gaze and its power relations in his discussion of landscape painting.

In the late 1980s and through the 1990s, with the *cultural turn* in human geography (discussed in the main Introduction), the attention of some cultural geographers turned increasingly to the issues of language and representation as these are worked through the landscape. Theory and methods developed in linguistics, literary criticism, and semiotics – fields that emphasize the construction in meaning through symbols, symbolic systems, and languages – were utilized by cultural geographers to read the landscape as a sort of text. Selections by Denis Cosgrove on the symbolic aspects of landscape, and James Duncan's close reading of the Kandyan landscape, provide examples of this approach. These geographers emphasize that, though one of the primary functions of landscape is to fix the meanings encoded within, as with other texts, landscape is an unstable medium and as such open to various interpretations and reworkings. Thus Duncan's work excerpted here interprets the meanings encoded in the Kandyan landscape within the historical context of the advent of a powerful kingship, analyzing how elements of the built environment worked to build support for a powerful ruler in the eyes of the subjects. There exists a mutually informative relationship between physical landscapes and their representation, whereby the representation constitutes much of the *meaning* of landscape. In the words of Stephen Daniels and Denis Cosgrove, "To understand a built landscape, say an eighteenth-century English park, it is usually necessary to understand written and verbal representations of it, not as 'illustrations', images standing outside it, but as constituent images of its meaning or meanings. And of course, every study of a landscape further transforms its meaning, depositing yet another layer of cultural representation."[3] Thus the study of *representations* of landscape is every bit as important to a complete understanding as physical immersion in *literal* landscapes, as with field-based exploration.

As with their more traditionally critical colleagues discussed above, these scholars also emphasized the power relations encoded in landscape. Typically, the dominant groups are those who are empowered to leave their mark – literally and figuratively – on society. But, given the fluidity of social and spatial systems alike, it is no surprise that unstable landscapes are also participants in contesting and reworking power relations. An important difference between those geographers who approach landscapes as crystallizing historical and material power relations, and those of a more textually inspired bent, is the focus of the latter on *representation*: both in the sense discussed above, of landscapes as depicted in art, literature, film, and photography; and in the sense that paradigmatic landscape representations have had a powerful role in shaping how we see and interpret the world around us. Indeed, the so-called "new" cultural geographers have at times been criticized for going too far in their emphasis on representation, and forgetting about the literal landscape or even denying the existence of a "real" landscape outside of representation. Though the differences between more traditionally critical cultural geographers adopting Marxist-inspired or feminist approaches, and the work of those focusing on language and representation, have been highlighted in some relatively recent cultural geography – for instance in Don Mitchell's piece included here – in fact, these approaches have a great deal in common, and over time the antagonism among various sub-groups has subsided considerably.

With the new millennium, some cultural geographers have consciously attempted to steer away from landscapes as representational. These cultural geographers instead pursue what is coming to be termed *non-representational* landscapes: in other words, landscapes that exist beyond humans and their dominant interpretive filters (particularly vision). These geographers suggest that landscapes may be understood as quite fluid constructs that are continually in the process of cohering and collapsing as we move through space. Thus rather than constituting fixed, static, material entities whose character is primarily visual, non-representational approaches see landscape as a sort of performance that is enacted much as is music or theater. This has broadened the focus on landscape beyond that "portion of the earth's surface that can be comprehended at a glance" to include the non-visual, non-human, and relational. British cultural geographers have been particularly active in this vein, with non-representational approaches to living, performing, and doing cultural geography. Michael Bull's selection here on personal stereo use is illustrative of this approach; in addition the Nature section of this *Reader* includes work – such as the selection by Cloke and Jones (see p. 232) – that is considered non-representational in approach.

Though the landscape idea can be traced through a series of historical currents, in reality scholarship is seldom packaged in such neat boxes. Rather, and as the essays included in this part illustrate well, there is often substantial overlap among periods and perspectives. For example, Gillian Rose's exploration of landscape draws on a critical feminist tradition as well as on the visual methods informed by art history. In addition, landscape is a term that is hardly exclusive to cultural geographers. Indeed, cultural geographers have long been in conversation with non-geographers, community activists, artists, and planners. Don Mitchell, whose discussion of California's landscape is included here, has been informed by his deep engagement with community activism. Finally, despite the various definitions of and approaches to landscape, the common thread that holds most cultural geographers who work on landscape together is the importance of field-based research. You will find that, for all of the extracts included in this section, their authors are deeply immersed in the primary materials they discuss, whether these are the paintings analyzed by W.J.T. Mitchell, the history and language of the Kandyan landscape examined by James Duncan, or Michael Bull's ethnographic interaction with users of personal stereos in British cities.

NOTES

1 J. Berger, *Ways of Seeing* (1972), p. 83.
2 Page 99 in Y. Tuan, "A view of geography," *Geographical Review* 81, 1 (1991): 99–107.
3 Page 1 in S. Daniels, and D. Cosgrove, "Introduction: Iconography and Landscape", in D. Cosgrove and S. Daniels (eds.) *The Iconography of Landscape* (1998), pp. 1–10.

"The Word Itself"

From *Discovering the Vernacular Landscape* (1984)

John Brinckerhoff Jackson

Editors' introduction

Though "J.B.", or just "Brinck", Jackson (1909–1996) was born in France, his work is renowned for embodying an essential American-ness. Throughout his long career as a scholar, writer, and artist, he contrasted what he called the *vernacular landscapes* built by everyday people meeting their needs through what was locally available with the *official landscapes* planned by governments. His work tended to glorify the rural elements symbolized by farms, country roads, and front yards, while displaying skepticism toward big cities, highways, and monumental construction. The contrast between rural-oriented *folk geographies* and urban-based *popular geographies* has shaped American and British cultural geography throughout much of the twentieth century.

These dichotomies resonate with contrasts in Jackson's own life. During his relatively privileged youth, Jackson spent time in Washington, D.C., Switzerland, New England, and New Mexico. He attended Harvard University in the early 1930s. He enlisted in the U.S. Army in 1940. His fluency in French and German made Jackson a natural for stationing in Europe during World War II. Jackson recounts browsing through the books of an occupied Norman chateau library, and spending a long winter in Germany's Huertgen Forest, reading and becoming intrigued with the work of notable European geographers such as Paul Vidal de la Blache (see p. 90). After his military service was concluded in 1946, Jackson went on to found the journal *Landscape* in 1951, where he remained as editor until 1968. Jackson also became a beloved teacher at the University of California, Berkeley. Despite his Harvard education and abiding love for French and Swiss cooking, he lived out his last years on an unassuming ranch in New Mexico with his dog.

Jackson viewed the landscape as a faithful record of man's presence, stating that "landscape is history made visible." His emphasis on reading the meaning of landscape from its material elements places him squarely among the "old" cultural geographers that Don Mitchell describes (see p. 159). In defining landscape as "a portion of the earth's surface that can be comprehended at a glance" Jackson reveals his Sauerian leanings (see p. 96). In fact, Carl Sauer and J.B. Jackson knew one another well, being colleagues at the University of California Berkeley's Department of Geography. Sauer contributed work to Jackson's journal, *Landscape*.

Contrary to many of the pieces in this section, Jackson's analysis did not spring from a critical concern with gender, labor or property ownership, or a theory-driven interest in landscape as representation or perspective, but rather from a sharp eye for detail and an abiding love of things rural, working-class, and everyday. Yet Jackson was hardly oblivious to societal power relations. His overt skepticism of the Establishment, big government, and growth for growth's sake put Jackson squarely in the corner of the "little guy".

In "The word itself", Jackson details the shifting meanings of the word "landscape." He traces the term's etymology through European languages, and its early use in agricultural traditions, administrative divisions, painting, and theater. Jackson expresses dismay at the increasingly metaphorical use of the term "landscape," as evidenced by the suffix "scape" being employed to mean any literal or figurative space. Rather, he calls for a more substantive inquiry into the shift of the term's usage away from the narrow circles of landscape painting and architecture. Though he declines to advance a new definition for the term "landscape," he does argue for a return to an understanding of landscape as a "concrete, three-dimensional, shared reality."

The relationship between nature and society as mediated through landscape painting is approached by Kenneth Clark in his canonical *Landscape into Art* (1949), in a fashion critiqued by many of those whose work is included in this section, including here by Jackson. Yi-Fu Tuan, a cultural geographer who has written many volumes including *Space and Place: The Perspective of Experience* (reprinted in 2001) evokes Jackson's insistence on firsthand experience of everyday events and places in constructing a meaningful landscape. "The word itself" prefigures the concerns of Kenneth Olwig, who wrote about landscape's role in theater, agriculture, and administration in Northern Europe, and how these early uses shifted in important ways toward statecraft, in *Landscape, Nature, and the Body Politic: From Britain's Renaissance to America's New World* (2002).

Key works from J.B. Jackson include *A Sense of Place, a Sense of Time* (1996), *Discovering the Vernacular Landscape* (1996), and *The Necessity for Ruins and other Topics* (1980). Many of the essays and sketches he contributed to the journal *Landscape*, some written under inventive pseudonyms, are gathered and reprinted in Helen Lefkowitz-Horwitz's *Landscape in Sight: Looking at America* (1997). One such essay, "Living outdoors with Mrs. Panther," is reprinted here (see p. 220). Lefkowitz-Horowitz's introduction to this collection is a detailed biography of Jackson's life and works, titled "J.B. Jackson and the discovery of the American landscape" (pp. ix–xxxi). Paul Starrs has also written in detail about Jackson's life and works in "Brinck Jackson in the realm of the everyday," in *Geographical Review* 88, 4 (1998): 492–506. Finally, Chris Wilson and Paul Groth edited a collection inspired by Jackson's approach, titled *Everyday America: Cultural Landscape Studies after J.B. Jackson* (2003).

Why is it, I wonder, that we have trouble agreeing on the meaning of *landscape*? The word is simple enough, and it refers to something which we think we understand; and yet to each of us it seems to mean something different.

What we need is a new definition. The one we find in most dictionaries is more than three hundred years old and was drawn up for artists. It tells us that a landscape is a "portion of land which the eye can comprehend at a glance." Actually, when it was first introduced (or reintroduced) into English it did not mean the view itself, it meant a *picture* of it, an artist's interpretation. It was his task to take the forms and colors and spaces in front of him – mountains, river, forest, fields, and so on – and compose them so that they made a work of art.

There is no need to tell in detail how the word gradually changed in meaning. First it meant a picture of a view; then the view itself. We went into the country and discovered beautiful views, always remembering the criteria of landscape beauty as established by critics and artists. Finally, on a modest scale, we undertook to make over a piece of ground so that it resembled a pastoral landscape in the shape of a garden or park. Just as the painter used his judgment as to what to include or omit in his composition, the landscape gardener (as he was known in the eighteenth century) took pains to produce a stylized "picturesque" landscape, leaving out the muddy roads, the plowed fields, the squalid villages of the real countryside and including certain agreeable natural features: brooks and groves of trees and smooth expanses of grass. The results were often extremely beautiful, but they were still pictures, though in three dimensions.

The reliance on the artist's point of view and his definition of landscape beauty persisted throughout the nineteenth century. [The nineteenth-century American landscape architect Frederick Law] Olmsted and his followers designed their parks and gardens in "painterly" terms. "Although three-dimensional composition in landscape materials differs from two-dimensional landscape painting, because a garden or park design contains a series of pictorial compositions," the *Encyclopaedia Britannica* . . . informs us, "nevertheless in each of these pictures we find the familiar basic principles of unity, of repetition, of sequence and balance, of harmony and contrast." But within the last half-century a revolution has taken place: landscape design and landscape painting have gone their separate ways. Landscape architects no longer turn to [painters] Poussin or Salvator Rosa or Gilpin for inspiration; they may not even have heard of their work. Knowledge of ecology and conservation and environmental psychology are now part of the landscape architect's professional background, and protecting and "managing" the natural environment are seen as more important than the designing of picturesque parks. Environmental designers, I have noticed, avoid the word *landscape* and prefer *land* or *terrain* or *environment* or even *space* when they have a specific site in mind. *Landscape* is used for suggesting the esthetic quality of the wider countryside.

As for painters, they have long since lost interest in producing conventional landscapes. Kenneth Clark, in his book *Landscape into Art*, comments on this fact. "The microscope and telescope have so greatly enlarged the range of our vision," he writes, "that the snug, sensible nature which we can see with our own eyes has ceased to satisfy our imaginations. We know that by our new standards of measurement the most extensive landscape is practically the same as the hole through which the burrowing ant escapes from our sight."

This does not strike me as a very satisfactory explanation of the demise of traditional landscape painting. More than a change in scale was responsible. Painters have learned to see the environment in a new and more subjective manner: as a different kind of experience. But that is not the point. The point is, the two disciplines which once had a monopoly on the word – landscape architecture and landscape painting – have ceased to use it the way they did a few decades ago, and it has now reverted, as it were, to the public domain.

What has happened to the word in the meantime? For one thing we are using it with much more freedom. We no longer bother with its literal meaning – which I will come to later – and we have coined a number of words similar to it: roadscape, townscape, cityscape, as if the syllable *scape* meant a space, which it does not; and we speak of the wilderness landscape, the lunar landscape, even of the landscape at the bottom of the ocean. Furthermore, the word is frequently used in critical writings as a kind of metaphor. Thus we find mention of the "landscape of a poet's images," the "landscape of dreams," or "landscape as antagonist" or "the landscape of thought," or, on quite a different level, the "political landscape of the NATO conference," the "patronage landscape." Our first reaction to these usages is that they are far-fetched and pretentious. Yet they remind us of an important truth: that we always need a word or phrase to indicate a kind of environment or setting which can give vividness to a thought or event or relationship; a background placing it in the world. In this sense, landscape serves the same useful purpose as do the words climate or atmosphere, used metaphorically. In fact, landscape when used as a painter's term often meant "all that part of a picture which is not of the body or argument" – like the stormy array of clouds in a battle scene or the glimpse of the Capitol in a presidential portrait.

In the eighteenth century, *landscape* indicated scenery in the theater and had the function of discreetly suggesting the location of the action or perhaps the time of day. As I have suggested elsewhere, there is no better indication of how our relation to the environment can change over the centuries than in the role of stage scenery. Three hundred years ago Corneille could write a five-act tragedy with a single indication of the setting: "The action takes place in the palace of the king." If we glance at the work of a modern playwright, we will probably find one detailed description of a scene after another, and the ultimate in this kind of landscape, I suppose, is the contemporary movie. Here the set does much more than merely identify the time and place and establish the mood. By means of shifts in lighting and sound and perspective, the set actually creates the players, identifies them, and tells them what to do: a good example of environmental determinism.

But these scenic devices and theater landscapes are mere imitations of real ones: easily understood by almost everyone, and shared. What I object to is the fallacy in the metaphorical use of the word. No one denies that as our thoughts become complex and abstract, we need metaphors to give them a degree of reality. No one denies that as we become uncertain of our status, we need more and more reinforcement from our environment. But we should not use the word *landscape* to describe our private world, our private microcosm and for a simple reason: a landscape is a concrete, three-dimensional, shared reality.

LANDS AND SHAPES

Landscape is a space on the surface of the earth; intuitively we know that it is a space with a degree of permanence, with its own distinct character, either topographical or cultural, and above all a space shared by a group of people; and when we go beyond the dictionary definition of landscape and examine the word itself, we find that our intuition is correct.

Landscape is a compound, and its components hark back to that ancient Indo-European idiom, brought out of Asia by migrating peoples thousands of years ago, that became the basis of almost all modern European languages – Latin and Celtic and Germanic and Slavic and Greek. The word was introduced into Britain sometime after the fifth century AD. by the Angles and Saxons and Jutes and Danes and other groups of Germanic speech. In addition to its Old English variations – *landskipe*, *landscaef*, and others – there is the German *Landschaft*, the Dutch *landscap*, as well as Danish and Swedish equivalents. They all come from the same roots, but they are not always used in the English sense. A German *Landschaft*, for instance, can sometimes be a small administrative unit, corresponding in size to our ward. I have the feeling that there is evolving a slight but noticeable difference between the way we Americans use the word and the way the English do. We tend to think that landscape can mean natural scenery only, whereas in England a landscape almost always contains a human element.

The equivalent word in Latin languages derives in almost every case from the Latin *pagus* – meaning a defined rural district. The French, in fact, have several words for landscape, each with shades of meaning: *terroir, pays, paysage, campagne*. In England the distinction was once made between two kinds

of landscape: woodland and champion – the latter deriving from the French *champagne*, meaning a countryside of fields. That first syllable, *land*, has had a varied career. By the time it reached England it signified *earth* and *soil* as well as a portion of the surface of the globe. But a much earlier Gothic meaning was *plowed field*. Grimm's monumental dictionary of the German language says that "*land* originally signified the plot of ground or the furrows in a field that were annually rotated" or redistributed. We can assume that in the Dark Ages the most common use of the word indicated any well-defined portion of the earth's surface. A small farm plot was a land, and so was a sovereign territory like England or Scotland; any area with recognized boundaries was a land. Despite almost two thousand years of reinterpretation by geographers and poets and ecologists, *land* in American law remains stubbornly true to that ancient meaning: "any *definite* site regarded as a portion of the earth's surface, and extending in both vertical directions as defined by law."

Perhaps because of this definition, farmers think of land not only in terms of soil and topography but in terms of spatial measurements, as a defined portion of a wider area. In the American South, and in England too, a "land" is a subdivision of a field, a broad row made by plowing or mowing, and horse-drawn mowers were once advertised as "making a land of so-and-so many feet." In Yorkshire the reapers of wheat take a "land" (generally six feet wide) and go down the length of the field. "A woman," says the *English Dialect Dictionary*, "would thus reap half an acre a day and "a man an acre." . . .

This is very confusing, and even more confusing is the fact that to this day in Scotland a *land* means a building divided into houses or flats. I confess that I find this particular use of the word hard to decipher, except that in Gaelic the word *lann* means a building divided into houses or flats. Finally, here is an example – if it can be called that – of *land* meaning both a fraction of a larger space and an enclosed space: infantrymen know that a land is an interval between the grooves of a rifle bore.

I need not press the point. As far back as we can trace the word, *land* meant a defined space, one with boundaries, though not necessarily one with fences or walls. The word has so many derivative meanings that it rivals in ambiguity the word *landscape.* Three centuries ago it was still being used in everyday speech to signify a fraction of plowed ground no larger than a quarter-acre, then to signify an expanse of village holdings, as in grassland or woodland, and then finally to signify England itself – the largest space any Englishman of those days could imagine; in short, a remarkably versatile word, but always implying a space defined by people, and one that could be described in legal terms.

This brings us to that second syllable: *scape.* It is essentially the same as shape, except that it once meant a composition of *similar* objects, as when we speak of a fellowship or a membership. The meaning is clearer in a related word: *sheaf* – a bundle or collection of similar stalks or plants. Old English, or Anglo-Saxon, seems to have contained several compound words using the second syllable *-scape* or its equivalent – to indicate collective aspects of the environment. It is much as if the words had been coined when people began to see the complexities of the man-made world. Thus *housescape* meant what we would now call a household, and a word of the same sort which we still use – *township* – once meant a collection of "tuns" or farmsteads.

Taken apart in this manner, *landscape* appears to be an easily understood word: a collection of lands. But both syllables once had several distinct, now forgotten meanings, and this should alert us to the fact that familiar monosyllables in English – house, town, land, field, home – can be very shifty despite their countrified sound. *Scape* is an instance. An English document of the tenth century mentions the destruction of what it called a "waterscape." What could that have been? We might logically suppose that it was the liquid equivalent of landscape, an ornamental arrangement, perhaps, of ponds and brooks and waterfalls, the creation of some Anglo-Saxon predecessor of Olmsted. But it was actually something entirely different. The waterscape in question was a system of pipes and drains and aqueducts serving a residence and a mill.

From this piece of information we can learn two things. First, that our Dark Age forebears possessed skills which we probably did not credit them with, and second, that the word *scape* could also indicate something like an organization or a system. And why not? If *housescape* meant the organization of the personnel of a house, if township eventually came to mean an administrative unit, then landscape could well have meant something like an organization, a system of rural farm spaces. At all events, it is clear that a thousand years ago the word had nothing to do with scenery or the depiction of scenery.

We pull up the word *landscape* by its Indo-European roots in an attempt to gain some insight into its basic meaning, and at first glance the results seem disappointing. Aside from the fact that, as originally used, the word dealt only with a small fraction of the rural environment, it seems to contain not a hint of the esthetic and emotional associations which the word still has for us. Little is to be gained by searching for some etymological line between our own rich landscape and the small cluster of plowed fields of more than a thousand years ago.

Nevertheless, the formula *landscape as a composition of man-made spaces on the land* is more significant than it first appears, for if it does not provide us with a definition, it throws a revealing light on the origin of the concept. For it says that a landscape is not a natural feature of the environment but a *synthetic* space, a man-made system of spaces superimposed on the face of the land, functioning and evolving not according to natural laws but to serve a community – for the collective character of the landscape is one thing that all generations and all points of view have agreed upon. A landscape is thus a space deliberately created to speed up or slow down the process of nature. . . . [I]t represents man taking upon himself the role of time.

A very successful undertaking on the whole, and the proof, paradoxically enough, is that many if not most of these synthetic organizations of space have been so well assimilated into the natural environment that they are indistinguishable and unrecognized for what they are. The reclamation of Holland, of the Fens in England, of large portions of the Po Valley are familiar examples of a topographical intervention producing new landscapes. Less well known are the synthetic landscapes produced simply by spatial reorganization. Historians are said to be blind to the spatial dimension of history, which is probably why we hear so little about the wholesale making of agricultural landscapes throughout seventeenth-century Europe.

It is not a coincidence that much of this landscape creation took place during a period when the greatest gardens and parks and the most magnificent of city complexes were being designed. A narrow and pedantic taxonomy has persuaded us that there is little or nothing in common between what used to be called civil engineering and garden or landscape architecture, but in fact from a historical perspective their more successful accomplishments are identical in result. The two professions may work for different patrons, but they both reorganize space for human needs, both produce works of art in the truest sense of the term. In the contemporary world, it is by recognizing this similarity of purpose that we will eventually formulate a new definition of *landscape*: a composition of man-made or man-modified spaces to serve as infrastructure or background for our collective existence; and if *background* seems inappropriately modest, we should remember that in our modern use of the word it means that which underscores not only our identity and presence but also our history.

It is not for me to attempt to elaborate on this new definition. My contribution would in any event be peripheral, for my interest in the topic is confined to trying to see how certain organizations of space can be identified with certain social and religious attitudes, especially here in America. This is not a new approach, for it has long been common among architectural and landscape architectural historians; and it leaves many important aspects of the contemporary landscape and contemporary city entirely unexplored. But it has the virtue of including the visual experience of our everyday world and of allowing me to remain loyal to that old-fashioned but surprisingly persistent definition of landscape: "A portion of the earth's surface that can be comprehended at a glance."

"California: The Beautiful and the Damned"

from *The Lie of the Land: Migrant Workers and the California Landscape* (1996)

Don Mitchell

Editors' introduction

Landscape is hard work. What is apparent to the eye, particularly for those landscapes that are known for their majestic qualities, often masks a significant amount of toil and exploitation behind the scenes. In "California: the beautiful and the damned," Don Mitchell explores the dark underside of the picturesque California landscape, rooted in the exploitative labor relations of California's agricultural industry. Mitchell uses the story of the Joad family from John Steinbeck's classic tale *The Grapes of Wrath* to introduce how landscape at once facilitates, and hides, the exploitative relations of production that shape it.

Mitchell distinguishes between what he calls "old" and "new" cultural geographers. On the one hand, the "old" cultural geographers, epitomized by Carl Sauer (see p. 96) and others who pursued a quasi-scientific, descriptive approach to landscape analysis, held the cultural landscape to be primarily a collection of its material elements that together displayed how human cultures had inscribed the physical world. On the other hand, "new" cultural geographers have emphasized landscape's representational and symbolic aspects. Denis Cosgrove and Stephen Daniels's edited collection titled *The Iconography of Landscape: Essays on the Symbolic Representation, Design and Use of Past Environments* (1988) is exemplary here, as is James Duncan's piece in this part, "From discourse to landscape: a kingly reading" (see p. 186). While the distinction between old and new is a bit over-simplified in this selection, Mitchell's concern is that the attention of critical cultural geographers to power relations and their contestation not be eclipsed by representational approaches that, taken to an extreme, render all inequality as "socially constructed": "[T]he abandonment of the material world as an object of study in order to focus exclusively on the politics of reading, language, and iconography represents a dangerous politics."

Mitchell demonstrates that though it is the laborer who does the work of physically shaping the landscape, under a capitalist system the laborer neither owns the land nor benefits in full measure from its products. Indeed, part of the "work" done by landscape is to hide this basic inequality both from the laborer, and the larger society of which he or she is a part. Thus attention to the very material power dynamics of capitalist society must be at the forefront of landscape analysis.

Mitchell presents us with a Marxist-influenced understanding of landscape. It is not surprising that Mitchell directs the "People's Geography Project" at Syracuse University in New York, where he is a Professor of Geography in the Maxwell School. The "People's Geography Project" aims to make critical geographic analysis of everyday life in US society accessible to ordinary people, by working directly with school-age youth and community members. Indeed, a number of critical cultural geographers strive to make their scholarship relevant to the communities – at diverse scales – of which they are a part. Mitchell's participatory, community-based action research is an excellent example of this commitment to social change.

Though Mitchell is a cultural geographer, he has expressed a healthy skepticism concerning the use of the term "culture" to stymie critical analysis. He has published this argument in a succinct piece titled "There is no such thing as culture: towards a reconceptualization of the idea of culture in geography," in *Transactions of the Institute of British Geographers* 19 (1995): 102–116.[1] Mitchell is also known for his work on issues of access to urban spaces, focusing on homelessness, protests, and public parks. This work can be found, among other places, in his book *The Right to the City: Social Justice and the Fight for Public Space* (2003).

Don Mitchell's work is inspired by that of other Marxist scholars who have examined at the social construction of urban space. In particular, the work of geographer David Harvey is frequently invoked by Mitchell and other critical geographers; see for example *The Urban Experience* (1989). The title of Mitchell's *The Right to the City* harkens to a term – *the right to the city* – originally employed by the French Marxist philosopher and urban sociologist Henri Lefebvre in his foundational work on cities; see Lefebvre's *Writings on Cities* (1995). Finally, Marxist literary critic Raymond Williams wrote critically and insightfully about the shifting images associated with British rural and urban landscapes in *The Country and the City* (1975) (see pp. 15 and 207).

NOTE

1 This piece attracted a great deal of critical response; see Peter Jackson, "The idea of culture: a response to Don Mitchell"; Denis Cosgrove, "Ideas and culture: a response to Don Mitchell"; James Duncan and Nancy Duncan, "Reconceptualizing the idea of culture in geography: a reply to Don Mitchell"; and Don Mitchell, "Explanation in cultural geography: a reply to Cosgrove, Jackson and the Duncans," an exchange appearing in 1996 in *Transactions of the Institute of British Geographers* 21, 3: 572–82.

After abandoning their farm in Oklahoma and joining the exodus across the desert to California, after seeing their family torn apart by the forced mobility of modernity, the Joads reach the top of Tehachapi Pass and gaze out over California's San Joaquin Valley. All of a sudden, the power and promise of the California landscape reveal themselves in a startling vista of color and pattern, instantly erasing the disillusionment that had accompanied the family all along their journey. In *The Grapes of Wrath*, John Steinbeck reduces this view to a list of characteristics, as if describing a painting: "The vineyard, the orchards, the great flat valley green and beautiful, the trees set in rows, and the farm houses." The Joads have at last reached the American apotheosis. "Pa sighed, 'I never knowed they was anything like her.' The peach trees and walnut groves, the dark green patches of oranges. And red roofs among the trees, and barns – rich barns . . ." The beauty and the wonder of the scene before them overwhelm the Joads: "And

then they stood, silent and awestruck, embarrassed before the great valley. The distance was thinned with the haze, and the land grew softer in the distance. A windmill flashed in the sun, and its turning blades were like heliograph, far away. Ruthie and Winfield looked at it, and Ruthie whispered, 'It's California.'"

This is a complex scene in which all the standard characteristics of landscape painting are present – a constructed, formal beauty, perspective represented by the thinning haze, a sense of proprietorship in the embarrassed gaze, a near complete absence of visible labor. It serves to represent California as dream, as spectacle, as a view to behold and perhaps to own. It shows California as a culmination of the American Dream – perhaps not a shining city on a hill, but a prosperous, rural, Jeffersonian, yeoman, countryside ideal. But Steinbeck is a wise writer, and he knows that to show this landscape as America, one must truly show it as an image, as a dream. All that has led the Joads to the top of this hill tells us that the perspective from there hides something, that the beauty of the place can only be an image constructed by hiding what makes it. The California Dream, the American Apotheosis that is California, can only be seen from afar. The dream itself is impossible without a certain haze that closes off perspective, that hides the struggle that goes into making landscape. Steinbeck thus has the Joads come down off the mountain, and he thereby opens up the view to show how it is constructed.

Hidden in the bushes along the creeks and irrigation ditches is the other side of the California Dream, a side that has been there all along, but that is easy to overlook from atop the hill: the invisible army of migrant workers who make the landscape of beauty and abundance that awed the Joads. Supposedly quiet, pliable, unorganized, they exist and reproduce themselves in landscapes of the most appalling deprivation. . . . Both indispensable as a class and completely expendable as individuals, it is quite clear that it is farmworkers who actively make what is visible as a landscape. The two landscapes – the broad, perspectival, aesthetic view from atop the hill, and the ugly, violent, dirty landscape of workers' everyday lives – are intimately linked.

. . . . [S]uch violence has in fact been *necessary*, not just to the construction of the American Dream, but to the workings of the economic system itself. Moreover, such violence has been mediated through the landscape itself: in all its complexity the landscape, as both more general view and more local, constructed environment, is an important player in the drama of capitalist development in California. Steinbeck had it right in two essential aspects. First, landscape must be understood as an interconnected relationship between view and production, between the aesthetic pleasure the Joads find on Tehachapi Pass and the reality of hobo jungles, Hoovervilles, labor camps, and skid rows they find down below. Second, in some very fundamental senses, it is the workers themselves who, in their struggle to make lives for themselves within and against a ruthless political economy, make the landscape – and it is they who are the glue that binds its two aspects.

For making these connections, for exposing the underbelly of the California Dream, Steinbeck saw his book banned and burned in Bakersfield (where

the Joads buried Granma after they came down off the hill), and he was roundly denounced by agribusiness and industrial concerns throughout the state as un-American. But these are precisely the connections that need to be explored if we are to understand both how the agricultural economy is continually reproduced despite its obvious unjustness and why the landscape looks the way it does. . . .

IMAGINING THE AMERICAN APOTHEOSIS

Members of the radical Industrial Workers of the World (IWW or Wobblies) in the first decades of the twentieth century liked to talk of "California, the Beautiful – and the Damned" precisely because they were continually forced to make the sorts of connections between landscape imagery and landscape reality that Steinbeck has the Joads make. Their phrase catches precisely the bloody irony of the California landscape. It is beautiful because it is damned. . . .

Most commentators on the California landscape, however, have been little interested in showing the connection between both sides of the landscape, and how these sides are dependent on each other. . . . Until recently, ignoring the blood and turmoil, the split heads and ruined lives, that allow the landscape to look as it does is an honored tradition in social-scientific, historical, and literary discourse on the California landscape. This discourse seems to imply, in the words of geographer James Parsons, that the landscape "is morally neutral." As neutral, both people and landscape may be transformed in their mutual encounters, but the moral content of the landscape remains fixed and imperturbable. It just is. The landscape is thus often understood in two interrelated ways; it is a relict rather than an ongoing construction; and it is organic, natural, and aesthetic. In the first case, the landscape is understood to be immutable at least in terms of the normal human life span. Rather than being molded directly by people, the landscape's immutability allows it to shape humans. In the second case, the landscape is something to be passed through and admired along the way.

[. . .]

. . . *Only* by erasing – or completely aestheticizing – the workers who made that way of life is its celebration possible. *Only* by seeing California purely as a landscape view can we see beauty without understanding the lives of the damned who are an integral part of that beauty. And that move, erasing the traces of work and struggle, is precisely what landscape imagery is all about.

[. . .]

Much of the work in geography on landscape-as-ideology and -representation has developed as a reaction, and thus in partial opposition, to the older landscape-as-morphology school. If a clear fault with the older landscape school in geography was its inability and unwillingness to adequately theorize its objects of study, to take them too much for granted, the primary fault of the newer landscape-as-ideology school has been to move too far away from the study of morphological production. . . . [M]uch of what gets called the "new cultural geography" has moved rather to a nearly exclusive study of (seemingly) disconnected images. And the most extreme forms of the "new cultural geography" have abandoned all interest in the world outside language and symbolic structure, outside representation. This has led to some theoretical positions that are hardly supportable. . . . [T]o see and understand a place *as* a landscape requires distance both from the place and from the labor that makes it. Landscape is thus not just ideology, it is *visual* ideology. "Landscape" is not so much experienced as *seen*. . . . [T]his ignores the fact that "landscape" is a relation of power, an *ideological* rendering of spatial relations. Landscapes transform the facts of place into a *controlled* representation, an imposition of order in which one (or perhaps a few) dominant ways of seeing are substituted for all ways of seeing and experiencing. . . . [T]he abandonment of the material world as an object of study in order to focus exclusively on the politics of reading, language, and iconography represents a dangerous politics.

[. . .]

Despite the shortcomings of both "new" and "old" cultural geographies, geographers should be able to build on the tools of both traditions to begin to explicate the nature of the connections between representations and materiality. "Landscapes" are produced in two ways. On one hand, there is labor – the work of shaping the land. This labor, of course, is organized not just locally but within a spatial division that cuts across myriad scales. On the other hand, the re-presentation of the products of labor *as* a landscape represents an attempt to naturalize and harmonize the appropriation of that labor and to impose a system of domination, consent, control, and order within the view. . . . Landscape is thus a unity of materiality and representation, constructed out of the contest between various social groups possessing varying amounts of social, economic, and political power. . . . There is, as "new cultural geographers" insist, an iconography of landscape, but that iconography must be constructed within the context of the form that landscape takes. Moreover, the morphological landscape is usually not produced in order to be read; rather it develops as both a product of and a means for guiding the social and spatial practices of production and reproduction in an area. . . . Landscapes, and landscape representations, are therefore very much a product of social struggle, whether engaged over form or over how to grasp and read that form. . . .

[. . .]

PRODUCING LANDSCAPE

For Steinbeck, the answers . . . start with the work of common people, and they proceed with an evaluation of how that work is organized. . . . The connection between local morphology and the representations through which those morphologies are ordered and sent into circulation is, simply, labor. This is neither far-fetched nor over-reductionist. . . . Under capitalism, however, the fruits of labor are alienated from those who make them. The shape of the land is the product of people, but it is not necessarily owned or controlled by them. While the appropriation process that structures landscape is certainly one of legal ownership of the land, it is also one of advancing and appropriating meanings in a way that tries to make the alienation of labor from the landscape seem at once natural and incontestable.

Landscape is thus quite a complex concept. A theory that seeks to explore the connections between landscape production and representation, it seems to me, must fulfill three basic requirements. . . . First, a theory of landscape representation and production must tell us what landscape is (how we understand "landscape" and what its relations are to the material world). Second, it must explain how "landscape" is produced as part of socially organized systems of production and reproduction (for landscapes in no way exist external to the functioning of society). Finally, landscape theory must specify the processes by which material landscapes and their representations function in society (which is a different question than the second).

What Landscape is

We have already spent a good deal of time discussing what landscape is, at least as far as geographers of differing perspectives have understood it. We can now go a step further. . . . Social struggle makes the landscape, and the landscape is always in a state of becoming: it is never *entirely* stable. Yet landscape is also a totality. That is, powerful social actors, as we have already suggested, are continually trying to represent the landscape as a fixed, total, and naturalized entity – as a unitary thing. Landscape is thus best understood as a kind of produced, lived, and represented space constructed out of the struggles, compromises, and temporarily settled relations of competing and cooperating social actors: it is both a thing (or suite of things), as Sauer would have it, and a social process, at once solidly material and ever changing. As a produced object, landscape is like a commodity in which evident, temporarily stable, form masks the facts of its production, and its status as a social relation. As both form and symbol, landscape is expected by those who attempt to define its meanings to speak unambiguously for itself.

. . . [T]he landscape is no simple reflection of the needs and desires of the domineering classes. Rather, it represents an important social contradiction within a unity of form: the reproduction of inequality and supposed powerlessness that is codified and naturalized in the landscape carries with it the seeds of revolt. Subordinate social actors can and do develop contestatory readings of landscape and can and do continually seek to impose a different, perhaps more equitable, suite of spaces and landscape forms in the place of the imposed architecture of social class. Yet if *productive* landscapes are to be maintained under the conditions of inequality that make capitalism possible, then revolt must be minimized, and threatening social groups must be neutralized. Powerful social actors

thus seek to build elements of landscape as a means of mediation, as a means of insuring neutralization – either by subverting subversion itself through cooptational blandishments (substituting better housing for the unjust social and economic conditions that make bad housing "acceptable," for example), or by seeking to reinforce the landscape as a representation of what is "natural."

The very *form* of the landscape incorporates the give-and-take of this process, now becoming solidified one way, now another, depending on the array of power at any given moment. The landscape itself, as a compromised unity, is therefore even more of a contradiction, held in an uneasy truce as ongoing and everyday social struggle forms and reforms it. In the midst of (as well as before and after) these struggles, social actors of all types continually seek to represent the landscape to themselves and to others in order to make sense of the struggles in which they are engaged. Landscape is thus a fragmentation of space *and* a totalization of it. People make sense of their fractured world by seeing it as a whole, by seeking to impose meanings and connections. But since social struggle is strategic, compromises often gain the appearance of stability: landscapes become naturalized; they become quite unremarkable.

How Landscape is Produced

[. . .]

An embodied set of processes that gains shape through struggle and contest (and is represented as self-evidently true), the landscape . . . is a social product that becomes naturalized through the very struggles engaged over its form and meaning. It is *enacted* in the process of struggle . . . [T]he shape of the landscape gives rise to new (social) realities. New battles are begun as soon as one shape is settled. The look of the land becomes at least partially determinate in the struggles that are to follow.

[. . .]

[Historian of science Bruno Latour] calls the resulting artifacts *quasi-objects* to suggest that they are not only material reality, but also an embodiment of the relations that went into building them. Similarly, a landscape may be seen as a *quasi-object*, embodying all the multifarious relations, struggles, arguments, representations, and conclusions that went into its making – even if it often appears as only an inert, or "natural," thing. As a *quasi-object* ... landscape structures social reality; it represents to us our relationships to the land and to social formations. But it does so in an obfuscatory way. Apart from knowing the struggles that went into its making (along with the struggles to which it gives rise), one cannot know a landscape except at some ideal level, which has the effect of reproducing, rather than analyzing or challenging, the relations of power that work to mask its function.

How Landscape Functions

Landscapes are produced and represented within specific historical conditions. While the development of a generalized theory of landscape production has been necessary, it is just as necessary to recall that agricultural California developed as (and remains) a part of an expanding capitalist economy. The promise of Eden that the Joads saw from Tehachapi Pass, and the reality of the Hoovervilles and unemployment that awaited them down below, were both part of a general process of capitalist development and of the local conditions within which that development occurred. Hence it is necessary to understand both how landscapes in general and the particular landscapes of rural California function within capitalism. We need now to examine the role that landscape plays in reproducing capitalist agriculture, and the social relations that allow the agricultural system to work. ... Landscape production ... is a moment in overall processes of uneven development. The "seesaw" motion of capital, restlessly searching out new opportunities for the production of surplus value, seeks differentials not just in land rent or locational advantage, but also in the ... needs and tendencies of labor. ...

... Labor qualities can be devalued or labor surpluses created (so that quantity substitutes for quality). The real wages of laborers can be driven down by lessening social needs, provided, of course, that labor is in no condition to press demands for its own improvement. The production of landscape, by objectifying, rationalizing, and naturalizing the social, has often had just this effect. If ... the landscape of capitalism is often a barrier to further accumulation and has to be creatively destroyed or otherwise overcome, then it is just as true that the landscape is often a great facilitator to capital (by helping to determine the "nature" of labor in a particular place). As this happens, workers must overcome not just conditions of inequality and the oppressive work of power, but the stabilized landscape itself. They must destabilize not just the relations of place, but the very ground upon and within which those relations are situated and structured.

Landscape is thus an uneasy truce between the needs and desires of the people who live in it, and the desire of powerful social actors to represent the world as they assume it should be. Landscape is always both a material form that results from and structures social interaction, and an ideological representation dripping with power. In both ways, landscapes are acts of contested discipline, channeling spatial practices into certain patterns and presenting to the world images of how the world (presumably) works and who it works for.

[...]

[N]o matter how beautiful, no matter how seemingly immutable, no matter how much it appears as a simulacrum, landscape is certainly not neutral. Nor are aesthetics ever free of the blood that goes into their making. In California, at least, there can be no beauty without a simultaneous damning.

"Imperial Landscape"
from *Landscape and Power* (1994)

W.J.T. Mitchell

Editors' introduction

In "Imperial landscape", University of Chicago Professor of English and Art History W.J.T. Mitchell questions three of the basic assumptions of traditional landscape studies: that landscape representation is a Western practice, a modern concept, and primarily pictorial. W.J.T. Mitchell's main point in "Imperial landscape" is that *imperialism* – the practice of exerting political and/or economic influence over foreign territories – is closely associated with the practice of depicting landscapes in particular ways. By approaching landscape thusly, the three assumptions of traditional landscape studies that Mitchell identifies – that it is a Western tradition, that it is modern, and it is a faithful reflection of reality – are called into question. Non-Western imperial powers also have strong landscape traditions, people throughout recorded history have probably enjoyed the beauty of their natural surroundings, and landscape painting does not simply reflect what is "out there" in nature. Rather, landscape is, from the start, a stylized form of communication that both encodes and conceals power relations in the societies from which they arise.

Though many scholars of landscape are themselves British, and focus their studies on British landscape traditions, European landscape painting did not originate in Britain, having its roots instead in modern-day Italy and Holland. Denis Cosgrove explored the origins of European landscape painting as located in emerging capitalist property relations, in his *Social Formation and the Symbolic Landscape* (reissued in 1998) (see also p. 176). Though deeply indebted to Cosgrove's arguments, W.J.T. Mitchell expands on his ideas to claim that most, if not all, societies that built empires developed traditions of landscape paintings. Thus Chinese landscape painting is as bound up with the rise and fall of China's Asian empire as English landscape painting is bound up with the rise and fall of Britain's own empire. Indeed, imperialism as a specific spatial formation of power is a focus of much critical cultural geography, and its study is not limited to landscape analysis.

Much of the scholarship that W.J.T. Mitchell draws upon to make his argument in "Imperial landscape" focuses on landscape traditions developed during the British imperial period. Important examples include John Barrell's *The Dark Side of Landscape: The Rural Poor in English Painting, 1730–1840* (1980) and Ann Bermingham's *Landscape and Ideology: The English Rustic Tradition, 1740–1860* (1986). These scholars emphasize the relationship between Britain's evolving industrial capitalism, changing labor relations and landholding patterns, and the artistic representation of the English landscape. By the early 1920s, Britain's far-flung empire reached far beyond its national borders, ruling one out of every four human beings on the planet, and covering nearly the same amount – 25 per cent – of the earth's territory. The phrase "the empire on which the sun never sets" was probably literally true, as at any given point in a twenty-four hour time span at least one of Britain's colonies was illuminated. Because of its geographic reach and longevity, Britain's imperial presence was bound to have a deep influence on the cultural fabric of many places. This is certainly the case with landscape painting as an artistic genre, and as a way of seeing more broadly understood.

Mitchell is the longtime editor, since 1978, of the internationally renowned journal *Critical Inquiry*. He has published extensively on cultural politics, political culture, and art history. Other publications by Mitchell include *What do Pictures Want? The Lives and Loves of Images* (2005), *The Last Dinosaur Book: The Life and Times of a Cultural Icon* (1998), and *Picture Theory: Essays on Verbal and Visual Representation* (1994).

THESES ON LANDSCAPE

1 Landscape is not a genre of art but a medium.
2 Landscape is a medium of exchange between the human and the natural, the self and the other. As such, it is like money: good for nothing in itself, but expressive of a potentially limitless reserve of value.
3 Like money, landscape is a social hieroglyph that conceals the actual basis of its value. It does so by naturalizing its conventions and conventionalizing its nature.
4 Landscape is a natural scene mediated by culture. It is both a represented and presented space, both a signifier and a signified, both a frame and what a frame contains, both a real place and its simulacrum, both a package and the commodity inside the package.
5 Landscape is a medium found in all cultures.
6 Landscape is a particular historical formation associated with European imperialism.
7 Theses 5 and 6 do not contradict one another.
8 Landscape is an exhausted medium, no longer viable as a mode of artistic expression. Like life, landscape is boring; we must not say so.
9 The landscape referred to in Thesis 8 is the same as that of Thesis 6.

[. . .]

Recent criticism of landscape aesthetics – a field that goes well beyond the history of painting to include poetry, fiction, travel literature, and landscape gardening – can largely be understood as an articulation of a loss of innocence. . . . "We" now know that there is no simple, unproblematic "we," corresponding to a universal human spirit seeking harmony, or even a European "rising" and "developing" since the Middle Ages. What we know now is what critics like John Barrell have

shown us, that there is a "dark side of the landscape" and that this dark side is not merely mythic, not merely a feature of the regressive, instinctual drives associated with nonhuman "nature" but a moral, ideological, and political darkness that covers itself with . . . innocent idealism. . . . Contemporary discussions of landscape are likely to be contentious and polemical. . . . They are likely to place the aesthetic idealization of landscape alongside "vulgar" economic and material considerations. . . .

I might as well say at the outset that I am mainly in sympathy with this darker, skeptical reading of landscape aesthetics and that this essay is an attempt to contribute further to this reading. . . . My aim in this essay, however, is not primarily to add to the stock of hard facts about landscape but to take a harder look at the framework in which facts about landscape are constituted – the way, in particular, that the nature, history, and semiotic or aesthetic character of landscape is constructed in both its idealist and skeptical interpretations.

As it happens, there is a good deal of common ground in these constructions, an underlying agreement on at least three major "facts" about landscape: (1) that it is, in its "pure" form, a western European and modern phenomenon; (2) that it emerges in the seventeenth century and reaches its peak in the nineteenth century; (3) that it is originally and centrally constituted as a genre of painting associated with a new way of seeing. These assumptions are generally accepted by all the parties in contemporary discussions of English landscape. . . .

The agreement on these three basic "facts" – let us call them the "Western-ness" of landscape, its modernity, and its visual/pictorial essence – may well be a sign of just how well founded they are. If critics of radically different persuasions take these things for granted, differing mainly in their explanations of them, then there is a strong presumption that they are true. . . .

[. . .]

There are two problems with these fundamental assumptions about the aesthetics of landscape: first, they are highly questionable; second, they are almost never brought into question, and the very ambiguity of the word "landscape" as denoting a place or a painting encourages this failure to ask questions. But the blurring of the distinction between the viewing and the representation of landscape seems, on the face of it, deeply problematic. Are we really to believe . . . that the appreciation of natural beauty begins only with the invention of landscape painting? Certainly the testimony of poets from Hesiod to Homer to Dante suggests that human beings did not . . . acquire a "new sense" sometime after the Middle Ages that made them "utterly different from all the great races that have existed before." Even the more restricted claim that landscape *painting* (as distinct from perception) has a uniquely Western and modern identity seems fraught with problems. The historical claim that landscape is a postmedieval development runs counter to the evidence . . . that Hellenistic and Roman painters evolved a school of landscape painting. And the geographic claim that landscape is a uniquely western European art falls to pieces in the face of the overwhelming richness, complexity, and antiquity of Chinese landscape painting. The Chinese tradition has a double importance in this context. Not only does it subvert any claims for the uniquely modern or Western lineage of landscape, the fact is that Chinese landscape played a crucial role in the elaboration of English landscape aesthetics in the eighteenth century, so much so that *le jardin anglo-chinois* became a common European label for the English garden.

The intrusion of Chinese traditions into the landscape discourse I have been describing is worth pondering further, for it raises fundamental questions about the Eurocentric bias of that discourse and its myths of origin. Two facts about Chinese landscape bear special emphasis: one is that it flourished most notably at the twilight of Chinese imperial power and began to decline in the eighteenth century as China became itself the object of English fascination and appropriation at the moment when England was beginning to experience itself as an imperial power. Is it possible that landscape, understood as the historical invention of a new visual/pictorial medium, is integrally connected with imperialism? Certainly the roll call of major originating movements in landscape painting – China, Japan, Rome, seventeenth-century Holland and France, eighteenth- and nineteenth-century Britain – makes the question hard to avoid. At a minimum we need to explore the possibility that the representation of landscape is not only a matter of internal politics and national or class ideology but also an international, global phenomenon, intimately bound up with the discourses of imperialism.

This hypothesis needs to be accompanied by a whole set of stipulations and qualifications. Imperialism is clearly not a simple, single, or homogeneous phenomenon but the name of a complex system of cultural, political, and economic expansion and domination that varies with the specificity of places, peoples, and historical moments. It is not a one-way phenomenon but a complicated process of exchange, mutual transformation, and ambivalence. It is a process conducted simultaneously at concrete levels of violence, expropriation, collaboration, and coercion, and at a variety of symbolic or representational levels whose relation to the concrete is rarely mimetic or transparent. Landscape, understood as concept or representational practice, does not usually declare its relation to imperialism in any direct way; it is not to be understood, in my view, as a mere tool of nefarious imperial designs, nor as uniquely caused by imperialism. Dutch landscape, for instance, which is often credited with being the European origin of both the discourse and the pictorial practice of landscape, must be seen at least in part as an antiimperial and nationalistic cultural gesture; the transformation of the Netherlands from a rebellious colony into a maritime empire in the second half of the seventeenth century indicates at the very least how quickly and drastically the political environment of a cultural practice can change, and it suggests the possibility of hybrid landscape formations that might be characterized simultaneously as imperial and anticolonial.

Landscape might be seen more profitably as something like the "dreamwork" of imperialism, unfolding its own movement in time and space from a central point of origin and folding back on itself to disclose both utopian fantasies of the perfected imperial prospect and fractured images of unresolved ambivalence and unsuppressed resistance. In short, the posing of a relation between imperialism and landscape is not offered here as a deductive model that can settle the meaning of either term, but as a provocation to an inquiry.

THE "RISE" OF LANDSCAPE

[. . .]

When does landscape first begin to be perceived? Everything depends, of course, on how one defines the proper or pure experience of landscape. Thus, Kenneth Clark dismisses the landscape paintings that adorned Roman villas as "backgrounds" and "digressions," not representations of natural scenery in and for itself. Landscape perception proper is possible only to "modern consciousness," a phenomenon that can be dated with some precision. . . . [However] long before Petrarch and long before St. Augustine, people had succumbed to the temptation of looking at natural wonders "for their own sake."

Numerous other originary moments in the viewing of landscape might be adduced, from Jehovah's looking upon his creation and finding it good to Michelet's French peasants running out of doors to perceive the beauties of their natural environment for the first time. The account of landscape contemplation that probably had the strongest influence on English painting, gardening, and poetry in the eighteenth century was Milton's description of Paradise, a viewing, we should recall, that is framed by the consciousness of Satan, who "only used for prospect" his vantage point on the Tree of Life. The "dark side" of landscape that Marxist historians have uncovered is anticipated in the myths of landscape by a recurrent sense of ambivalence. Petrarch fears the landscape as secular, sensuous temptation; Michelet treats it as a momentary revelation of beauty and freedom bracketed by blindness and slavery; Milton presents it as the voyeuristic object for a gaze that wavers between aesthetic delight and malicious intent. . . .

This ambivalence, moreover, is temporalized and narrativized. It is almost as if there is something built into the grammar and logic of the landscape concept that requires the elaboration of a pseudohistory, complete with a prehistory, an originating moment that issues in progressive historical development, and (often) a final decline and fall. The analogy with typical narratives of the rise and fall of empires becomes even more striking when we notice that the rise and fall of landscape painting is typically represented as a threefold process of emancipation, naturalization, and unification. . . . Landscape painting is routinely described as emancipating itself from subordinate roles like literary illustration, religious edification, and decoration to achieve an independent status in which nature is seen for its own sake. Chinese landscape is prehistoric, prior to the emergence of nature enjoyed for its own sake. . . .

 . . . . [T]he emancipation of landscape as a genre of painting is also a *naturalization*, a freeing of nature from the bonds of convention. Formerly, nature was represented in highly conventionalized or symbolic forms; latterly, it appears in naturalistic transcripts of nature, the product of a long evolution in which the vocabulary of rendering natural scenery gained shape side by side with the power to see nature as scenery. This evolution from subordination to emancipation, convention to nature has as its ultimate goal the *unification* of nature in the perception and representation of landscape. . . .

Each of these transitions or developments in the articulation of landscape presents itself as a historical shift, whether abrupt or gradual, from ancient to modern, from classical to Romantic, from Christian to secular. Thus, the history of landscape painting is often described as a quest, not just for pure, transparent representation of nature, but as a quest for pure painting, freed of literary concerns and representation. . . . One end to the story of landscape is thus abstract painting. At the other extreme, the history of landscape painting may be described as a movement from "conventional formulas" to "naturalistic transcripts of nature." Both stories are grail-quests for purity. On the one hand, the goal is nonrepresentational painting, freed of reference, language, and subject matter; on the other hand, pure hyperrepresentational painting, a superlikeness that produces natural representations of nature.

As a pseudohistorical myth, then, the discourse of landscape is a crucial means for enlisting "Nature" in the legitimation of modernity, the claim that "we moderns" are somehow different from and essentially superior to everything that preceded us, free of superstition and convention, masters of a unified, natural language epitomized by landscape painting. . . .

THE SACRED SILENT LANGUAGE

The charming landscape which I saw this morning, is indubitably made up of some twenty or thirty farms. Miller owns this field, Locke that, and Manning the woodland beyond. But none of them owns the landscape. There is a property in the horizon which no man has but he whose eye can integrate all the parts, that is, the poet. This is the best part of all these men's farms, yet to this their land-deeds give them no title. (Emerson, *Nature*, 1836)

I have been assuming . . . that landscape is best understood as a medium of cultural expression, not a genre of painting or fine art. It is now time to explain exactly what this means. There certainly is a genre of painting known as landscape, defined very loosely by a certain emphasis on natural objects as subject matter. What we tend to forget, however, is that this subject matter is not simply raw material to be represented in paint but is always already a symbolic form in its own right. The familiar categories that divide the genre of landscape painting into subgenres – notions such as the Ideal, the Heroic, the Pastoral, the Beautiful, the Sublime, and the Picturesque – are all distinctions based, not in ways of putting paint on canvas, but in the kinds of objects and visual spaces that may be represented by paint.

Landscape *painting* is best understood, then, not as the uniquely central medium that gives us access to ways of seeing landscape, but as a representation of something that is already a representation in its own right. Landscape may be represented by painting, drawing, or engraving; by photography, film, and theatrical scenery; by writing, speech, and presumably even music and other sound images. Before all these secondary representations, however, landscape is itself a physical and multisensory medium (earth, stone, vegetation, water, sky, sound and silence, light and darkness, etc.) in which cultural meanings and values are encoded, whether they are *put* there by the physical transformation of a place in landscape gardening and architecture, or *found* in a place formed, as we say, "by nature." The simplest way to summarize this point is to note that it makes Kenneth Clark's title, *Landscape into Art* quite redundant: landscape is already artifice in the moment of its beholding, long before it becomes the subject of pictorial representation.

Landscape is a medium in the fullest sense of the word. It is a material means (to borrow Aristotle's terminology) like language or paint, embedded in a tradition of cultural signification and communication, a body of symbolic forms capable of being invoked and reshaped to express meanings and values. As a medium for expressing value, it has a semiotic structure rather like that of money, functioning as a special sort of commodity that plays a unique symbolic role in the system of exchange-value. Like money, landscape is good for nothing as a use-value, while serving as a theoretically limitless symbol of value at some other level. At the most basic, vulgar level, the value of landscape expresses itself in a specific price: the added cost of a beautiful view in real estate value; the price of a plane ticket to the Rockies, Hawaii, the Alps, or New Zealand. Landscape is a marketable commodity to be presented and re-presented in packaged tours, an object to be purchased, consumed, and even brought home in the form of souvenirs such as postcards and photo albums. In its double role as commodity and potent cultural symbol, landscape is the object of fetishistic practices involving the limitless repetition of identical photographs taken on identical spots by tourists with interchangeable emotions.

As a fetishized commodity, landscape is what Marx called a "social hieroglyph," an emblem of the social relations it conceals. At the same time that it commands a specific price, landscape represents itself as beyond price, a source of pure, inexhaustible spiritual value. "Landscape," says Emerson, "has no owner," and the pure viewing of landscape for itself is spoiled by economic considerations: "you cannot *freely* admire a noble landscape, if laborers are digging in the field hard by.". . . . "Landscape" must represent itself, then, as the antithesis of "land," as an "ideal estate" quite independent of "real estate," as a poetic property, in Emerson's phrase, rather than a material one. The land, real property, contains a limited quantity of wealth in minerals, vegetation, water, and dwelling space. Dig out all the gold in a mountainside, and its wealth is exhausted. But how many photographs, postcards, paintings, and awestruck sightings of the Grand Canyon will it take to exhaust its value as landscape? Could we fill up Grand Canyon with its representations? How do we exhaust the value of a medium like landscape?

Landscape is a medium not only for expressing value but also for expressing meaning, for communication between persons – most radically, for communication between the Human and the non-Human. Landscape mediates the cultural and the natural, or Man and Nature, as eighteenth-century theorists would say. It is not only a natural scene, and not just a representation of a natural scene, but a *natural* representation of a natural scene, a trace or icon of nature *in* nature itself, as if nature were imprinting and encoding its essential structures on our perceptual apparatus. Perhaps this is why we place a special value on landscapes with lakes or reflecting pools. The reflection exhibits Nature representing itself to itself, displaying an identity of the Real and the Imaginary that certifies the reality of our own images.

The desire for this certificate of the Real is clearest in the rhetoric of scientific, topographical illustration, with its craving for pure objectivity and transparency and the suppression of aesthetic signs of style or genre. But even the most highly formulaic, conventional, and stylized landscapes tend to represent themselves as true to some sort of nature, to universal structures of Ideal nature, or to codes that are wired in to the visual cortex and to deeply instinctual roots of visual pleasure associated with scopophilia, voyeurism, and the desire to see without being seen.

[. . .]

. . . We say "landscape is nature, not convention" in the same way we say "landscape is ideal, not real estate," and for the same reason – to erase the signs of our own constructive activity in the formation of landscape as meaning or value; to produce an art that conceals its own artifice, to imagine a representation that breaks through representation into the realm of the nonhuman. That is how we manage to call landscape the "natural medium" in the same breath that we admit that it is nothing but a bag of tricks, a bunch of conventions and stereotypes. Histories of landscape, as we have seen, continually present it as breaking with convention, with language and textuality, for a natural view of nature, just as they present landscape as transcending property and labor. . . .

These semiotic features of landscape, and the historical narratives they generate, are tailor-made for the discourse of imperialism, which conceives itself precisely (and simultaneously) as an expansion of landscape understood as an inevitable, progressive development in history, an expansion of culture and civilization into a "natural" space in a progress that is itself narrated as "natural." Empires move outward in space as a way of moving forward in time; the prospect that opens up is not just a spatial scene but a projected future of development and exploitation. And this movement is not confined to the external, foreign fields toward which empire directs itself; it is typically accompanied by a renewed interest in the representation of the home landscape, the "nature" of the imperial center. The development of English landscape conventions in the eighteenth century illustrates this double movement perfectly. At the same time as English art and taste are moving outward to import new landscape conventions from Europe and China, it moves inward toward a reshaping and re-presentation of the native land. The Enclosure movement and the accompanying dispossession of the English peasantry are an internal colonization of the home country, its transformation from what Blake called "a green & pleasant land" into a landscape, an emblem of national and imperial identity. . . .

[. . .]

"Looking at Landscape: The Uneasy Pleasures of Power"

from *Feminism and Geography: The Limits of Geographical Knowledge* (1993)

Gillian Rose

Editors' introduction

Feminism and Geography: The Limits of Geographical Knowledge was a landmark publication because of its encompassing critique of the masculinism of human geography. Like the other pieces in Part Three, Gillian Rose notes the predominantly visual quality of landscape. Rose argues that landscape is not just imbued with the power relations of *labor* in a capitalist society, as Don Mitchell has argued (see p. 159), but that it is also imbued with the power relations of *gender*. It is not accidental that landscapes are so often depicted as feminine forms, argues Rose. Rather, it is at the heart of geography as an enterprise that the domination of knowledge about landscapes is, at the same time, a domination of the feminine Other that haunts, and bedevils, cultural geography. Looking at landscape is a gendered act of power on the part of male geographers, one which is part and parcel of the masculine gaze that bestows ownership and control on that which is gazed upon. This uneasy relationship with the feminine is an unacknowledged yet fundamental aspect of much of cultural geography.

Gillian Rose is by no means alone amongst those who have "looked at landscape" through feminist eyes. Annette Kolodny's classic *The Lay of the Land: Metaphor as Experience and History in American Life and Letters* (1975) explores the long-standing associations of the landscape of the American West and the feminine. Art historian Griselda Pollock examines representations of women in landscape painting in *Vision and Difference* (2003).

Gillian Rose is a Professor of Cultural Geography at the Open University in London. Rose's primary interest as a cultural geographer is in the field of visual culture. She draws on a long tradition, mentioned throughout this section on landscape, of visual studies. In this selection, she utilizes John Berger's brief but highly readable and revealing work on power and visual representation, *Ways of Seeing* (1972). Rose's research on family photographs has revealed a complex dynamics of childhood, parenting, and the domestic; see "'Everyone's cuddled up and it just looks really nice': the emotional geography of some mums and their family photos," in *Social and Cultural Geography* 5 (2004): 549–564. Rose's book, *Visual Methodologies: An Introduction to Interpreting Visual Materials*, second edition (2007) provides a resource for using visual materials, such as film, photographs, and painting, in scholarly analysis. The method of using photographs in cultural geographic scholarship is further explored in this volume in the selection by Alan Latham (see p. 68).

Though it is not tremendously apparent in the selection that appears here, Rose's work in *Feminism and Geography* and beyond, has been deeply influenced by feminist psychoanalytic theorists. Feminists in a variety of disciplines have critically examined some of the key ideas of Sigmund Freud and Jacques Lacan, particularly about sex, gender, how we form identities, and the roles played by family and society in the shaping of the self. A classic text is Simone de Beauvoir's *The Second Sex* (1952); more challenging explorations of these topics are provided by Luce Irigaray's *This Sex which is Not One* (trans. Catherine Porter, 1985); and Judith Butler's *Gender Trouble: Feminism and the Subversion of Identity* (1999). Liz Bondi, who is a feminist geographer, has written on psychiatric counseling from a spatial perspective; see "Making connections and thinking through emotions: between geography and psychotherapy" in *Transactions of the Institute of British Geographers* 30, 4 (2005): 433–448.

As is apparent from the titles of some of the publications previously cited in this introduction, the geography of emotions is a area that is closely related to other topics discussed here. Examples from this emerging field include Kay Anderson and Susan Smith's editorial titled "Emotional geographies" in *Transactions of the Institute of British Geographers* 26 (2001): 7–10; Fernando Bosco, "The Madres of the Plaza de Mayo and three decades of human rights activism: embeddedness, emotions, and social movements" in the *Annals of the Association of American Geographers* 96, 2 (2004): 342–365; and Joyce Davidson, Liz Bondi, and Mick Smith (eds.) *Emotional Geographies* (2005).

Landscape is a central term in geographical studies because it refers to one of the discipline's most enduring interests: the relation between the natural environment and human society, or, to rephrase, between, Nature and Culture. Landscape is a term especially associated with cultural geography, and although "literally [the landscape] is the scene within the range of the observer's vision," its conceptualization has changed through history. By the interwar period, for its leading exponents, such as Otto Schluter in Germany, Jean Brunhes in France and Carl Sauer in the USA, the term "landscape" was increasingly interpreted as a formulation of the dynamic relations between a society or culture and its environment. . . . The interpretation of these processes depended in particular on fieldwork, and fieldwork is all about looking. . . . Just as fieldwork is central not only to cultural geography but also to the discipline as a whole, however, so too the visual is central to claims to geographical knowledge: a president of the Association of American Geographers [John Fraser Hart] has argued that "good regional geography, and I suspect most good geography of any stripe, begins by looking." The absence of knowledge, which is the condition for continuing to seek to know, is often metaphorically indicated in geographical discourse by an absence of insight, by mystery or by myopia; conversely, the desire for full knowledge is indicated by transparency, visibility and perception. Seeing and knowing are often conflated.

More recent work on landscape has begun to question the visuality of traditional cultural geography, however, as part of a wider critique of the latter's neglect of the power relations within which landscapes are embedded. Some cultural geographers suggest that the discipline's visuality is not simple observation but, rather, is a sophisticated ideological device that enacts systematic erasures. They have begun to problematize the term "landscape" as a reference to relations between society and the environment through contextual studies of the concept as it emerged and developed historically, and they have argued that it refers not only to the relationships between different objects caught in the fieldworker's gaze, but that it also implies a specific way of looking. They interpret landscape not as a material consequence of interactions between a society and an environment, observable in the field by the more-or-less objective gaze of the geographer, but rather as a gaze which itself helps to make sense of a particular relationship between society and land. They have stressed the importance of the look to the idea of landscape and have argued that landscape is a way of seeing which we learn; as a consequence, they argue that the gaze of the fieldworker is part of the problematic, not a tool of analysis. Indeed, they name this gaze at landscape a "visual ideology," because it uncritically shows only the relationship of the powerful to their environment. . . .

Questions of gender and sexuality have not been raised by this . . . work, however. This seems an important omission.... A consequence has been that, historically, in geographical discourse, landscapes are often seen in terms of the female body and the beauty of Nature.... This feminization of what is looked at does matter, because it is one half of ... the dominant visual regime of white heterosexual masculinism.... This particular masculine position is to look actively, possessively, sexually and pleasurably, at women as objects.... [T]he feminization of landscape in geography allows many of the arguments made about the masculinity of the gaze at the nude to work in the context of geography's landscape too, particularly in the context of geography's pleasure in landscape.... [G]eography's look at landscape draws on not only a complex discursive transcoding between Woman and Nature ... on a specific masculine way of seeing: the men acting in the context of geography are the fieldworkers, and the Woman appearing is the landscape. This compelling figure of Woman both haunts a masculinist spectator of landscape and constitutes him.

The pleasures that geographers feel when they look at landscape are not innocent, then, but nor are they simple. The pleasure of the masculine gaze at beautiful Nature is tempered by geography's scientism.... The gaze of the scientist has been described ... as part of masculinist rationality, and to admit an emotional response to Nature would destroy the anonymity on which that kind of scientific objectivity depends.... [W]hen Descartes discovered that the eye was a passive lens, in order to retain an understanding of the accession to knowledge as active he was forced to separate the seeing intellect from the seeing eye. This was one aspect of the split between the mind and the body so much associated with his work, and it rendered the objects of the gaze separate from the looking subject.... Such disembodiment separated knowing from desire, and protected men's scientific neutrality from Woman's wild nature.... [G]eographers are constituted as sensitive as well as objective scientists in their approach to Nature and landscape. This contradiction produces a conflict between desire and fear in visual forms. It creates a tension between distance from the object of the gaze and merger with it, which is at work both in the conflict between knowledge and pleasure – a conflict between "a highly individual response" and "a disinterested search for evidence"

– and also within the pleasured gaze. These complex contradictions between and within (social-) scientific objectivity and aesthetic sensitivity disrupt cultural geography's claim to know landscape.... I argue that the structure of aesthetic masculinity which studies landscape is inherently unstable, subverted by its own desire for the pleasures that it fears.

[...]

LANDSCAPE AS VISUAL IDEOLOGY

[...]

.... Merchants often commissioned paintings of their newly acquired properties, and in these canvases, through perspective, they enjoyed perspectival as well as material control over their land ... It is argued then ... that landscape is meaningful as a "way of seeing" bound into class relations....

This is an extremely important critique of the ideologies implicit in graphical discourse. Its strengths are evident in the interpretation, shared by cultural geographers, of the mid-eighteenth-century double portrait of Mr and Mrs Andrews, by the English artist Thomas Gainsborough. In their discussions of this image, geographers concur that pleasure in the right-hand side of the canvas – those intense green fields, the heaviness of the sheaves of corn, the English sky threatening rain – is made problematic by the two figures on the left, Mr and Mrs Andrews. [John] Berger, whose discussion of this painting geographers follow, insists that the fact that this couple owned the fields and trees about them is central to its creation and therefore to its meaning.... Their ownership of land is celebrated in the substantiality of the oil paints used to represent it, and in the vista opening up beyond them, which echoes in visual form the freedom to move over property which only landowners could enjoy. The absence in the painting's content of the people who work the fields, and the absence in its form of the signs of its production by an artist working for a fee on a commission, can be used to [demonstrate that] landscape painting is a form of visual ideology: it denies the social relations of waged labour under capitalism. "Mr and Mrs Andrews," then, is an image on which geographers are agreed: it is a symptom of the capitalist

"Mr and Mrs Andrews," *c.* 1750, by Thomas Gainsborough, English landscape painter (1727–1788) *Source* Reproduced by courtesy of the National Gallery, London

property relations that legitimate and are sanctioned by the visual sweep of a landscape prospect.

However, the painting of Mr and Mrs Andrews can also be read in other ways. In particular, it is possible to prise the couple – "the landowners" – apart, and to differentiate between them. Although both figures are relaxed and share the sense of partnership so often found in eighteenth-century portraits of husband and wife, their unity is not entire: they are given rather different relationships to the land around them. Mr Andrews stands, gun on arm, ready to leave his pose and go shooting again; his hunting dog is at his feet, already urging him away. Meanwhile, Mrs Andrews sits impassively, rooted to her seat with its wrought iron branches and tendrils, her upright stance echoing that of the tree directly behind her. If Mr Andrews seems at any moment able to stride off into the vista, Mrs Andrews looks planted to the spot. This helps me to remember that, *contra* Berger, these two people are *not* both landowners – only Mr Andrews owns the land. His potential for activity, his free movement over his property, is in stark contrast not only to the harsh penalties awaiting poachers daring the

same freedom of movement over his land (as Berger notes), but also to the frozen stillness of Mrs Andrews. Moreover, the shadow of the oak tree over her refers to the family tree she was expected to propagate and nurture; like the fields she sits beside, her role was to reproduce, and this role is itself naturalized by the references to trees and fields. . . . [T]his period saw the consolidation of an argument that women were more "natural" than men. Medical, scientific, legal and political discourses concurred, and contextualize the image of Mr and Mrs Andrews in terms of a gendered difference in which the relationship to the land is a key signifier. Landscape painting then involves not only class relations, but also gender relations. Mr Andrews is represented as the owner of the land, while Mrs Andrews is painted almost as a part of that still and exquisite landscape: the tree and its roots bracketing her on one side, and the metal branches of her seat on the other.

. . . [M]y interpretation of the figure of Mrs Andrews stresses her representation as a natural mother. Obviously, her representation also draws on discourses of class and even nation. I emphasize her femininity, however, because there are feminist arguments which offer a critique not just of the discourses that pin Mrs Andrews to her seat, but also of the gaze that renders her as immobile, as natural, as productive and as decorative as the land. Such arguments consider the dynamics of a masculine gaze and its pleasures. . . .

Woman, landscape and nature

. . . The massive social, economic and political upheavals in [Europe and North America] during [the nineteenth century] – upheavals which included the colonial explorations through which geography developed as a discipline – meant that many of the schema previously used by artists to represent the world seemed increasingly outmoded, and new iconographies were sought to articulate the changes producing and reproducing the lives of art's audience, the bourgeoisie. By the mid nineteenth century, the emergence of this new public for paintings was fuelling a vigorous debate about the role of art: art was drawn into debates about social, political and moral standards which might structure the emerging modern world and, as feminists have remarked, central to these wider issues was the figure of Woman – fallen, pure, decadent, spiritual. . . . Woman becomes Nature, and Nature Woman, and both can thus be burdened with men's meaning and invite interpretation by masculinist discourse. . . . It should be emphasized that the "naturalization" of some women is asserted more directly than that of others: allegorical figures especially, but also, in bourgeois and racist society, working-class and black women. Thus the visual encoding of nineteenth century Western hegemonic masculinist constructions of femininity, sexuality, nature and property are at their most overtly intertwined in the landscapes with figures set in the colonies of Europe and America. . . . I suggest that, as well as contextualizing stories of geography's beginnings, the conflation of Woman and Nature can also say something about contemporary cultural geography's visual pleasure in landscape.

[. . .]

. . . The female figure represents landscape, and landscape a female torso, visually in part through their pose: paintings of Woman and Nature often share the same topography of passivity and stillness. The comparison is also made through the association of both land and Woman: with reproduction, fertility and sexuality, free from the constraints of Culture. Incorporating all of these associations, both Woman and Nature are vulnerable to the desires of men. Armstrong examines this vulnerability by arguing that if Art and the spectator constitute both Woman and Nature as what they work on and interpret, they do so especially by looking at both in a similar manner. Both are made to invite the same kind of observation. Rarely do the women in landscape images look out from the canvas at the viewer as an equal. Their gaze is often elsewhere: oblivious to their exposure, they offer no resistance to the regard of the spectator. Perhaps they will be looking in a mirror, allowing the viewer to enjoy them as they apparently enjoy themselves. If they acknowledge the spectator/artist, they do so with a look of invitation. The viewer's eye can move over the canvas at will, just as it can wander across a landscape painting, with the same kin of sensual pleasure. Here is another parallel between Woman and landscape: the techniques of perspective used to record landscapes were also used to map female nudes, and the art genre of naked women emerged in the same period as did landscape painting.

[. . .]

. . . . [T]he sensual topography of land and skin is mapped by a gaze which is eroticized as masculine and heterosexual. This masculine gaze sees a feminine body which requires interpreting by the cultured knowledgeable look; something to own, and something to give pleasure. The same sense of visual power as well as pleasure is at work as the eye traverses both field and flesh: the masculine gaze is of knowledge and desire.

This discussion of the visual representation of women and landscape concentrated on the complex construction of images of "natural" Woman as the objects of male desire. I have argued that Nature and Woman are represented through masculinist fantasies, and that makes looking pleasurable. Women are seen as closer to Nature than men because of the desirable sexuality given to them in these images and other discourses . . . Pleasure in landscape, it appears, is for straight men's eyes only.

[. . .]

"Geography is Everywhere: Culture and Symbolism in Human Landscapes"

from *Horizons in Human Geography* (1988)

Denis Cosgrove

Editors' introduction

Denis Cosgrove grew up in Liverpool, England. Reflecting upon his childhood in an interview, he remarked upon how important Sunday family walks along the docks, with its landscape of ships from faraway ports, were to his early interest in cultural geography. Today, Denis Cosgrove holds the prestigious Humboldt Chair in Geography at the University of California at Los Angeles. One of the key figures in the landscape-as-text approach, Cosgrove asks us to rethink the established technique of "reading" the cultural landscape. Cultural geographers have long been encouraged to examine the visible, material landscape around them for clues to the cultures that fashioned them from nature. In the oft-quoted words of Peirce Lewis, "our human landscape is our unwitting autobiography, reflecting our tastes, our values, our aspirations, and even our fears in tangible, visible form.... All our cultural warts and blemishes are there, and our glories too; but above all, our ordinary day-to-day qualities are exhibited for anybody who wants to find them and knows how to look for them."[1] In other words, landscapes are a faithful mirror in which we can see ourselves reflected. The task for the cultural geographer is to observe as carefully as possible, as a detective might, for those clues that might otherwise go unnoticed (see also Hoskins, p. 105).

In the 1980s, particularly amongst British cultural geographers, this task of "reading" the landscape underwent a profound transformation. If the landscape is indeed a text, theoretical developments in interpreting literary texts could surely be extended to how we read the cultural landscape. Specifically, post-structural approaches in literary criticism and related fields led some cultural geographers to emphasize that landscapes could be read in multiple ways. Rather than possessing one unitary meaning that the cultural geographer painstakingly uncovers, diverse individuals and groups in society might well read the same landscape in profoundly different ways. To draw an example from the Cosgrove excerpt reproduced below, the shopping center he frequents with his family is interpreted and used in a plethora of ways by different people. Some, like the unemployed youths he mentions, are even shut out of participating in it. In Cosgrove's words, the shopping center is "a highly-textured place, with multiple layers of meaning ... a symbolic place where a number of cultures meet and perhaps clash."

Particularly important to this theoretically informed landscape-as-text approach in cultural geography is *symbolic representation*. Dominant culture has the upper hand (as in most matters) in deciding the content of the landscape: what (and who) will be included, and what (or who) will be excluded. In other words, social power is reproduced through the landscape. As Kenneth Foote has explored in his book, *Shadowed Ground: America's Landscapes of Violence and Tragedy* (1997), the marking of significant events in a nation's history – or failing to mark them, or even refusing to do so – says quite a bit about what the nation wishes to project about itself to the outside world, as well as to its own citizenry. For example, Civil War battlefields in the United States are well marked landscapes of national reverence, while events that still loom shameful in the U.S. national conscience – such as the Manzanar concentration camp where Japanese-Americans were interned during World War II, or the site of the Salem witch executions – remain unmarked.

In addition, the link between the symbol (called the "signifier" in linguistics) and what it represents (the "signified") is neither natural nor unchangeable. Rather, it is socially constructed, and as such, it can be contested and changed. In the Foote example above, as the United States comes to terms with episodes of racialized violence that are a part of its past, the sites of this violence are slowly but surely becoming more visible on the landscape. Plaques, monuments, and other symbols of memorialization are placed there. They are made discrete from the surrounding landscape, through fencing or other techniques, and their grounds are tended. They appear on maps. Thus reading a landscape had become, in the work of many critical cultural geographers, an analysis of social power relations in all of their dynamic complexity. Cosgrove's *The Iconography of Landscape*, co-edited with Stephen Daniels (1988), is a landmark text in the study of landscape and representation.

The core of Denis Cosgrove's scholarly work examines the evolution of landscape representation and practice in Europe, particularly in Venice and northern Italy. As he explores at length in his now classic *Social Formation and the Symbolic Landscape* (1984; reissued with a new introductory chapter in 1998), the evolving socio-economic relations of capitalism, with its emphasis on privately held land and wage labor, is reflected, encoded, and contested through the European landscape and its representations, particularly in landscape architecture and painting. Cosgrove's more recent publications have focused even more intently on the use of symbolism. Using the paradigmatic "blue marble" image of the earth seen from space, Cosgrove has argued that from the time this image appeared in 1972 our understanding of the earth and our place in it has changed profoundly. In *Apollo's Eye: A Cartographic Genealogy of the Earth in the Western Imagination* (2001), Cosgrove notes that the blue marble image has come to symbolized human unity. Yet human attempts to represent the Earth have a long history, and Cosgrove links these representations to changing notions of Western identity. Most of these images have been represented primarily through maps, a notion Cosgove explores still further in his edited collection titled *Mappings* (1999).

NOTE

1 Page 12 in P. Lewis, "Axioms for reading the landscape: some guides to the American scene," in D.W. Meinig (ed.) *The Interpretation of Ordinary Landscapes* (1979).

I. MEANINGS AND LANDSCAPES

On Saturday mornings I am not, consciously, a geographer. I am, like so many other people of my age and lifestyle, to be found shopping with my family in my local town-centre precinct. It is not a very special place, artificially illuminated under the multi-storey car park, containing an entirely predictable collection of chain stores . . . fairly crowded with well-dressed, comfortable family consumers. The same scene could be found almost anywhere in England. Change the names of the stores and then the scene would be typical of much of Western Europe and North America. Geographers might take an interest in the place because it occupies the peak rent location of the town, they might study the frontage widths or goods on offer as part of a retail study, or they might assess its impact on the pre-existing urban morphology. But I'm shopping.

Then I realise other things are also happening: I'm asked to contribute to a cause I don't approve of; I turn a corner and there is an ageing, evangelical Christian distributing tracts. The main open space is occupied by a display of window panels to improve house insulation – or rather, in my opinion, to destroy the visual harmony of my street. Around the concrete base of the precinct's decorative tree a group of teenagers with vividly coloured Mohican haircuts and studded armbands cast the occasional scornful glance at middle-aged consumers. I realise that, unemployed as they almost certainly are and of an age when home is the least comfortable environment, they will "hang around" here until this space is closed off by the steel barriers that enclose it at night.

The precinct, then, is a highly textured place, with multiple layers of meaning. Designed for the consumer, to be sure, and thus easily amenable to my retail geography study, nevertheless its geography stretches way beyond that narrow and restrictive perspective. The precinct is a symbolic place where a number of cultures meet and perhaps clash. Even on Saturday morning I am still a geographer. Geography is everywhere.

Culture and symbolism are words that today do not slip easily or frequently off the tongues of most human geographers in Britain. By and large we rather pride ourselves on our down-to-earth practicality and relevance. We prefer to handle tangible, empirical materials, to interpret the world in the precise and measurable terms of practical necessity. Since the 1960s British human geographers have tended to work with certain unstated assumptions about how they should set about explaining patterns of human occupance and activity, assumptions which tend to exclude from consideration culture and symbol. . . .

These assumptions are in no sense dishonourable. But they do result in excluding from our agenda much that human geography could potentially study in the realms of human spatial activity and its environmental expressions. Further, they produce a deep contradiction within the subject. If our intentions are morally founded and the outcome of our work supposedly of value to humankind, while our materials remain exclusively empirical and our interpretations of human motivation resolutely utilitarian, we deny ourselves a language for framing the very goals we seek: the making of a better human world. . . .

Firstly, lost on the tide of earnest practicality and among the shingles of demonstrable fact is the real magic of geography – the sense of wonderment at the human world, the joy of seeing and reflecting upon the richly variegated mosaic of human life and of understanding the elegance of its expressions in the human landscape. This is the experience that still makes the *National Geographic* one of the most popular journals in the world. Geography, after all, is everywhere. . . . One of the tasks of geographers is to show that geography is there to be enjoyed. Too often we have been more successful in dulling rather than enhancing that pleasure.

Secondly, what we also lose in the utilitarian functionalism of so much geographical explanation is the recognition of human motivation other than the narrowly practical. Banished from geography are those awkward, sometimes frighteningly powerful motivating passions of human action, among them moral, patriotic, religious, sexual and political. We all know how fundamentally these motivations influence our own daily behaviour, how much they inform our response to places and scenes, even the shopping precinct. Yet in human geography we seem to wilfully ignore or deny them, refusing to explore how such passions find expression in the worlds we create and transform. Consequently our geography misses much of the meaning embedded in the human landscape, tending to reduce it to an impersonal expression of demographic and economic forces. The idea of applying to the human landscape some of the interpretative skills we deploy in studying a novel, a poem, a film or a painting, of treating it as an intentional human expression composed of many layers of meaning, is fairly alien to us. Yet this is what I propose to explore, and to suggest ways of treating geography as a *humanity* as much as a social science.

Such an approach has begun to emerge among a small number of human geographers since the early 1970s. . . . As with all shifts in the direction of geographical research, this change is related to broader social movements: protests against environmental exploitation and pollution, unease with megascale planning and the anonymous landscapes of urban redevelopment, the growing voice of organised women challenging the dominance of male culture and the failure of the post-war social and political consensus have all played their part in nudging human geography towards *humanistic* geography. But the idea of human geography as a *humanity* is scarcely a mature or fully developed one. So what follows must be a personal assessment of possibilities. I will approach this through a discussion of three terms – landscape, culture and symbolism – and lead on to some examples of interpreting the symbolism of cultural landscapes.

Landscape

Landscape has always been closely connected in human geography with culture, the idea of *visible* forms on the earth's surface and their composition. Landscape is in fact a 'way of seeing', a way of composing and harmonising the external world into a 'scene', a visual unity. The word landscape emerged in the Renaissance to denote a new relationship between humans and their environment. At the same time cartography, astronomy, architecture, land surveying, painting and many other arts and sciences were being revolutionised by the application of formal mathematical and geometrical rules derived from Euclid. Such rules, it was believed, would return the arts and sciences to their classical perfection. Perhaps the most striking of all these 'mechanical arts' from the point of view of space relations was the invention of linear perspective. Perspective allows us to reproduce in two dimensions the realistic illusion of a rationally composed three-dimensional space. A consistent order and form can be imposed intellectually and practically across the external world. Little wonder that in the same period landscape painting appeared for the first time in Europe as a popular style, paralleled by a blossoming art of landscape in poetry, drama, garden and park design. This was also the age when terrestrial space was being mapped rationally on to the graticules of sophisticated map projections, while rational human landscapes were being constructed in capital cities like Rome, Petersburg and Paris, and written across newly reclaimed lands in northern Italy, Holland and East Anglia, or on the enclosed estates of progressive landowners and over the vastnesses of overseas colonial territories.

Landscape is thus intimately linked with a new way of seeing the world as a rationally ordered, designed and harmonious creation whose structure and mechanism are accessible to the human mind as well as to the eye, and act as guides to humans in their alteration and improvement of the environment. In this sense landscape is a complex concept of whose implications I want to specify three: (i) a focus on the *visible* forms of our world, their composition and spatial structure; (ii) unity, coherence and rational order or design in the environment; (iii) the idea of human intervention and control of the forces that shape and reshape our world. Such intervention, it should be stressed, is not a mindless, exploitive or destructive relationship but one which should harmonise human life with the inherent order or pattern of nature itself. This point is crucial, for as we can see from even the merest acquaintance with landscape representation in painting, poetry or drama, the most powerful themes are those which comment on the ties between human life, love and feeling and the invariant rhythms of the natural world: the passage of the seasons, the cycle of birth, growth, reproduction, age, death, decay and, rebirth; and the imagined reflection of human moods and emotions in the aspect of natural forms.

For these reasons landscape is a uniquely valuable concept for a humane geography. Unlike place it reminds us of our position in the scheme of nature. Unlike environment or space it reminds us that only through human consciousness and reason is that scheme known to us, and only through technique can we participate as humans in it. At the same time landscape reminds us that geography is everywhere, that it is a constant source of beauty and ugliness, of right and wrong and joy and suffering, as much as it is of profit and loss.

Culture

I claimed above that landscape in human geography has long been associated with culture. This is particularly so in American human geography, where Carl Sauer's teaching and writings gave birth to a school of landscape geography focusing on humans' role in transforming the face of the earth. The emphasis was mainly on technologies: for example the use of fire, the domestication of plants and animals, hydraulics, but also to some extent on non-material culture (that is religious belief, legal and political systems and so on). Attention centred on pre-modern societies or their evidence in the contemporary landscape, for example the evidence in the American scene of the various Indian, African and European cultures that have shaped it.

Cultural geography in this tradition concentrated on the visible forms of landscape – farmhouses, barns, field patterns and town squares – although in Britain a similar tradition examined such non-visible phenomena as place names for evidence of past cultural influences. Culture itself was regarded as a relatively unproblematic concept: a set of shared practices common to a particular human group, practices that were learned and passed down the generations. Culture seemed to work through people to achieve ends of which they seemed but dimly aware. Critics have called this 'cultural determinism', and have stressed the need for a more nuanced cultural theory (in geography, particularly) if we are to treat contemporary landscapes and sophisticated modern culture.

A revived cultural geography seeks to overcome some of these weaknesses with a stronger cultural theory. It would still read the landscape as a cultural text, but recognises that texts are multilayered, offering the possibility of simultaneous and equally valid different readings. . . .

[. . .]

Symbol

To understand the expressions written by a culture into its landscape we require a knowledge of the 'language' employed: the symbols and their meaning within that culture. All landscapes are symbolic, although the link between the symbol and what it stands for (its referent) may appear very tenuous. A dominating slab of white marble inscribed with names, surmounted by a cross and decorated with wreaths and flags standing at the heart of a city is a powerful symbol of national mourning for fallen soldiers, although there is no link between the two phenomena outside the particular code of military remembrance. The birthplace of a great national figure may be an ordinary house, yet it bears enormous symbolic meaning for the initiated.

Much of the symbolism of landscape is far less apparent than either of these examples. But it still serves the purpose of reproducing cultural norms and establishing the values of dominant groups across all of a society. Take for example the municipal park of an English provincial town. Normally it occupies ten to fifteen acres in the Victorian inner suburbs, accessible on foot from the town centre. Surrounded by green or black painted railings, it still maintains its nineteenth-century design of mown lawns, carefully edged, serpentine paths winding past herbaceous borders, chromatic summer beds and shrub plantations with perhaps a small lake and scattered deciduous trees. In one corner is a children's playground, carefully fenced off.

Anyone entering the park knows instinctively the boundaries of behaviour, the appropriate codes of conduct. In general one should walk or rather stroll along the paths. Running is only for children and the grass for sitting on or picnics. Ducks may be fed, but the pool neither paddled nor fished in. Trees should not be climbed, nor should music be played except by the uniformed brass band on the wrought iron bandstand. In sum, behaviour should be decorous and restrained. When these codes are transgressed, as they are, by music centres, BMX bikers, over-amorous couples or bottle-toting tramps, then the fact is observed, and disapproval clearly registered by those who, although perhaps numerically a minority, nevertheless have the moral symbolism of the whole designed landscape on their side. There is little need for signs, although the unread printed park regulations peeling at the entrance would confirm the interpretation of the righteous guarantors of propriety.

Despite the enormous social changes that have occurred since its Victorian origins, the codes of behaviour still have legitimacy in the park because the landscape itself, the organisation of space, the selection of plants, the use of colour and the mode of maintenance will remain largely unchanged. They communicate a specific set of values. If we trace the history of such parks we find that the declared aim of their founders was moral and social control. With the intention of improving the physical and spiritual welfare of the labouring classes (whose dissolution cut into profits) the Victorian middle class actively discouraged traditional pastimes: tavern drinking, cockfighting and common-land festivals or fairs. They substituted the public park, writing the rules of conduct within it most precisely. Despite the passage of time, these characteristic slices of English urban landscape still symbolise ideals of decency and propriety held by the Victorian bourgeoisie.

All landscapes carry symbolic meaning because all are products of the human appropriation and transformation of the environment. Symbolism is most easily read in the most highly-designed landscapes – the city, the park and the garden – and through the representation of landscape in painting, poetry and other arts. But it is there to be read in rural landscapes and even in the most apparently unhumanised of natural environments. These last are often powerful symbols in themselves. Take for example the polar landscape, whose cultural significance derives precisely from its apparent savage unconquerability by humans. During the period of the great polar expeditions at the turn of the century the landscape of ice, crevice, snowstorm, polar bear and green seas became the very paradigm of a *Boys' Own* world, the setting for a British upper-class male cultural fantasy. Scott's death in 1912 made a corner of Antarctica 'forever England'. Imperial themes of military heroism taking strength from a barren and hostile environmental setting were revived in 1982, as British troops "yomped" across the South Atlantic islands during the Falklands-Malvinas war.

Reading Symbolic Landscapes

The many-layered meanings of symbolic landscapes await geographical decoding. The methods available for this task are rigorous and demanding, but not fundamentally esoteric or difficult to grasp. Essentially they are those employed in all the humanities. A prerequisite is the close, detailed reading of the text, for us the landscape itself in all its expressions. Geographers have always recognised, at least by lip service, the centrality of a deep and intimate knowledge of the area under study. The two principal routes to this are via fieldwork, map-making and interpretation. In developing such personal knowledge a highly individual response is inevitably generated. This is a response, or responses, of which we need to be conscious, not in order to discount them in the search for 'objectivity', but rather so that they may be reflected upon and honestly acknowledged in the writing of our geography.

At the same time we seek 'critical distance', a disinterested search for evidence and a presentation of that evidence free from conscious distortion. By evidence I mean any source that can inform us of the meanings contained in the landscape, for those who made it, altered it, sustain it, visit it and so on, and evidence that may challenge our predilections and theories just as its very collection

will be informed by those predilections and theories. It is important to realise that what is proposed here does not presuppose profound or specialised knowledge, only a willingness to look, to ask the unexpected question and be open to challenges to taken-for-granted assumptions. Very often it is children, so much less acculturated into conventional meanings, who can be the best stimulus to recovering the meanings encoded into landscape. The kind of evidence that geographers now use for interpreting the symbolism of cultural landscapes is much broader than it has been in the past. Material evidence in the field and cartographic, oral, archival and other documentary sources all remain valuable. But often we find the evidence of cultural products themselves – paintings, poems, novels, folk tales, music, film and song – can provide as firm a handle on the meanings that places and landscapes possess, express and evoke as do more conventional 'factual' sources. All such sources present their own advantages and limitations, each requires techniques to be learned if it is to be handled proficiently. Above all, a historical and contextual sensitivity on the part of the geographer is essential. We must resist the temptation to wrench the landscape out of its context of time and space, while yet cultivating our imaginative ability to get 'under its skin' to see it, as it were, from the inside. Finally, in such a geography *language* is crucial. The results of our study are communicated primarily through the texts that we ourselves produce. The text of a geographical landscape interpretation is the means through which we convey its symbolic meaning, through which we *re-present* those meanings. Inevitably our understanding is informed by our own values, beliefs and theories, but it is grounded in the pursuit of evidence according to the acknowledged rules of disinterested scholarship. In the act of representing a landscape written words and maps, themselves symbolic codes, are the principal tools of our trade.

Decoding Symbolic Landscapes: Some Examples

I suggested earlier that from the perspective of culture as power we could speak of dominant, residual, emergent and excluded cultures, each of which will have a different impact on the human landscape. I will use that threefold typology as the framework for exemplifying the approach to landscape that a 'humane' geography might adopt. I make no claim for the inclusiveness or objective validity of the classification. It serves as a useful organising device, no more.

Landscapes of dominant culture

By definition dominant culture is that of a group with power over others. By power I do not mean only the limited sense of a particular executive or governing body, rather the group or class whose dominance over others is grounded objectively in control of the means of life: land, capital, raw materials and labour power. In the final analysis it is they who determine, according to their own values, the allocation of the social surplus produced by the whole community. Their power is sustained and reproduced to a considerable extent by their ability to project and communicate, by whatever media are available and across all other social levels and divisions, an image of the world consonant with their own experience, and to have that image accepted as a true reflection of everyone's reality. This is the meaning of ideology.

To take a specific example: during the years immediately following the French Revolution there was considerable fear among the English ruling class, still dominated by landed interests, that English agricultural labourers, the largest single group of workers, might become 'infected' by the revolutionary spirit of liberty, equality and fraternity. From the perspective of an English squire such an outcome would be disastrous for the whole social order, because the harmonious balance which it suited him to believe existed between all classes in his justly governed realm would be shattered and anarchy would take its place. All sorts of appeals to patriotism and the ancient liberties of freeborn, well-fed English yeomen appeared, together with caricatures of emaciated French peasants starving in their liberty.

Another, probably only dimly conscious, response was the popularity among connoisseurs of painting – themselves landowners and ruling class members – of painted landscapes showing peaceful rural scenes with contented labourers gathering abundant harvests or resting with their families at the cottage door. Such scenes, however distant from rural realities, were recognisably English in topography and reassuringly peaceful socially. Only by looking at such landscape images in their context can we begin to uncover one of their key cultural meanings: that for the English squirearchy God was in his heaven and all was well with the world. They also give us a purchase on one of the most enduring images of English landscape, an image still reproduced today in the landscapes we seek to conserve in picturesque villages and well-regulated fields of hay and corn, as well as on our post cards and tourist posters.

In terms of existing landscapes, of course, we are most likely to see the clearest expression of dominant culture at the geographical centre of power. In class societies, just as the surplus is concentrated socially so it is concentrated spatially, in country houses and their parks for example, but above all in the city. It is instructive to observe how historically consistent has been the use of rational, geometrical forms in the design of cities: the circle, square and axial orthogonal or grid-iron road system all recur. Such geometry is radically different from the curves and undulations of natural landscape. It represents human reason, the power of intellect. Euclidian geometry as the foundation of urban form is to be found in ancient Greek, Roman, Renaissance, Baroque and Victorian city plans, even in the apparently benevolent landscape of Ebenezer Howard's garden city design, as well as in Chinese, Indian and Mayan urban form. Modernist city landscapes are equally exercises in applied geometry, whether we are considering Le Corbusier's Radiant City or the cubes of Manhattan or Dallas skylines.

To take one specific example of this theme of power and geometrical landscape, consider the capital city of the USA. Built upon 'virgin land' handed to the federal government by Virginia and Maryland and named after the first President, Washington DC was to be the seat of power for the first new nation of modern times and the centre of a territory larger than all of Europe. In its Declaration of Independence and Constitution the white, Europeanised, patrician founders of the United States had declared their vision of a new and perfect society and democracy. It was their cultural ideals that were celebrated in the designed landscape of Washington DC. The French architect

L'Enfant composed the plan of two simple geometrical designs: the orthogonal radiating pattern traditionally favoured by European monarchs exercising an absolute power which radiated from their persons and their courts, and the infinitely repeatable grid pattern which had become the basis for every colonial town, a democratic and egalitarian form that gives no single location a privileged status.

Here, inscribed in the very street pattern of the nation's capital, is the American resolution of European centralism and colonial localism, of federalism and states' rights. . . . [T]he plan . . . produces fifteen nodes, one for each existing state of the Union (thirteen former colonies plus Kentucky and Tennessee), and . . . symbolic buildings are [centrally] located. The White House and Capitol, the two balanced powers of executive and legislature under the American Constitution, stand at the ends of a great L at whose corner rises the Washington Monument commemorating the founding hero of the revolution, located on the bank of the Potomac river where nature and culture meet. White House and Capitol are joined directly by the line of Pennsylvania Avenue, named after the 'keystone state'. Washington's urban landscape can thus be 'read' as a declaration of American political culture written in space.

Such symbolic landscapes are not merely static, formal statements. The cultural values they celebrate need to be actively reproduced if they are to continue to have meaning. In large measure this is achieved in daily life by the simple recognition of buildings, place names and the like. But frequently the values inscribed in the landscape are reinforced by public ritual during major or minor ceremonies. Each year the British monarch 'opens' Parliament, an occasion of elaborate ritual at the Palace of Westminster. Much of the ritual is highly public and employs London's landscape. The monarch in a state coach accompanied by a retinue of the military and civil establishment processes from Buckingham Palace down the Mall and through Admiralty Arch – through a gate opened only for the passage of the Crown – passing Trafalgar Square with its monuments to British military victories and down Whitehall to Parliament. Crown and Parliament are thus conjoined via a ceremonial route and the passage marked by elaborate and impressive public ritual. Here, and at other such rituals,

such as Trooping the Colour, State visits, royal weddings and victory parades, urban space combines with (often invented) tradition and patriotic references in order to celebrate 'national' values and present them as the common heritage of all citizens. It is instructive to compare the routes taken by such official cultural events with those followed by other ceremonial users of the urban landscape: trades union processions, nuclear protesters or West Indian carnivals for example. A similar analysis could be applied at different scales to the design and use of space in any community from the largest city to the smallest village with its symbolic locations of war memorial, church, square, British Legion Hall or working men's club. Each of these landscapes has its ritual uses as well as its symbolic design. To examine and decode them allows us to reflect upon our own roles in reproducing the culture and human geography of our daily world.

Alternative landscapes

By their nature alternative cultures are less visible in the landscape than dominant ones, although with a change in the scale of observation a subordinate or alternative culture may appear dominant. Thus most English cities today have areas which are dominated by ethnic groups whose culture differs markedly from the prevailing white culture. This can produce a disjuncture between the formal built environment of inner city residential areas, constructed before the post-war wave of immigration from former imperial territories and still bearing the symbols appropriate to that time, and the informal uses and new meanings and attachments now introduced in a plural society. The former tram depot may be a mosque, bright paintwork, reggae rhythms and evangelical posters may be layered over a street of Victorian bye-law terraces. But however locally dominant an alternative culture may be it remains subdominant to the official national culture. At this latter scale I divide alternative cultures into residual, emergent and excluded.

Residual. Many landscape elements have little of their original meaning left. Some may be devoid of any meaning whatsoever to large numbers, as for example the concrete pyramids that can still be found near British coasts scattered over flat terrain and half overgrown – relics of symbolic wartime protection against invading German tanks. Geographers have long taken an interest in relict landscapes, generally using them as clues for the reconstruction of former geographies. But as with all historical documents, the meaning of such features for those who produced them is difficult to recover, and indeed the interpretations we make of them tell us as much about ourselves and our cultural assumptions as about their original significance.

A case in point is Stonehenge. Set starkly on the Wiltshire downs it is a dominating symbol, not merely because of its size and age but because its original cultural meaning lies beyond reasonable hope of recovery. Inigo Jones, the seventeenth-century architect, believed it was the ruin of a Roman theatre, discounting existing theories that it had been a Druid temple or the magic setting for Arthurian deeds created by Merlin's wand. Later theorists have claimed it as a giant observatory, a calendar device and the focal point of a sacred ley-line system whose influence still exists. Each of these interpretations indicates the role of residual landscape symbols in revealing contemporary alternative cultures.

The most ubiquitous residual landscape element in Britain is the medieval church building. From great gothic cathedral to village steeple, nearly every settlement has its ancient church, however altered by later accretions and renovations. In location, architecture and scale these are still powerful symbolic statements in our landscape, and their surrounding graveyards trace the cultural history of their community in layout, headstone design, lettering and funerary inscription. A gothic pointed arch is still recognised by the least religious of us as a sacred symbol. Yet the role of the church in contemporary English life cannot in any sense be called dominant. Indeed, one indication of its residual status is the difficulty architects have in finding a style appropriate to the cultural role of the church in modern life. Ancient church buildings become discotheques and cheap supermarkets while new church buildings look like discos and cheap supermarkets! There is much interesting work to be undertaken on landscapes of the past and their contemporary meanings, and their apparent re-creation in museums and theme parks is a good point of departure.

Emergent. Emergent cultures are of many kinds, some being very transient and having relatively little permanent impact on the landscape as, for example, the hippie culture of the late 1960s with its associated communes, alternative food shops and organic smallholdings. Yet they all have their own geography and their own symbolic systems. It is in the nature of an emergent culture to offer a challenge to the existing dominant culture, a vision of alternative possible futures. Thus their landscapes often have a futuristic and utopian aspect to them, as for example the geodesic domes so favoured by commune dwellers in America during the 1970s. But precisely because of this utopian strain emergent cultures very often deal in blueprints – paper landscapes. They are no less interesting or relevant to geographical study for that, because every utopia is as much an environmental as a social vision. There *is* a geography of 1984, of Brave New World and of Things To Come, as well as of every science fiction book, comic or film. To study that geography tells us much about the links between human society and environment.

We should not scorn the study of imaginative geographies, nor the use of real landscapes to anticipate future cultures and social relations. The New York skyline, for example, has been used since the days of King Kong and Superman to present an image of future urban society and its sophisticated yet precarious culture, tottering always on the edge of destruction by overwhelming forces of evil. There is also the landscape of sport, particularly international and Olympic sport, which remains a utopian vision of human concord even though its landscape expression has consistently been subverted by nationalistic culture, from Nuremburg in 1936 to Los Angeles in 1984. Contrasting landscape symbols of the future are rarely as poignantly juxtaposed as they are in the few hundred yards that separate the grey, regimented nuclear silos and the sprawling domestic anarchy of the Peace Camp at Greenham Common.

Excluded. By the time this essay appears in print one of those two emergent landscapes may well have disappeared. The particular culture promoted in the women's Peace Camp may have been officially excluded. In general women represent the largest single excluded culture, at least as far as impact on the public landscape is concerned.

Female culture is evident in the home, perhaps in the domestic garden. But the domestic landscape is one that geographers, significantly, have avoided studying. The organisation and use of space by women presupposes a very different set of symbolic meanings than by men, and in the past decade some important beginnings have been made in revealing the significance of gender in the attribution and reproduction of landscape symbolism. This has largely been the work of anthropologists. The maleness and femaleness of public landscape remains largely an excluded subject for geographical investigation, for no other reason than that the questions have never been put.

The same is very largely true for other excluded cultures, apart from the occasional study, itself usually treated as either of marginal interest or mildly suspicious. But the human landscape is replete with the symbols of, and symbolic meaning for, excluded groups. The symbolic space of children's games and their imaginative use of everyday places to create fantasy landscapes, the gypsy caravan site, the marks left by tramps to indicate the character of a neighbourhood as a source of charity, the graffiti of street gangs, the discreet notices and landscape indicators of such varied groups as gays or freemasons or prostitutes, are all coded into the landscape of daily life and await geographical study. It is fascinating to compare the official landscape meanings of the public park discussed earlier with its symbolic geography for various excluded cultures.

The taken-for-granted landscapes of our daily lives are full of meaning. Much of the most interesting geography lies in decoding them. It is a task that can be undertaken by anyone at the level of sophistication appropriate to them. Because geography is everywhere, reproduced daily by each one of us, the recovery of meaning in our ordinary landscapes tells us much about ourselves. A humane geography is a critical and relevant human geography, one that can contribute to the very heart of a humanist education: a better knowledge and understanding of ourselves, others and the world we share.

"From Discourse to Landscape: A Kingly Reading"

from *The City as Text: The Politics of Landscape Interpretation in the Kandyan Kingdom* (1990)

James S. Duncan

Editors' introduction

This selection by James Duncan falls squarely into the landscape-as-text approach. Unlike some of the other selections in this part, "From discourse to landscape: a kingly reading" is not an explanation of the author's theoretical framework regarding landscape. Rather, it is an illustration of how approaching the landscape as a text can be applied to a case study. Thus it works well with the previous selection by Denis Cosgrove, as it illustrates how a cultural geographer *actually uses* the approach described by Cosgrove in an applied landscape analysis.

Duncan scrutinizes the landscape of Kandy, a major city in the highlands of (what is today) Sri Lanka. In this selection, he looks particularly at the king's approach to shaping the elements of the capital city in ways that reinforce his leadership role as a *cakravarti*, or strong king. The physical layout of the city, and its buildings and grounds, work together to convey specific meanings about the king's changing role in Kandyan society. Familiar myths and symbols are encoded into the very elements of the built landscape such that the king's subjects are constantly reminded of his divine status. Thus the landscape is not a neutral backdrop against which society plays out its dramas. Rather, it plays a leading role in shaping those dramas. Landscape, in other words, is ideological.

The period chosen by Duncan for examination, the late eighteenth and early nineteenth centuries, encompassed several important changes for the Kandyan kingdom. Sri Vikrama, the last king of Kandy, assumed the throne in 1798 and reigned until 1815, when the kingdom was conquered by the British. During the course of his reign, leadership was undergoing a transition from the Asokan model of kingship which held that the king was a benevolent ruler in the Buddhist tradition, to a Sakran model of kingship in which the king was seen as divine: both more powerful and more active than in the earlier period. The transition in leadership style was part of a tension in the region between Buddhist (Asokan) ideologies associated with the Sinhalese people, and Hindu (Sakran) principles associated with the Tamil people. You may be aware that, even today, tension between Sri Lanka's Sinhalese majority and Tamil minority populations is at the root of ongoing civil conflict.

Duncan emphasizes that landscapes are definitely *not* unproblematic documents that can be read much as a book can be read. Rather, Duncan's point is that landscapes are many-layered entities full of erasures, silences, and struggles for power. Furthermore, landscapes are never merely passive records of society's struggles. Rather, they are active participants in waging those struggles. Duncan illustrates the active role that the landscape played in imposing and reinforcing the institution of a strong kingship in late eighteenth century Kandy. Duncan's method is to connect the elements of the built landscape to the larger narratives, myths, and symbols that promoted strong kingship.

The attention to language, representation, and relative theoretical sophistication of this work shared important parallels with similar developments in cultural anthropology, as evidenced by the popularity of James Clifford and George Marcus's edited collection, *Writing Culture* (1986) amongst the "new" cultural geographers at the time. Historian Simon Schama has explored the historic workings of European nationalism through landscape representation in his very readable *Landscape and Memory* (1995). More recently, geographers Lily Kong and Brenda Yeoh centralize the symbolic aspects of Singapore's ideological landscape in their critical exploration of nation building, in *The Politics of Landscape in Singapore* (2003).

James S. Duncan is a Reader in Cultural Geography at the University of Cambridge in the United Kingdom. Duncan made an early mark in cultural geography with his critique of the so-called "superorganic" approach to culture; in other words, the reluctance to problematize the concept of culture itself that was prevalent in the work of Carl Sauer and his disciples (particularly Wilbur Zelinsky, see p. 113); see "The superorganic in American cultural geography," in the *Annals of the Association of American Geographers* 70, 2 (1980): 181–198. Other work by Duncan on the colonial landscape of Kandy includes "Embodying colonialism? Domination and resistance in nineteenth century Ceylonese coffee plantations," in *Journal of Historical Geography* 28, 3 (2002): 317–338; and "The struggle to be temperate: climate and 'moral masculinity' in mid-nineteenth century Ceylon," in *Singapore Journal of Tropical Geography* 21 (2000): 34–47. Duncan has also explored suburban landscapes in New York; see for example *Landscapes of Privilege: The Politics of the Aesthetic in an American Suburb* (2004). This book was co-written with Nancy Duncan, who is a well recognized cultural geographer in her own right. As a team, James and Nancy Duncan have published widely on substantive cultural geography topics, as well as on broader concerns of the field, as with "Culture unbound," in *Environment and Planning A* 36 (2004): 391–403. James Duncan is also a co-editor, with Nuala Johnson and Richard Schein, of *A Companion to Cultural Geography* (2004), which gathers contributions by the leading scholars in the field on a broad range of contemporary cultural geography's concerns.

THE KING'S READING OF THE LANDSCAPE

[. . .]

. . . I will thicken the description [of Kandyan landscape elements] by offering a reading of the royal city of Kandy, its sacred and profane spaces, buildings, and architectural detail, which I suggest was the king's reading – one that he hoped the people and especially the nobles would accept. I will argue that this landscape is a text, written in the language of the concrete, and that it communicated the governing ideas of political and religious life. By tacking back and forth between the landscape text and various written works – religious scriptures, architectural manuals, political and historical texts as well as court poetry – I will attempt to reconstruct the king's reading: how it served to link the city of Kandy with an ideal landscape in order to legitimate his claims to political power.

What do I imply when I say that this was the king's reading? First, it was not a personal reading, not the idiosyncratic reading of a particular king; rather it was a kingly reading generated by a particular model of kingship within a general discursive field on kingship which can be traced back through Sinhalese and Indian texts. Although the king could emphasize one model of kingship or the other, he could not stray outside the wider discursive field and remain effective. Second, it implies that there were other possible readings of the city, readings of the nobles or of the ordinary citizens. . . .

It would appear that the landscape of Kandy which Sri Vikrama inherited in 1798 represented in concrete form the history of a compromise between the Asokan and Sakran philosophies of kingship. After his defeat of the British in the early nineteenth century, Sri Vikrama undertook a re-creation of the landscape which spoke more forcefully of Sakran kingship. His building program was designed to reinforce his claims to Sakran kingship, while the buildings themselves provided a more fitting backdrop for his civic ceremonies.

There are two principal ways in which the landscape and the king's quest for political power were intertwined. The first was in his attempt to employ the magic of parallelism to strengthen his political power, in this case to create an homology between the landscape of Kandy and the landscapes of the cities of the gods, and thereby to partake of the power of the gods.

The second way was implicit in the first. The king attempted to stun his subjects with the sheer magnificence of his surroundings. This was not simply a form of elaborate impression management, however, as all concerned – the king as well as the nobles, citizens, and monks – also believed in the real causal efficacy of spatial parallels and the power of symbols. Nevertheless the king took a calculated risk in over-emphasizing the Sakran self-aggrandizement to the detriment of the *sangha* [Buddhist clergy] and the people's welfare. But for now our concern is with the role the landscape played in the effort of this king to portray himself as divine. . . .

. . . [T]he Sakran discourse is based on two principal intertwined narratives. I will refer to the first of these as "The world of the gods." This narrative can be in turn subdivided into three subnarratives. The first was the story of the cities of the gods, especially the city of Sakra. This served as the model of an ideal capital in which the king was omnipotent. The second was the story of the Ocean of Milk, with its reference to the creation of the world and the renewal of the world's fertility. This served as a reminder of the fertility which emanated from the capital of a righteous king. The third subnarrative was of the cosmic axis which located the capital at the center of the world, assured its stability and allowed it to serve as a conduit between the worlds of the humans and the gods. The principal motifs in this were Mount Meru and the cosmic tree.

The second principal narrative I refer to as "The world of the *cakravarti*." It was also subdivided into three subnarratives. The first of these centered around the *cakravarti*'s control over the whole world, the second his control over his kingdom; while the third concerned the cities of the hero-kings of Lanka. . . . All three of these subnarratives spoke of a mythic time when the power of kings was, in theory at least, far less circumscribed.

These narratives were expressed in multiple media. The first medium was concrete, and its representation was iconic. It included various landscape features such as walls, ponds, canals, architectural detail, and the spatial relation of structures within the landscape. The second medium was language and its representation was metonymic. Objects within the landscape were denominated just as they were in the world of the gods. It is important to note that such iconic and linguistic representation was similar, for both allegorically transformed myth into landscape.

The third medium was behavior and its representation was ritualistic. Here the king, his entourage, and the common people emulated the world of the gods or of the *cakravarti*. They reproduced the allegory in rituals acted out in the landscape, itself an allegorical representation of these narratives. Thus, repeatedly composed in these multiple media was a powerful statement about an allegedly powerful king. . . .

The mechanisms by which the Sakran narratives were communicated included two important tropes. The first was synecdoche and the second recurrence. Synecdoche . . . is a metonymic device by which a single element out of a series . . . is made to stand for the whole of which it is a part. The wholes in this case were composed of elements drawn from the divine order of existence and expressed in the world of humans. Within the context of Kandy, these synecdoches were elements of the above-mentioned narratives which stood for the whole narrative. These synecdoches were found in different media; for example, there was an iconic representation in the wave-shaped wall around the lake in Kandy which stood for the waves raised during the churning of the cosmic ocean at the time of creation. Others were linguistic, such as the metonymic reference to the king's palace as the palace of Sakra. Some were ritualistic, such as the king's ascent of the square coronation stone which represented his ascension to the square cities of the gods on the top of Mount Meru. . . .

[. . .]

THE MYTH OF THE FOUNDING OF KANDY

There is no way of knowing whether the town of Kandy was founded in the manner suggested by the foundation myth; for our purposes its instrumentality and not its veracity is at issue. Its social function was allegorical; it told the inhabitants of Kandy a story about the founding of their city that elevated the city out of the realm of ordinary cities. As such it told a story not only about the city, but about its kings and its people. As we shall see, this myth derives from a larger tradition of foundation myths.

The *Rajavaliya* . . . a late-seventeenth-century Sinhalese document, describes the founding of the city of Kapilawastupura near Benares in northern India by the four sons of King Suta of the Tritlya Okkaka people. These princes were important to the Sinhalese for it was they who had founded the Sakya dynasty into which the Buddha was born. It was said that the four princes:

roamed the forest, seeking a site in its midst to fell and clear, with a view to construct tanks and dams, making fields and gardens, and build a city. There they found Bodhisattva [a future Buddha] who in his birth as the hermit Kapila was practicing severe austerities at the foot of a tree in the vicinity of a lake in the midst of a forest. He, seeing the Princes walking through the forest, asked them, "Princes, what seek ye in this forest?" They replied that they had left their country and were in search of a site whereupon to build a city. On learning this the Bodhisattva examined the nature of the site eighty cubits upwards and eighty cubits downward and said, "Princes, if you would build a city, take the site of my *pansala*: when foxes chasing after hares come to this place, the hares turning back chase the foxes; when cobras chasing after rats and frogs come to this place, these turn round and pursue the cobras; and when tigers hunting deer come to my *pansala* premises, they chase the tigers. A person who will hereafter live in this place will be kindly treated by the gods and Brahmas. Take, therefore, this *pansala* ground of mine; even if an army of *Cakravarti* should come [here] it would be defeated: therefore take ye this site and build a city: the only favor I ask is that ye call the city Kapila-wastu-pura, after my name, when ye have completed the building of it." Accordingly, the four princes when they completed the city gave it the name of Kapilawastupura.

. . . . [T]he king asked the sage what the sign meant and was told that this was victorious ground that the gods had ordained for the establishment of his kingdom. "You will be well protected in this place and instead of fleeing before thine enemies thou wilt turn and put them to flight". . . . [The legend] spoke of a weak kingdom that was insecure in the face of stronger enemies. The Kandyans in the mountain kingdom were the rabbits or cobras while their enemies the Sinhalese of the coastal kingdoms and the Europeans were the jackals and mongooses.

The foundation myth of Kandy justified the choice of its location in several different ways. First, it showed that it was a place chosen by the gods. As a place where the normal order of the mundane world was reversed it was liminal [existing between heaven and earth], an *axis mundi* where the worlds of humans and gods mingled and merged. Second, in this place that had received the favor of the gods, weakness prevailed over strength. . . .

THE CITY AS AN ALLEGORICAL LANDSCAPE

The very form of the city suggests that the king conceived of it as a cosmic capital. Kandy was composed of two rectangles . . . the sacred shape of the cosmic cities of the gods. As such the very outline of the city was a powerful iconic synecdoche standing for the two central allegories that I have identified: "The world of the gods" and "The world of the *cakravarti*." But the city . . . was composed of several different parts and I will now interpret each in turn.

The western rectangle

The western rectangle was the location of both the residences of the nobles and the houses and shops of the common people. The city was divided into four quarters by two major streets running north–south and east–west. Of these two streets, the one running east–west was the more important, for it divided the city into its two administrative units. The northeast and northwest quarters of the city were under the jurisdiction of the king's first *adikar* [chief officer], while the southeast and southwest quarters were under the second *adikar.* Division of the city into two parts was metaphoric, as it mirrored the division of the kingdom itself between the first *adikar* who had responsibility for the north and east of the kingdom and the second *adikar* who had responsibility for the south and the west. Through the power of like numbers, the western rectangle stood for the kingdom as a whole. This parallelism was thought to be efficacious, extending the power of the *adikar* spatially.

The number four, and multiples of four, are highly symbolic throughout Indian Asia as they represent totality, the four cardinal directions which are synecdoches for the four quarters of the world. . . . [T]he typical kingdom in Lanka was conceived of as being, at least in theory, composed of four quarters. . . . In keeping with this practice we can see the recurrence of the number four within the city. For example, as I stated above, the city was divided into four quarters, there were four shrines to the gods in Kandy, four gates to the city, four great festivals, and four ferries to bring people across the river into the city. These many references to the four quarters served as recurrent synecdoches affording the king symbolic power over the kingdom and the world beyond. To have power over the four quarters is to be a world ruler. Thus the number four occurring throughout the city in various media makes a clear reference to the narrative of "The world of the *cakravarti.*" In other words, the city becomes a microcosm of the kingdom, the world, and beyond that the macrocosmos. It is . . . a cosmopolis – a city that mirrors a world.

But this theme of the microcosmic reduction of the world and the kingdom also recurred in many other synecdoches. For example . . . in the late eighteenth and early nineteenth centuries the Kandyan kingdom was composed of twenty-one administrative units. These units appear to have been arrayed in the shape of a sacred cosmic diagram or *mandala* around the capital. . . . In the center was the city of Kandy surrounded by an inner circle of nine counties or *rata.* The outer ring was composed of twelve provinces or *disa*, which is the Pali term for a direction point of the compass. The term for governor is *disava*. . . . The governors, therefore, were the lords of the compass points and in Kandy there were four major and eight minor *disavas*, mirroring in the bureaucracy the four and eight points of the compass.

Until the reign of the last king of Kandy, the four quarters of the city were further subdivided by streets into sixteen squares constituting what one mid-eighteenth-century Kandyan text called a "properly divided street pattern". . . . [T]his is the number of squares (4×4) into which a properly designed capital should be divided.

Sri Vikrama, the last king, added two streets and extended three others. This increased the number of squares in the city to twenty-one and in the process made the western part of the city a more perfectly shaped rectangle. . . . [T]here are two reasons why these additions to the city can be better understood as a systematic attempt by the last king to reinforce his power.

First, by adding five squares to his city he raised their number to twenty-one, which is the number of administrative units in the kingdom. The kingdom, therefore, symbolically recurred within the city; the macrocosmos was reduced to the microcosmos. Second, there is strong evidence that the king was attempting a magical solution to the problems besetting his kingdom when he reshaped the city into a perfect rectangle. As a more faithful representation of the heavenly city of the gods it might, through the power of parallelism, partake of the potency of a heavenly city. . . . Furthermore, by assigning each province a square in the city, he was also able, through the metonymic power of the synecdoche, to bring the whole kingdom into the city. Thus he could magically control the kingdom by controlling the city. Because the city was a liminal place, the power of the gods manifest in the power of the king could be deployed against such irksome banalities as the kingdom's budget deficit.

The streets forming the borders of the twenty-one squares contained shops providing services for the king. Here also were the *valavvas*, the mansions of the governors of the twenty-one administrative units of the kingdom, where the families of the nobles were kept hostage as guarantors of their patriarch's loyalty. Evidently, whatever power parallelism may have had to preserve his kingdom, the king was not above more practical precautions.

. . . . [T]he western rectangle was the profane portion of the city which, in relation to the eastern rectangle of the city, stood as does the earth to the heavens . . . .

The eastern rectangle

To more fully understand the role of urban form in the legitimation of power one must be able to interpret the eastern rectangle, the so-called "sacred rectangle," the real locus of ritual power in the kingdom where the temples and the palace were located. Many royal cities in India were composed of two rectangles, the first for the palace and the temples and the second for the citizens.

One way to unlock the mythic structure of a landscape, the hermeneutic circle of landscape and myth, is to begin with one element of the landscape and securely anchor it through synecdoche to a set of narratives. Having done this one can then move on to relate that landscape element to other elements of the landscape. These can be explained in turn through synecdoches that refer back to the initial narratives. This process entails simultaneous reconstruction of a landscape text and a myth system, with a constant tacking back and forth between elements of each.

Although the major features of the landscape remain the same today, in the description that follows I will use the past tense. While there is evidence in contemporary reports, paintings, and maps that the architecture and layout of the town was as I describe it, there is no evidence of what the frescoes on the walls inside the Temple of the Tooth Relic were like during the reign of Sri Vikrama or whether they were as they are today. My description of these frescoes is based on my own observations of the temple. Whereas the paintings may have changed, there is reason to assume that the symbols would be essentially the same, as their purpose was to reconfirm the same sets of religious narratives found in the architecture and spatial configurations of buildings, streets, and monuments.

THE LAKE AS THE COSMIC OCEAN

As a point of entry into the circle of landscape and narrative I have chosen a landscape element that is unambiguously allegorical. I will first uncover the synecdoches which link it to the narrative and then proceed outward in the field of other landscape elements to those whose allegorical connections are perhaps less obvious, that is, those which require a more indirect method of decoding. The lake in Kandy, which lay to the south of the sacred rectangle, was an unambiguous element. By its very name, "Kiri Muhuda", the Ocean of Milk, it was linguistically secured to the narrative "The world of the gods."

However, before continuing it is important to pause and enquire as to the purpose of this lake. Clearly this large lake served no agricultural purpose; in fact some paddy fields were removed from production when it was dug. Furthermore, the capital was well watered by the Tingol Kumbura stream, the various channels that had been cut by prior kings, and by Bogambara Lake within the western limits of the city. Why then was the lake constructed? Although, as I have said, the answer that most historians give is that Sri Vikrama was an aesthete, who constructed the lake in order to beautify the capital . . . we might wish to take [this] hunch one step further. . . .

During [the eighteenth century] the court became increasingly Hinduized and kings strove to portray themselves as gods. Sri Vikrama's building program must be seen in this context. He wished to show his subjects that he was a god like Sakra, a *cakravarti* and a future Buddha. This was to be accomplished largely through ritual and environmental symbolism. His capital already had the lake called Bogambara which was a representation of the mythical Lake Anotatta. What it lacked was the much larger body of water, the cosmic Ocean of Milk. Sri Vikrama accomplished this with the lake that he named the Kiri Muhuda, the Ocean of Milk. He now had symbolically reproduced within his capital both Anotatta, the mountain lake near the center of the world, and Kiri Muhuda, the cosmic ocean which surrounds Mount Meru. In doing so he captured the universe; he had reduced the macrocosm to the microcosm. How better to symbolize that he was a universal monarch? The island in the lake, with its white pleasure house, also acted as a powerful synecdoche for both the allegories of "The world of the *cakravarti*" and "The world of the Gods". . . . Sri Vikrama had not, then, engaged in a frivolously aesthetic project; he had, through the power of metonymy, more firmly placed his capital, and by extension himself, at the center of the universe.

But let us now return to a consideration of the implications of naming the lake "The Ocean of Milk", the ... links between the lake and other elements in the landscape, and the power of this name to assemble disparate elements into an allegorical text or a landscape, that spoke of divine power. The name alluded to and embodied a greater complex of ideas just as its creator, the king, was the embodiment of divinity and all the glory associated with the gods.

... [T]he Ocean of Milk was the name given in the sacred texts to the cosmic ocean which lies at the foot of Mount Meru at the center of the universe. It forms one of the three subnarratives of the narrative "The world of the Gods," an important part of which is the churning of the Ocean of Milk. According to the *Visnu Purana*, the Ocean of Milk was churned by the gods and demons with Vasuki, the cosmic serpent, wrapped around Mount Mandara, itself balanced on a tortoise that was Visnu incarnate. From this churning arose *soma*, the potion of immortality; the milk-white elephant, who became the mount of Sakra, the king of the gods; the milk-white horse and milk-white cow, which Sakra also took as his own; and the *kapruka*, the gift-giving tree with the milky sap, which he took for his garden.

The presence of this artificial lake named the Ocean of Milk with its allegorical references reinforced the power of other linguistic, iconic, and behavioral synecdoches which also allegorically referred to the same subnarrative. For example, consider the color of the Ocean of Milk. White is held to be a "natural symbol" of fertility, symbolizing, in the South Asian tradition, both milk and semen. Furthermore, it is the color associated with Sakra who, in addition to being the king of the gods, is the god of rain and hence of fertility. White was therefore the official color used by the king of Kandy to symbolize his claim to be an incarnation of Sakra and a guarantor of fertility throughout the kingdom.

Like Sakra, the kings of Kandy possessed the gifts symbolizing fertility – the elephant, the horse, the cow, the tree – that arose out of the churning of the Ocean of Milk. The kings owned white, or light-colored state elephants and were known to have frequently requested European ambassadors to send white horses, which were unavailable locally. The kings also possessed small herds of sacred white cattle and kept a *kapruka* in the Temple of the Tooth. By naming the lake in Kandy

the "Ocean of Milk" the last king simultaneously established a link between the lake and the world of the gods, and by extension between himself and Sakra. Furthermore, this act also vivified these other synecdoches evoking the subnarrative of the churning of the Ocean of Milk. Here, both iconically and linguistically, he possessed the Ocean of Milk itself. No longer did the people have only a white horse or white cattle to remind them of the churning. Now they had before their very eyes the Ocean of Milk itself. Collectively these synecdoches transformed the landscape into concrete evidence of the king's role as a god-like, creative agent.

However ... the lake was constructed with forced labor, and this building project imposed real hardships and provoked abiding resentment within the kingdom. In the eyes of the people, *rajakariya* (forced labor due the king) was legitimately due only when employed in building a "proper" capital for the king or for religious projects. In an attempt to mollify those who suspected that his excessive city building represented and sanctioned engrossment of regal power, the king apparently sought to define the lake construction as a religious project. This is apparent not only in the verbal associations which we have just outlined, but I will argue in the supplement of an important new component to the city's foundation myth. ...

According to this version of the foundation myth, King Vikrama Bahu IV of Gampola, having decided to found a new capital, sent an old man to search for an auspicious place (*jaya bhumi*). At the spot where the Temple of the Tooth was eventually located the old man saw a squirrel defeat a rat snake. Later others were sent to discover the meaning of this. They saw a frog defeat a rat snake on the same spot. When he was asked for the meaning of these two events, the king's *adikar* interpreted them as favorable portents and invited the king to examine the place himself. The king brought his astrologer ... who agreed that it was indeed an auspicious spot. The king remained unconvinced. He scanned the dubious terrain and asked, "Why should I leave Gampola for a place so surrounded by marshes and hills?" With this the king ordered his astrologer to consult the oracle for forty-eight hours. At the end of the two days the astrologer made his prediction. He ordered that the king's men begin digging at the *jaya bhumi* and said that they would first find milk-white clay, then a layer of sand and finally water. After the king's men had found these layers just as had been predicted, the astrologer asked for a pure-white cloth predicting that a milk-white tortoise would be found; as they dug in the mud the tortoise did indeed appear and was wrapped in the cloth. Delighted with the success of the predictions, the king ordered that a city be built around this lucky site. He intended to build his palace directly upon the *jaya bhumi*, but the astrologer said: "This is too good a place for a palace, it is a place for a temple." The king subsequently abandoned his capital at Gampola and moved to Kandy, building the town around a temple on the lucky spot. The white tortoise was given a small pool at the eastern end of what is now the Kandy Lake. This pool was called the Ocean of Milk (Kiri Muhuda) and the tortoise was served food from the king's kitchen. Later this land was converted into paddy fields for the king.

What is interesting about this foundation myth is that it took the basic myth, which we reviewed earlier, of the weak overcoming the strong in an auspicious site, and appended an allusion to the sub-narrative of the Ocean of Milk. This allusion to the Ocean of Milk, I would argue, is a rather desperate attempt by the last king to justify the construction of the lake by linking it to the founding of the city. The suggestion is that latent in that spot there was *always* an Ocean of Milk. It was for the last king and his subjects to realize this potentiality. If accepted, such a claim would, he hoped, extinguish the unrest. For who could object to fulfilling a plan laid by the gods?

We have seen how Sri Vikrama attempted to transform his lake into the Ocean of Milk though linguistic parallelism. A mundane landscape was made sacred through naming. But naming was not sufficient in and of itself. Names are used to establish a metonymic relation, a bridge across which meaning flows like electricity between two poles. But the poles are necessary. The positive pole in this case was the cosmic ocean in mythic time and the negative pole or ground was the lake in Kandy. By naming the lake in Kandy the Ocean of Milk, the connection was made and the symbolic charge flowed from mythic time to real time, from the ocean to the lake. Without the negative pole, the concrete synecdoche of the landscaped ground, the connection could not have been made, mythic time and place could not have been realized.

[. . .]

The whole basis of this kingly reading of the city was metaphoric and metonymic. Through the magic of parallelism, synecdochic elements in the landscape stand for and attract to themselves the power of the larger allegorical whole. These relationships which are established symbolically are highly complex. They are not only metaphorical in an especially efficacious way, but through the important religious concept of liminality they are metonymic or syntagmatic, for there is a kind of contiguity established between heavenly and earthly landscapes through such mechanisms as the cosmic axis. Also, important syntagmatic relationships were established through the spatial sequencing and juxtapositioning of iconic, linguistic, and behavioral symbols in the landscape. These relations of contiguity and similarity, along with relations of difference, such as sacred versus profane spaces, all joined to transform the landscape of Kandy into a highly complex, intertextual, and multivocal system of communication.

"Reconfiguring the 'Site' and 'Horizon' of Experience"

from *Sounding out the City: Personal Stereos and the Management of Everyday Life* (2000)

Michael Bull

Editors' introduction

In most of these selections, landscapes are approached as primarily *visual* entities. The *act of seeing* is thus paramount to constructing, experiencing, and understanding landscapes. Other senses, such as touch, smell, or taste are not taken into account. Painting and other visual representations such as photography and film are by far the most important media for landscape as an image that is meant to be seen.

In "Reconfiguring the 'site' and 'horizon' of experience" Michael Bull explores landscape as an aural entity. Bull uses interviews with personal-stereo users to argue that landscapes can be made of music: soundscapes. Aural landscapes are both a key element in contemporary urban culture and an important way that listeners craft their individual identities, as well as their relationship to the places and people around them. Music can be used to cocoon the listener by blocking out unwanted noise, inhibiting interaction with others, and allowing the listener to focus his or her thoughts. Alternatively, music can be used to connect the listener through shared experiences of listening, by recalling past events or places where the music was heard, or by establishing a fictive rapport with the musicians. Personal-stereo users are thus able to create the sort of connection – or disconnection – that they desire with their surroundings.

For practically all of us, our mundane everyday activities – shopping, commuting to work, conversing with friends, relaxing – are *mediated*, or experienced through, technological devices and their associated images and sounds. Television, cellphones, and the Internet are just three examples of ubiquitous technologies that mediate our daily lives. Bull uses the generic term 'personal stereo' to refer to portable audio-cassette players. (He was prevented by Sony from using the trademarked name 'Walkman'.) You might be more familiar today with the digital music players such as Apple's iPod. Bull's point, however, is the same: everyday technologies create sensory landscapes that shape our identity, our experience of place, and our connections to others. You have but to look around you at your classmates moving between classes, studying, and socializing on any university campus to realize the importance of individual aural landscapes created through cellphones and iPods.

Michael Bull is a Reader in Media and Film Studies at the University of Sussex. Though not formally trained as a geographer, he works closely with geographers on topics concerning urban space and technology. Bull's recent publications include *Sound Moves: iPod Culture and Urban Experience* (2007), and co-editorship of *The Auditory Culture Reader* (2003).

Works that have centralized the aural landscape of the everyday include Tia DeNora's *Music in Everyday Life* (2000); Ben Anderson's "Recorded music and practices of remembering" in *Social and Cultural*

Geography 5 (2004): 3–20; and Karin Bijsterveld's "'The city of din': decibels, noise and neighbors in the Netherlands, 1910–1980" in *Osiris* 18 (2003): 173–93.

In emphasizing landscapes that are constructed and reconstructed by listeners as they move through space, "Reconfiguring the 'site' and 'horizon' of experience" invokes several important themes in contemporary cultural geography. First is *non-representational theory*, an approach that attempts, among other things, to get beyond the visuality of so much of human geography and emphasize instead the emotional, performative, and multi-sensory nature of being-in-the-world. Nigel Thrift's *Spatial Formations* (1996) provides a good introduction to non-representational theory in cultural geography, while an overview of recent publications in this spirit is provided in Hayden Lorimer's review article titled "Cultural geography: the busyness of being 'more than representational'" in *Progress in Human Geography* 29 (2005): 83–94.

Second, the theme of the body has, since the 1980s, become an increasingly important site of research and theory in cultural geography (see also the introduction to Part Seven). "Reconfiguring the 'site' and 'horizon' of experience" treats the body moving through space as being simultaneously enveloped and extended by music. Thus the body becomes an integral part of the landscape. Nigel Thrift's exploration of the body in motion through dance in "The still point: resistance, expressive embodiment, and dance," pp. 124–151 in Steve Pile and Michael Keith (eds.) *Geographies of Resistance* (1997) echoes this approach.

Third, walking through city streets is a long-standing theme for urban geographers and sociologists of a cultural bent. Marxist literary critic Walter Benjamin's important work, conducted before World War II but published only posthumously in *The Arcades Project* (1999), centralizes the *flâneur*, or wealthy gentleman who strolled the streets of late nineteenth century Paris much as today's iPod listener might be seen to do. Henri Lefebvre, like Benjamin a twentieth century European intellectual of a Marxist persuasion, wrote of human movement through the city as in part constructive of the urban landscape, a theme particularly evident in *Rhythmanalysis* (2004). Theodore Adorno, who was greatly influenced by Walter Benjamin's interpretation of Marx, also wrote of contemporary urban culture and centralized music in his analysis; see his *Philosophy of Modern Music* (2003).

■ ■ ■ ■ ■ ■

Personal stereo use reorientates and re-spatializes the users' experience with users often describing the experience in solipsistic and aesthetic terms. Personal stereos appear to provide an invisible shell for the user within which the boundaries of both cognitive and physical space become reformulated. . . .

> I don't necessarily feel that I'm there. Especially if I'm listening to the radio. I feel I'm there, where the radio is, because of the way, that is, he's talking to me and only me and no one else around me is listening to that. So I feel like, I know I'm really on the train, but I'm not really . . . I like the fact that there's someone still there. (Mandy)

Personal-stereo users often describe habitation in terms of an imaginary communion with the source of communication. Mandy is twenty-one. She spends four hours each day travelling across London and uses her personal stereo throughout this time. She likes to listen both to the radio and to taped music on her machine. She listens to music habitually, waking up to it and going to sleep to it. Her description of listening sheds some light upon the connections between technology, experience and place. Using a personal stereo appears to constitute a form of company for her whilst she is alone, through its creation of a zone of intimacy and immediacy. This sense of intimacy and immediateness . . . appears to be built into the very structure of the auditory medium itself. The headphones of her machine fit snugly into the ears to provide sound which fills the space of cognition. The 'space' in which reception occurs is decisive, for just as the situation of the television in the home changes the structuring of experience there, so the use of a personal stereo changes the structuring of experience wherever it is used. Mandy describes herself as being where the music or the DJ is. She constructs an imaginary journey within a real journey each day. The space of reception becomes a form of mobile home as she moves through the places of the city. The structuring of space through personal-stereo use is connected to other forms of communication strategies enacted through a range of communication technologies. Users live in a world of technologically mediated sounds and images. . . . This is demonstrated in the following remark by Mandy:

I can't go to sleep at night without my radio on. I'm one of those people. It's really strange. I find it very difficult. I don't like silence. I'm not that sort of person. I like hearing things around me. It's like hearing that there's a world going on sort of thing. I'm not a very alone person. I will always have something on. I don't mind being by myself as long as I have something on. (Mandy)

Mandy goes on to describe her feeling of centredness, of being secure with her personal stereo by excluding the extraneous noises of the city or at least her ability to control this:

Because I haven't got the external sort of noises around me I feel I'm in a bit of a world of my own because I can't really hear so much of what is going on around me. (Mandy)

The use of a personal stereo either creates the experience of being cocooned by separating the user from the outside world or alternatively the user moves outwards into the public realm of communication culture through a private act of reception and becomes absorbed into it. . . . The user does not perceive herself as being alone but understands that neither is she 'really there'. Using a personal stereo makes her feel more secure as it acts as a kind of boundary marker for her.

Her use of a personal stereo transforms her experience of place and social distance. Through use, the nature and meaning of being 'connected' within a reconfiguration of subject and object itself becomes problematic. The very distinction between them appears to be blurred. The following description of situatedness is typical in which the user describes use as filling:

The space whilst you're walking . . . It also changes the atmosphere as well. If you listen to

music, like, and you're feeling depressed it can change the atmosphere around you. (Sara)

The auditory quality of listening is described as being all-engulfing. The site of experience is transformed from the inside out. Effectively it is colonized. Habitable space becomes both auratic and intimate:

Because when you have the Walkman it's like having company. You don't feel lonely. It's your own environment. It's like you're doing something pleasurable you can do by yourself and enjoy it. I think it creates a sense of a kind of aura sort of like. Even though it's directly in your ears you feel like it's all around your head. You're really aware it's just you, only you can hear it. It makes you feel individual . . . (Alile)

Listening also constitutes 'company':

If there's the radio there's always somebody talking. There's always something happening. (Alice)

This is contrasted with the observation that nothing is happening if there is no musical accompaniment to experience. The auratic space of habitation collapses. . . . When the personal stereo is switched off the 'we-ness' falls away and the user is left in an experiential void often described with various degrees of apprehension or annoyance. Left to themselves with no distractions, users often experience feelings of anxiety. This is apparent in the many users who either put their personal stereos on to go to sleep or alternatively go to sleep with sound or music from their record players or radios. The activity is of course pleasurable in its own right:

I like something to sing me to sleep. Usually Bob Marley because I don't like silence. It frightens me. If it's silent and it's dark as well. It helps me think. Because I have trouble sleeping so if I have a song I like; it's sort of soothing. It's like your mum rocking you to sleep. I like someone to sing me to sleep. (Jana)

I don't like silence. I hate it at night. I suppose it's at night and you're on your own. I just don't like being alone. I just have to have someone with me or if not with me some type of noise. That's why I have music on for. It kinds of hides it. It just makes me feel comfortable. (Kim)

Just having the noise. If it's not music I have the TV. If there's the radio there's always someone talking. There's something happening. (Sara)

These responses contextualize the role of personal stereos to other forms of communication technologies that also act as forms of 'we-ness'. Dorinda, a thirty-year-old mother, describes using her personal stereo whilst cycling. For her the state of 'being with' is very specific. She plays one tape for months on end on her personal stereo. At present it is *Scott Walker sings Jacques Brel*. The tape has personal connotations for her and whilst listening she describes feeling confident, as if she's 'with' the singer. The sense of security she gains from this imagined familiarity is conveyed in the following remark:

Yeh. It's me and Scott [Walker] on the bike. (Dorinda)

Other users also describe this in terms of a feeling of being protected. Their own space becomes a protected zone where they are 'together' with the content of their personal stereo:

If I'm in a difficult situation or in new surroundings then I think nothing can affect you, you know. It's your space. (Paul)

Use appears to function as a substitute for company in these examples. Instead of company, sound installs itself, usually successfully. Jade, a habitual user, describes his relationship with his personal stereo in interpersonal terms in which the machine becomes an extension of his body. Users often describe feeling more comfortable when they touch or are aware of the physical presence of their personal stereo. These users normally don't like other people to use their machine:

It's a little like another person. You can relate to it. You get something from it. They share the same things as you do. You relate to it as if it's another person. Though you can't speak to it.

The silence is freaky for me. That is kind of scary. It's almost like a void if you like. (Jade)

The above extract is also indicative of the feeling of being deserted when the music stops. This feeling might also be described in terms of communication technology enhancing the space and the time of the user. As such it becomes both taken for granted and everyday in terms of the user's experience. Experience without it is seen as either void or at least inferior to experience through it. The spacing of experience becomes transformed, as the following group of teenagers testifies:

It fills the space whilst you're walking. (Rebecca)

It also changes the atmosphere as well. If you listen to music you really like and you're feeling depressed it can change the atmosphere around you. It livens everything up. (Sara)

The invigoration and heightening of the space of experience enacted through use collapses the distinction between private mood or orientation and the user's surroundings. The world becomes one with the experience of the user as against the threatened disjunction between the two. Using a personal stereo colonizes space for these users, transforming their mood, orientation, and the reach of their experience. The quality of these experiences is dependent upon the continued use of the personal stereo. This is graphically demonstrated by the following seventeen-year-old respondents who were asked in a group interview to describe how the atmosphere changes with the switching off of their personal stereos:

An empty feeling. (Kayz)

Got nothing to do. (Zoe)

Just sitting there and get bored. (Donna)

It's like when you're in a pub and they stop the music. It's an anticlimax. Everyone just stops. You don't know what to say. (Sara)

Switching off becomes tantamount to killing off their private world and returning them to the diminished space and duration of the disenchanted and mundane outside world. . . . The heightening and colonizing nature of personal-stereo use is clearly brought out in the following examples of holiday use. Personal stereos are a popular holiday companion for users:

> I use it lying on the beach. You need music when you're tanning yourself. There's the waves and everybody's around. You just need your music. On the plane we were listening to Enigma and things like that. It fitted in . . . Not bored, it livens everything up. Everything's on a higher level all the time. It makes it seem a bit busier. You get excited. Everything's happening. (Donna)

Donna isn't describing use as an antidote to boredom but as a form of harmonizing the environment to herself. Using a personal stereo enhances her experience, helping her to create a 'perfect' environment. Use allows her to experience the environment through her mediated fantasies. The holiday brochure might also come to life through use, as Jay's description demonstrates:

> I use it on the beach. I feel that I'd be listening to my music. I have the sea, I have the sand. I have the warmth but I don't have all the crap around me. I can eliminate that and I can get much more out of what the ocean has to offer me. I can enjoy. I feel that, listening to my music, I can really pull those sun's rays. Not being disturbed by screaming kids and all that shouting which is not why I went there. I went to have harmony with the sea and the sun . . . The plane journey, flying out and back and you listen to different music, but it just helps me to still my mind and to centre myself and I feel that by taking this tape with me I'm carrying that all day and I feel that I'm able to take more from the day and give more to the day. Whether that's right or wrong I don't know but that's how I feel. (Jay)

The environment is re-appropriated and experienced as part of the user's desire. Through her privatized auditory experience the listener gets more out of the environment, not by interacting with it but precisely by not interacting. Jay focuses on herself as personally receiving the environment via her personal stereo. There is only the sun and the user's body and state of mind.

Actual environments, unadorned, are not normally sufficient for personal-stereo users. It is either populated with people (Jay) or merely mundane (Donna). Music listened to through the personal stereo makes it 'what it is' for the user and permits the recreation of the desired space to accord with the wishes of the user. This is achieved by the user repossessing space as part of, or constitutive of their subjective desire. Personal-stereo users thus tend to colonize and appropriate the here-and-now as part of the re-inscribing of habitable space through the colonizing of place.

PERSONAL-STEREO USE: HOME AND AUDITORY MNEMONICS

Just as representational space is transformed, so is the user's experience of habitable space. As personal-stereo users traverse the public spaces of the city they often describe the experience in terms of never leaving home, understood either symbolically or sometimes literally. The aim here is not to reach outwards into a form of 'we-ness' but rather to negate distance enabling the user to maintain a desired sense of security. Using a personal stereo is often described in terms of a feeling of being surrounded or enveloped. This is what users frequently mean when they refer to feelings of being at home. . . .

> I like to have a piece of my own world. Familiar and secure. It's a familiarity. Something you're taking with you from your home. You're not actually leaving home. You're taking it with you. You're in your own little bubble. You're in your own little world and you have a certain amount of control and you don't have so much interruption . . . What it evokes for me is that I didn't really have to worry about it at all because there's someone there who'll take care of me. In a sense like when you're little and you have your mum and dad. So that's what it would evoke for me, a feeling of security that it will be all right . . . I don't like it [the urban] to totally take over. I have to have a piece of my own world. (Jay)

Jay listens to tapes that she associates with her own world and memories. She does not visualize this sense of home literally in terms of concrete memories but rather relates to it in terms of a sense of well-being and security. In this sense, she does not demonstrate an interest in an ongoing communicative process with a socially constructed public state of 'we-ness'. Rather, certain tunes or songs give her a heightened sense of well-being reminding her of childhood and family.

Other users describe travelling back into their own narratives by visualizing situations or re-experiencing the sensation of pleasurable situations whilst listening to their personal stereos in discounted public spaces. Their imaginary journey takes precedence over their actual physical journey and their actual present is overridden by their imaginary present. Whilst daydreaming is a common activity, users appear to have great difficulty conjuring up these feelings and images of home and narrative without using their personal stereos. As such, daydreaming becomes mediated, constructed and constituted through the technological medium of personal stereos and music.

The control exerted over the external environment through use is also described in terms of clearing a space for thoughts or the imagination. The random nature of the sounds of the street does not produce the correct configuration or force to successfully produce or create the focusing of thoughts in the desired direction. For users who are habitually accompanied by music there arises a need for accompaniment as a constituent part of their experience. The world and their biography is recollected and accompanied by sound. This construction of a space or clearing for the imagination to, either function in, or be triggered by personal-stereo use appears to be connected to the habitualness of use rather than the type of environment within which the experience takes place. It often makes little difference to the user whether they are walking down a deserted street or travelling on a congested train in terms of the production of the states of 'being' discussed here.

Home and narrative appear to be closely connected in the lifeworld of users. Personal stereos can be construed as functioning as a form of auditory mnemonic in which users attempt to construct a sense of narrative within urban spaces that have no narrative sense for them. The construction of a narrative becomes an attempt to maintain a sense of pleasurable coherence in those spaces that are perceived to be bereft of interest. Users describe a variety of situations relating to this point:

The music sparks off memories. Just like that. As soon as you hear the tunes. (Kim)

I'll remember the place. I'll be there. I'll remember what I was doing when I was listening to that music. (Jana)

If I'm listening to Ben E. King's 'Stand by me' I can imagine myself walking down Leicester Square because that's where I heard it with that guy. (Mandy)

Sometimes it brings back memories. Like how you felt. Some types of music and songs like, you only listen to them at certain times with certain people, so you listen to them on your own and it brings back memories ... atmospheres. (Sara)

Every time you listen to music it takes you back ... I visualize it. Like if I heard a certain song at a party or something and when I heard it again on my Walkman I'd just be at that party again with my friends doing what I was doing. (Rebecca)

Especially here, where I don't have such a big network of social connections. It's like ... having a photo of old friends. (Magnus)

Personal stereo use therefore represents one form of biographical travelling. The narrative quality that users attach to music permits them to reconstruct these narrative memories at will in places where they would otherwise have difficulty in summoning them up. ... Sound appears as the significant medium here as users rarely describe constructing narratives out of television-watching for example, at least whilst alone and in public areas.

PLACE AS BODY IN PERSONAL-STEREO USE

The use of personal stereos also helps users to reconceptualize their experience of the body as the site of action. The relationship of sound to the body also demonstrates the dual nature of the auditory. It is both a 'distance' sense, as is sight, as well as a 'contact' sense together with touch and taste. The physicality of sound is brought out admirably by the following user's description:

> You hear things not just through your eardrums, but through your whole bones. Your whole body is vibrating. I suppose it cancels out the vibration from the traffic around you. (Karin)

Users often describe feelings of being energized. The following account of cycling to the sounds of the personal stereo is typical:

> It's like when you've got music on and you're on your bike. It's like flying in a way. You're kind of away from things and you're not having any other contact with people. So flying above everything I suppose. You're more aware of cycling. Of the physical action of cycling. (Dorinda)

The experience of cycling is thus transformed. A heightened sense of the body as the site of action is commonly described by users, especially those who use them for physical activity. This type of use often results in an emptying out of thoughts from the body together with a greater awareness of the body as the site of action:

> I'd enjoy the feeling of my body working hard. It made me more concentrated on that. I enjoyed the feeling. It was channelling you in on that feeling. . . . Certain tracks, get into a rhythm, follow the bass line. It's always dance music. It's got energy. It's like clubbing or dancing. I get the same energy working or cycling. You become part of the bicycle. (Ben)

In these descriptions the physical body becomes the centre of action. This can be understood as a form of 'de-consciousnessing' by which I mean the giving over of oneself to the body as the site of action. The closest analogy would be the experience of extended dancing at 'rave' evenings. . . . [T]he body is experienced as merging with the activity of cycling. As such the body tends to lose its weight and resistance, becoming consumed in the present, thus banishing time. For users to successfully produce this experience the personal stereo must normally be played loudly in order to preclude the intrusive sounds of the world that would otherwise threaten to diminish the experience. Users are often aware of the possibility of sound encroaching into their world and respond by varying the sound level of their personal stereo appropriately, thus maintaining the hermetically sealed nature of their listening experiences.

Users' relations to representational space are transformed, enabling them to construct forms of 'habitable' space for themselves. In doing so users can be described as creating a fragile world of certainty within a contingent world. Users tend not to like being left to their own thoughts, not for them the reveries of a Rousseau who liked nothing more than walking in the solitude of the countryside in order to be alone with his own thoughts. Personal-stereo users prefer to be 'alone' with the mediated sounds of the culture industry. . . .

PART FOUR

Nature

Courtesy of Alex Dorfsman

INTRODUCTION TO PART FOUR

You may be asking yourself, "Why even read about nature? I'm studying cultural geography." If so, you've hit upon one of the key questions for cultural geographers. For it is common practice to separate "humans" from "nature"; furthermore, it is usual to consider "culture" the exclusive purview of human societies (see also the introduction to Part One). Non-human animals, such as primates, dolphins, insects, and birds, may well have relatively complex ways of communicating with one another, surprisingly sophisticated social hierarchies, and the ability to express emotions such as grief upon the death of a mate. Yet conventional wisdom holds that it is only humankind that truly engages in higher-order acts that together constitute *culture*: using fire to cook food, forge metal, or ward off cold; having an awareness of one's own mortality; constructing abstract symbolic communication systems; and so on. *Nature*, by contrast, is typically conceptualized as the non-human world that surrounds us. Nature is composed of both living beings, such as animals, trees, and microbes; and non-living entities, such as rocks, water, and clouds. In other words, *nature is all that culture is not*. Thus, cultural geography by definition would preclude the study of nature, if we are to abide by such distinctions between nature and culture.

In fact, the border between nature and culture is far from sharply drawn. Moreover, it never has been. Though prevailing notions of just what should be included in the category *nature* have changed over time, there has always been some contention over its definition. In particular, the place of human beings *vis-à-vis* nature has posed a particularly intriguing dilemma, addressed throughout the ages by a variety of theological, literary, scientific, and philosophical perspectives. Do human beings exist outside of nature, not subject to the natural laws that affect other living beings? Do human beings have a right, perhaps even a mandate, to utilize and modify nature for our survival and pleasure? Was the natural world in fact created by God (or the gods) for humans, or is nature itself a god (or goddess)? Are human beings simply another element in nature, subject to the same laws and impulses as non-human beings? What right do humans have to consume non-human beings, utilize them for work, or dominate them for companionship? Is there any proof that humans possess superior intelligence, sensitivity, or durability when compared to non-human beings? Are the actions of human beings in fact destroying the earth's life support systems – water, soil, atmosphere, animal and plant life – with which we humans are so intimately intertwined; and if so, does it not behoove us to recognize that nature and culture are, at some level, inseparable?

The shifting contours of nature and culture are closely bound up in the diverse and changing ways that language is used, and how this changes over time. This is a point that Raymond Williams's selection, "Nature", makes quite clear. The three main uses of the term 'nature' – all still in use today – are at odds over whether, for example, nature is an inherent force that emanates from within, or does nature refer to external qualities? Does nature reference the divine, or is it restricted to the material world? Most important, does nature include or exclude humans, and what are the implications of this for us as human beings? Are we above nature? Do we have a right or even a mandate to dominate nature? Or are we part of it, and thus perhaps part of God's divine architecture, or the divine reason of science? In Williams's words, caution is in order, "since *nature* is a word which carries, over a very long period, many of the major variations of human thought – often, in any particular use, only implicitly yet with powerful effect on the character of the argument – it is necessary to be especially aware of its difficulty."

To go a bit further, we might ask if there is even such a thing as "nature" *per se*. Or, as with so many other taken-for-granted terms – such as gender, race, nation, and so on – is nature, too, a social construction? This contention can be understood at a conceptual level, to mean that how we define nature says much more about *who* is doing the defining than it does about *what* is being defined. In other words, the contents of nature are potentially so varied that, ultimately, what gets defined as nature is a reflection of social power relations (and, in turn, can act to shape those social power relations). Think, for example, of animal rights activists and their highly publicized clashes with the fashion industry over the use of animal fur and skins in clothing. If humans are indeed superior to animals, and furthermore have a God-given right to utilize them for our survival and pleasure, then there is no problem with wearing a fur coat or leather shoes. If, however, one's definition of nature grants animals the same – or even superior – status to humans, animals have rights that preclude their killing by other animals (humans), such that wearing an animal is amoral. Or consider racism and colonialism, both of which hold that some humans are in fact animals – part of nature – and thus their domination, exploitation, and enslavement are legitimate. Such clashes over contending definitions of nature, and how these are transposed to human interactions such as with racialized categories, are discussed in Elder, Wolch, and Emel's selection titled "*Le Pratique sauvage*."

The contention that nature is a social construction can also be understood in a more literal fashion. The imprint of human modification on the earth is inescapable; indeed, this realization was at the heart of Carl Sauer's morphological approach to understanding landscape (see the introduction to Part Three). Today you would be hard pressed to find any corner of the earth, however remote, that remains utterly unaltered by human influence. In the selection titled "Creating a Second Nature," Clarence Glacken argues that even in the ancient world, what humans conceptualized as nature was in fact a *second nature*, profoundly altered by human civilization. The selection by "Ajax," one of the many pen names utilized by J.B. Jackson, provides an acerbic take on the separation of humans from nature in the 1950s in his piece "Living outdoors with Mrs. Panther."

Over the course of the nineteenth century, the discipline of geography was just becoming established, in Europe as well as the United States, as a legitimate field of study in which one could obtain a university degree. The question of nature and geography's relationship to nature proved to be a pressing issue then, as well as today. In geography's early decades, there was no sharp distinction between human and physical geographers, as there is today. Rather, the prevailing sentiment was that geography's purpose as a discipline was to integrate, or bridge the gap, between the natural and the human sciences. This has proven to be an enduring rationalization of the discipline of geography that can be readily heard today: geography explores the interface between nature and culture. Yet this interface has always been rather one-sidedly skewed toward the human side of the equation. As Sarah Whatmore has noted, "as human geographers set about trafficking between culture and nature, a fundamental asymmetry in the treatment of the things assigned to these categories has been smuggled into the enterprise."[1]

Through the 1930s, one thesis on the relationship between humans and nature held that processes such as natural selection shaped not just the *evolution* of species (human and non-human), but determined their *character* as well. The notion that nature determines human character and potential can be traced to Hippocrates' essay "Airs, Waters, Places," written in the fifth century BCE. Hippocrates asserted that the human body was comprised of four humors or fluids: blood, black bile, yellow bile, and phlegm. The prevalence of one humor over the others was determined by climate; for instance, residents of cooler, moist climes possessed an excess of phlegm (hence the adjective 'phlegmatic' to describe the unemotional, rational folks of northern regions), while residents of warm, hot climes tended toward blood (hence the adjective 'sanguine' to describe the quick-tempered inhabitants of the Torrid Zone). The notion that climate determines human character and potential reached an apogee in the *environmental determinism* of the nineteenth and early twentieth centuries, as reflected in the work of the German geographer Carl Ritter, and the American geographer Ellen Churchill Semple. Environmental determinism lent itself readily to racist and imperialist endeavors, through positing the immutable inferiority of those residing in the tropics. It was also a simplistic approach that was unable to accommodate in a scientific fashion the complexity of human–nature interactions, and fell out of favor with geographers (see also the introduction to Part Two).

Over the course of the mid-twentieth century, the discipline of Geography became more firmly divided between those physical geographers who study the natural world, narrowly defined as our non-human surroundings and processes (e.g., geomorphologists, biogeographers, oceanographers, and hydro-geographers), and human geographers who study the human world (e.g., social geographers, political geographers, economic geographers, and cultural geographers). Until surprisingly recently, human geo-graphers as a group did not take non-human nature into much account as an explicit object of study, with the notable exception of environmental geographers. Much less did they regard nature and humankind's place in nature as something to be problematized or critically analyzed. In cognate fields, such as environmental studies and nature writing, this relationship was problematized earlier, which has allowed cultural geographers to draw from their insights.

By the 1980s, however, some human geographers turned their focus to these very questions. As discussed in the introduction to the selection by Glacken, the notion of the *social production* of nature was taken on board by Marxist geographers. Rather than that which is outside of the human sphere, nature was repositioned squarely *within* the realm of human production. In a landmark publication in this vein, Neil Smith remarks that "Nature is generally seen as precisely that which cannot be produced; it is the antithesis of human productive activity. . . . But with the progress of capital accumulation and the expansion of economic development, this material substratum is more and more the product of social production . . ."[2] Think for a moment about the food that you eat. While in some ways natural, much of what we consume today is raised in a factory farming setting or in gigantic fields of genetically modified monocrops. There are even attempts under way to grow meat in labs, in order to avoid the economic, environmental, and moral costs of raising livestock, slaughtering it for meat, and transporting the meat to market.

In a related, but somewhat later move, nature as a *social construction* became the focus of some contemporary cultural geographers. Alexander Wilson's selection here, from his book titled *The Culture of Nature*, exemplifies this approach. In it, Wilson explores how the landscaping of post-war American suburbs in fact destroyed indigenous vegetation and replaced it with a stylized combination of non-native species and technology (in the form of fertilizers, pesticides, irrigation, and mechanical grooming) designed to provide a constitutive ground upon which to stage a specific form of modern suburban subjectivity. Wilson's work puts a contemporary spin on Glacken's notion of second nature. More broadly, it under-scores (along with "*Le Pratique sauvage*") how the social construction of nature is tightly bound to the social construction of other societal categories, such as gender and race. These approaches share their genesis with others in cultural geography in the 1990s, particularly landscape, which derive from the broader cultural turn in the social sciences (see also the introduction to Part Three). Thus they too focus on *representations of nature* in language, image, and symbol, and how representations act to natural-ize broader social relations of power in society.

Critics of these production and construction of nature approaches note that for all their insights, they ultimately reassert the primacy of humans in the nature–culture relationship, whereby nature is absorbed into the human side of affairs as simply a product of human activities, rather than possessing an exis-tence independent of the realm of the human. These geographers understand humans to be just one of many actors involved in complex networks composed of animals, plants, and the earth's life support systems of soil, water, and air. In this approach, referred to as *actor network theory*, humans are not privileged; rather they are regarded simply as partners with non-human actors in a delicate, place-based interchange, as Owain Jones and Paul Cloke discuss in their selection, "Orchard." Rather than consti-tuting an oppositional pair of categories, the distinction between culture and nature itself is questioned, as is the pervasive focus on the human side of things. Non-humans can be actors, possess agency and intentionality, and hold equal if not more power than humans do.

Whether informed by Marxist, feminist, postmodern, actor network, or non-representational theories, contemporary cultural geographers have brought nature back into the spotlight in exciting ways.

Notes

1 Page 165 in S. Whatmore, "Introduction: more than human geographies," in K. Anderson, M. Domosh, S. Pile, and N. Thrift (eds.) *Handbook of Cultural Geography* (2003) pp. 165–167.
2 N. Smith, *Uneven Development: Nature, Capital, and the Production of Space*, (1984), p. 32.

"Nature"

from *Keywords: A Vocabulary of Culture and Society*, revised edition (1983)

Raymond Williams

Editors' introduction

In this selection from his seminal work, *Keywords*, Williams explores the lineage of "perhaps the most complex word in the [English] language": nature. Though *Keywords* is styled much like a dictionary or an encyclopedia, with terms arranged and discussed in alphabetical order, Williams's intent is not to provide a concise or even a "correct" definition for the reader. Rather, *Keywords* is meant to be an inquiry into the shared meanings of terms and concepts that, in Williams's estimation, form the bedrock of English-speaking culture and society. For years, scholars have turned to *Keywords* not for definitions or historical summaries of important English words and concepts, but instead for clues to the *relationships between those words and broader patterns* of social and cultural change. Language is dynamic, always adapting and changing in response to, or in anticipation of, broader changes in society. Williams highlights that the meanings of terms and concepts change because of larger changes that are occurring in society. However, Williams insists that language does not simply *reflect* social change and historical process, but that these changes and processes themselves occur in part *within* language itself. In such movements, problems, meanings, and relationships are worked out in the confusions and ambiguities of language itself. Culture and society are in a continuous process of change, and that change occurs most fundamentally at the level of language. Nor is change a straightforward process of the old giving way to the new. Old meanings linger in language, just as they do in other aspects of our everyday lives. In short, Williams is keenly conscious of the central role played by language in broader processes of social change. Words are key.

All of the entries in *Keywords* dig behind terms that are often assumed to possess stable, straightforward, uncontroversial meanings. What Williams does is to construct a family tree of sorts – a genealogy – composed of the historic forebearers of contemporary words, their often quite divergent meanings, and their lingering imprint on the term we know today. In the entry for nature, for example, Williams notes how the word comes to contemporary English via Old French, and previous to that, Latin. The root of "nature" lies in the Latin verb *nasci*: to be born. Thus nature shares a common origin, and hence a common meaning, with other words, such as *nation*, *native*, and *innate*.

In addition to tracing the genealogy of the meanings of nature, Williams also offers a way of thinking through the complexities of the term without surrendering to the desire for a simple definition that will resolve ambiguity. This is an extremely important, yet subtle message. While noting that it is important for any discipline to clarify its terminology, Williams argues that "in general it is the range and overlap of meanings that is significant." He illustrates how the early use of nature, in the thirteenth century, indicated the essence or quality of something specific. For example, one might say "The nature of our discussion was friendly." In this example, "nature" refers to the *quality* or the *essence* of the specific discussion in question. Over the 300 years from the fourteenth century to the seventeenth, however, "nature" assumed two additional, more abstract, meanings: first, "the inherent force that directs . . . the world"; and second, "the material world itself". Whether either of these two uses of nature includes humans was, as Williams points out, debated at the time (as it still is). All three uses of nature are still in force today. As discussed in the introduction to this part, defining the contours of what counts as nature, and determining whether humans and human cultures are part of nature or separate (and superior) from nature, is an enduring question that philosophers, historians, writers, theologians, geographers, and others have pondered across the ages. Williams does not settle this debate for us; to do so would violate his intent in *Keywords*.

Of all the terms Williams discusses in *Keywords*, he considers nature (along with culture, see p. 15) to be so difficult, and important, because of its centrality to human identity. He states: "Any full history of the uses of nature would be a history of a large part of human thought." It is thus not surprising that so many cultural geographers who write about nature begin with Raymond Williams's exploration of the term. In a wonderfully accessible overview of nature seen from a geographer's perspective, Noel Castree starts off by noting the polysemy, or multiple meanings, of the term nature, and uses Williams's work to help craft his discussion; see *Nature* (2005). David Harvey, in his book *Justice, Nature, and the Geography of Difference* (1996), engages deeply albeit critically at times with Williams's discussion of nature and other keywords. Harvey, who is sympathetic to Williams's Marxist approach, notes the importance of labor as the way in for most of us to experience and define nature, and our place in (or out) of it: through work. Historian William Cronon edited *Uncommon Ground: Toward Reinventing Nature* (1995), and his observations have become key for cultural geographers looking at nature. In keeping with Williams's observations, Cronon centralizes the idea that "the way we describe and understand [the nonhuman world] is so entangled with our own values and assumptions that the two can never be fully separated. What we mean when we use the word 'nature' says as much about ourselves as about the things we label with that word" (p. 25). Finally, though Williams was no feminist, his observations on the gendered treatment of nature – as goddess, as mother – are taken up by feminist cultural geographers (see Rose, p. 171).

Raymond Williams (1921–1987) was born into a working-class Welsh family. He served in the British army during World War II, engaging in combat in France and Germany. After the war, Williams worked as an adult education instructor. In 1961, he was invited to join the faculty at Cambridge University, eventually becoming a Professor of Drama. Williams is known for his wide-ranging interests, as a literary and media critic, political analyst, dramatist, novelist, and social historian. The author of over twenty books, Williams is perhaps best known for *Culture and Society* (1958), *The Long Revolution* (1961), and *Marxism and Literature* (1977). Perhaps his most geographical work of non-fiction was *The Country and the City* (1973), but Williams's short stories and novels – such as *Border Country* (1960) – are also rich in geographical themes. Williams sought to avoid a deterministic Marxist approach, emphasizing instead the rich interrelation between language, culture, and material structures, particularly as these played out in the realm of the everyday. Williams was active in the so-called New Left social activism of the 1960s, and in close conversation with the founding members of the British cultural studies movement at the University of Birmingham. Along with Stuart Hall (see p. 264) and E.P. Thompson, Raymond Williams founded the radical journals *New Left Review* and *The New Reasoner*.

Nature is perhaps the most complex word in the language. It is relatively easy to distinguish three areas of meaning: (i) the essential quality and character *of* something; (ii) the inherent force which directs either the world or human beings or both; (iii) the material world itself, taken as including or not including human beings. Yet it is evident that within (ii) and (iii), though the area of reference is broadly clear, precise meanings are variable and at times even opposed. The historical development of the word through these three senses is important, but it is also significant that all three senses, and the main variations and alternatives within the two most difficult of them, are still active and widespread in contemporary usage.

Nature comes from [the immediate forerunner] *nature*, [Old French] and *natura*, [Latin], from a root in the past participle of *nasci*, [Latin] – to be born (from which also derive nation, native, innate, etc.). Its earliest sense, as in [Old French] and [Latin], was (i), the essential character and quality *of* something. **Nature** is thus one of several important words, including *culture*, which began as descriptions of a quality or process, immediately defined by a specific reference, but later became independent nouns. The relevant [Latin] phrase for the developed meanings is *natura rerum* – the nature of things, which already in some [Latin] uses was shortened to *natura* – the constitution of the world. In English sense (i) is from the thirteenth century, sense (ii) from the fourteenth century, sense (iii) from the seventeenth century, though there was an essential continuity and in senses (ii) and (iii) considerable overlap from the sixteenth century. It is usually not difficult to distinguish (i) from (ii) and (iii); indeed it is often habitual and in effect not noticed in reading.

> In a state of *rude* nature there is no such thing as a people . . . The idea of a people . . . is wholly artificial; and made, like all other legal fictions, by common agreement. What the particular nature of that agreement was, is collected from the form into which the particular society has been cast.

Here, in [Edmund] Burke, there is a problem about the first use of **nature** but no problem – indeed it hardly seems the same word – about the second (sense (i)) use. Nevertheless, the connection and distinction between senses (i), (ii) and (iii) have sometimes to be made very conscious. The common phrase **human nature**, for example, which is often crucial in important kinds of argument, can contain, without clearly demonstrating it, any of the three main senses and indeed the main variations and alternatives. There is a relatively neutral use in sense (i): that it is an essential quality and characteristic of human beings to do something (though the something that is specified may of course be controversial). But in many uses the descriptive (and hence verifiable or falsifiable) character of sense (i) is less prominent than the very different kind of

statement which depends on sense (ii), the directing inherent force, or one of the variants of sense (iii), a fixed property of the material world, in this case 'natural man'.

What has also to be noticed in the relation between sense (i) and senses (ii) and (iii) is, more generally, that sense (i), by definition, is a specific singular – the **nature of** something, whereas senses (ii) and (iii), in almost all their uses, are abstract singulars – the **nature of** all things having become singular **nature** or **Nature**. The abstract singular is of course now conventional, but it has a precise history. Sense (ii) developed from sense (i), and became abstract, because what was being sought was a single universal 'essential quality or character'. This is structurally and historically cognate with the emergence of *God* from *a god* or *the gods*. Abstract **Nature**, the essential inherent force, was thus formed by the assumption of a single prime cause, even when it was counterposed, in controversy, to the more explicitly abstract singular cause or force *God*. This has its effect as far as sense (iii), when reference to the whole material world, and therefore to a multiplicity of things and creatures, can carry an assumption of something common to all of them: either (a) the bare fact of their existence, which is neutral, or, at least as commonly, (b) the generalization of a common quality which is drawn upon for statements of the type, usually explicitly sense (iii), '**Nature** shows us that . . . this reduction of a multiplicity to a singularity, by the structure and history of the critical word, is then, curiously, compatible either with the assertion of a common quality, which the singular sense suits, or with the general or specific demonstration of differences, including the implicit or explicit denial of a common effective quality, which the singular form yet often manages to contain.'

Any full history of the uses of **nature** would be a history of a large part of human thought. . . . But it is possible to indicate some of the critical uses and changes. There is, first, the very early and surprisingly persistent personification of singular **Nature**: Nature the goddess, 'nature herself'. This singular personification is critically different from what are now called 'nature gods' or 'nature spirits': mythical personifications of particular natural forces. 'Nature herself' is at one extreme a literal goddess, a universal directing power, and at another extreme (very difficult to distinguish from some non-religious singular uses) an amorphous but still all-powerful creative and shaping force. The associated 'Mother Nature' is at this end of the religious and mythical spectrum. There is then great complexity when this kind of singular religious or mythical abstraction has to coexist, as it were, with another singular all-powerful force, namely a monotheistic God. It was orthodox in medieval European belief to use both singular absolutes but to define God as primary and Nature as his minister or deputy. But there was a recurrent tendency to see Nature in another way, as an absolute monarch. It is obviously difficult to separate this from the goddess or the minister, but the concept was especially used to express a sense of fatalism rather than of providence. The emphasis was on the power of natural forces, and on the apparently arbitrary or capricious occasional exercise of these powers, with inevitable, often destructive effects on men.

As might be expected, in matters of such fundamental difficulty, the concept of **nature** was usually in practice much wider and more various than any of the specific definitions. There was then a practice of shifting use, as in Shakespeare's *Lear*.

Allow not nature more than nature needs,
Man's life's as cheap as beasts'...

... one daughter
Who redeems nature from the general curse
Which twain have brought her to.

That nature, which contemns its origin,
Cannot be border'd certain in itself...

... All shaking thunder,
Crack nature's moulds, all germens spill at once,
That make ungrateful man...

... Hear, nature hear; dear goddess, hear...

In these examples there is a range of meanings: from nature as the primitive condition before human society; through the sense of an original innocence from which there has been a fall and a curse, requiring redemption; through the special sense of a quality of birth, as in the rootword; through again a sense of the forms and moulds of nature which can yet, paradoxically, be destroyed by the natural force of thunder; to that simple and persistent form of the goddess, Nature herself. This complexity of meaning is possible in a dramatic rather than an expository mode. What can be seen as an uncertainty was also a tension: nature was at once innocent, unprovided, sure, unsure, fruitful, destructive, a pure force and tainted and cursed. The real complexity of natural processes has been rendered by a complexity within the singular term.

There was then, especially from the early seventeenth century, a critical argument about the observation and understanding of nature. It could seem wrong to inquire into the workings of an absolute monarch, or of a minister of God. But a formula was arrived at: to understand the creation was to praise the creator, seeing absolute power through contingent works. In practice the formula became lip-service and was then forgotten. Paralleling political changes, nature was altered from an absolute to a constitutional monarch, with a new kind of emphasis on natural laws. Nature, in the eighteenth and nineteenth centuries, was often in effect personified as a constitutional lawyer. The laws came *from* somewhere, and this was variously but often indifferently defined; most practical attention was given to interpreting and classifying the laws, making predictions from precedents, discovering or reviving forgotten statutes, and above all shaping new laws from new cases: nature not as an inherent and shaping force but as an accumulation and classification of cases.

This was the decisive emergence of sense (iii): nature as the material world. But the emphasis on discoverable laws –

Nature and Nature's laws lay hid in night;
God said, Let Newton be! and all was light!
([Alexander] Pope)

– led to a common identification of Nature with Reason: the object of observation with the mode of observation. This provided a basis for a significant variation, in which Nature was contrasted with what had been made of man, or what man had made of himself. A 'state of nature' could be contrasted – sometimes pessimistically but more often optimistically and even programmatically – with an existing state of society. The 'state of nature', and the newly personified idea of Nature, then played critical roles in arguments about, first, an obsolete or corrupt society, needing redemption and renewal, and, second, an 'artificial' or 'mechanical' society, which learning from **Nature** must cure. Broadly, these two phases were the Enlightenment and the Romantic movement. The senses can readily be distinguished, but there was often a good deal of overlapping. The emphasis on law gave a philosophical basis for conceiving an ideal society. The emphasis on an inherent original power – a new version of the much older idea – gave a basis for actual regeneration, or, where regeneration seemed impossible or was too long delayed, an alternative source for belief in the goodness of life and of humanity, as counterweight or as solace against a harsh 'world'.

Each of these conceptions of Nature was significantly static: a set of laws – the constitution of the world, or an inherent, universal, primary but also recurrent force – evident in the 'beauties of nature' and in the 'hearts of men', teaching a singular goodness. Each of these concepts, but especially the latter, has retained currency. Indeed one of the most powerful uses of nature, since the late eighteenth century, has been in this selective sense of goodness and innocence. **Nature** has meant the 'countryside', the 'unspoiled places', plants and creatures other than man. The use is especially current in contrasts between town and country: **nature** is what man has not made, though if he made it long enough ago – a hedgerow or a desert – it will usually be included as **natural**. **Nature-lover** and **nature poetry** date from this phase.

But there was one further powerful personification yet to come: nature as the goddess, the minister, the monarch, the lawyer or the source of original innocence was joined by nature the selective breeder: natural selection, and the 'ruthless' competition apparently inherent in it, were made the basis for seeing nature as both historical and active. Nature still indeed had laws, but they were the laws of survival and extinction: species rose and flourished, decayed and died. The extraordinary accumulation of knowledge about actual evolutionary processes, and about the highly variable relations between organisms and their environments including other organisms, was again, astonishingly, generalized to a singular name. Nature was doing this and this to species. There was then an expansion of variable forms of the newly scientific generalization: 'Nature teaches . . .', 'Nature shows us that . . .' In the actual record what was taught or shown ranged from inherent and inevitably bitter competition to inherent mutuality or cooperation. Numerous **natural** examples could be selected to support any of these versions: aggression, property, parasitism, symbiosis, co-operation have all been demonstrated, justified and projected into social ideas by selective statements of this form, normally cast as dependent on a singular **Nature** even while the facts of variation and variability were being collected and used.

The complexity of the word is hardly surprising, given the fundamental importance of the processes to which it refers. But since **nature** is a word which carries, over a very long period, many of the major variations of human thought – often, in any particular use, only implicitly yet with powerful effect on the character of the argument – it is necessary to be especially aware of its difficulty.

[. . .]

"Creating a Second Nature"

from *Traces on the Rhodian Shore: Nature and Culture in Western Thought from Ancient Times to the End of the Eighteenth Century* (1967)

Clarence J. Glacken

Editors' introduction

Do humans of necessity impose order upon nature in order to survive, progress, or improve upon what is before them; or is nature already divinely ordered such that no human has the ability or right to alter its perfection? Is the earth a finite being with a youth, middle age, old age, and eventual death? Has mankind improved, or fallen from grace, since a supposed golden age of yore? Is there such a thing as nature unmediated by humans; and for that matter, humans unmediated by nature?

Clarence J. Glacken's *Traces on the Rhodian Shore* was, and continues to be, a foundational contribution to addressing these long-standing questions. In *Traces on the Rhodian Shore*, Glacken ponders historical perspectives on three encompassing, interrelated hypotheses with respect to the relationship between humankind and the earth: that the earth was divinely designed for humankind's benefit and enjoyment; that the environment influences the character, occupations, and health of human beings residing in different places; and that humankind has long played a determining role in shaping and modifying the natural world. Though its focus is on the past – from classical antiquity through the eighteenth century – in many ways *Traces on the Rhodian Shore* was ahead of its time. For Glacken poses the now widespread notion that nature is not *natural*; rather, it is always mediated by the influence of humans. In other words, what we see before us is a *second nature*. The notion that nature is socially produced became central to Marxist geographers' understandings of human–nature interactions, as evidenced by Neil Smith's *Uneven Development: Nature, Capital, and the Production of Space* (1984).

In this selection, Glacken ponders the ancients' understandings of human endeavors on the landscapes of Egypt and Greece. He contends that even in the Egyptian civilizations that pre-dated Greece's rise to prominence in the ancient world, there was awareness that man's interaction with the natural world, for the purposes of raising crops, domesticating animals, and constructing cities, had modified the environment such that no such thing as a pristine or "untouched" nature existed. Debates that might strike us as particularly contemporary – for example, climate change due to human activities such as deforestation, erosion of land, human desires to dam or re-route rivers for agricultural purposes, the depletion of the earth's resources – were the subject of heated debate in the ancient world. A predecessor of Glacken's perspectives can be found in George Perkins Marsh, whose *Man and Nature; or, Physical Geography as Modified by Human Action* (1864) noted the decisive, often destructive, role of human modification of nature.

A California native, Glacken worked during the years of the Great Depression not as an academic, but in public service to the state's needy. He was employed by agencies created under the auspices of the California

Resettlement Administration, the Works Progress Administration, and the California State Relief Administration. These labors brought Glacken into contact with those impoverished Dust Bowl-era migrants that Don Mitchell invokes in his *Lie of the Land* (see p. 159). It was not until age forty that Glacken decided to seek his PhD in geography from the Johns Hopkins University. This followed an eleven-month trip around the world, and a stint in the US army during World War II, where he received training in Japanese culture and language and was posted to Korea. In the then common "old boy" network of the time, Carl Sauer (see p. 96) offered Glacken an academic position in Berkeley's Department of Geography, where he remained for the duration of his career.

Clarence Glacken's personal history was not a wholly happy one, particularly during the last two decades of his life. He chaired the Geography Department at the University of California at Berkeley in the late 1960s, a period that coincided with often violent clashes on university campuses in the United States. On the heels of these upheavals on the Berkeley campus, Glacken went into a deep depression, suffered a heart attack, and never fully recovered his former emotional or physical vitality.

The cornerstone of Glacken's publications, *Traces on the Rhodian Shore* was initially meant to be an introductory chapter to a more comprehensive volume; however, it grew to a substantial 763 page tome of its own accord. Glacken mostly completed, but never published, a second volume on nineteenth and twentieth century art, science, and philosophy. Other works by Clarence Glacken include *The Great Loochoo: A Study of Okinawan Village Life* (1955); "Man and nature in recent Western thought," pp. 163–201 in Michael Hamilton (ed.) *This Little Planet* (1970); and an autobiographical essay titled "A late arrival in Academia," pp. 20–34 in Anne Buttimer (ed.), *The Practice of Geography* (1983).

ON ARTISANSHIP AND NATURE

If the apparent unity and order of nature led men to a belief that behind it was a plan, a purpose in which human beings were deeply involved, if differences among peoples were perceived as a matter of everyday observation in the Eastern Mediterranean, and if these were ascribed to custom or to nature, there was also an awareness of the novelty that men could create in nature, of differences brought about by art and by the power derived from the control over domestic animals. Man was a creator of order, an agent of control, a possessor of the unique skill of the artisan. Long before the Greeks there was impressive evidence of these skills in the metallurgy, mining, and building of the older civilizations, especially of Egypt. It has been said by many that Greek science, unlike modern science, did not lead to the control of nature but the occupations, crafts and the skills of everyday life were evidences that changes were possible that either brought order, or more anthropocentrically, produced more orderly accessibility to things men needed. If by control over nature one means its modem sense, the application of theoretical science to applied science and technology

. . . there was no such control in the ancient world. Conscious change of the environment need not, however, rest on complex theoretical science. . . . The power of mind was acknowledged in the analogy of the creator-artisan and in its potentials for rearrangement of natural phenomena, such as in the establishment of a village, the discipline of animals by men, the indirect control over wildlife with weapons, snares, and the like.

Finally there is the mythology of the celestial archetypes of territories and temples, of which their worldly counterparts are copies. . . . This is why, when possession is taken of a territory – that is, when its exploitation begins – rites are performed that symbolically repeat the act of Creation: the uncultivated zone is first "cosmicized," then inhabited. Thus, "Settlement in a new, unknown, uncultivated country is equivalent to an act of Creation."

Myths of this kind strongly suggest that man is an orderer of nature. In the literature interpreting the changes that men make in their environment, in the attempts to bestow meaning on these changes, there are, as we shall see, recurrent themes of man as a finisher of the creation, of man bringing order into nature, and after the age of discovery, of European man discovering new lands, which despite the presence of primitive peoples, are considered to be unchanged since the creation and awaiting his transforming hand. Did men become aware of themselves as modifiers of nature, as creators of a new environment because of the distinctions they made between themselves and the animals – mainly, higher intelligence and upright carriage – because they had a sense of creating . . . an order, because their artisanship enabled them to bring about this cosmos, and because through their power over plants and animals they were able to maintain and perpetuate it? Early Greek writings on the subject, few as they are, suggest that these awarenesses did exist.

In reading the comments of the ancient authors regarding the changes which man has made in the physical environment, one has two impressions: there was a recognition of man as an active, working, achieving being, despite the seeming stability that might be implied from the dominance of environmental influences . . . and that the living nature that these men observed – and often loved – was, as we now know, a nature already greatly altered by man.

In the ancient world, there was a lively interest in natural resources and how men could exploit them: in mining, in ways of obtaining food, in agricultural methods, in canals, in maintaining soil fertility, in drainage and grazing and many other economic activities which – even if they produced only a partial philosophy of man as a part of nature which he was engaged in changing – are eloquent proof of his busyness, his incessant restlessness in changing the earth about him. The preoccupation with technology is clear in the literature related to primitivism, whether the individual thinkers looked back to a happier, less complicated period or approved of the amenities of their own civilization. The golden age of the past was often an age of simplicity and one in which the soil required no cultivation but supported life spontaneously rather than by tillage and ordered plantings; if there had been a moral decline to the hard realities of the contemporary iron age, it owed much to the advances of the arts and sciences and to applied technology. . . .

Although many of these thinkers had traveled widely, the environment they knew best and about which they wrote with greatest affection was that of the Mediterranean basin. In the fifth century B.C.,

it was known that the history of its settlement was already a long one. Hippocrates had said that the present ways of living, unlike those – and the crude foods – of an earlier age, had been discovered and elaborated over a long period of time. They were accustomed to surroundings full of evidences of change and of human activity. . . .

One feels that to these writers – Greek and Roman alike – the vineyards, the olive orchards, the irrigation ditches, the grazing goats on the rocky summits, the villages, and the villas were inseparable from the landscape of the dry parched hills of the Mediterranean summer, the winds for which there were so many local names, the deep blueness of the sea, and the bright Mediterranean skies. It was an altered landscape, upon which they gazed and whose beauties they loved.

[. . .]

ENVIRONMENTAL CHANGE IN THE HELLENISTIC PERIOD

Although the notices from pre-Hellenistic times reveal an awareness of environmental change, they are isolated. In the ancient world as a whole, there is no lack of evidence regarding change, but interpretations of it are few. One learns of grafting, fertilizing, the laying-out of towns, but for the most part the facts are stated, and that is all. Occasionally it is possible to infer an attitude from the spirit of the writing or the spirit behind the activities described. Excellent illustrations come from Ptolemaic Egypt, such as *The Tebtunis Papyri*, the correspondence of Apollonius and Zenon, the reclamation work of Cleonand Theodorus in the Fayum (i.e., Lake Moeris, about fifty miles southwest of Cairo). All of them suggest the fervor with which the Greek colonists went about their tasks in Egypt, implying a philosophy of activity, optimism, and desire for land improvement.

When Hieron of Syracuse engages in shipbuilding and Archimedes superintends it, and boats are launched with the windlass he constructed, one receives . . . the impression that men consciously seek to change their environment, whether by building cities or ships or by introducing plants for their own purposes.

On February 16, 256 BC, Apollonius, the minister and landholder, approves of an order which

Zenon had given that olive and laurel shoots should be planted in the park at Philadelphia where Zenon had now gone or was going to reside as superintendent of Apollonius's property. In a letter dated December 27, 256 BC, Zenon is ordered to take from Apollonius's own garden and from the palace grounds in Memphis pear shoots and young plants – as many as possible – and to get some sweet-apple trees from Hermaphilos; all are to be planted in orchards at Philadelphia. In another letter of the same date, Apollonius orders Zenon to plant at least three hundred fir trees all over the park and around the vineyard and the olive trees. "For the tree has a striking appearance and will be of service to the king"; it will provide him timber for his ships and be an ornament to his estate. On January 7, 255 BC, Apollonius reminded Zenon that it was time to plant vines, olives, and the other shoots; Zenon should send to Memphis for them and give orders to begin planting. Apollonius promises to send from the Alexandria district more vine shoots and whatever other kinds of fruit trees may be useful. On October 8, 255 BC, Apollonius orders Zenon to take at least three thousand olive shoots from his park and from the gardens at Memphis. Before the fruit is gathered, he is to mark each tree from which he intends to take shoots. And he is to choose above all the wild olive and the laurel, for the Egyptian olive is suitable only for parks and not for olive groves.

[. . .]

The prevailing mood of the Eastern Greeks in early Hellenistic times . . . was one of buoyant optimism; they had confidence and faith, supported by the leading philosophical schools, "in the unlimited capabilities of man and his reason". . . . Agriculture and related occupations such as cattle-breeding were the most important sources of wealth in the ancient world. Intensification of such economic activity is favorable to landscape changes visible to the eye. Canals appear, swamps vanish, river courses change. If, as seems probable from reading the classical writers on agriculture, the judging of soils empirically was a primordial skill, the good soils had long since been known and further improvement could come only from the acquisition of new land. Land reclamation during this period was based on the science of mechanics, and on practical experience with canal-digging, irrigation, and swamp drainage. The purpose of one

famous scheme: the drainage of Lake Copais in Boeotia under the supervision of Crates, a mining engineer in Alexander's army, apparently was to increase the cultivated area of Greece. Similar projects were undertaken in the Eastern Hellenistic monarchies and in Egypt.

. . . . In the conscious development of the natural resources of Ptolemaic Egypt, about which far more is known than of the other large areas of the Hellenistic world, the purpose was to make the country self-sufficient, and to create in modern terminology a favorable balance of trade. Here . . . environmental change is a product of conscious government policy. In carrying out this policy, the solicitude of the Ptolemies for the Greek settlers led to visible changes in the appearance of the land, an apt illustration of the influence of national tastes and diet which are exported to another land. The Egyptian drink was beer, but the Greeks liked wine, and soon there were extensive vine plantings in Ptolemaic Egypt. It was the same with the indispensable olive. So vineyards and olive groves became witnesses of the Greek presence as did the fruit trees and the sheep. (It was not that such plantings were unknown in Egypt before, but they were few and not very successful.)

A history of attempts at plant acclimatization, especially in Egypt, would have in it a chapter on Greek taste in food and clothes. Experiments were not confined to Egypt, for Harpalus attempted to acclimatize pines in Mesopotamia. Theophrastus says Harpalus tried repeatedly to plant ivy in the gardens of Babylon and failed. The Greeks liked wool for their clothing, and sheep in Ptolemaic Egypt became important. Foreign sheep were imported and efforts made to acclimatize them. . . .

If one could have taken a series of photographs of Ptolemaic Egypt at suitable intervals, one could probably see, at least through the earlier period, the different crops, the new devices, and the introductions that created a more variegated landscape.

It is tantalizing to speculate on the policy of the Hellenistic monarchs toward deforestation, because this practice probably more than any other in a preindustrial society changes the ecosystem and the appearance of the land. The rulers of Egypt had given careful attention to tree planting and to cutting, but it is not known if they were interested in conservation.

During the Hellenistic period. . . . [there was a] special place of architects and engineers because of the immense amount of building, especially in the principal islands and the great commercial cities along the coasts of Asia Minor, the Straits, and the Propontis: remodeling harbors, replanning and rebuilding of such cities as Miletus, Ephesus, Smyrna, and lesser cities of Asia Minor. New cities and new temples were built, and others already in existence were rebuilt to make life within them easier, through drainage and the construction of aqueducts. Building obviously was also closely related to the exploitation of mines, quarries, and forests where they existed. War and military construction played a vital part too. . . . There seemed to be closer alliance between building construction and military engineering, and science and art, than between practice and theory in agriculture, in the absence of scientifically conducted agricultural experiments. The technical innovation that occurred was not revolutionary; it was based partly on scientific discoveries, partly on the interchange of long-established methods among the constituent nations of the Hellenistic world.

[. . .]

IS THE EARTH MORTAL?

Theories of soil exhaustion were also related to the idea of senescence in nature, an application of the organic analogy to the earth itself. The theory is ably expressed and refuted by Columella, who lived, probably, in the first century AD; although he does not mention Lucretius by name, it is his doctrine which Columella is attacking.

This idea of the senescence of the earth survived through the Middle Ages and into modern times; it was one consideration in the quarrel between the ancients and the moderns. Men like George Hakewill, Jon Jonston, John Evelyn, discuss it; and Montesquieu, arguing in the *Persian Letters* that the populations of modern times are less than those of ancient times, asks, through a letter his Persian Rhedi wrote from Venice to Usbek in Paris in 1718: "How can the world be so sparsely populated in comparison with what it once was? How can nature have lost that prodigious fertility of primitive times? Could she be already in her old age, and will she fall into her dotage?"

Lucretius believes . . . that the earth is a mortal body; it will grow old gradually, and ultimately it will die. Nor is it sacrosanct. It is nonsense, he says, to think that "the glorious nature of the world" has been fashioned by the gods according to a divine plan for the sake of man; it is foolish to think of a divine artisan who has created an eternal and immortal abode for him. . . . In modern times the point of view has been urged against the diversion of rivers, the digging of canals, and of course in medicine, the administration of anesthesia being an outstanding example. If the Lord had intended these things, they would have been created by Him in the beginning. . . .

No, Lucretius continues, the universe is too full of imperfections, the earth too full of land which cannot be used, to admit the possibility of its creation for man by divine power. Furthermore, the earth is older than it was. The greater fertility of the golden age is ascribed to the youth of the earth. The strength of man and his oxen is worn down; the plow can scarcely turn the soil of the grudging fields. The ploughman compares his ill fortune with the blessings of his fathers, who won a living from the soils so much more easily. "So too gloomily the planter of the worn-out, wrinkled vine rails at the trend of the times, and curses the age, and grumbles to think how the generations of old, rich in piety, easily supported life on a narrow plot, since afore time the limit of land was far less to each man. Nor does he grasp that all things waste away little by little and pass to the grave fordone by age and the lapse of life."

Columella attacks a similar idea, apparently widely accepted among the administrators of the state. . . . Leading men of the state complain about the lack of soil fertility and bad climatic years as being responsible for poor crops, basing their complaints "as if on well-founded reasoning, on the ground that, in their opinion, the soil was worn out and exhausted by the overproduction of earlier days and can no longer furnish sustenance to mortals with its old-time benevolence." Speaking more plainly, Columella continues:

For it is a sin to suppose that Nature, endowed with perennial fertility by the creator of the universe, is affected with barrenness as though with some disease; and it is unbecoming to a man – of good judgment to believe that Earth, to whose lot was assigned a divine and everlasting youth, and who is called the common mother of all things – because she has always brought forth all things and is destined to bring them forth continuously – has grown old in mortal fashion.

Columella does not mean that the soils cannot be exhausted, that they are everlastingly productive, but that their failures may have a human cause.

The comparison of Mother Earth with a human mother, he says, is a false one. After a certain age, even a woman can no longer bear children; her fertility once lost cannot be restored, but this analogy does not apply to soil which has been abandoned, for when cultivation is resumed "it repays the farmer with heavy interest for its periods of idleness." Soil exhaustion is not related to the age of the earth but to agricultural practices . . .

[. . .]

INTERPRETING ENVIRONMENTAL CHANGES WITHIN A BROADER PHILOSOPHY OF CIVILIZATION

Among the works of writers whose interpretations of environmental change are part of a broader philosophy, those of Cicero, the Hermetic writers (who probably derive their ideas on this matter from the Stoics), Lucretius, Varro, and Virgil are the most instructive. Despite differences in approach, each of them either implicitly or explicitly assumes that cultural history has at least in part been the history of environmental change and that the development of the arts and sciences has brought about changes in the physical environment.

In the Stoic philosophy, man's technological achievements, his inventions, the changes he brings about in nature, are combinations of the skill of the hand, the discoveries of the mind, observations of the senses; he has his share of the artisanry and reason which permeate the world, the earth being particularly suited to him, as witness the arrangements of external nature like the Nile, the Euphrates, and the Indus, that exist for his preservation and care.

Environmental change by man, the creation of a "second nature" within the world of nature, is explained in essence by the basic qualitative dif-

ference between human and animal art. Man is a reasoning creature, whose cumulative experience through time permits innovation and invention; he participates in the creative life and spirit pervading the whole world.

The naturalistic view of Lucretius presents an alternative interpretation. . . . Men by their struggles add to what is already provided by nature. Tilled lands are better than untilled ones; they produce more. Earning a livelihood is enmeshed in a physical and human cycle: rain from the skies ultimately brings food to the towns; later the water of the streams returns to the ocean to be lifted again to the sky. Lucretius is deeply aware of the physical difficulty men have in maintaining the environments they create; with failure, carelessness, or laziness, the thorns, coppice, and weeds will again invade the tilled field.

[. . .]

Men in the past, though hardier than those of today, did not spend their energies at the plow, for they knew nothing of plowing, planting, or pruning. Like people of the golden age, they accepted freely the spontaneous gifts of the earth. The invention of fire was a great step forward in the conquest of nature; lightning, or possibly the friction of tree branches with one another, first made it available. Then, in lessons from the sun and its effects on earthly substances, men learned how to cook. With the invention of fire, the next step was the discovery of metallurgy.

Lucretius' theory of the origin of metallurgy reveals how conscious he was of the activities of man: the discovery of the metals (copper, gold, iron, silver, lead) he ascribes to great forest fires which may have been started by lightning, by warring men who started fires against one another, or by those who desired to increase their arable lands and pastures at the expense of forests or who wished to kill off wild beasts. "For hunting with pit and fire arose first before fencing the grove with nets and scaring the beasts with dogs." The forest fire, whatever its cause, burned so fiercely that the melted streams of silver, gold, copper, and lead flowed into the hollows of the earth's surface, and men, attracted by the luster and polish of the metals, could see from their odd shapes that they could be molded. They could now make tools to clear forests and work up lumber, and to till the fields, first with copper tools and later with the iron plow.

Taught by the model of nature and in imitation of her, men planted and grafted plants, and experimented with various types of cultivation. With gentle care, they brought the wild fruits under human protection and cultivation, and following the suggestions of nature, they widened areas of change, substituting a domesticated environment for the pristine.

And day by day they would constrain the woods more and more to retire up the mountains, and to give up the land beneath to tilth, that on hills and plains they might have meadows, pools, streams, crops, and glad vineyards, and the grey belt of olives might run between with its clear line, spreading over hillocks and hollows and plains; even as now you see all the land clear marked with diverse beauties, where men make it bright by planting it here and there with sweet fruit-trees, and fence it by planting it all round with fruitful shrubs.

... In the passages just quoted, [Lucretius] is clearly describing, in poetical language and without any suggestion of decay or death, the manner in which a people transforms the landscape.

Man's progress in the arts has its effects on his environments as well; he has learned by imitation, by using his mind, and he has increased his knowledge by practice and experience; he has saved many animal species; he has domesticated plants, has cleared and drained land, and the landscape about him is, at least in part, a result of his own creativity.

[...]

Man is thus a part of nature; he shares his creative endowment with the whole cosmos but his arts are in a different realm of being than are those of the animals. With his hands, his tools, his intelligence, he has changed the earth by creating arts and techniques of agriculture, fishing, animal domestication, by mining, clearing, and navigation.... [N]ature has given man opportunities, such as the life-giving floods of the Nile, the Euphrates, and the Indus, with the idea that man in turn has not only preserved but improved animals and plants which would become extinct without his care; that nature has given man hands, a mind, and senses, the basic endowments of his art: the mind to invent, the senses to perceive, the

hands to execute. By human hands much of nature has been both controlled and changed. Our foods are a result of labor and cultivation; wild and domesticated animals are put to many uses; the mining of iron is indispensable to tillage; clearings are made for fire, cooking, house- and ship-building.

[...]

CONCLUSION

The thinkers of antiquity developed conceptions of the earth as a fit environment for human life and human cultures whose force was still felt in the nineteenth century. The conception of a designed earth was strongest among the Academic and the Stoic philosophers, but even among the Epicureans there could exist a harmony between man and nature, orderly even if not a product of design. Geographically, it was a most important idea: if there were harmonious relationships in nature ... of which man was a part, the spatial distribution of plants, animals, and man conformed to and gave evidence of this plan; there was a place for everything and everything was in its place. It assumed the adaptation of all forms of life to the arrangements of nature found on the earth.

Furthermore, this conception was hospitable to our two divergent if not contradictory ideas: the influence of the environment on man, and man's ability to change it to his own uses. The first could be accommodated by pointing to evidence of design in the different climates of the earth and the peoples, plants, and animals living in them and adapted to them. So could the second. Man, as the highest being of creation, changes nature – even improves it – through art and invention; his habitats, in Strabo's words, show that art is in partnership with nature. His environments may be those of art – the towns and cities, centuriation, clearings, irrigation works, farming and viticulture – but they are really products of his divinely endowed intelligence; his inventions, tools, and techniques spring from a higher creative source as he improves and brings the pristine earth to a finished state.

Equally important was the utilitarian bias of these speculations, especially in those thinkers who saw the creation as serving the uses of man, and who, interpreting the past by observation of the present, saw in the usefulness of the grains, of beasts of burden, of the dog, the sheep, and the goat the reasons for their creation. These domestications took place in the past for purposes illustrated by the uses to which they are put in the present. And lastly, if one may speak of ancient thought in modern language, the idea of design was antidiffusionist in character; the idea of a design with all parts well in place and adapted to one another in an all-embracing harmony implied stability and permanence; nature and the human activity within it were a great mosaic, full of life and vigor, conflict and beauty, its harmony persisting among the myriads of individual permutations, an underlying stability.

Since classical times, this conception of a designed earth has been but a part of a wider teleology and a philosophy of final causes, but one should not forget that it is the beauty, the utility, the productivity of nature on earth that, with proper selectivity and avoidance of the harsh and unproductive, provided convincing evidence of purpose in the creation, and in turn a traditional proof for the existence of God. The conception of the earth developed by the classical thinkers and the moderns who followed them was no abstract natural law. It could be enriched with lovely, often poetic, descriptions of nature itself. It owes its force and its influence to its all-embracing character; all ideas could be fitted into it and this hospitality was the reason for its failure: anything that existed, any relationship could be explained as part of the design, if one ignored (as Lucretius refused to do) certain characteristics of the earth as a habitable planet that were hard to explain as products of purpose and design. . . .

. . . The history of theories based on situation is derived from multiple sources, a result both of the diversity of Mediterranean life and of relief and site in the Mediterranean basin and in less-known peripheral areas. Generalizations emerged from the role of the sea in Greek history, the rise of Rome to become the cosmopolitan capital of an empire, and of the effects of Greek and Roman civilization on the barbarian peoples living adjacent to them. . . .

. . . If the earth was divinely ordered for life, man's mission on earth was to improve it. Such an interpretation found room for triumphs in irrigation, drainage, mining, agriculture, plant breeding. If this interpretation of man serving as a partner of God in overseeing the earth were correct, understanding man's place in nature was not difficult. When, however, unmistakable evidences that undesirable changes in nature were made by man began to accumulate in great volume in the eighteenth and nineteenth centuries, the philosophical and theological underpinnings of the classical, and later of the Christian, idea of stewardship were threatened. For if man cleared forests too rapidly, if he relentlessly killed off wildlife, if torrents and soil erosion followed his clearings, it seemed as if the lord of creation was failing in his appointed task, that he was going a way of his own, capriciously and selfishly defiant of the will of God and of Nature's plan; but castigations of this kind do not appear until the eighteenth and nineteenth centuries, reaching their culmination in Marsh's *Man and Nature*.

[. . .]

There is a sharp contrast between ancient and modern literature on the modifications of the earth by human agency. If the surviving works from the ancient world are representative, the contrast is a measure not only of the vast increase in the amount and rate of change in modern times, but also of an awareness of change, accumulating in the Middle Ages, advancing rapidly in the seventeenth, eighteenth, and nineteenth centuries, rising to a crescendo in our own times, and for which we are still seeking explanations that rise above description, technical solutions, and naive faith in science.

"Living Outdoors with Mrs. Panther"

from *Landscape* 4, 2 (1954–1955): 24–25

"Ajax"

Editors' introduction

"Ajax" is one of the many pseudonyms adopted by the American cultural geographer J.B. Jackson (1909–1996; see also p. 153). In 1951 Jackson founded the journal *Landscape*, from which the selection featured here, "Living outdoors with Mrs. Panther," is drawn. In the journal's early years, Jackson himself wrote most of the articles, opinion pieces, and sketched the illustrations as well. He adopted a variety of pen names – G.A. Feather, A.W. Conway, H.G. West (or H.G.W.), P.G. Anson, and of course the picaresque Ajax – a tactic which allowed him to (somewhat) disguise the fact that he was the sole author of all of the journal's pieces. More importantly, the pseudonyms gave Jackson license to express a variety of perspectives in his writing.

In "Living outdoors with Mrs. Panther" Jackson adopts an ironic, at times acerbic, and always tongue-in-cheek style to mimic an interview that might have appeared at the time in a popular magazine on modern life and architecture, such as *Architectural Record* or *Progressive Architecture*. Highly critical of city folk who purport to "live simply," Jackson points out time and again that the fictive "Babs Panther" has – to put it mildly – a hypocritical relationship with nature. Though Mrs. Panther may say that her family desires nothing more than to live a simple life far away from the bustle of the big city, it is clear that her actions contradict her ideals at every turn. From the "temperatrolled" house to the windows that cannot be opened without the help of an engineer, to her obsession with maintaining a germ- and insect-free environment, to the artificial coloring in the swimming pool, the Panther family could not be more nature-phobic. Jackson's distain for the big-city socialites he mocks in the form of Mrs. Panther is barely disguised: Babs is portrayed as vain, silly and infantile. The artists and architects referred to throughout the piece – architect Mies Van der Rohe, sculptor Henry Moore, painter Georges Braque, artist Alexander Calder – are key figures in the modern design movement of which Jackson was extremely critical. Jackson's writings on the topic pre-date Robert Venturi's similar criticisms; see Venturi's *Complexity and Contradiction in Architecture* (1966).

Babs Panther, and the so-called "young moderns" of whom she is an archetype, envisions herself as blurring the divide between the natural and the artificial worlds created by humans. Yet, as Jackson demonstrates in this piece, Mrs. Panther has in fact sharpened the line between the artificial and the natural; furthermore, she has chosen to live exclusively on the artificial side of the divide. Certainly some of the hypocrisy Jackson exposes in this piece from the mid-1950s resonates with contemporary ironies. Think, for example, of the trendy "back to basics" ethos extolled by the now popular consumption of organic foods, ecotourism, and the restoration of historic houses. These activities are billed as allowing ordinary people to cut out the excesses of modern living and get in closer touch with nature and, thus, with their true selves. In reality, to participate in these activities requires a great deal of disposable income and free time. They are available only to the wealthy, those whose income and free time may well be enabled by a disingenuous complicity with the excesses and exploitations both of nature and of other human beings. Jennifer Price, in her book titled *Flight Maps: Adventures with Nature in North America* (1999), makes similar observations about the hypocrisy behind such practices as the rising popularity of the plastic yard flamingo at the same time that living species of birds were becoming extinct, or the oxymoronic term "natural company" that is so prevalent today.

Key works by J.B. Jackson are detailed on p. 153. With respect to Jackson's critical engagement with modernity and architecture, see particularly "Review of built in the U.S.A." (written under the pseudonym H.G. West), in *Landscape* 3, 1 (1953): 29–30; "Hail and farewell," in *Landscape* 3, 2 (1953–1954): 5–6; and statement in "Whither architecture? Some outside views," in *AIA Journal* 71 (1982): 205–206. Jackson's positive assessment of the vernacular, often commercially oriented, design of small-town US houses and towns can be found in "The almost perfect town," in *Landscape* 2, 1 (1952): 2–8; "The westward-moving house," in *Landscape* 2, 3 (1953): 8–21; and "Other-directed houses," in *Landscape* 6, 2 (1956–1957): 29–35. These essays and more are gathered and reprinted in Helen Lefkowitz-Horwitz's *Landscape in Sight: Looking at America* (1997). An overview of Jackson's life and works from an architectural historian's perspective is found in Marc Treib's "The measure of wisdom: John Brinckerhoff Jackson (1909–1996)," in *Journal of the Society of Architectural Historians* 55, 4 (1996): 380–381 and 490–491.

"The immediate experience of nature": How many of us really know what that means? Well, plenty of Young Moderns do, and Mr. and Mrs. Jeffrey Panther – he's the New York publisher, of course – have gone about proving it in a smart, typically Young American way.

Quite simply, quite casually, entirely without fanfare, the Panthers have decided to live out of doors. Not in a tent like their pioneer ancestors – Mrs. Panther, incidentally, is a direct descendant of Clara Peabody Newell – No; in a house specially designed by famed Modernist Mies van der Rohe. On a small ten-acre lot in Connecticut's Fairfield County there has recently been built a gay little $100,000 home for the enterprising young family of four.

"Jeff and I call the whole thing an experiment in Modern Living," Mrs. Panther laughingly explained when we telephoned her one day last summer. "But do come and see what fun we're having."

So we did, and because we found the Panther home so excitingly modern (in the wholesome American sense of the word), we want to tell the readers of *Landscape* all about it.

ART BELONGS IN THE MODERN HOME!

Enchantingly sleek and simple in appearance – a long white box perched on stilts – the Panthers'

house is situated in a grove of wonderfully natural-looking trees. Mrs. Panther – "Babs" to her many friends in the World Federalists and on the Community Forum Committee – meets us at the door. She is wearing black velvet toreador tights, ballet slippers, and a divine yellow linen shirt. With her blonde hair in a horsetail, she looks for all the world like a little girl. "This is my year-round costume," she explains later; "I never wear anything different. You see, the house is temperatrolled."

We glance, fascinated, into the enormous living room – or, as the Panthers call it, the play space. We are speechless with delight: one entire wall is occupied by a vast window (of Sanilite glass, of course, which lets in only the health-giving rays) reaching fifteen feet from brick floor to ceiling. Outside is a charmingly unspoilt view of trees and rocks and underbrush. "Here we sit, like Hansel and Gretel, Jeff and I, right in the heart of the woods! We even have a tree here in the middle of the play space!" And so they have: the slender trunk of a maple rises out of the floor and then disappears through the ceiling. "We love our tree," she says softly, laying her hand on the trunk. "The texture of the bark is so exciting. Mies van der Rohe was a lamb and let us have it." And how wonderfully *right* it is! It lends just that simple sophisticated touch to the decor of the room. The natural form is repeated by a small but important piece of Henry Moore sculpture on the floor; a witty Calder mobile twinkles overhead. "Don't you adore our tiny little art collection? These two," Mrs. Panther says, "and a sweet little Braque are all we could afford; we saved and saved and *saved* to buy them." A gay little smile admits us to her confidence. "But we simply *had* to have them," she continues, "because if you love plants and animals and birds the way Jeff and I do, you just have to have that kind of art – like nature."

THE AMERICAN HOME AT ITS SIMPLE BEST

Despite the summer heat out of doors and the bright clean light streaming through the uncurtained windows, the play space (living room) is wonderfully cool and fresh. And that, of course, is because of the temperatroll. Pointing to an instrument panel, Mrs. Panther explains this modern miracle, this triumph of the American will to live beautifully and wholesomely – and with simplicity. "Oh, we scraped pennies in order to have our own special climate. Jeff and I are essentially outdoor people – like all our younger and more stimulating friends," she adds. "We simply couldn't *stand* living in an old-fashioned Victorian house with all that absurd closing and opening of doors and windows. We want to live indoors just the way we would live outdoors: freely and informally and spaciously. If you know what I mean."

So there is no need to open any of the windows in the Panther house. Not that it's complicated to do so; a telephone call to the local ventilating engineer, and out he comes at once with his special equipment, and in no time the great windows are opened outwards.

"We spend hours in the outdoor play space," Mrs. Panther remarks. "That's where we have our swimming pool and the squash court. Of course we had to cut down on the cost of the house in order to have them. But Mies was a darling about it."

For a moment she disappears to spread a special insect-repellent suntan oil on herself, and to get her special sunglasses and a shade hat, before taking us out into the garden. What a fabulous spot it is! Small, but so natural and so modern in feeling! No prim flowerbeds and tiresome hedges; a stretch of that chic Brazilian gravel which is so popular in California this season; a few potted jub-jub

trees ("They were flown in from Hawaii") and a casual array of Chinese ivy in pots. That's all. Italian beach furniture and gaily striped parasols are grouped in front of a wall of the stylish split-beech French fence. We catch a glimpse of the pool beyond: a spot of turquoise in a free-form basin. "Of course," she confesses, "the color is artificial. But the water has been thoroughly tested; it is chlorinated and filtered and kept at the correct temperature. The children splash about in it like savages! I *do* dislike old-fashioned restrictions, don't you? Ronny and Jody" (those are the two Panther children) "can do anything they like, provided they take their multivitamin shots and never eat anything except what comes out of the kitchen or pick wild plants or fondle stray animals or play with children who might be dirty or socially maladjusted.

"That's why Jeff and I won't have pets around." (The squash court building contains a shower and an air-conditioned exercise room with a marvelous family-size sunlamp.) "We don't believe in interfering with nature. We spray the trees, disinfect the soil, and change the potted plants every two months – and then we let things take their course."

We find this admirable; we like this forthright rejection of pruning and clipping and transplanting. Do the Panthers have a vegetable garden?

"No; but we're trying hydroponics in the guest bathroom so that the children will have a feel for growing things."

SCIENCE PLUS AMUSING INFORMALITY IS THE WATCHWORD

Nor do the Panthers sleep or eat out of doors. "Jeff, poor darling, is allergic to practically everything that grows in Connecticut – or anywhere else, for that matter. He has to have an air-conditioned room all of his own." As for eating outside: "Well," Mrs. Panther says with a delightful smile, "I think I prefer to keep the outdoors for the very simplest kind of pleasure. And I adore my work area" (kitchen, in old-fashioned parlance) "and spend a great deal of time there. When we have company I open some cans and toss a salad; we have a bottle of French wine, some cheese, and then sit around on cushions and discuss McCarthyism and how we dislike it. I've become quite a cook," she adds proudly.

The children? They have their own rooms – sound-proof and out of the way. "Besides, they spend most of their time at the Play Clinic in town, where there's a marvelous psychiatric guidance expert."

Yes, we reflect, as Mrs. Panther leads us back into the house, this typical American family leads a natural life for Young Moderns. The artificialities of city existence are far, far removed from the quiet little eight-room house out there on stilts in the Connecticut woods. Nightclubs, traffic jams, dirt, and confusion are no part of their life. Excitement? A casual little concert on recorders, or a new wine-and-shallot sauce Babs discovers, or waking up on a winter's morning to see the Japanese-printlike effect of snow on the black branches– these comprise the Panthers' happiest moments. The Panthers, by the way, have an automatic snow-melting system from the garage door to the road a hundred yards distant, so that Jeff need not shovel snow like his Victorian forebears. What's more, it disposes of the melted snow so that no ice is ever formed on the driveway. "Let it snow," says Babs in the words of the once popular song. She turns up the thermostat, adjusts the temperatroll to suit her toreador tights and yellow shirt and little-girl hairdo; and once the children have been called for by the school bus, she settles down with a volume of her favorite author, André Gide, to enjoy a winter's day in the country. "I'm afraid," she laughingly tells us, "that I wouldn't know how to behave in the city any more. But we Young Moderns are like that: we want to live abundantly, the way Jeff and I do: in a simple kind of house with this immediate kind of experience of Nature." She thoughtfully caresses the Henry Moore composition. "Or do you think I'm utterly barbaric?"

Well, frankly, Mrs. Panther, since you ask. . . .

"Nature at Home"

from *The Culture of Nature: North American Landscape from Disney to the Exxon Valdez* (1991)

Alexander Wilson

Editors' introduction

Throughout Part Four, one of the key questions – if not *the* key question – has been "What is humankind's relationship to nature?" In this selection by Alexander Wilson, "Nature at Home," the focus is on the twentieth-century American suburb. In this particular place and time, the answer to the question about humans and nature has, by and large, been that humans and their technology stand outside of and dominate nature. Wilson provides a keen observation on how American suburban development is predicated on the obliteration, and subsequent recreation, of "natural" vegetation. Rather than attempting to reinstate what was there before the developers came, however, the post-war landscaping industry introduces a stylized ideal of vegetation that is intended to provide a specific backdrop for human habitation. Thus Wilson illustrates how the human drive to shape nature to our purposes has, in post-war suburban America, had profound consequences for what we think of as the normal ("natural") and desirable. This applies not only to the non-human world of vegetation (and the absence of wildlife in suburbs), but also to post-war gender relations.

Shaping nature to human needs and desires is by no means limited to twentieth century North American suburbs. Indeed, as Clarence Glacken argues in "Creating a Second Nature" (see p. 212), the routine cultivation of vegetation for human consumption and pleasure is a tradition that goes back at least as far as ancient Mediterranean civilization (and certainly farther, in the non-Western world). In the selection by Jones and Cloke in this part, "Orchard" (see p. 232), we get a complementary contemporary example from rural Britain of a botanical landscape consciously directed toward human consumption and pleasure. Thus landscape design as an active and professionalized undertaking that shapes nature to human needs and wants in American suburbs, British orchards, and ancient gardens might well be seen together as varieties of *working landscapes*, much in the way that this is discussed by Don Mitchell in the context of California's agricultural industry (see p. 159).

What is particularly distinctive, and troubling, about post-war American suburban landscape design is its highly technological character. Pesticides, fertilizers, large amounts of water used for irrigation, machinery that runs on fossil fuels, and the desire for showy non-native plants that tend to be less hardy in their new environments all add up to potential ecological damage. Rachel Carson, in her landmark book *Silent Spring* (1962) made an early and forceful case for the destructive effects of household and garden chemical use on wildlife, particularly birds.

But Wilson's prognosis is far from grim. Indeed, he asserts that American landscape design has a longer history than just its post-war suburban manifestation. In the early work of landscape architects, such as Frederick

Law Olmstead, who designed New York City's Central Park, or the quintessential American architect Frank Lloyd Wright – himself a supporter of suburbs – we see an enduring concern with the aesthetic intertwining of human habitation with vegetation. Wilson concludes this piece with an optimistic discussion of *restoration ecology*, which he sees as a promising trend that attempts to come to terms with the fact that humans must intervene in nature, but can do so in friendlier ways that we have been accustomed to doing. Restoration ecology "nurtures a new appreciation of working landscape, those places that actively figure a harmonious dwelling-in-the-world" which was very much in line with Wilson's own horticultural practice. An intriguing exploration of harmonious human habitation of working landscapes can be appreciated in the work of Scottish artist Andy Goldsworthy. In videos such as *Rivers and Tides* (2001), Goldsworthy works with materials found around him, such as minerals, flower petals, or twigs, to create ephemeral large-scale sculptures or more permanent constructions such as walls and cairns. Though the intent of much of Goldsworthy's work is for the creation to dissipate without leaving a trace, some of his projects – such as the draping of low stone walls with sheep's wool bunting – blanket a working landscape in beauty crafted from ordinariness.

Contemporary cultural geographers who have studied the balance (or lack thereof) between humans and the non-human natural world include Roderick P. Neumann, whose work on wilderness preserves in Tanzania questions the shifting, and highly politicized, boundaries between natives and nature in *Imposing Wilderness: Struggles over Livelihood and Nature Preservation in Africa* (2002); and Richard A. Schroeder's *Shady Practices: Agroforestry and Gender Politics in the Gambia* (1999), which can be read for some interesting parallels to, as well as striking divergences from, Wilson's observations on the ways that gender roles and relationships to the environment change together.

Alexander Wilson (1953–1993) was born in the United States, and grew up in Oakland, California. In his twenties, he moved to Toronto, Canada. An active scholar, Wilson also engaged in community activism and practiced landscape design. He designed the landscaping of the AIDS Memorial, in Cawthra Park (located in a predominantly gay neighborhood of Toronto). Sadly, Wilson did not live to see his plans carried out, as AIDS claimed his own life shortly thereafter at the age of forty. In 1998, five years after his death, the Alex Wilson Community Garden was established in Toronto in his memory.

A SOCIAL ECOLOGY OF POSTWAR LANDSCAPE DESIGN

We don't just talk and dream about our relations with the non-human world. We also actively explore them in the real places of our streets, gardens, and working landscapes. By crossing to the sunny side of the road on a winter's day, or by arranging some flowers in a vase, we both respond to and address the animals and plants, rocks and water and climate that surround us. Those working landscapes – the ordinary places of human production and settlement – are enormously complex places. Their history is in part a history of engineering – of how we build bridges, contain water, prune trees, and lay sidewalks. But it is also an aesthetic history. It is about shaping, defining, and making the world beautiful in a way that makes sense to us in the time and place that we live.

Throughout the twentieth century, landscape design ("landscaping" as opposed to landscape) has expanded into new spheres. Regional planning agencies have built new towns and reorganized entire watersheds, all of which require landscaping. In addition to traditional sites such as public parks and private estates, landscaping is now done alongside freeways and in industrial parks. We see landscaping at airports and outside restaurants and shopping centres, as well as inside buildings. Some of these sites either didn't exist before or weren't typically planted and tended by humans.

There have also been changes in the way people have come to make their domestic spaces fit their ideas of – or felt needs for – nature. In the twentieth century, millions of North Americans left rural communities and settled in cities and suburbs, disrupting their traditional physical relationship with the non-human world. Yet in the construction of suburban yards, victory gardens, and, later, shopping malls, community parks, and "wild gardens," people have addressed and replicated nature in other ways, developing new aesthetics in the process.

Changes in North American settlement patterns have been slow and uneven, and they have had complex social and geographical repercussions. City and country can no longer be thought of as the two poles of human settlement on the land. As agriculture was industrialized and the economy shifted its centre to the city over the course of the last century, many people abandoned rural areas, leaving whole regions of the continent both socially and economically impoverished. By the 1960s, when this trend peaked, more than two-thirds of North Americans lived within the rough boundaries of urban agglomerations. But those boundaries have gradually become indistinct. In the postwar years, regional planners directed most population growth to the new geography of the suburb, which took over rural lands on the margins of cities. By 1970 almost 40 per cent of U.S. citizens lived in the suburbs, which became, ideologically at least, the dominant land form on the continent.

Yet the next twenty years brought further changes. Many people moved back to rural areas, or to more intact examples of the small towns that were engulfed by the rapidly expanding cities of the postwar years. In the 1960s the back-to-the-land movement . . . was merely one symptom of a much more systematic development that brought about an increasing interaction of urban and rural economies. Rural areas became very different places than they were two decades earlier. Agriculture, for its part, became closely (and perhaps fatally) linked with urban money markets. In legitimated scenic areas, the leisure industry . . . propelled itself into existence through the mass marketing of raw land, recreational communities, resort condominiums, and second homes.

As the nature of the capitalist economy shifted towards information and commodity production, production was decentralized. Now, many industrial activities no longer rely on concentrated workforces or physical proximity to resources or markets. Data processing centres and small more specialized industries have parachuted themselves into forests and fields well away from metropolitan areas, giving rise to new kinds of exurban settlements that some commentators have called

"technoburbs." All of these developments have intensified the reinhabitation of rural space.

These complex displacements and resettlements . . . have contributed to a jumble of landscape design styles. . . .

In recent years a great many critical and alternative landscaping practices have emerged. Some of these try to combine modernist forms with an environmentalist ethic – by using conservation and wildlife plantings, for example. Some, like urban agriculture projects, insist on integrating horticulture with local economies. "Natural landscaping" and wild gardens attempt to reintroduce indigenous land forms to horticulture and to reanimate the city. Current trends in horticulture suggest a movement away from concentrating on individual species and towards the creation of whole communities of plants, of habitat.

All of this work challenges the orthodoxies of postwar landscaping, the culture of golf courses and petrochemicals and swimming pools that many of us grew up aspiring to. In the best of this work . . . we can see the re-emergence of a pre-modern relationship with nature, a relationship that is not about domination and containment. We can begin again to imagine nature as an agent of historical forces and human culture.

THE PLANTING OF THE SUBURB

The postwar suburb has had an enormous influence on modern landscaping practice and its aesthetic continues to influence human geographies the world over. . . .

Mobility is the key to understanding contemporary landscape design, because in the last forty years planners and builders have organized most land development around the automobile. This has had enormous effects on how most of us see the landscape. It has also changed the look and feel of the land itself. The car has encouraged – indeed, insisted on – large-scale development: houses on quarter-acre lots, giant boulevards and expressways that don't welcome bicycles or pedestrians, huge stores or plazas surrounded by massive parking lots.

The mass building techniques practised in North America both require and promote uniformity. To build on land, property owners first have to clear and level it. Everything must go. Once they put up the structures they replant the land. Biological life is allowed to reassert itself, but it is always a life that corresponds to prevailing ideas about nature. Obviously, building contractors cannot restore the land to its former appearance – an impossible task, because they've had the topsoil removed and heavy machinery has compacted the remnant subsoils. But it is also ideologically impossible. A suburban housing development cannot pretend to look like the farm, or marsh, or forest it has replaced (and often been named after), for that would not correspond to popular ideas of progress and modernity, ideas based more on erasing a sense of locale than on working with it. By and large, contemporary design and materials strive towards universality. Regional character . . . is now a matter of choice rather than necessity. When buildings were made of local stone, wood, and clay, they had an organic relationship to the soils and plants of the region.

We can get a direct sense of these changes by considering what has been planted in the suburban landscape. First, the plantings have had to be species able to survive the harsh conditions of most North American suburbs: aridity, soil compaction, salt spray from roads, and increasingly toxic air and water. Where I live, the plants that "naturally" grow in such places are pioneer species like dandelion, sumach, tree of heaven, and brambles of various kinds – plants that, ironically, are usually considered weeds. Yet instead of recognizing the beneficial functions of these opportunistic species, university horticulture departments spent much of the 1950s and 1960s breeding properly decorous plant varieties and hybrids able to tolerate the new urban conditions. The plants had to be fast growing, adaptable to propagation in containers, and, perhaps above all, showy. By definition these requirements preclude most native North American species – for the showy very often means the exotic. Unfortunately, with so much effort put into breeding the top of the plant for appearance's sake, the resultant hybrid invariably has a shallow, weak root system, a bare base, and needs frequent pruning, fertilizing, and doses of pesticides during its short life.

Evergreens became another common feature of the suburban aesthetic. The junipers, spruces, yews, and broadleaf evergreens planted throughout the temperate regions of the continent constantly say "green" and thus evoke nature over and again. The implication is that nature is absent in the leafless winter months (or perhaps all too present), because by some oversight she does not produce green at that time of year. So evergreens are massed around the house as a corrective.

But what are the economic strategies of the culture in remaking the domestic landscape? Certainly some already existing ideas were carried over to the postwar suburbs. Many people planted fruit trees and vegetable gardens when they moved to the suburbs, and indeed, some even brought their pigs and chickens – at least until municipalities passed anti-husbandry legislation in the name of sanitation. Yet the backyard could not serve as a displaced farmyard. Too much had intervened. The suburb quickly became locked into a consumer economy in which agriculture, energy, transportation, and information were one consolidated industry. Sanitation and packaging technologies further mediated relations with the environment. So while suburban hedges and fences could recall the now ancient enclosures of farm and range, for example, they also promoted reinvigorated ideologies of private property and the nuclear family.

Most of the North American suburb was built quickly in the years following the Second World War. One result of such an immense undertaking was a standardization of landscape styles. Several extant styles were drawn upon to create an aesthetic that everywhere is synonymous with modernity and that until very recently dominated landscaping practice. In its caricatured form, the most prominent feature of the modern suburban aesthetic is the lawn, in which three or four species of exotic grasses are grown together as a monoculture. Native grasses and broadleaf plants are eradicated from the lawn with herbicides, and the whole is kept neatly cropped to further discourage "invasion" by other species, a natural component of plant succession. Massive doses of pesticides, synthetic fertilizers, and water are necessary to keep the turf green. . . .

The aesthetic value of the lawn is thus directly proportional to the simplicity of its ecosystem, and the magnitude of inputs. The "byproducts" of this regime are now familiar: given the intensive inputs of water and fossil fuels, there's a related output of toxins that leach into the water table.

Typically, the suburban lawn is sparsely planted with shade trees and occasionally a small ornamental tree bred to perform for its spectators: it either flowers or is variegated or somehow contorted or stunted. These species are planted to lend interest to an otherwise static composition. The house is rung with what are called foundation plantings, very often evergreen shrubs planted symmetrically or alternated with variegated or broad-leafed shrubs. These are usually clipped into rounded or rectangular shapes. The driveway and garage otherwise dominate the front of the lot. A hard-surfaced area for outdoor cooking and eating is off to the rear or side of the house and a bed for vegetables or flowers is usually at the far side of the backyard. The house's positioning on the lot has little to do with the movement of the sun or any other features of the place. The determinants of the design are more often the quantifiable ones: number of cars per family (the industry standard is 2.5 cars, plus recreational vehicles and lawn-mowers), allowable lot coverage, and maximum return on investment. Such is the suburban garden as it has been planted in countless thousands of communities up, down, and across the continent.

[. . .]

MEN AND WOMEN IN THE SUBURBAN GARDEN

In postwar North America, patterns of management and domination suffused popular culture. The pastoral lawn, for example, not only predominates in suburban front yards, but also stretches across golf courses, corporate headquarters, farmyards, school grounds, university campuses, sod farms, and highway verges. For such enormous expanses of this continent to be brought under the exacting regime of turf management, an entire technological infrastructure had to be in place. There had to be abundant sources of petroleum and electricity to provide for an increasingly mechanized horticulture. Power mowers, clippers and edgers, weed whips, leaf blowers, sod cutters, fertilizer spreaders, and sprayers brought nature under control. Hedges and shrubbery were closely clipped. Each housing lot needed its own driveway (a large one, to accommodate the 2.5 cars). In colder climates this often necessitated the purchase of a snow

plough or blower. In the 1950s, the new petro-chemical industry introduced chlorinated hydrocarbon pesticides as virtual miracle products that would liquidate unwanted weeds, insects, or fungi. Popular horticultural literature reduced the soil . . . to a lifeless, neutral medium that did little more than convey water-soluble fertilizers and help plants stand up. As a site of mediation between humankind and nature, the postwar garden had become technologized.

While contemporary garden chores may still be a source of pleasure, the chores themselves have changed. Many people talk fondly today about climbing on to a tractor mower and cutting an immense lawn – not unlike the way a combine harvests a field of grain. This is an activity that ends up integrating the human body into a mechanistic view of nature. The idea of the body as machine has been around since the Enlightenment and the beginnings of industrial capitalism; gardening had also begun to be mechanized by the early nineteenth century. But in postwar North American culture, a great many people became gardeners for the first time, for street trees and parks were no longer the only horticultural presence in the city. The space that surrounded the suburban tract home was of a new kind, however. It was neither the kitchen garden and barnyard familiar to women nor the rural field or urban street that was most often the domain of men.

As gardening became both less exacting and more technologized – in other words, as it came to be synonymous with turf management – it was increasingly an enterprise carried out by men. Previously, for men technics had always been confined to the workplace. The home, and the symbolic clearing in which it stood, had been thought of as a refuge from the world of alienated labour. But changes in the economy brought changes in the relationship between work and home. In some ways the workplace has been demasculinized as industry has shifted away from primary production towards what are called "services." As consumption, rather than production, came to dominate Western economies in the second half of the twentieth century, men often took up more exacting "hobbies" to compensate for the loss of physical labour. Care of the garden was one such hobby.

That's not to say that women stopped gardening, any more than they stopped cooking when men began to preside over the backyard barbecue. But women's presence in the garden tended to become associated even more with everything that could be generalized as "flowers": perennial borders, herb gardens, arbours and trellises, window boxes, bedding plants, and greenhouses. The landscape profession often dismisses this horticultural work (and horticulture is not a strong tradition in North America) as being too fussy or labour-intensive, when it is perhaps better thought of as evidence of a keen awareness of and interest in the other communities of the biophysical world. For women, the domestic spheres of food and sanitation had also gradually become mechanized; flower beds remained one of the few household locations not mediated by technology. Men wielded a lawnmower over the grass; women dug into the soil with a trowel.

The suburb was a new form of human settlement on the land, a new way of living. Often far from friends and kin, and "independent" of neighbours (as the suburb was supposed to be independent of city and country), the nuclear family of the 1950s clung to newly revived ideologies of togetherness. Yet the suburban form itself accentuated the feeling of absence at the centre of middle-class family life. The new houses replaced fireplace and kerosene stove with central heating, thus dissipating social experience throughout the home. A fridge full of "raidables" and supper-hour TV programs broke down the pattern of meal-times. Separate bedrooms for all or most of the children and the evolution of men's spaces like the workshop and the "yard" further encouraged rigid gender distinctions. At the same time, communal experiences within the family often became more a matter of choice than necessity. The growing independence that children felt from their parents and siblings opened up the possibility for an affective life outside the confines of the nuclear family for both men and women. . . .

The suburb stands at the centre of everything we recognize as "fifties culture." Beneath its placid aesthetic appearance, its austere modernism, we can now glimpse the tensions of a life that for many had no precedent. Until these tensions were brought to the surface in the 1960s, the suburb was a frontier. There were no models for a family newly disrupted by commodity culture, any more than there were for garden design in a place that

had never existed before. It was as if nature and our experience of it were in suspension. Things were unfamiliar in the suburb, and it's no surprise that people who could afford it fled whenever they could. Weekends and summer holidays were often spent not in the ersatz idylls of Don Mills, Levittown, or Walnut Creek, but in what was imagined to be nature itself: newly created parks and lakes and recreation areas. Here, at last, out the car window or just beyond the campsite or cottage, was an experience of nature that was somehow familiar. In fact it seems that this holiday place – and not the suburb – was nature.

But the idea of nature that was invented by postwar suburban landscaping was not a unitary one. The distinction I've made between "lawn" and "flowers" – and the parallels with gender roles – were and continue to be refuted by many people's gardening habits. Organic gardening, for example, is a very old practice that allowed many people to resist the technological incursions of the 1950s. And technology was resisted in more obvious ways, too. The mass movement against the bomb was perhaps the earliest expression on this continent of modern environmentalism.

Outside of the suburbs, in the older settled areas of the cities themselves, other forms of resistance gathered strength. The social movements whose beginnings we casually ascribe to the "sixties" – civil and human rights, feminism, peace, free speech, sexual liberation, as well as environmentalism – were in part struggles over the nature and use of urban land. Urban activism developed its own very different ideas about landscape design – ideas that are now more influential than ever.

[. . .]

THE ECOLOGICAL IMPERATIVE

The suburban landscaping of the immediate postwar years is still the spatially predominant model, but it has come to mean something different today. As modernity itself is being questioned right across the culture, we experience its expressions with much more ambivalence. Consider these examples: the "no-maintenance" garden of coloured gravel that was once popular in Florida and the U.S. Southwest is on the wane. Its matrix was the Japanese-Californian work of the early

1960s, and when well done it was striking. But it turned out that no-maintenance meant that you got rid of weeds with regular doses of [fertilizer] or a blast with a blow torch or flame thrower. It's unlikely that in a culture that has been through Vietnam and the Love Canal such a regime can have quite the cachet it once did. Likewise with "growth inhibitors" that you spray on hedges so they don't need to be clipped. These are landscaping strategies that deny change and the presence of life.

In recent years, ecological science has begun to change the way North Americans think about and work their gardens. Ideas of ecosystem and habitat have become new models for landscape work. There is new interest in native plants and wildflower gardens, in biological pest control and organic foods, as well as in planting for wildlife. These are all symptoms of a new understanding of urban land as animated, dynamic, and diverse.

These issues are now often forced into the open. Many North American cities mandate water conservation, for example. The city of Santa Barbara, California, forbids people to water their lawns with municipal water. Marin County, California, pays residents to remove their lawns and replace them with drought-tolerant plants. In many parts of the western United States, new land development is contingent on no net increase in water use, forcing communities to investigate composting toilets, the reuse of grey water (non-sewage waste water), and what is now called "xeriscaping," water-conserving planting schemes. Sometimes these schemes mean drawing strictly from the region: cactus and rock landscapes in Arizona, for example. But they can also mean working with composites of native plants and plants from similar bioregions elsewhere. In southern California this means rejecting the tropical and subtropical plant species that have been so long associated with Los Angeles and drawing instead from the chaparral and dry woodland plant communities of the Mediterranean regions of the world: southern France, central Chile, South Africa, Australia, and of course southern California itself. All of this work gives the places we live a sense of regional integrity.

[. . .]

Questions of place and values resonate differently across generations, classes, and political cultures. But some landscape work is able to galvanize both communities and professions. A promising example is ecological restoration, an emerging discipline – and movement – dedicated to restoring the Earth to health. Restoration is the literal reconstruction of natural and historic landscapes. It can mean fixing degraded river banks, replanting urban forests, creating bogs and marshes, or taking streams out of culverts. Since the early 1980s, this work – a great deal of it carried out by people working for free in their spare time – has been going on in forest, savannah, wetland, and prairie ecosystems all over North America. The Society for Ecological Restoration was founded in 1987 to co-ordinate the endeavours of its disparate practitioners: farmers, engineers, gardeners, public land managers, landscape architects, and wildlife biologists, among many others.

Restoration ecology is multidisciplinary work, drawing on technical and scientific knowledge for a generalist pursuit. It is more than tree planting or ecosystem preservation: it is an attempt to reproduce, or at least mimic, natural systems. It is also a way of learning about those systems, a model for a sound relationship between humans and the rest of nature. Restoration projects actively investigate the history of human intervention in the world. Thus they are at once agriculture, medicine, and art. . . .

These are not new ideas, but they are ideas newly current in the culture. . . . The recirculation of these ideas has led to some fascinating philosophical and political debates. What is an authentic landscape? What is native, or original, or natural? These are cultural questions, and it's refreshing to see them raised within a technical – even scientific – profession.

Restoration actively seeks out places to repair the biosphere, to recreate habitat, to breach the ruptures and disconnections that agriculture and urbanization have brought to the landscape. But unlike preservationism, it is not an elegiac exercise. Rather than eulogize what industrial civilization has destroyed, restoration proposes a new environmental ethic. Its projects demonstrate that humans must intervene in nature, must garden it, participate in it. Restoration thus nurtures a new appreciation of working landscape, those places that actively figure a harmonious dwelling-in-the-world.

What we see in the landscaping work of the late twentieth century is residues of many traditions: romantic, modernist, environmentalist, pastoral, countercultural, regionalist, agrarian, and, now, restorationist. The suburban aesthetic was able to accommodate some of those traditions, but today suburbia is clearly a landscape that can no longer negotiate the tensions between city and country – much less those posed by the many people and movements already busy making new relationships with the non-human world.

Changing environmental and cultural circumstances have brought changing aesthetics. If these changes have left the landscape profession (and the landscape) in disarray, they have also allowed large numbers of people to become involved in shaping the physical world as never before. As landscaping ideas have been reinterpreted and reversed, the boundaries of the garden have become less distinct. Much recent work attempts to reintegrate country and city, suggesting that what was once nature at home may soon become nature as home.

"Orchard"

from *Tree Cultures: The Place of Trees and Trees in their Place* (2002)

Owain Jones and Paul Cloke

Editors' introduction

Can non-human entities, such as trees, possess *agency*: the capacity for instrumental action? Or is agency reserved for humans alone? *Actor-network theory*, or ANT, is one way scholars have addressed this question. In this selection, Owain Jones and Paul Cloke utilize ANT to assert that trees exert creative agency, thereby placing trees at the heart of the interconnected human and non-human practices that together comprise their research site: West Bradley orchard. ANT deeply challenges the traditional division between nature and culture which puts humans in charge of non-humans. In a provocative example, the authors drive this point home using the example of pruning. Pruning both *shapes* the tree through the action of humans using tools, but it is equally importantly *shaped by* the tree itself as a unique, and powerful, orchard dweller. The interaction of tree and human is intimate, patterned, and binds them together in place. As Jones and Cloke illustrate, the orchard gathers a variety of human and non-human "actants" (a term preferred to "actors" by some ANT theorists), as well as a mixture of tradition and modernity. Bees, trees of many varieties and ages, laborers, farm animals, machinery ranging from old-fashioned poles used to knock ripe cider apples to the ground to specialized tractors, and weekenders picking their own apples, all mingle in the orchard in relatively regular, though never static, patterns. Jones and Cloke argue that the orchard is far from a closed system; rather, it is in continual flux as streams of inputs such as fuel, labor, pesticides, and knowledge flow in, while specialized apple products flow out. In other words, the orchard is not a typical static landscape involving a fixed view of an enframed scene; rather, it is a fluid and ever-changing node of productive flows: a *taskscape*. Thus Jones and Cloke's view of the orchard as place is reminiscent of Doreen Massey's notion of place as an open, dynamic coalescence of social relations (see p. 275). It can also be contrasted in interesting ways to Don Mitchell's view of agricultural labor and landscape (see p. 159).

Central to "Orchard" is the notion of *dwelling*. In English, "to dwell" is used interchangeably with "to reside," or "to live" (in a place). The act of dwelling, however, refers to a deeper connection to place: a rootedness that involves a life-giving connection of humans to the earth. Dwelling was considered at length by the German philosopher Martin Heidegger; in particular in his essay "Building dwelling thinking", originally written in 1951 translated and published in English in *Poetry, Language, Thought* (1971). In Heidegger's view, dwelling is enacted over generations through repeated traditions and customs. True dwelling is incompatible with the flux of modern urban life, according to Heidegger. Jones and Cloke note that Heidegger's romanticized take on rural life and traditions seeps into contemporary views of orchards like the West Bradley orchard they study. Indeed, cultural geography on both sides of the Atlantic has had a long-standing focus on rural landscapes and so-called *folk cultures*, a focus that has come under critical scrutiny from a variety of perspectives. Jones and Cloke argue for a notion of dwelling that accommodates tradition and modernity, human and non-human actants intertwined in ways that do not presuppose human dominance over nature, and for a notion of dwelling that is open to the dark as well as the romantic side of places like orchards.

An important contribution to the cultural geography literature on dwelling and nature is found in Tim Ingold's *The Perception of the Environment: Essays in Livelihood, Dwelling, and Skill* (2000). Further reading on ANT is available in Bruno Latour's *Reassembling the Social: an Introduction to Actor-Network Theory* (2005). For an additional example from a cultural geographer, see Russell Hitchings's "People, plants, and performance: on actor network theory and the material pleasures of the private garden," in *Social and Cultural Geographies* 4, 1 (2003): 99–114.

Owain Jones is a Research Fellow at the School of Geography, Archaeology and Earth Resources at the University of Exeter. His research examines biodiversity and food production. Jones is also researching the geographies of children and childhood. He is associate editor of the journal *Children's Geographies*. Jones's recent publications include (with M. Williams, L. Wood, and C. Fleuriot), "Investigating new wireless technologies and their potential impact on children's spatiality: a role for GIS," in *Transactions in GIS* 10, 1 (2006): 87–102; "Non-human Rural Studies," pp. 185–200 in P. Cloke, T. Marsden and P. Mooney (eds.) *Handbook of Rural Studies* (2006).

Paul Cloke is a Professor of Human Geography at the University of Exeter. His research interests include geographies of rurality, nature–society relations, geographies of homelessness, and landscapes of spirituality. He is the founding editor of the *Journal of Rural Studies*. Cloke's publications include *International Perspectives on Rural Homelessness*, co-edited with Paul Milbourne (2006); and *Handbook of Rural Studies*, co-authored with Terry Marsden and Patrick H. Mooney (2005).

FOUR

West Bradley is a sixty-five-acre orchard in the Glastonbury area of the county of Somerset in south-west England. . . .

West Bradley is privately owned. The owner lives in a house on the edge of the orchard and takes an active role in the strategic management and development of the orchard in conjunction with a manager who also oversees day-to-day operations and a small, flexible workforce. The orchard is 'drawn' or marked as a place in multiple ways. For example, it has an overall perimeter hedge which physically demarks it; it is mapped on legal deeds of ownership; identified as orchard on Ordnance Survey maps; has signs proclaiming itself, and is classified as orchard in local authority surveys of agricultural land-use and orchard-cover. It is well known locally in a number of ways: for its farm shop, as a source of seasonal casual work, as a place to visit for PYO [pick your own] apples, as a place of spectacle in blossom and fruiting times, as a place that keeps up the local and regional traditions of orchards, and as a place of orchard practice for other local producers (there are local producer associations).

The orchard produces a range of predominately apple 'products', although it also includes a small number of pear trees, and recently a few walnut trees have been planted for future harvesting. The apples which are grown are routed into three main production/consumption streams: retail, cider and processing, each of which is associated with a range of products and outlets. Thus the orchard produces dessert and culinary apples which are sold on site at a small farm shop and via seasonal PYO weekends, supplied to local shops, and passed on into larger cooperative wholesale systems which supply major food retail chains. The cider apples grown in the orchard are sold both to local concerns and to further-flung markets for cider production. The orchard's process apples are shipped to specific companies to be used for making juice and baby food products.

To achieve these outputs, a complex mix of people, organic entities, technologies, and knowledges are present in the orchard, at the centre of which stand the trees. The production practices range from 'traditional', long-standing orchard practices (e.g. hand-picking and pruning) to modern commercially developed practices (such as the use of pesticides and modern fruit varieties). This complex collective is maintained by a stream of inputs into the site such as new root stock, fertilizers/pesticides, and information from commercial research bodies, human labour, hardware, and fuel for machinery.

The orchard is divided into a number of areas – named The Bees, The Park, The Wilderness – dedicated to the various production streams and defined by hedges, ditches and tracks. A combination of tree varieties and attendant forms is spread throughout these areas, and different management regimes work with them. For example, the old cider orchard comprises well established standard ('full-size') trees and is hedged, and has cattle running in it at certain times of year. The pick-your-own section and the trees which produce many of the dessert and culinary apples are mainly half-standard ('half full-size') trees, while some newer varieties for the cider and culinary market, and for the more recent process apple market, are in the form of bush (small) trees. The variety of apple and tree types grown is designed to not only feed different apple qualities into the various markets supplied but also to do so at different times of the season. . . .

The creativity which enables such precise product is a relational achievement from the orchard collective, with the apple tree varieties playing a key role *as agents*. The density and texture of all this at work *in place* makes the orchard *a place*. So the orchard can be understood as being contemporaneously both an achievement woven by a complex set of networks and a place marked by different imaginative and material articulations. . . . [W]e shall discuss how the orchard can be considered as a dwelling place, and as being woven into wider ideas of place, and we shall show that this orchard identity is connected with all manner of cultural resonances which fold into the place milieu in fluid multidimensional performative ways. We shall also show how obviously, yet fundamentally, the presence of the trees and their creative abilities are at the heart of this whole achievement.

THE ORCHARD AS DWELLING

In our research at West Bradley, we became fascinated by its potential as a grounded example of the concept of dwelling. . . . [D]welling suggests a rich, intimate and ongoing togetherness of beings and materials which constitute and reconstitute landscapes and places. These conceptual togethernesses seemed to come alive at West Bradley orchard. Our account of the orchard as a tree-place, then, focuses on issues relating to the interconnections between trees, place, landscape and dwelling. . . .

In West Bradley's 1999 publicity leaflet, there is more than a hint of the notion of dwelling in the description of the orchard's situation in the landscape:

> Our orchards are situated three miles due east of Glastonbury, at the edge of the Somerset moors and tucked under the shelter of Pennard Hill. The combination of this shelter, soil type, and the gentle climate of south-west England, gives us a flavour which cannot be easily matched.

There is an implicit assertion here of nature and culture coming together harmoniously, and of an authenticity and rightness which resonates of dwelling as it has been articulated from Heidegger onwards. But as we shall show West Bradley cannot be seen simply as a traditional, authentic orchard landscape, ideas seemingly so significant in ideas of dwelling. It has adopted 'modern' practices such as modern fruit types and pesticide systems which, being elements of globalized industrial fruit-growing practice, could be said to be anti locally embedded 'dwelling'. We shall argue that dwelling is a more fluid notion than this which can incorporate 'modern' practice and ideas of networks within dynamic notions of place, and that the orchard illustrates this well.

A closer look at the orchard reveals a deep hybridity of people, nature, and technology – new and old – which is embedded in a complex array of networks, but which also has a time-thickened, place-forming dimension, and the trees are at the creative centre of all this. . . . [T]he fruit trees at West Bradley are at its heart as a place, as a network mode, or however else it is constructed. . . .

In the orchard the human 'actants' engage with the non-cider trees with great intimacy, pruning, painting (covering the pruning cuts to prevent infection), thinning (reducing clusters of young growing apples to two so the remaining apples will grow bigger and have more space to develop), and picking. This intimate relationship in not just networking, it is also the stuff of dwelling. (And this is not to say that this work might not often be regarded as boring, a grind, low-paid, insecure.) This is because it is about temporal materiality expressed through repeated rounds of doing. There is science, abstract 'objective knowledge' (of nature) here with regard to how and where the branches of the tree are cut, at what time of year, and with what objective in mind, and there is 'art' too. Those pruning the trees assured us that it was an 'art form' and every tree had to be approached as 'an individual'. Each tree presents a unique pattern of branches, through its own disposition for growth and previous rounds of pruning, which in turn 'the art of pruning' engages with year on year. It might even be suggested that each tree is an individual taskscape developed over time. These kinds of intimate skill and relationship are part of the idea of dwelling and, in the more conventional senses, account for the cultural attraction of orchards. . . .

At West Bradley this attraction is indeed particularly articulated through the larger, older trees. The owner, manager, and workers all admitted to some sadness when these were grubbed out and replaced with new bush trees. These older trees were seen as 'unique characters'. Their form, created by rounds of pruning and growth, is an example of the materialization of place narrative. . . . These older trees then contribute to the unfolding of the place (as well as the network) in this way and also others. There is concern when they have to be replaced, and a determination to keep at least some of the standard and half-standard trees in place to preserve the image of the orchard, and in fact local council landscape grants are dedicated to supporting the areas of older cider trees. These larger trees are closely connected to the traditional orchard culture of the region and the continued practice of local ceremonial customs. Being taller than people, these trees make the spaces of the orchard enclosed and intimate, and give the rows and paths a maze-like quality. The new bush trees, being about the same height as a human adult, do not produce this effect to the same extent, and are not so visually prominent in the landscape.

Surrounding these intimate processes of human–tree interaction and their cultural accretions, there are all manner of other components to the orchard taskscape. To aid pollination, beehives are kept and crab apple trees dispersed throughout the orchard. To protect the trees and crop, rabbit guards are placed around the trees, fences maintained (against deer), kite bird-scarers are flown, and various chemical insect pest- and disease-control systems are employed. There is a paraphernalia of technology such as tractors, mowers, ladders, sprayers, stakes, crates, and an infrastructure of packing sheds, grading tables and cold stores. These are all deployed in different combinations within the three different areas of production.

The preceding account is by no means intended as a comprehensive depiction of the orchard and its processes which represent very complex and detailed hybrid networks. . . . Rather, our aim is to give *an impression* of the intimate mix of nature, humans, and technology which make up the taskscape therein and the network elements which thread through it. These relational agencies can be seen in terms of actants and networks, and such a perspective could easily be enhanced by tracing more precisely the particular interconnections which constitute particular chains relating to production and consumption. However, to do so would be to stray from the place-related togetherness of the orchard. Our analysis of West Bradley, therefore, is as a dwelling place of contextualized lived practices, a taskscape which articulates practices of dwelling.

[. . .]

It is the weaving together across the nature–culture divide represented in this history which produces the incredibly rich cultural/natural ecology. . . . [T]o see nature as a pure realm which can only be depleted through interaction with the human realm is to completely misread the nature–society relationship. The ecological/cultural diversity exemplified by this example is the outcome of non-human/human relational agency at work in particular formations. This is how the idea of dwelling should develop, as a way of seeing the intimacy of these interconnections as they perform the diverse world, but it needs to avoid certain static views of the local, of landscape and place as we shall set out below.

THE TREES AS CREATIVE AGENTS

We want to emphasise that a dwelling perspective can make room for the *creative* presence, the non-human agency of 'things', in its accounts. At West Bradley orchard, from an ANT perspective, the trees are enrolled by human actors, and all manner of other actants are deployed in performing the production of fruit. However, we emphasize that the trees bring to this process the unique creativity of being able to produce fruit in the first place. We . . . argue that the creativity of fruit trees is obviously essential to the networks and place-characteristics of West Bradley. Furthermore the different types of tree produce different kinds of fruit which stream off into the three main markets supplied by the orchard. Thus there is both a general creativity of producing apples and a more specific creativity of producing particular types of apple with particular properties, which is at the heart of this relational achievement. *It is precisely this unique creative ability of fruit trees which is mourned, and considered a potential economic and scientific loss, when particular rare varieties of fruit are lost.* Thus there are 'rare-breed collections' working to preserve such creative wellsprings. At present, at least, this is a creativity which humans cannot re-create, a fact which suggests that there is a form of agency at work here which is beyond that of humans.

Moreover, it is not just this key creative ability that can be termed as non-human agency in a relational achievement. Many of the different management and production techniques which are deployed to sustain and nurture this creativity are also relational achievements. A good example of this is the pruning of the trees which is carried out to maximize, specialize and control the growth, shape and fruiting habits of the trees. This may seem to be merely a form of human control over, and imposition on, the trees. But acquiring of pruning techniques and knowledges is a relational achievement developed over time. The nature of the trees – how and when they grow, form branches, fruit and leaves – *has shaped pruning as a practice*. It works with and within the active capacities of the trees. In other words the trees have 'shaped' the art of pruning (and the forms of pruning equipment), just as pruning shapes the trees.

[. . .]

ORCHARD DWELLING AS AUTHENTICITY

Many . . . have suggested that modernism in its many forms is destructive to the practice of dwelling. . . . Such an argument assumes a relationship between authenticity and dwelling which poses important questions about nature and landscape under contemporary conditions which we will explore in the context of the orchard.

In Heidegger, authenticity is a critical element . . . and seems near impossible under the conditions of modernity . . . The obvious concern is that . . . dwelling has been obliterated by alienating modernity and becomes an impossibility. . . . So should the notion of dwelling be abandoned as a useful conceptual view of the world and of landscape because of this problem of a lack of authenticity in the terms described above? . . . [A] number of writers show that the notion of authenticity is itself a modem construction. As soon as authenticity is prescribed, or preserved, or even re-created within modernity, we can begin to stray into the world of simulacra.

The view of authenticity of being as some original (natural) form, some blessed state, can certainly be found in writings on orchards. . . . Once local varieties are moved from 'their home' and 'industrial practices' take over, we are apparently on the steady downward slope towards the modern inauthentic apple. . . .

Taken to their extreme, these arguments lead to a view of true nature, or authentic landscapes, or communities, as consisting of diminishing pockets of harmonious authentic dwelling in an ever encroaching sea of alienation. This seems a deeply flawed view and one which would make the deployment of dwelling as a view of landscape, place and nature redundant. . . .

Any notion of the 'authenticity' of West Bradley orchard is problematic. Here, traditional practices (such as pruning and the keeping of bees to aid pollination) merge with, and are interspersed by, more modem forms and practices such as the use of mechanical harvesters and state-of-the-art chemical fungicides and pesticides. But of course those traditional methods would themselves have been innovations in the development of orchard practice. . . . [T]raditions of orchard cultivation stretch back to ancient Roman and Greek civilizations and beyond. This time-depth which challenges notions of simple authenticity is seen in the types of apple grown at West Bradley. Of the fifteen or so types of culinary and eating apple grown at the orchard, the oldest type – Blenheim Orange – dates from 1740, while Fiesta and Jonagold are modern apple types developed by commercial plant suppliers. Some areas of the orchard are of traditional standard trees, while since 1990 only small-bush trees have been planted. Does this, and the use of tractors and other technologies, mark the orchard as compromised in terms of an authentic landscape? Our view is that such simple concepts of authenticity do not sit well with the notion of dwelling wherein landscape can be seen as being temporally complex, with the past being co-present with the future through both material and imaginative processes. At West Bradley we do not uncover a sterilized museum of past landscape and dwelling, somehow untouched by change or even current technologies and practices. Instead we see a series of practices which have evolved over time, and changes which are constantly informed by shifting economic, technical and cultural formations, and a place that is not conducive to fixed-point notions of authenticity.

West Bradley was bought in 1858 by the grandfather of the last-but-one owner. Through the later part of the nineteenth century it was not an orchard at all but a small dairy farm in the Somerset tradition. The farm was planted as an orchard at the turn of the century because the son who took over from his ill father was allergic to cows, so he immediately sold the cows and started planting apple trees. So this switch in land use was pragmatic. . . .

Cider apples were at first the only crop. A cider-making business was soon built up; a small-scale enterprise selling 4.5 gallon barrels made to particular requirements of customers, such as sweet or dry, clear or dark. In later years the production set-up was quite advanced, but continued to supply small private customers rather than supplying the retail trade as the larger cider companies did. As the dynamics of cider production shifted, other markets were pursued. The orchard was one of the earliest to plant and grow Bramley (cooking) apples on a commercial basis, sending apples in barrels to wholesalers in Leeds by train. These developments were all bound up with innovative modern commercial and technological practice (such as the cold storage of the crop), but throughout these changes, we suggest, the authenticity of the orchard as an orchard has been maintained in two ways.

First, over time, in this and other orchards, series of innovations and changes (such as the use of modern fruit types, and agrichemicals) constantly weave together with some older threads (such as old apple varieties, pruning, the keeping of bees) creating new *hybrid forms and practices* which are neither authentic nor inauthentic. It does, though, remain distinctive because, secondly, the new technologies which have been brought in to 'modernize' production cannot be seen purely as abstracted, undifferentiated modernity being imposed upon and obliterating 'traditional' practice. The new technologies adopted carry the marks of orchardness. For example the tractor, imported from France, is of a special narrow design for moving up and down in between the rows of trees.

In contrast to the 'eaters' which are still traditionally (and laboriously) picked from the tree by hand, cider apples are picked up from the ground after being knocked off the trees (with long poles – an old practice), by modern machinery hand-guided around and under the trees. The mechanized cider-apple harvester also has appleness and orchardness embedded in its materiality through the way it is designed to pick up apples and move through the orchard. This may seem an obvious point, but it is crucial in that the new techniques and equipment are bound into, even enrolled into, the continuation of a form of orchard identity. Orchardness may shift over time, but it retains some form of dynamic identity as it migrates though economic, technological and cultural space. The authenticity of dwelling, then, should be seen as a form of dynamism, of ongoing freshness, rather than anything static, but which at the same time retains an identity.

The individual trees at West Bradley are routinely grubbed out and replaced, either with new trees of the same type, or often with new types of trees for new markets, or with better yields, or lower related production costs. The cider trees are left longest, but no 'working' trees have lived through the life-span of the orchard. It is the orchard itself, an ongoing presence of trees, which is the ongoing taskscape. The ongoing rich mixture of nature, technology and humans retains a form of oneness which is bound together in some form of cohesion, which perhaps can be seen as 'authentic', but only in a dynamic time-embedded sense, rather than in comparison to any fixed time-point referencing. . . .

We need first to re-emphasize here . . . that there is no *necessary* equation of dwelling with goodness, morality or aesthetic benefit (as there seems to be in Heidegger). Secondly, it is also important to note that the 'authenticity' relayed by West Bradley is not confined to its spatial boundaries, but rather is projected on different scales, and this adds a layer of complexity to understanding the orchard as landscape or place.

[. . .]

DWELLING AND SPATIAL BOUNDEDNESS

The degree to which the togetherness of dwelling relies on (harmonious) spatial boundedness is obscure and problematic. [In Heidegger's] notion of dwelling . . . there is a move from a gathering (of things) to nearness, to dwelling, which is always 'dwelling in nearness'. 'To be in a place is to be near to whatever else is in that place, and pre-eminently the things that are co-located there.' In his vision of the farmhouse in the landscape of the Black Forest, Heidegger depicts a place where the material and design of the house, and the topography of the land (the house is placed in the lee of the hill for shelter), permit divinities and mortals and things to enter a 'simple oneness'. . . . Oneness implies rootedness where people and landscape become joined. In this kind of oneness and being rooted in the landscapes . . . there is a correspondence between community, landscape and place. There is a fixedness of the space in terms of a bounded local space. This is the kind of idyll . . . where communities and their corresponding landscapes are closed, and have a pointed temporal dimension in terms of the purity of the space being projected in the future. . . . Closed intimate spatial boundedness is a key way in which such stable familiar idylls or dwellings are imagined to be formed.

This kind of oneness and rootedness, then, like authenticity, to which it is closely linked, has a powerful appeal and intuitively seems best delivered within intimate, stable local sets of relations. . . . To be rooted is to have a localness; to be rooted in a local space that is distinct. . . . This is, in effect, the local taskscape: the particular dynamic of dwelling formed of rich, dense local relations between people and environment.

As with the concept of authenticity, such a view of dwelling as a local spatially bound distinctiveness of nearness is highly problematic. In part, such problems stem from the sinister (nationalist) rustic romanticism which pervades Heidegger's ideas. . . . If dwelling is to be a serviceable concept for contemporary landscapes, it needs to shed this reliance on local boundedness and instead reflect a view of space and place which is dynamic, overlapping and interpenetrating.

In the case of West Bradley, as for most modern places and landscapes, this idea of oneness and simple rootedness is a redundant vision. The owner, those who work at West Bradley, those who visit it, and those who encounter it in other ways all live spatially complex lives which take them through all manner of spaces both practically and imaginatively. Through these people and those who know West Bradley through other means – such as the County Council Tree Officer (who takes particular interest in orchards), the members of the local cider apple growers' association, the commercial suppliers, and those who are supplied by the orchard – West Bradley is clearly being engaged with sporadically, partially, and through widely differing socio-cultural constructions. The meanings of West Bradley as an orchard cannot in any way be seen as confined to the space itself. As we have outlined, meanings and materials flow in and out of its space in complex ways.

One major flow is the concern for the loss of orchards in Somerset and in Britain, set within wider concerns about environmental decline and the destruction of the countryside more generally. Another major flow is the notion of Somerset as a place of orchards. The material and cultural environment in which West Bradley is immediately set is marked with constant reminders of the orchardness of Somerset. Local cider is advertised and sold in shops. The local radio station is 'Orchard FM'. Pubs bear names and images of traditional orchard culture. The local media constantly use apples as visual icons and cover local orchard stories. Such cultural discourses are more or less consciously, and differently, carried into West Bradley. For example, some of those engaging in the PYO weekends were clearly doing so in part as a ceremonial partaking in the regional apple culture.

Moreover, there is also an awareness of Englishness at work. Apples produced by the orchard are marketed as 'English apples'. Many of those who come to the orchard to buy the produce from the farm shop, and particularly those who come to pick-your-own, are also more or less reflexively aware of larger scales of landscape and production dynamics. Many are concerned to support local orchards and concerned to support the English landscape in the context of competition with and hostility toward France and the EU in particular, it seemed.

. . . . [R]epresentations as well as practices are important here. Representations will actively render any spatially bounded notion of dwelling permeable to the cultural flows of ideas, meanings, significations and symbols operating on different scales. Secondly, practices themselves are multiple, suggesting multiple taskscapes associated with anyone dwelling. To return to the example of the taskscape . . . is it reasonable to consider it as a single taskscape? The paths are described as being worn by countless journeys of the community, but those journeys are likely to have been markedly different in their nature. The labourers, the owners, the priest, the village officials, the women, the men, the children, the sad, the lonely, the happy, the poor, the wealthy, will have walked those paths doing different tasks in different ways and constructing the landscape differently. . . . We want to argue therefore that dwelling's oneness is formed of a complex multiplicity of practice and representation. Further, in the context of both practice and representation, spatial proximity alone cannot map the boundedness of dwelling encountered at West Bradley orchard.

At West Bradley the experiences and constructions of the owner, the manager, the casual labourers do have differing, often contested, representations of the place. The present-day labour relations of this landscape and wider labour relations of agriculture are part of the elements contained here and should not be glossed over by some organic harmonious vision. Yet they remain bound together in a complex material and imaginative taskscape by all manner of forces, which range from the material boundedness of the place itself to common cultural constructions, and to the disciplines of the networks which flow to and from the place.

DWELLING AND THE FRAMING OF LANDSCAPE

[. . .]

West Bradley, being a sixty-four-acre orchard laid out on flat land, is not readable as a landscape as a framed view at all. It presents itself in many, many ways. It is trees showing over the lane as you drive or walk past; trees which may be in flower, or in full leaf, or in fruit, or in winter bareness. It is glimpses through gateways and into rows of trees. It is being on the main paths, looking along, where the end of each row going off at right angles is marked by the end tree. It is looking up one of the rows of trees. In many positions trees may blank out any depth of view at all, their foliage filling your field of vision. Or in the area of old standard cider trees, in summer, you are in a wondrous space under the canopy. Where the small bush trees are planted you can see over them and into the surrounding landscape. All these views change significantly through the seasons. Sounds and smells emphasize your being in the landscape, and as you walk your orientation changes, and your head and eyes move about. It is an embodied embeddedness.

You may (or may not) be carrying, and be more or less consciously engaging with in your mind, imaginative constructions of trees, orchards, Somerset, England, countryside, freshness, supermarkets, militant French farmers, EU bureaucrats, pesticide residues, bullfinches, and so on. Images of the orchard at differing times of the year may flash into your mind as seasonal comparisons and preferences. In other words, dwelling is an embodied and an imaginative embeddedness in landscape. These combine to create complex sensory and imaginative, dynamic collages of being-in-this place. The view is never the same twice, even for any one person, yet the place can and/or does remain deeply familiar. The orchard may be framed imaginatively as a whole, for example as somewhere owned, as somewhere where there are so many apples to be picked before the weather turns, as a source of casual labour, as an element of the orchard landscape of Somerset, as an example of a working orchard with certain working practices, as a place to go and see in blossom, or to go and pick your own fruit. These are imaginative dwellings, which interact with the dynamic spatial/temporal process of viewing it as described above. Dwelling cannot be happily represented or understood in terms of a fixed gaze upon a framed landscape. Rather it should suggest an embodied, practised, contextualized melange of experience within that landscape. This view of dwelling has much more chance of doing justice to the rich experience of being in place than does the fixed view. . . .

In many ways, we are aware that our interpretation of West Bradley orchard still chimes rather too neatly with . . . romantic overtones. . . . Orchards, after all, seem to be deeply appealing landscapes. . . . We have asked ourselves whether the concept, and the equally romanticized notion of taskscape, would be as applicable in harsher conditions where everyday practices involved more industrialized or socially or ethnically regulated procedures. Would dwelling prove to be as appealing a concept among the huge industrialized orchards of the American state of Washington, or as an aid to understanding of the taskscapes of Black labour in Apartheid (or even post-Apartheid) South African agriculture or viticulture?

The conceptual appeal of dwelling is not necessarily negated by such questions. In our view, it offers an important acknowledgement of how human actants are embedded in landscapes, how nature and culture are bound together, and how landscape invariably has time-depth which relates the present to past futures and future pasts. . . .

Landscapes of conflict clearly can be just as rich, intimate and hybrid, even if all the qualities are terrible in form. We see dwelling as concerned with this rich intimate mixing, which are all in one way parts of networks at work, but which also fold and hold space into particular forms and characters that can become places of some kind or other.

However, it is clear that the conceptualization of dwelling requires a new and more complex imagination in order to lift interpretative horizons beyond limited local and fixed-point expectations. Dwelling can only be a useful concept if it can adapt to a world where views of authenticity as some form of idealized past original stable state are clearly unhelpful; to the complex interpenetration of places with other places, and to the flows of ideas, people and materials which co-constitute and co-construct those places; and to the need for dynamic rather than fixed ways of understanding embodied engagement with landscapes.

[. . .]

"*Le Pratique sauvage*: Race, Place, and the Human–Animal Divide"

from *Animal Geographies: Place, Politics, and Identity in the Nature–Culture Borderlands* (1998)

Glen Elder, Jennifer Wolch, and Jody Emel

Editors' introduction

As discussed in the introduction to Part Four, one of the fundamental questions for the field of cultural geography involves the *limits of culture*. It is commonly assumed that culture is exclusively the purview of human beings. To be sure, non-human animals do communicate with one another, and even with humans, using language-like systems. If they live in groups, they often establish elaborate social hierarchies. But non-human animals are not typically granted the same order of cultural magnitude as humans. What are the implications of this assumption?

Some cultural geographers, such as Glen Elder, Jennifer Wolch, and Jody Emel, have examined the social construction of human-animal divide. As they detail in the selection presented here, "*Le Pratique sauvage*: race, place, and the human–animal divide," there is no hard-and-fast distinction from culture to culture, place to place, or across time about what counts as human and, by extension, what species are understood to fall into the category of "animal." Likewise, distinctions between companion animals, food animals, and working animals differ greatly across place, culture, and time. Even within cultures, there can be highly emotional conflicts over these distinctions. A familiar case in point is E.B. White's *Charlotte's Web* (1952), a popular children's story about a pig named Wilbur and the dramatic contention over his ultimate fate: pet or roast pork loin? Though so-called "animal geographies" have experienced rising popularity since the 1990s, Yi-Fu Tuan's *Dominance and Affection: The Making of Pets* (1984) was an early contribution to the cultural geographic literature on the unstable divide between humans and animals.

What is consistent across cultures, however, is the assumption of human superiority over animals. In some contexts, this leads to a legitimation of human dominance of non-human beings, even a license to do violence to animals in ways that would be unimaginable toward a fellow human being. In "*Le Pratique sauvage*," Elder, Wolch, and Emel consider the cultural clashes occurring in the United States today when immigrant populations confront norms different from those they are accustomed to. Using stories that made the news because they seemed so shocking to mainstream America – a puppy killed by a Laotian immigrant to Fresno, California, in order to appease evil spirits afflicting the man's wife; the consumption of a German Shepherd puppy by Cambodian immigrants in Long Beach, California; horse-tripping as rodeo entertainment performed by *vaqueros* – Elder, Wolch, and Emel suggest that immigrants are vilified in part by what is deemed to be their *inhuman* treatment of animals. In this way, the immigrants themselves come to be seen as *less than human*, and thus are open to being treated in ways that fellow humans would not be treated. This, argue the authors, is one way that racism operates in contemporary Western societies. Allan Pred has explored racism in Sweden, noting the slippage between animal and racialized humans in contemporary and historic times, in *The Past is not Dead: Facts, Fictions, and Enduring Racial Stereotypes* (2004).

Glen Elder is an Associate Professor of Geography at the University of Vermont. His research focuses on sexual and racial identities. He has examined this topic in the context of South Africa under apartheid. Elder's publications include *Hostels, Sex and the Apartheid Legacy: Malevolent Geographies* (2003).

Jennifer Wolch is a Professor of Geography at the University of Southern California. Her research explores the worlds of homeless and public service-dependent people in American cities, the impacts of welfare reform, and the relationships between animals and people. Wolch's publications include *Malign Neglect: Homelessness in an American City* (1994), co-authored with Michael Dear.

Jody Emel is a Professor of Geography at Clark University. Her research interests encompass the social construction of animals and animal–society relations, environmental activism centering on the gold mining industry, and water resource use in cotton production. Emel's publications include *North American Llano Estacado: Environmental Transformation and Potential for Sustainability* (2000), co-authored with E. Brooks. Wolch and Emel co-edited *Animal Geographies* (1998), from which the selection featured here, "*Le Pratique sauvage*: race, place, and the human–animal divide," is drawn.

■ ■ ■ ■ ■ ■

INTRODUCTION

[. . .]

Animal practices are extraordinarily powerful as a basis for creating difference and hence racialization. This is because they serve as defining moments in the social construction of the human–animal divide. While universally understood in literal terms, the divide is a shifting metaphorical line built up on the basis of human–animal interaction patterns, ideas about hierarchies of living things (both human and nonhuman), and the symbolic roles played by specific animals in society. Certain sorts of animals (such as apes, pets, or revered species) become positioned on the human side of this metaphorical line, rendering some practices unacceptable. But other harmful practices are normalized, to reduce the guilt (or at least the ambivalence) associated with inflicting pain or death, and to justify them as defensible behaviors differentiated from the seemingly wanton violence observed in nonhuman nature.

Norms of legitimate animal practice are neither consistent nor universal. Instead, codes for harmful animal practices are heavily dependent on the immediate context of an event. Here, the critical dimensions of context include the animal species, human actor(s), rationale for and methods of harm, and site of action involved in the practice. And because animal practices emerge over long periods of time as part of highly variable cultural landscapes,

place is also implicated in constructing the human–animal divide. When distinct, place-based animal practices are suddenly inserted into new locales by immigrants and are thus decontextualized conflict erupts. Those newcomers who violate or transgress the many-layered cultural boundary between people and animals become branded as "savage," "primitive," or "uncivilized" and risk dehumanization, that is, being symbolically allocated to the far side of the human-animal divide.

Driven by anxiety over declining global hegemony, economic and social polarization, and growing population diversity that threatens the country's image as "white," dominant groups in the US are waging an intense battle to maintain their positions of material and political power. Moreover, they seek to protect a socially constructed national identity built upon some particular . . . categories of people and places in part defined in contradistinction to others. In this situation, racialization of those immigrants whose darker skin color feeds into entrenched racial ideologies, stereotypes, and discursive practices serves to demarcate the boundaries of national culture and belonging to place, and to exclude those who do not "fit." Conflicts over animal practices, rooted in deep-seated cultural beliefs and social norms, fuel ongoing efforts to racialize and devalue certain groups of immigrants. . . .

Our readings of the links between race, place, and animals imply that violence done to animals and the pain inflicted on them are inevitably interpreted in culturally and place-specific ways. It is therefore both difficult and inappropriate to characterize one type of harm or death as more painful or humane than another. This categorically does not imply, however, that animal suffering, agony, and death are mere social constructs; *they are only too real.* Indeed, our ultimate purpose is to stimulate a profound rethinking of all "savage practices" toward animals as well as toward "othered" people. As our title suggests, we promote a "wild practice" (or *pratique sauvage*) in which heterogeneous others use their marginality as a position from which to pursue a radically open, anarchic, and inclusive politics. We conclude by raising the possibility that a truly inclusive *pratique sauvage* could encompass animals, the ultimate other.

POSTCOLONIAL ANIMAL STORIES

We launch our arguments by telling a series of stories drawn from recent events in the US. Unlike colonial animal stories such as Babar, in which the animals are representations of colonists and "natives," these postcolonial stories focus on the treatment of animals by subaltern groups and the ways these practices are used to devalue them. Their practices, interpreted as "out of place" by dominant groups, serve to position them at the very edge of humanity – to racialize and dehumanize them through a complicated set of associations that measure their distance from modernity and civilization and the ideals of white America.

The rescue dog

Late in 1995, a three-month-old German shepherd puppy was beaten to death in a residential neighborhood of Fresno, one of the fastest growing urban regions in California's vast Central Valley. The puppy death created a public furor. Neighbors complained to local authorities, and the man responsible for the dog's death was taken into custody on felony charges of animal cruelty. Later these charges were reduced to misdemeanor cruelty, to which the defendant pleaded guilty. The man charged in the case was Chia Thai Moua, a Hmong immigrant from Laos who had come to

the United States in the 1970s. Moua was also what the press reports termed a "shaman." Curiously, his shaman's logic in turning to the puppy was precisely that of so many others who use dogs to serve people: he was trying to rescue another human (in this case, his wife). He explained that he had killed the dog in order to "appease an evil spirit" that had come to plague her in the form of diabetes. The sacrifice could drive out the spirit and effect a cure. According to Hmong beliefs, "a dog's night vision and keen sense of smell can track down more elusive evil spirits and barter for a sick person's lost soul." Other animals, such as chickens and pigs, are sacrificed first, but if the killing of such animals does not solve the problem, then, according to Moua, "If it is a serious case . . . I have no other choice" but to "resort" to a dog. Moua stated that each year he performs a special ceremony to release the souls of all the animals who have helped him, so that they can be reborn. Thus, according to Moua, Hmong people from the highlands of Laos "are not cruel to animals. . . . We love them. . . . Everything I kill will be reborn again."

Moua's reliance on the Hmong conception of the human–animal border and the appropriate uses for certain animals puts him at odds with mainstream American ideas on the subject. He killed a dog. His reasons for doing so had no resonance or legitimacy for members of the dominant culture, who only sanction a limited number of contexts for dog killing. Dogs can be "laboratory workers" and "give" their lives to science, or they can be "entertainment workers" and be legitimately killed when no longer "employable" – witness the large numbers of "surplus" racing greyhound dogs that are killed each year. (Note that some forms of entertainment such as dog fighting, in which the *purpose* of the event, rather than the result, is dog injury and death, are strictly illegal.) But neither canine "lab workers" nor "entertainment workers" can be pets: dogs are usually purpose-bred for both the laboratory and the track.

Because Moua killed the puppy in his home, the dog was automatically a pet (and a pet of a revered breed at that). People are expected to dote on pet puppies in their homes, lavishing on them toys, tidbits, and attention. Barring unfortunate accidents, humans are not supposed to kill pets, except for veterinarians or euthanasia technicians in an animal shelter. Moua was neither. Worse, instead of using medicalized instruments such as the scalpel or syringe, to be wielded in the name of science or "kindness," Moua used a method (bludgeoning) widely seen as "inhuman" – a gross act of physical force that suggests a deeply disturbing animality.

An insightful head investigator for Fresno's Humane Society claimed that he could "count on my hand the actual cases [of Hmong dog sacrifices] I know about. . . . A lot of the false complaining is racism, pure and simple." Nonetheless, the publicity around Moua's deed and arrest did nothing to resolve ethnic tensions between the Anglo population of Fresno and the sizable Hmong population, which continue to fester.

[. . .]

The bowser bag

Two Long Beach men were charged with cruelty to animals for allegedly killing a German shepherd puppy and eating the dog for dinner on a March evening in 1989. A Los Angeles area judge ruled that there was no law against eating dogs, and that the animal had not been killed in an inhumane fashion. The charges were therefore dropped.

The case did not die, however. Rather, it spurred the introduction of a law, signed by then-Governor George Deukmejian, making pet-eating a criminal misdemeanor, punishable by a six-month jail term and a $1,000 fine. Pets are defined in this statute as any animal commonly kept as a pet. Killing and eating wildlife, poultry, livestock, fish, or shellfish remain legal since these sorts of creatures fall beyond accepted definitions of "pet."

But all this is beside the point, which is that Americans eat hot dogs, not dogs. In fact, given the status of most pet dogs and cats as quasi-human members of the family, eating a dog or cat is much too close to cannibalism for comfort. Indeed, the puppy involved was killed in an apartment complex, at home, it was all in the family. But the two men above were not "American," they were refugees from Cambodia. Trying to minimize the backlash against his community, the head of the Cambodia Association of America claimed that "Cambodians don't eat dogs," but it is widely known that many people from various parts of Asia do. (Isn't this how chow-dogs got their name?) And some Asians eat cats as well; civet cats, for example, are eaten in many parts of China and Southeast Asia. But in the Asian context, dogs and cats are "specialty" meats, considered "delicacy" foods. While most people see nothing wrong with eating many animals for food (including baby animals) and even taboo animals under conditions of duress, killing a cute helpless puppy for a luxury meal is another story – an act guided by self-indulgence, not the hand of necessity.

As initially drafted, the pet-protection bill only covered cats and dogs. Protests by Asian civic organizations led to an extension of the killing ban to all animals "commonly kept as pets." Curiously, however, the law still disregards pet turtles, rabbits, and pigeons, which are commonly eaten by Anglos. As Vietnamese-born editorial writer Andrew Lam claimed, the legislation implied that "[t]he yellow horde is at it again, that the eating habits of South East Asians, specifically the Vietnamese, are out of control" while "[i]t remains chic in a French restaurant to eat squab, as it is an accepted ritual for American fraternity boys to swallow live goldfish. And rabbit is nice in red wine."

Horses heading for a fall

Several localities and states have recently banned horse tripping, an event traditionally performed in *charreadas* or Mexican-style rodeos. *Charreadas* have been staged throughout Mexico for several centuries and are also frequently held throughout the southwestern United States. In this event, the legs of a horse that is galloping across the rodeo arena are lassoed by men who are pursuing on horseback. Once the legs are encircled by the lasso, the rope is pulled tight, throwing the horse to the ground. It is not uncommon for horses felled in this fashion to suffer injuries or even death.

The spreading efforts to ban horse tripping are grounded on the argument that the event is inhumane. But more to the point, horse tripping violates the deeply contradictory human–animal borders in force within dominant Anglo culture. It is difficult to underestimate the importance of horses to Anglo-European culture, including Hispanic-origin societies. But in the US today, horses are seen both as pets (the number of working horses is now vanishingly small) and as perhaps the premier animal symbol of freedom, nobility, beauty, grace, and power. While it is acceptable to derive money from equine suffering and death (after all, no one is seriously trying to ban horse racing, picket the horse slaughterhouses that supply Alpo or Purina, or prevent the export of horse flesh to France), how could civilized people derive entertainment pleasure from watching such a glorious animal be thrown violently to the ground? Also, the method – tripping an innocent, noble, unsuspecting individual – is so sneaky and underhanded. It might be OK for cattle to be "hazed" (that is, roped, thrown, and hog-tied), but then again, they're *cattle*.

The people who perform the horse tripping are *charros* or *vaqueros*. Historically, *vaqueros* were simply Mexican cowboys who worked throughout the western borderlands. But as the Anglo land grab of the frontier proceeded, they were displaced by American cowboys who went on to become the most revered figures of the American West. Hollywood subsequently recast the *vaquero* in racialized and heavily masculinized terms, to become the image of a cruel, macho Mejicano, a mustachioed bandit figure digging his razor-sharp spurs into his horse's sides until they bleed.

[. . .]

POSTCOLONIAL RACIALIZATION AND THE HUMAN–ANIMAL DIVIDE

Our cases illustrate how, in the contemporary US, racialization of others is fostered by postcolonial interpretations of the human–animal boundary or divide, under time–space conditions of post-modernity. Many forms of racialization have, in fact, long relied upon a discourse about human–animal boundaries, namely the dichotomous division of sentient beings into categories of "human" and "animal." The most basic and durable criteria used to fix the boundary have involved differences in *kind*. But although humans and animals do manifestly differ (a point that is universally recognized), the interspecific divide is not solely a behavioral or biologically determined distinction. Rather, like so many other common categorizations (such as race or ethnicity), it is also a place-specific social construction subject to renegotiation over time. Moreover, the reasons for assigning one human group to one side of the boundary or another may also change between times and places.

From its earliest beginnings, Christian theology identified the soul as the defining feature of humanity. Even with the advent of Enlightenment ideas about animals, such as Descartes's identification of animals with machines, the boundary rested on the presence/absence of souls. With the rise of a more secular Western science, the key differences in kind became biological and behavioral characteristics; criteria such as language or intentionality were employed to maintain the borders. But Darwin's theory of evolution cast a fundamentally new light on the issue. The boundary distinguishing humans and animals was reinterpreted in the West to involve not only differences in kind but also differences in *progress* along an evolutionary path. This path began with "lower" life forms, proceeded through intermediate stages inhabited by "higher" animals, and reached its pinnacle with (white) "man."

This scientific, evolutionary recasting fit squarely within an interconnected set of understandings about the human geography of the colonial world, in which the "discovery" of "races" raised complex questions of human taxonomy. Categorizing exotic-looking peoples from distant lands as lower on the evolutionary scale and thus closer to animals echoed and relied upon a myriad of similar divisions used to separate some humans from others: primitive versus modern, civilized versus savage, heathen versus Christian, cannibal versus non-cannibal. In turn, the human–animal division construed as a continuum of *both* bodily form/function and temporal stage in evolutionary progress was used to reinforce these intra-human categorizations and interpret them in temporal, evolutionary terms rather than in solely social or geographic ways. The stubborn and threatening heterogeneity of the colonies was contained and disciplined not only by branding them socially or geographically different from Europe but also . . . temporally different. . . .

In postcolonial, Western capitalist space, the idea of a human–animal divide as reflective of both differences in kind and in evolutionary progress has retained its power to produce and maintain racial and other forms of cultural difference. The dominant uses of human–animal distinctions during the colonial epoch relied upon representations of similarity to animals to dehumanize and thus racialize particular cultural groups. Contemporary arguments, in contrast, are primarily characterized by a focus on animal practices employed by subdominant cultural groups as cruel, savage, criminal, and *inhuman*: the literal blood-letting of animals, the slicing up of their bodies. . . .

[. . .]

ANIMALS AND THE BODY POLITIC

[. . .]

In general, animal bodies can be used to racialize, dehumanize, and maintain power relations in three key ways. First, animals serve as absent referents or models for human behavior. Being treated "like an animal" is typically interpreted as a degrading and dehumanizing experience, and such treatment is therefore a powerful tool for subjugation of others. The specific "treatments" in mind here are not the many loving forms of human–animal interaction, but rather involve abuse or violation, physical and/or emotional. The key aspect of such violent treatment that makes it dehumanizing, however, is not just the abuse or violation: it is the fact that victims are *objectified* and used like animals, who are commonly objectified and used without second thought. Abusive treatment of slaves by masters, for example, was modeled on how people use animals without consideration of their subjectivity.

Second, people are dehumanized by virtue of imputed similarities in behavior or bodily features and/or associations with the animal world in general or certain animals in particular. . . . Imputations are often made on the basis of associational representations of both humans and the animals to which they are being linked: colonial images of Africans as "ape-people" come readily to mind. . . . [P]eople of color (especially Africans) were historically situated by Westerners as lower on the "chain of being" and thus in closer evolutionary and behavioral proximity to nonhuman animals (especially the great apes). Colored bodies were thus both more primitive and uncivilized, and closer to animals and their unbridled biological urges and passions. Such associations persist and are often made explicit; in contemporary pornography, for example, it is most often people of color depicted in sex scenes involving intercourse with animals.

The third and least explored manner in which animals play a role in the social construction of racial difference . . . involves specific human practices on animal bodies. . . . [T]aboos about which animal bodies to eat (and which body parts) are common amongst contemporary peoples, with the result that outsider groups not observing such taboos may be viewed with disgust and distain. The many other sorts of practices on animal bodies – such as those described in our animal stories – that can constitute powerful weapons for the devaluation and dehumanization of people of color have been less remarked. We turn now to an analysis of why certain animal bodies and body practices are taken up in this fashion.

ANIMAL PRACTICES AND DEHUMANIZATION

What makes one animal practice acceptable and another a potent symbol of savagery that can be used to dehumanize those who engage in it? We have argued that every human group defines the boundary between humans and other animals in part on the basis of their treatment of animal bodies or animal practices. Specific forms of human–animal interactions, legitimized and rationalized over time, are part and parcel of the repertoire of "civilized" behavior that defines the human–animal divide. Those who do not stay within this field fall over the human–animal boundary or at least into the netherworld of "savagery"; if the practices are too far over the line, they can be interpreted as cannibalism, the ultimate act of inhumanity. Policing the human–animal boundary through the regulation of animal practices is necessary to maintain identity as humans and, not coincidentally, to sustain the legitimacy of animal practices of dominant groups.

It is widely recognized that in most societies certain types of animal practices are taboo. Taboo practices involve sexual relations with animals (bestiality is rarely sanctioned, although sometimes tolerated). Beyond bestiality, the killing and eating of the "wrong" species or categories of animals (especially totemic species or those seen as too similar to humans) can also be forbidden. For example, the consumption of apes is widely interpreted as tantamount to cannibalism, since simians occupy an ambiguous position along the human–animal boundary. They are not fully inside the human camp: one would not marry King Kong or have sex with Bonzo (even at bedtime)! But apes are seen almost literally as "inferior" humans because of their physiological similarity to humans. Eating them is thus strictly taboo. Similarly, in societies where pets are perceived to be members of the family and household, they can also come to occupy ambiguous or intermediate positions. Eating them, like the Cambodian men in our story did, becomes out of the question for civilized people.

Despite the importance of animal species or category in determining which animal practices fall beyond the bounds of humanity in any given society, practices are rarely considered (un)acceptable on the basis of species alone. . . .

Specifically . . . there are at least four other key elements of context which define the human–animal borderline. One is reason or rationale for harm. Was a specific harmful practice necessary for survival or to minimize human or animal pain/death? Few humans raise objections to killing and eating taboo animals if the alternative is starvation; the most commonly stated reason for killing laboratory animals (even "pet" species such as dogs and cats) is to prevent suffering or death; and "euthanasia" of companion animals is justified as a way to reduce animal suffering. When the rationale for harm is seen as unnecessary or irrational, or the results are defined as damaging, however, practices may be condemned. Just what is unnecessary or irrational or damaging varies from group to group.

Another important aspect of context is the social location of the perpetrator: was the person(s) involved in the harmful practice "appropriate"? For example, if an animal was killed for purposes of human consumption, did a butcher or slaughterhouse worker perform the act? Or if a companion animal was killed, was a veterinarian presiding? As our cases illustrate, problems arise when the human actor does not have the role and/or training deemed necessary by the dominant group to legitimize the act. Religious functionaries, for example, are no longer normally linked with animal sacrifices: Christian clergy are trained to deal in immortal souls, not corporeal affairs; and rabbis only serve to insure that kosher methods of killing are used. Thus, as religious specialists, neither Hmong shamans nor Santeria priests are seen to have the credentials to sacrifice food animals, much less companion animals. Similarly, where the actual killing of animals has become industrialized, professionalized, and removed from the course of everyday life, lay people (such as the Cambodian men charged with pet eating at home) have no legitimacy as animal killers.

A further contextual element revolves around the means or methods of harm: How was the harm inflicted? What techniques or tools were utilized, and did they fall within the range of local convention? Or were methods seen as archaic, barbaric, or brutally employed? A puppy can legitimately lose her head in a laboratory decapitator, but bludgeoning her to death is deemed too brutal. Similarly, bolt-guns are acceptable for dispatching a lamb led to (professional) slaughter, but the kitchen knife is no longer seen as civilized or humane. Certainly "twisting off the heads" of small birds is completely beyond the pale, and hunting to injure rather than swiftly kill is apt to be defined not only as cruel and inhumane but both unmanly and unsporting.

Lastly, the site of harm is perhaps the most crucial aspect of context in determining the legitimacy of an animal practice. Was an animal killed in a slaughterhouse or in the backyard barbecue pit next to the pool? Were rats killed in the lab or were they disemboweled in the living room? The issue of site has two dimensions. One is whether the harmful action is carried out in purpose-built quarters or reserved places (slaughterhouses, labs, shelters, forests during hunting season) or "out of site" in unspecialized spaces more typically used for other purposes or banned for the animal practice in question (residential areas, posted lands). A second site-related issue is whether the action occurs "out of sight" in abattoirs or factory farms banished from the city or in labs behind locked doors, or in highly visible places of everyday life such as homes, street corners, or church. Although in traditional societies the killing and death of individual animals was (and in many places remains) a quotidian experience, keeping mass, mechanized, and industrialized violence toward animals "out of sight" is necessary to legitimize suffering on the vast scale required by the mass market's demand for meat and medicine.

PLACE AND THE BORDERS OF HUMANITY

Human–animal borders and human practices on animals vary according to place. In representational politics that seek to dehumanize people by associating them with certain animals, place is often used to reinforce such associations. Places are imbued with negative characteristics because they harbor (or are thought to harbor) certain feared or disliked animals, and then these places are linked to people who take on the dirty, polluted, or dangerous aspects of the place (and its animals). For example, "jungles" are dangerous places in the Western popular imagination, conjuring up images of dense foliage beneath which poisonous snakes slither and vicious beasts wait to pounce on unsuspecting humans. More concretely, marginalized groups such as gypsies are often relegated to residual places in urban areas (such as dumps), often inhabited by "dirty" and "disease-ridden" animals, for example, rats. Thus a "dirty–unsafe–rats–gypsies" association arises, linking a so-called pest-species to a particular subaltern group. This associational process has long been used to connect poor people, "dirty" animals, and dirt more generally.

In the case of animal practices, however, place plays both more straightforward and more nuanced roles. At a basic level, specific repertoires of animal practices evolve and become normalized in *place*. Such repertoires are in part environmentally determined, since the diversity of animal species available to kill, eat, or otherwise use is shaped by environmental factors.... In addition, however, cultural ideas about animals (like other aspects of culture) evolve in place over time due to social or technological change generated within a society, or by externally driven events such as migrations or invasions. Thus values and practices concerning cosmological, totemic, or companionate relations between people and animals, and the material uses of animals as food or clothing, medicines or aphrodisiacs, shift as a result of social dynamics, technological change, or culture contact. The result is a shifting but place-specific ensemblage of animals, valued and used according to particular, legitimized codes. Transgressions of such place-specific codes or boundaries of practice *by definition* situate an individual or group as "outsider," "savage," or "subhuman."

What happens when the coding of animal bodies and the codes of animal practice shared by people dominant in one place are broken or challenged by people from another place, who do not share these codes but share the same space? When people are uprooted and brought to new places, they encounter different human–animal boundary constructions and if they persist in their indigenous practices are much more likely to transgress the border than locals. During much of (pre)history, the pace of such culture contact was relatively slow, allowing both host and newcomer groups to adjust; in earlier international migration waves to the US, origins of immigrants were sufficiently similar to host populations that conflict on the basis of animal practices does not appear to have been rife. With the economic globalization, escalating geopolitical instabilities and conflicts, and vast international population flows that characterize the postmodern condition, the "empire" has come home. Newcomers from a wide variety of radically different environments and cultural landscapes are suddenly living cheek by jowl. Typically, immigrants must move into the territories of a more powerful host community. Adjustment possibilities are foreshortened; for the largest immigrant groups, the need to adjust may be obviated by the emergence of relatively self-contained immigrant districts, such as "ethnoburbs." Thus in the contemporary US, immigrants whose indigenous animal practices clash with the codes of dominant society are at the greatest risk of racialization and dehumanization.

Nevertheless, non-immigrant people of darker (versus lighter) color can also be at risk on the basis of their animal practices.... Thus cock fighting among Native Americans or Chicanos, the adoption of Santeria on the part of many Chicanos and African Americans, or the keeping of aggressive, vicious dogs (or, worse, dog-fighting) among youth in inner-city communities of color can place such subaltern groups on the far side of the human–animal boundary. When problematic practices occur in racialized and marginalized places, such as "ghetto" areas that are already indirectly and sometimes even explicitly linked to Africa (by virtue of names like "The Jungle"), prospects of racialization on the basis of animal practices may rise still higher.

[...]

TOWARD *LE PRATIQUE SAUVAGE*

Our purpose in attempting to explicate the links between race, place, and animal practices has been to show how deeply engrained ideas about people and animals have been used to produce cultural difference and devalue subaltern groups. In the US, such differences play into a multifaceted and dynamic process of racialization in which immigrants who appear to threaten dominant cultural identities, are powerfully marked as outside the project of becoming American, and thus excluded from its associated benefits. This exploration reveals the extreme relativity of legitimate animal body codes and practices with respect to time, place, and culture. Ironically, however, our consideration also exposes the universality of human violence toward animals. We are left with a dual challenge: how to break the links between animals and racialization, and stop the violence done to people racialized on the basis of their animal practices; and how to make the links between animals and people, and stop the violence directed at animals on the basis of their nonhuman status. . . .

We maintain that making the links between animals and people requires a rejection of "dehumanization" as a basis for cultural critique. For the connotations of the very term "dehumanization" are deeply insidious. They imply human superiority and thus sanction mastery over animals and nature, and also suggest that violent or otherwise harmful treatment is acceptable as long as the targets are nonhuman beings. Thus dehumanization not only stimulates violence toward people, it implicitly legitimizes violence toward animals.

This does not mean that the human–animal boundary should simply be banished for good. For . . . the denial of difference can be as harmful as its production. Instead, difference – whether amongst humans or between humans and animals – must be respected and valorized. Stopping the violence means neither dismissing difference nor using it to legitimize harm or domination. Rather, in our view, stopping the violence requires adopting recipes for "wild practice" and extending them to embrace animals as well as people.

What changes in human thought and practice does *le pratique sauvage* imply? We see three basic shifts as necessary. One is that humans, especially dominant groups, accept rather than deny some of the vulnerability that animals have always known and reject the illusion that a devaluation of others (human or animal) either empowers or offers protection from harm. Another is that all humans need to abandon the drive for overarching control and instead choose a position of humility or marginality with respect to the Earth that balances needs for safety and security with consideration for the needs of other life-forms. Such marginality must be internally imposed (as opposed to the marginality that humans impose on each other to oppress or gain power) and its costs must be fairly borne. Finally, this sort of *pratique sauvage* implies that people must actively engage in a radically inclusive politics which considers the interests and positionality of the enormous array of animal life and lives, as well as the lives of diverse peoples. Neither human nor animal lives can ever be fully known, of course. We are obliged, however, to discern them as best we are able, through both the practices of interaction and exchange, and the exercise of all our powers of empathy and imagination.

PART FIVE

Identity and Place in a Global Context

Courtesy of Alex Dorfsman

INTRODUCTION TO PART FIVE

Philosophers, poets, theologians, and scholars of a humanistic bent have long remarked on the deeply felt human need for connections to place. Whether the connection between place and identity is an innate part of the individual psyche, or an adaptive cultural prerequisite for social belonging to a community, is certainly open to debate. What is clear, however, is that places need us as much as we seem to need places. Cultural geographer Yi-Fu Tuan coined a term for this: *topophilia*, or love of place. The very fact of our existence is place-dependent: we occupy discrete spaces with our physical bodies, our lives are lived in particular places, and we take in the very air and water and sustenance of place – if not of the place we are in at the moment, of *someplace*. To move a bit further from these literal examples, we also carry places with us in our memories, conjuring a favorite place-memory in times of sadness, for example, or dreaming fondly of childhood landscapes long after we have moved away. As Robert Sack has noted of the mutuality of this relationship, "Places cannot exist without us. But equally important, we cannot exist without places."[1]

Yet though we are profoundly *emplaced* creatures, we are not *place-bound*. Humans have always wondered about other places and the people that might inhabit them, and some have acted on this curiosity, generating travel narratives that date to ancient times yet whose fascination of discovery resonates with contemporary readers. For example, the dramatic account told in Bernal Díaz del Castillo's *The Discovery and Conquest of Mexico* is as thrilling a read for today's students as it was 500 years ago. Humans have also long used forced displacement as a punitive strategy, whether this involved the individual cast out from the village, or whole groups of people driven from their homeland because of ethnic or religious persecution. Indeed, the dark side of Díaz del Castillo's story of adventure and discovery involves the enslavement, massacre, and forced displacement of much of Mexico's indigenous population. Today, many people find themselves relocating several times over their life course, often living far away from family members. You yourself may be a voluntary migrant, on the move in search of educational opportunities, a better job, or part of a family whose primary decision-maker has decided to relocate. You may be a member of a *diasporic community* (this term refers groups displaced from their ancestral homeland), or perhaps a refugee; both are so-called *involuntary cosmopolitans* who have been forcibly displaced and relocated or scattered across the globe. In less traumatic experiences of globalization, we may sample cuisines, musical styles, fashions, and media, and thereby be tempted to try on different identities, or to invent new identities, with the elements assembled from a list of far-flung places.

As the number of people on the move voluntarily or involuntarily has reached a worldwide crescendo, what then of place? Does place even matter any more? Has the world become a series of homogeneous, nearly indistinguishable non-places? Or are people in fact more likely to highlight the distinctiveness of place, and their special attachment to place, as a strategy to defend against the erosion of identity threatened by globalization? There is no definitive answer to this question, but because of the strong spatial and cultural aspects of globalization, cultural geographers have a great deal of interest in the topic. This interest is shared closely with cultural anthropologists, a number of whose work is featured in this part.

One of the more confounding distinctions made by geographers is that between *space* and *place*. Though most people use these terms interchangeably, for cultural geographers they have quite distinct meanings. Of the two terms, space is the more abstract. *Space* suggests dimensionality (depth, volume, area), infinitude, and emptiness, as in "outer space." Yet, as contemporary cultural geographers are quick to point out, space is far from devoid of entanglement in social relations. Indeed, space is socially produced. *Place* is the less abstract of the two notions, invoking familiarity, finitude, and immediateness. Place is space infused with meaning. In its most simple expression, place is often equated with locality, as in "you are here." But place is understood by most contemporary cultural geographers to be more complex than mere locality. Place can be viewed as a fluid nexus of lived social relations on a variety of scales, from abstract to concrete, and from global to the local. Indeed, it is the very local scale of the body that has brought place under new scrutiny by critical human geographers (see also the introduction to Part Seven). Our dwelling or being-in-the-world (what Heidegger termed *dasein*) involves an intimately embodied engagement with our surroundings: place-making and meaning-making as conjoined activities. Particularly in these global times, this process involves connections across scale and space. As philosopher Edward Casey has noted, "The seemingly contracted locus of the lived body, which is always just *here*, has proved to be an effective basis for what has become an expansive vision of what place is all about, even when it is located over *there* and far away."[2]

Nowadays, however, as people, ideas, and products seem to be in constant motion across the earth's surface, the connection between place and identity is evermore open to question. For place can imply stability, familiarity, and belonging. Through this lens, place is made through the patterned repetition of behaviors in one location over generations. It is often proposed that the global proliferation of chain stores, fast food franchises, and American mass-produced architectural styles ranging from strip malls to suburban tract housing, means that landscapes – particularly in cities – have begun to look more alike the world over. The distinctiveness of place is becoming further and further eroded, and along with it, the stable sense of belonging that is – for some – at the heart of identity. Others, however, note that such notions of place are misguided. Place has never connoted stability, natural unity, or an experience of belonging that is shared by all. Quite to the contrary, the history of any one geographic location reveals an often tumultuous influx of outsiders, conflicts over belonging, displacement, and cultural mixture. Thus these global times can be understood as nothing more than more of the same, albeit on a larger scale and faster pace than before.

There are those who see a real value to preserving existing senses of place, or returning to them. In the selection by Arturo Escobar, it is argued that the Pacific coast Colombians' attempts to preserve the biodiversity of the region where they live provides a laudable illustration of resistance in the face of globalization. Keith Basso documents the importance of place-based attachments for preserving the history and identity of the Western Apache people he interviewed in New Mexico, noting that losing the land is tantamount to losing one's sense of self and community. Others have embraced the freedom from *place-bound* identities and practices (as opposed to *place-based* identities and practices) that is offered by the heightened levels of movement and exchange offered by globalization. The notions of *diaspora*, *hybridization*, and *creolization* – all terms that connote the ways that human movement involves the mixing of traditions, identities, and practices into something new – invoke the ability, sometimes born of need and sometimes of choice, to pick and choose from diverse cultural elements in fashioning novel and more resilient senses of self. Both Stuart Hall and Lisa Malkki note how reactionary notions of places, particularly nation-states, have served to draw boundaries that stigmatize those who fall, or are forced, outside of their limits. In this sense, understandings of place reveal a dark and exclusionary side.

In these discussions, the term *globalization* must be approached with some care. A concise (though perhaps insufficiently subtle) definition of globalization is the heightened level of connections and interchange amongst the world's people and places, leading to the sense that time is speeding up and distances are shrinking. Globalization is sometimes approached as a largely economic phenomenon, enacted by corporations and experienced on the ground through the effects of economic restructuring.

One of the contributions of cultural geographers has been to argue that globalization is at least as importantly a cultural as an economic process, though there is some disagreement as to the degree of independence culture can exert (see also the introduction to Part One). Another common misconception is that globalization is a fairly recent phenomenon, beginning some time in the mid-twentieth century. Yet if globalization is viewed more broadly, it is clear that cultural, political, and economic exchange of people, ideas, and goods has gone on for centuries. It is important to note that globalization is not limited to the industrialized West. Indeed, some of the most powerful agents and most innovative contributions to globalization come from non-Western societies. For example, the worldwide appeal of Japanese cartoon-inspired merchandise and media, such as Hello Kitty, Pokémon, and *animae*, is striking. As Doreen Massey reminds us, globalization is a *spatially and socially uneven* process. Much of what we hear about globalization involves liberation from the confines of borders, exposure to new cultures and ideas, and the freedom to choose new identities. Yet as many of the selections in this part discuss, people may well find themselves on the receiving end of globalization: impoverished, trapped in place, or forced to move against their will. It is in this context that the politics of place and identity become crucial.

Scale, or the level of spatial aggregation at which a thing exists or a process occurs, is also an important consideration. For most of modernity in most places in the world, the scale of the nation-state has been paramount. Some see a move away from the scale of the nation-state as a defining site of loyalty, citizenship, action, and identity as emblematic of a global era. The assumption of a stable link between place and identity is at the heart of the nation-state, a link that is progressively becoming undone through heightened levels of mobility and the formation of place attachments at scales other than the nation-state. Some scholars go so far as to question the very future of the nation-state, asserting that we now (or will soon) live in post-national times. In a post-national era country borders cease to matter, place-based loyalties are either forged at different scales (smaller or larger) than the nation-state, or not at all, and socio-spatial connections become far more fluid and fleeting than before.

In the face of this, at least two reactions are possible. On the one hand, defensiveness can set in as attempts to protect the eroding nation-state are undertaken. Borders are fortified, outsiders (refugees, immigrants, and foreigners) refused entry, and desperate attempts to purify the national space are undertaken through practices such as deportation, imposition of dress codes, and laws declaring an official language. As David Morley and Kevin Robins explore with regard to Germany, references to supposedly timeless homelands or pure ethnic roots can be revived and promoted in the media. It is important to note that the associations between place and identity at the national scale often involve storytelling, such as the narrative constructed in the films analyzed by Morley and Robins. Myths of origin, for example, can provide powerful claims to place and are often at the heart of national and ethnic identities. Narrative and language thus play an important role in nation-building. As Liisa Malkki's selection elaborates, *metaphor* is a rhetorical device that has often been used to link people to places in ways that suggest organic – thus natural – connections, such as those found between tree roots and soil. Thus, both media depictions and language itself, along with most of the symbols and performances associated with the nation-state (holidays, parades, swearing-in ceremonies, anthems, flag rituals, and costumes) work culturally to present the nation-state as a natural scale at which belonging and identity reside. In this sense, and as discussed in the Introduction to the *Reader*, the nation-state has correctly been described as an *imagined community*.

On the other hand, what Doreen Massey calls an *extroverted sense of place* can arise in the face of globalization. Places are – and have always been – the product of spatial interaction, through colonialism, conquest, exploration, and migration. Indeed, as Liisa Malkki points out, the notion of discrete countries with boundaries closing them off from other countries is a fiction: a powerful fiction, to be sure, but a fiction nonetheless, and one that can be rewritten in a number of ways. Massey suggests that understanding places not as things, but as nodes in networks of socio-spatial networks, provides an appreciation of how deeply connected our lives are to the lives of people elsewhere. Stuart Hall illustrates this using the example of transatlantic African diasporic musical forms.

In the United States, contention over the integrity of borders in light of the attacks of September 11, 2001, and the ongoing porosity of the border with Mexico, highlights the tensions between defensive and progressive senses of place. In Western Europe, the world region that has made the largest strides toward economic, cultural, and political integration, debates surrounding the deepening and widening of the European Union bring these tensions between defending the nation-state and embracing the new global order to the forefront.

But is it enough to try and keep outsiders outside? Or are nation-states always constructed upon suppressed internal differences? Women, homosexuals, racial and religious minorities have never been fully enfranchised by nation-states. Though nation-states may strive to construct the appearance of unity and inclusion, they are never fully unified or inclusive. As the selections by Keith Basso and Arturo Escobar argue, indigenous peoples and racialized minorities constitute examples of such outsiders-within. In both cases, place is vital to maintaining these communities. In the case of the Native Americans discussed by Basso, it is the intricate association of language and place that is responsible for keeping the historical record, and enforcing proper behavior in the face of pressures from mainstream American society. Escobar highlights Afro-Colombian political activism *vis-à-vis* global forces seeking to extract resources from the biodiverse Pacific coastal region where they live. Defense of place, seen in this light, opens to question whether a defensive stance with regard to place can in fact be politically progressive.

NOTES

1 R. Sack, "Place, Power, and the Good" in P. Adams, S. Hoelscher, and K. Till (eds.) *Textures of Place: Exploring Humanist Geographies* (2001), p. 233.
2 E. Casey, *The Fate of Place: A Philosophical History*, (1998), p. 332.

"A Global Sense of Place"
from *Space, Place, and Gender* (1994)

Doreen Massey

Editors' introduction

How can we understand the cultural dimensions of globalization from a geographic perspective? In "A global sense of place," Doreen Massey suggests that there are at least two sides to the issue. On the one hand, the speed at which people, ideas, and commodities flow might seem to make space less and less of a barrier, and place less and less relevant, in today's world. On the other hand, Massey notes that globalization is not experienced by all people everywhere in the same way. Rather, there exists what Massey calls the *power geometry* of space. Some people are indeed on the initiating end of the movement that characterizes globalization: sending faxes, traveling in airplanes, shopping online. But others are not. Refugees, for example, do move but they do not have much choice in the matter. Still others don't move at all, but are trapped in place by low wage jobs, lack of access to transportation, and larger forces such as economic restructuring that leave them behind.

Massey explores a theme common to many of the pieces included in this part namely, how to go about both conceptualizing, as well as effectively engaging with, the changes wrought by globalization. In "A global sense of place," Massey advances the notion of a *progressive sense of place*. Rather than seeing places as homogeneous entities that must be maintained pure in the face of globalization, Massey suggests we see places as sets of social relations linked into networks that cross space and scale. Under globalization, social relations are becoming more spatially stretched-out than before. Thus we are, and in many important ways have long been, closely connected to people in other places, and with processes at scales larger than the local. Indeed, we are often more closely connected to people geographically distant places than we are to people living just down the street from us.

Whether globalization is symptomatic of a *postmodern era*, an era beyond modernity where the speed of flows has led to a qualitatively different experience of being-in-the-world, or not, is a matter of some debate amongst geographers. David Harvey's *The Condition of Postmodernity: An Inquiry into the Dimensions of Cultural Change* (1989) and Edward Soja's *Postmodern Geographies: The Reassertion of Space in Critical Social Theory* (1989) are both important works on this topic by geographers. Though Harvey argues that we live in a late modern world, while Soja suggests that we've made the transition to postmodernity, Doreen Massey has critically engaged with both of them from a feminist perspective; see particularly her article titled "Flexible sexism," in *Environment and Planning D: Society and Space* 9, 1 (1991): 31–57.

Known for her incisive discussion of big issues like globalization, theories of space and place, and gender and economic restructuring, Doreen Massey is one of the most widely read living human geographers in the English-speaking world. Her most recent book, *For Space,* was published in 2005. A compendium of her work spanning several decades was published in 1993 under the title *Space, Place, and Gender.* Doreen Massey is a Professor of Geography at the Open University.

▪ ▪ ▪ ▪ ▪ ▪

This is an era – it is often said – when things are speeding up, and spreading out. Capital is going through a new phase of internationalization, especially in its financial parts. More people travel more frequently and for longer distances. Your clothes have probably been made in a range of countries from Latin America to South East Asia. Dinner consists of food shipped in from all over the world. And if you have a screen in your office, instead of opening a letter which . . . has taken some days to wend its way across the country, you now get interrupted by e-mail.

This view of the current age is one now frequently found in a wide range of books and journals. Much of what is written about space, place and postmodern times emphasizes a new phase in what Marx once called 'the annihilation of space by time'. The process is argued, or – more usually – asserted, to have gained a new momentum, to have reached a new stage. It is a phenomenon which has been called 'time–space compression'. And the general acceptance that something of the sort is going on is marked by the almost obligatory use in the literature of terms and phrases such as speed-up, global village, overcoming spatial barriers, the disruption of horizons, and so forth.

One of the results of this is an increasing uncertainty about what we mean by 'places' and how we relate to them. How, in the face of all this movement and intermixing, can we retain any sense of a local place and its particularity? An (idealized) notion of an era when places were (supposedly) inhabited by coherent and homogeneous communities is set against the current fragmentation and disruption. The counterposition is anyway dubious, of course; 'place' and 'community' have only rarely been coterminous. But the occasional longing for such coherence is none the less a sign of the geographical fragmentation, the spatial disruption, of our times. And occasionally, too, it has been part of what has given rise to defensive and reactionary responses – certain forms of nationalism, sentimentalized recovering of sanitized 'heritages'

and outright antagonism to newcomers and 'outsiders'. One of the effects of such responses is that place itself, the seeking after a sense of place, has come to be seen by some as necessarily reactionary.

But is that necessarily so? Can't we rethink our sense of place? Is it not possible for a sense of place to be progressive: not self-enclosing and defensive, but outward-looking? A sense of place which is adequate to this era of time–space compression? To begin with, there are some questions to be asked about time–space compression itself. Who is it that experiences it, and how? Do we all benefit and suffer from it in the same way?

For instance, to what extent does the currently popular characterization of time–space compression represent very much a western, colonizer's, view? The sense of dislocation which some see at the sight of a once well-known local street now lined with a succession of cultural imports – the pizzeria, the kebab house, the branch of the middle-eastern bank – must have been felt for centuries, though from a very different point of view, by colonized peoples all over the world as they watched the importation, maybe even used, the products of, first, European colonization, maybe British (from new forms of transport to liver salts and custard powder), later US, as they learned to eat wheat instead of rice or corn, to drink Coca-Cola, just as today we try out enchiladas.

Moreover . . . we also need to ask about its causes: what, is it that determines our degrees of mobility, that influences the sense we have of space and place? Time–space compression refers to movement and communication across space, to the geographical stretching-out of social relations, and to our experience of all this. The usual interpretation is that it results overwhelmingly from the actions of capital, and from its currently increasing internationalization. On this interpretation, then, it is time, space and money which make the world go round, and us go round (or not) the world. It is capitalism and its developments which are argued to determine our understanding and our experience of space.

But surely this is insufficient. Among the many other things which clearly influence that experience, there are, for instance, 'race' and gender. The degree to which we can move between countries, or walk about the streets at night, or venture out of hotels in foreign cities, is not just influenced by 'capital'. Survey after survey has shown how women's mobility, for instance, is restricted – in a thousand different ways, from physical violence to being ogled at or made to feel quite simply 'out of place' – not by 'capital', but by men. . . . A simple resort to explanation in terms of 'money' or 'capital' alone could not begin to get to grips with the issue. The current speed-up may be strongly determined by economic forces, but it is not the economy alone which determines our experience of space and place. In other words, and put simply, there is a lot more determining how we experience space than what 'capital' gets up to.

What is more, of course, that last example indicated that 'time–space compression' has not been happening for everyone in all spheres of activity. . . .

In other words, and most broadly, time–space compression needs differentiating socially. This is not just a moral or political point about inequality, although that would be sufficient reason to mention it; it is also a conceptual point.

Imagine for a moment that you are on a satellite, further out and beyond all actual satellites; you can see 'planet earth' from a distance and, unusually for someone with only peaceful intentions, you are equipped with the kind of technology which allows you to see the colours of people's eyes and the numbers on their numberplates. You can see all the movement and tune in to all the communication that is going on. Furthest out are the satellites, then aeroplanes, the long haul between London and Tokyo and the hop from San Salvador to Guatemala City. Some of this is people moving, some of it is physical trade, some is media broadcasting. There are faxes, e-mail, film-distribution networks, financial flows and transactions. Look in closer and there are ships and trains, steam trains slogging laboriously up hills somewhere in Asia. Look in closer still and there are lorries and cars and buses, and on down further, somewhere in sub-Saharan Africa, there's a woman – amongst many women – on foot, who still spends hours a day collecting water.

Now, I want to make one simple point here, and that is about what one might call the *power geometry* of it all; the power geometry of time–space compression. For different social groups, and different individuals, are placed in very distinct ways in relation to these flows and interconnections. This point concerns not merely the issue of who moves and who doesn't, although that is an important element of it; it is also about power in relation *to* the flows and the movement. Different social groups have distinct relationships to this anyway differentiated mobility: some people are more in charge of it than others; some initiate flows and movement, others don't; some are more on the receiving end of it than others; some are effectively imprisoned by it.

[There are] . . . those who are both doing the moving and the communicating and who are in some way in a position of control in relation to it – the jet-setters, the ones sending and receiving the faxes and the e-mail, holding the international conference calls, the ones distributing the films, controlling the news, organizing the investments and the international currency transactions. These are the groups who are really in a sense in charge of time–space compression, who can really use it and turn it to advantage, whose power and influence it very definitely increases. On its more prosaic fringes this group probably includes a fair number of western academics and journalists – those, in other words, who write most about it.

But there are also groups who are also doing a lot of physical moving, but who are not 'in charge' of the process in the same way at all. The refugees from El Salvador or Guatemala and the undocumented migrant workers from Michoacán in Mexico, crowding into Tijuana to make a perhaps fatal dash for it across the border to grab a chance of a new life. Here the experience of movement, and indeed of a confusing plurality of cultures, is very different. And there are those from India, Pakistan, Bangladesh, the Caribbean, who come half way round the world only to get held up in an interrogation room at Heathrow.

Or – a different case again – there are those who are simply on the receiving end of time–space compression. The pensioner in a bed-sit in any inner city in this country, eating British working-class-style fish and chips from a Chinese take-away, watching a US film on a Japanese television; and not daring to go out after dark. And anyway the public transport's been cut.

Or – one final example to illustrate a different kind of complexity – there are the people who live in the *favelas* of Rio, who know global football like the back of their hand, and have produced some of its players; who have contributed massively to global music, who gave us the samba and produced the lambada that everyone was dancing to last year in the clubs of Paris and London; and who have never, or hardly ever, been to downtown Rio. At one level they have been tremendous contributors to what we call 'time–space compression' and at another level they are imprisoned in it.

This is, in other words, a highly complex social differentiation. There are differences in the degree of movement and communication, but also in the degree of control and of initiation. The ways in which people are placed within 'time–space compression' are highly complicated and extremely varied.

But this in turn immediately raises questions of politics. If time–space compression can be imagined in that more socially formed, socially evaluative and differentiated way, then there may be here the possibility of developing a politics of mobility and access. For it does seem that mobility, and control over mobility, both reflects and reinforces power. It is not simply a question of unequal distribution, that some people move more than others, and at some have more control than others. It is that the mobility and control of some groups can actively weaken other people. Differential mobility can weaken the leverage of the already weak. The time–space compression of some groups can undermine the power of others.

This is well established and often noted in the relationship between capital and labour. Capital's ability to roam the world further strengthens it in relation to relatively immobile workers, enables it to play off the plant at Genk against the plant at Dagenham. It also strengthens its hand against struggling local economies the world over as they compete for the favour of some investment. The 747s that fly computer scientists across the Pacific are part of the reason for the greater isolation today of the island of Pitcairn. But also, every time someone uses a car, and thereby increases their personal mobility, they reduce both the social

rationale and the financial viability of the public transport system – and thereby potentially reduce the mobility of those who rely on that system. Every time you drive to that out-of-town shopping centre you contribute to the rising prices, even hasten the demise, of the corner shop. And the 'time–space compression' which is involved in producing and reproducing the daily lives of the comfortably-off in First World societies – not just their own travel but the resources they draw on, from all over the world, to feed their lives – may entail environmental consequences, or hit constraints, which will limit the lives of others before their own. We need to ask, in other words, whether our relative mobility and power over mobility and communication entrenches the spatial imprisonment of other groups.

But this way of thinking about time–space compression also returns us to the question of place and a sense of place. How, in the context of all these socially varied time–space changes do we think about 'places'? In era when, it is argued, 'local communities' seem to be increasingly broken up, when you can go abroad and find the same shops, the same music as at home, or eat your favourite foreign-holiday food at a restaurant down the road – and when everyone has a different experience of all this – how then do we think about 'locality'?

Many of those who write about time–space compression emphasize the insecurity and unsettling impact of its effects, the feelings of vulnerability which it can produce. Some therefore go on from this to argue that, in the middle of all this flux, people desperately need a bit of peace and quiet – and that a strong sense of place, of locality, can form one kind of refuge from the hubbub. So the search after the 'real' meanings of places, the unearthing of heritages and so forth, is interpreted as being, in part, a response to desire for fixity and for security of identity in the middle of all the movement and change. A 'sense of place', of rootedness, can provide . . . stability and a source of unproblematical identity. In that guise, however, place and the spatially local are then rejected by many progressive people as almost necessarily reactionary. They are interpreted as an evasion; as a retreat from the (actually unavoidable) dynamic and change of 'real life', which is what we must seize if we are to change things for the better. On this reading, place and locality are foci for a form of romanticized escapism from the real business of the world.

While 'time' is equated with movement and progress, 'space'/'place' is equated with stasis and reaction.

There are some serious inadequacies in this argument. There is the question of why it is assumed that time–space compression will produce insecurity. There is the need to face up to – rather than simply deny – people's need for attachment of some sort, whether through place or anything else. None the less, it is certainly the case that there is indeed at the moment a recrudescence of some very problematical senses of place, from reactionary nationalisms, to competitive localisms, to introverted obsessions with 'heritage'. We need, therefore, to think through what might be an adequately progressive sense of place. . . . The question is how to hold to that notion of geographical difference, of uniqueness, even of rootedness if people want that, without it being reactionary.

There are a number of distinct ways in which the 'reactionary' notion of place described above is problematical. One is the idea that places have single, essential, identities. Another is the idea that identity of place – the sense of place – is constructed out of an introverted, inward-looking history based on delving into the past for internalized origins. . . . A particular problem with this conception of place is that it seems to require the drawing of boundaries.

Geographers have long been exercised by the problem of defining regions and this question of 'definition' has almost always been reduced to the issue of drawing lines around a place. I remember some of my most painful times as a geographer have been spent unwillingly struggling to think how one could draw a boundary around somewhere like the 'east midlands'. But that kind of boundary around an area precisely distinguishes between an inside and an outside. It can so easily be yet another way of constructing a counterposition between 'us' and 'them'.

And yet if one considers almost any real place, and certainly one not defined primarily by administrative or political boundaries, these supposed characteristics have little real purchase.

Take, for instance, a walk down Kilburn High Road, my local shopping centre. It is a pretty ordinary place, north-west of the centre of London. Under the railway bridge the newspaper stand sells papers from every county of what my neighbours, many of whom come from there, still often call the Irish Free State. . . .

Thread your way through the often almost stationary traffic diagonally across the road from the newsstand and there's a shop which as long as I can remember has displayed saris in the window. Four life-sized models of Indian women, and reams of cloth. On the door a notice announces a forthcoming concert at Wembley Arena: Anand Miland presents Rekha, live, with Aamir Khan, Salman Khan, Jahi Chawla and Raveena Tandon. On another ad, for the end of the month, is written, 'All Hindus are cordially invited'. In another newsagent's I chat with the man who keeps it, a Muslim unutterably depressed by events in the Gulf, silently chafing at having to sell the *Sun*. Overhead there is always at least one aeroplane – we seem to be on a flight-path to Heathrow and by the time they're over Kilburn you can see them clearly enough to tell the airline and wonder as you struggle with your shopping where they're coming from. Below, the reason the traffic is snarled up (another odd effect of time–space compression!) is in part because this is one of the main entrances to an escape route from London, the road to Staples Corner and the beginning of the M1 to 'the North'.

This is just the beginnings of a sketch from immediate impressions but a proper analysis could be done of the links between Kilburn and the world. And so it could for almost any place.

Kilburn is a place for which I have a great affection; I have lived there many years. It certainly has 'a character its own'. But it is possible to feel all this without subscribing to any of the static and defensive – and in that sense reactionary – notions of 'place' which were referred to above. First, while Kilburn may have a character of its own, it is absolutely not a seamless, coherent identity, a single sense of place which everyone shares. It could hardly be less so. People's routes through the place, their favourite haunts within it, the connections they make (physically, or by phone or post, or in memory and imagination) between here and the rest of the world vary enormously. If it is now recognized that people have multiple identities then the same point can be made in relation to places. Moreover, such multiple identities can either be a source of richness or a source of conflict, or both.

One of the problems here has been a persistent identification of place with 'community'. Yet this is a misidentification. On the one hand, communities can exist without being in the same place – from networks of friends with like interests to major religious, ethnic or political communities. On the other hand, the instances of places housing single 'communities' in the sense of coherent social groups are probably – and, I would argue, have for long been – quite rare. Moreover, even where they do exist this in no way implies a single sense of place. For people occupy different positions within any community. We could counterpose to the chaotic mix of Kilburn the relatively stable and homogeneous community across the road from the newsstand (at least in popular imagery) of a small mining village. Homogeneous? 'Communities' too have internal structures. To take the most obvious example, I'm sure a woman's sense of place in a mining village – the 'spaces through which she normally moves, the meeting places, the connections outside' – are different from a man's. Their 'senses of the place' will be different.

Moreover, not only does 'Kilburn', then, have many identities (or its full identity is a complex mix of all these) it is also, looked at in this way, absolutely not introverted. It is (or ought to be) impossible even to begin thinking about Kilburn High Road without bringing into play half the world and a considerable amount of British imperialist history (and this certainly goes for mining villages too). Imagining it this way provokes in you (or at least in me) a really global sense of place.

And finally, in contrasting this way of looking at places with the defensive reactionary view, I certainly could not begin to, nor would I want to, define 'Kilburn' by drawing its enclosing boundaries.

So, at this point in the argument, get back in your mind's eye on a satellite; go right out again and look back at the globe. This time, however, imagine not just all the physical movement, nor even all the often invisible communications, but also and especially all the social relations, all the links between people. Fill it in with all those different experiences of time–space compression. For what is happening is that the geography of social relations is changing. In many cases such relations are increasingly stretched out over space. Economic, political and cultural social relations, each full of power and with internal structures of domination and subordination,

stretched out over the planet at every different level, from the household to the local area to the international.

It is from that perspective that it is possible to envisage an alternative interpretation of place. In this interpretation, what gives a place its specificity is not some long internalized history but the fact that it is constructed out of a particular constellation of social relations, meeting and weaving together at a particular locus. If one moves in from the satellite towards the globe, holding all those networks of social relations and movements and communications in one's head, then each 'place' can be seen as a particular, unique, point of their intersection. It is, indeed, a *meeting* place. Instead, then, of thinking of places as areas with boundaries around, they can be imagined as articulated moments in networks of social relations and understandings, but where a large proportion of those relations, experiences and understandings are constructed on a far larger scale than what we happen to define for that moment as the place itself, whether that be a street, or a region or even a continent. And this in turn allows a sense of place which is extroverted, which includes a consciousness of its links with the wider world, which integrates in a positive way the global and the local.

[. . .]

These arguments, then, highlight a number of ways in which a progressive concept of place might be developed. First of all, it is absolutely not static. If places can be conceptualized in terms of the social interactions which they tie together, then it is also the case that these interactions themselves are not motionless things, frozen in time. They are processes. One of the great one-liners in Marxist exchanges has for long been, 'Ah, but capital is not a thing, it's a process.' Perhaps this should be said also about places; that places are processes, too.

Second, places do not have to have boundaries in the sense of divisions which frame simple enclosures. 'Boundaries' may of course be necessary, for the purposes of certain types of studies for instance, but they are not necessary for the conceptualization of a place itself. Definition in this sense does not have to be through simple counterposition to the outside; it can come, in part, precisely through the particularity of linkage *to* that 'outside' which is therefore itself part of what constitutes the place. This helps get away from the common association between penetrability and vulnerability. For it is this kind of association which makes invasion by newcomers so threatening.

Third, clearly places do not have single, unique 'identities'; they are full of internal conflicts. Just think, for instance, about London's Docklands, a place which is at the moment quite clearly *defined* by conflict: a conflict over what its past has been (the nature of its 'heritage'), conflict over what should be its present development, conflict over what could be its future.

Fourth, and finally, none of this denies place nor the importance of the uniqueness of place. The specificity of place is continually reproduced, but it is not a specificity which results from some long, internalized history. There are a number of sources of this specificity – the uniqueness of place. There is the fact that the wider social relations in which places are set are themselves geographically differentiated. Globalization (in the economy, or in culture, or in anything else) does not entail simply homogenization. On the contrary, the globalization of social relations is yet another source of . . . geographical uneven development, and thus of the uniqueness of place. There is the specificity of place which derives from the fact that each place is the focus of a distinct *mixture* of wider and more local social relations. There is the fact that this very mixture together in one place may produce effects which would not have happened otherwise. And finally, all these relations interact with and take a further element of specificity from the accumulated history of a place, with that history itself imagined as the product of layer upon layer of different sets of linkages, both local and to the wider world.

In her portrait of Corsica, *Granite Island*, Dorothy Carrington travels the island seeking out the roots of its character. All the different layers of peoples and cultures are explored; the long and tumultuous relationship with France, with Genoa and Aragon in the thirteenth, fourteenth and fifteenth centuries, back through the much earlier incorporation into the Byzantine Empire, and before that domination by the Vandals, before that being part of the Roman Empire, before that the colonization and settlements of the Carthaginians and the Greeks . . . until we find . . . that even the megalith builders had come to Corsica from somewhere else.

It is a sense of place, an understanding of 'its character', which can only be constructed by linking that place to places beyond. A progressive sense of place would recognize that, without being threatened by it. What we need, it seems to me, is a global sense of the local, a global sense of place.

"New Cultures for Old?"
from Doreen Massey and Pat Jess (eds)
A Place in the World? Places, Cultures and Globalization (1995)

Stuart Hall

Editors' introduction

Stuart Hall was born in Kingston, Jamaica, in 1932, and moved to England in 1951. Hall is best known for his long association with the Centre for Contemporary Cultural Studies at Birmingham University, which he directed from 1968 to 1979. While his predecessor, Richard Hoggart, is widely credited with founding the so-called cultural studies movement in Britain, Hall oversaw the movement's flowering under his leadership of the Centre. The term *cultural studies* is rather loosely utilized throughout Anglophone academia to indicate the general study of cultural phenomena, particularly in the industrialized West. However, the meaning more closely associated with the Birmingham school of Hall's day was an overtly politicized one, emphasizing the power of the culture industry embodied in television, literature, fashion, film, and so on. Hall realized that popular culture and the culture industry that produces it hold incredible power in contemporary society. In consuming the products of the culture industry, people acquire their cultural identity, according to Hall. For Hall and his contemporaries, consumer capitalism was understood as controlling the mass production of cultural meaning.

In 1979 Hall was appointed to a professorship of sociology at the Open University, from which he retired in 1997. Like many of the scholars whose work is excerpted in this *Reader*, Hall's scholarship is derived from a Marxist tradition. In fact, Hall worked with two other noted Marxist cultural theorists featured in this *Reader*, E.P. Thompson (p. 20) and Raymond Williams (pp. 15 and 207), in the 1950s. Together they founded the radical journals *New Left Review* and *The New Reasoner*, among other projects. Other works by Stuart Hall include *The Popular Arts* (written with Paddy Whannel, 1964), *Resistance Through Rituals: Youth Subcultures in Post-war Britain* (1989), *Questions of Cultural Identity* (co-edited with Paul du Gay, 1996), and *Visual Culture: The Reader* (co-edited with Jessica Evans, 1999).

In this engaging selection, "New cultures for old?" we see Hall at his didactic best. Culture – which, as you know by now, is a slippery term – is succinctly defined as a set of shared meanings. It is within this flexible set of meanings that our identity as part of a community is formed. Like languages, cultures produce and convey meaning. Though Hall is not trained as a geographer, he is well aware that place plays a central role as an anchor of cultures and the identities associated with them. With contemporary globalization and its fast-paced flows of people, goods, and information, our sets of shared meanings have begun to shift away from a notion of bounded nation-states as providing the most important framework for identity. Hall is careful to note, however, that globalization has a long history, and that nation-states themselves are not natural constructs but must rather be conjured from amidst difference. Two responses are possible in the face of these cultural shifts, according to Hall. One is a *defensive* response, an attempt to draw sharp boundaries around a culture and its place, to maintain the imagined purity of both in the face of dramatic changes. The second response is one that emphasizes cultural *openness* through connections across place, which Hall approaches through the notion of diaspora.

The often difficult relationship between globalization, culture, and identity is the subject of much contemporary work in both academia and literature. With respect to the latter, Salman Rushdie's work is paradigmatic; see *The Satanic Verses* (1988) and *Imaginary Homelands* (1992). In academia, Hall's student at the Centre for Contemporary Cultural Studies at Birmingham University, Paul Gilroy, has written on the cultural politics of race in ways that inform Hall's discussion of transatlantic African diaspora musics; see *Black Atlantic: Modernity and Double-Consciousness* (1991); and with respect to black culture and identity in contemporary Britain, "*There ain't no Black in the Union Jack*": *The Cultural Politics of Race and Nation* (1987).

Hall's critical take on capitalist culture is strongly informed by the so-called Frankfurt school (from the German city of Frankfurt, where the scholars associated with this perspective worked at the University of Frankfurt). A key text is Max Horkheimer and Theodor Adorno's *The Dialectic of Enlightenment: Philosophical Fragments*, originally published in 1947, translated and reissued in 2002, which provides a discussion of the politically stupefying effects of capitalist mass culture. Other works that underpin Hall's analysis of communication, media, and cultural meaning include Roland Barthes's *Mythologies*, originally published in 1952 and reissued in 1972; Jean Baudrillard's seminal work on power, culture, and communication, see for example *The Consumer Society: Myths and Structures* (1970); and Marshall McLuhan (who coined the phrase "global village"), *The Medium is the Message: An Inventory of Effects* (1967).

[. . .]

By *culture* we mean the systems of shared meanings which people who belong to the same community, group, or nation use to help them interpret and make sense of the world. These meanings are not free-floating ideas. They are embodied in the material and social world. The term 'culture' includes the social practices which produce meaning as well as the practices which are regulated and organized by those shared meanings. Sharing the same 'maps of meaning' gives us a sense of belonging to a culture, creates a common bond, a sense of community or identity with others. Having a position within a set of shared meanings gives us a sense of 'who we are', 'where we belong' – a sense of our own identity. Culture is thus one of the principal means by which identities are constructed, sustained and transformed.

Cultures are usually thought of as relatively stable or fixed sets of meanings and practices which have achieved a settled continuity over time and place. To think of oneself as 'English' or 'British' is inevitably to place oneself within a set of meanings that have a long history and continuity. Cultures predate the individual. They seem to provide a frame of reference or a tradition which connects one's present mode of existence to the way of life of one's ancestors, thereby giving a culture a distinctive coherence and shape over time and making it internally homogeneous. Those who identify with a culture, who share a cultural identity, are assumed to be the same (identical) by virtue of this membership. Cultures are usually seen as well-bounded formations, clearly marked off from other cultures. This marking of cultural difference both increases the sense of community or group solidarity ('us') amongst those who belong and . . . sharpens the sense of difference from 'other cultures' ('them').

Given that cultures are defined and perceived in this way, it is hardly surprising that the impact of globalization is seen by many commentators as profoundly unsettling for cultures and cultural identity. With its accelerated flows of goods, peoples, ideas, and images, the 'stretching' of social relations, its time and space convergences, its migrating movements of people and cultures, globalization is calculated to disturb culture's settled contours. Established traditions and customary ways of life are dislocated by the invasion of foreign influences and images from the new global cultural industries which traditional communities find enormously seductive, impossible to reject, yet difficult to contain. Global consumerism, though limited by its uneven 'geography of power', spreads the same thin cultural film over everything – Big Macs, Coca-Cola and Nike trainers everywhere – inviting everyone to take on western consumer identities and obscuring profound differences of history and tradition between cultures. Migration, which is part of the same process, moves peoples of very different backgrounds and traditions into the same space and time-frame. Sometimes, cultures are caught between, on the one hand, the desire for the mobility and material rewards of modernity and, on the other, the nostalgia for a lost purity, stability and traditional coherence which the present no longer provides, and consequently they splinter and fragment. The consequences of globalization for culture and cultural identity are profound. They are also contradictory, moving in different directions, and difficult to understand or predict.

[. . .]

How, then, are we to understand the changing shape of the cultural map in the most recent phase of the age of globalization? How are the new linkages of 'global' and 'local' affecting cultures? Are cultural differences being strengthened or eroded? Are new local cultures emerging as the older local cultures decline? How should we think about or re-conceptualize cultural identity in these more global times?

[. . .]

My argument, briefly, is that globalization forces us to question many of our commonsense ideas about cultural identity and obliges us to conceptualize culture in new ways. We can start the process of dismantling some older notions of cultural identity, and putting some alternative notions in play, by revisiting . . . the concept of place. What is involved in thinking further about the changing relationships between culture, identity and place?

[. . .]

What do we mean when we say cultures are *systems of meaning?* How do they work? One way of understanding how such systems of meaning work, is to think of culture in terms of a model of language. Things – objects, events, people – don't have a fixed meaning, a single truth, which exists for all to see and which we simply reflect in the language we use about them. Rather, it is through language that we *give meaning* to the world. Snow does not know that it is 'snow'. It is we who agree, within the conventions of our language code, to call

that soft, cold, frozen white water from the sky 'snow'; and by doing so we give it a meaning by distinguishing it from two other closely related phenomena, 'rain' and 'hail'. The Inuit people, who have rather a lot of the stuff, are said to have many more words in their language for distinguishing between different kinds of 'snow'.

So cultures consist of different systems which produce meaning, which classify the world *meaningfully.* Thus, there is a very close relationship between the sorts of meanings and resonances embodied in the English language and English culture. Systems of meanings or cultures work like languages. They provide us with interpretive frameworks through which we make sense of the world. Using the same language code or system of meanings to make sense of the world places us English-speakers within what is sometimes called the same 'interpretive community'. Individuals can't change language – the meanings of words – by an act of will; nevertheless, language is constantly changing, historically. This suggests that language is a shared, a collective social construction, though each of us can use it individually to say what we mean.

In the same way, systems of meaning or cultures cannot be fixed since there is no way of insisting that events, practices, rituals or relationships mean only one thing or of preventing them, over time, or in different contexts, from taking on new meanings. In culture, as in language, each usage changes or inflects the meaning in new ways; and over time (in different periods or different contexts, in relation to new topics or situations) new meanings or inflections will establish themselves in common usage. . . . Some novel meanings will emerge by combining older meanings, while others will simply fall away or become archaic, useless for communicative exchange, through lack of use or relevance to a changing situation. We may try to use these systems of meaning like we use language, as accurately as we can, in order to say what we mean, to express ourselves perfectly. But we know that every statement will slip a little when it is appropriated or interpreted or translated into their own frame of reference by the persons with whom we are communicating. Meanings shift and are always open to interpretation – and other people never understand perfectly everything we say or catch every nuance of meaning we try to express. On the other hand, just as to communicate to a

Swahili speaker requires us to use a common Swahili language code, so to express a cultural meaning requires us to position ourselves within a shared cultural meaning system, and then to use it to say or mean something – even if the idea may not have been expressed quite like that before, and often goes beyond the established tradition or conventions of the culture. A reference to 'the cross' will carry a very different meaning in a Christian, as opposed to a Muslim, country. The Union Jack evokes very different sentiments in Dublin and London.

We have been using language as a model of how cultural systems work. But language, we must remind ourselves, is also itself one of the fundamental cultural systems. People who share a language can communicate with or 'make sense' about the world to one another. A shared language is something which helps to give a person a cultural identity – being a Gaelic or Basque or Standard English or Urdu or patois speaker is to *be* a certain kind of person. It places one on a particular part of the language map: the community of English or Gaelic or Basque or Urdu or patois speakers to which one 'belongs'. Speaking a language means that you are familiar with, even if you do not share, the values of other language-users. That is why the right to use a particular language has been, historically, so significant a part of the struggles for national independence; and why being obliged to speak someone else's language – the language of the conqueror or colonizer, for example – is such a powerful symbol of cultural subjugation.

However, language is only one of the systems of meaning which produce culture. A culture is composed of many such systems. For example, religion is a powerful bearer of shared meanings about the sacred which carries a great deal of symbolic meaning and authority in many cultures. Again, religion has traditionally been a powerful source of 'belongingness', and, perhaps for that very reason, also a source of division and the marking of difference and 'otherness' (for example, between Hindus and Muslims during the partition of India; or between Protestants and Catholics in Northern Ireland today). Another powerful source of cultural meanings and cultural identities is custom and tradition: the distinctive, traditional ways things are done in 'this' part of the country or world as opposed to 'that'; the everyday rituals and practices

which establish what the 'folkways' have been down the ages, or which mark special occasions (births, puberty, marriages, deaths, anniversaries), and connect present forms of life with the distinctive ways of life of one's ancestors. There are also shared traditions of representation: *genres* of painting or decoration; narratives about the past of the 'tribe'; sculptures or crafts which have a special significance in the working, familial or sacred life of the group; forms of dress and adornment; or stories (popular or highbrow) which maintain in collective memory the historical record of the group.

'Place' is another of those cultural systems. . . . We use a discourse of 'place' to give meaning to life and to position ourselves in certain definite ways within society and its belief system. I would argue that cultures are often understood as 'placed' in at least two senses:

1 First, we associate 'place' with an actual location where many different relationships have overlapped over time, producing a dense, richly textured sense of life. Until recently, it had been assumed that the idea of place was a significant, though not a necessary, element in the way we understand cultures. In fact, it is the case that shared meaning systems can develop between people who live in very different places – across time and space. Indeed, this tendency may be increasing under the most recent forms of globalization where, for example, transnational migrants maintain important linkages between place of origin and place of settlement. Modern communications systems give rise to shared 'communities of interest' (e.g. chess players), or shared 'communities of taste' (e.g. opera lovers, jazz enthusiasts, *Neighbours* or *Eastenders* addicts), or shared consumer tastes (e.g. youth fashions) amongst people who are widely separated, who do not actually share a place, and who may indeed have never met one another. Shared systems of meaning can also survive lengthy spatial separation. Think of the successful struggle to keep traditions and folkways alive among many exiled or migrant communities (e.g. black slaves in the New World, Orthodox diaspora Jews in Europe, Puerto Ricans or Sicilians in New York, Latvians and Rumanians in Milwaukee). Nevertheless, it is still common to *think* of cultures as if they depended

on the stable interaction of the same people, doing the same sorts of things, over and over again, in the same geographical location – a set of meanings which we traditionally associate with 'place'.

2   In other words, while not literally necessary to culture, 'place' seems to act as a sort of symbolic guarantee of cultural belongingness. It establishes symbolic boundaries around a culture, marking off those who belong from those who do not. . . . It ensures the continuity of patterns of life and of tradition amongst a gathered and interrelated population who have been together, living in the same spatial environment, since 'time immemorial'. Again, let us emphasize that it is possible to think of many cultures which are maintained by groups that are not settled in one place, and by cultural influences between groups that have never shared the same place. However, physical settlement, continuity of occupation, the long-lasting effects on ways of life arising from the shaping influence of location and physical environment, coupled with the idea that these cultural influences have been exercised amongst a population which is settled and deeply inter-related through marriage and kinship relations, are meanings which we closely associate with the idea of culture and which provide powerful ways of conceptualizing what 'culture' is, how it works, and how it is transmitted and preserved.

When shared meaning systems are underpinned by long, historical settlement of a population and 'shaping' in one physical environment, with strong kinship links as a result of continuous intermarriage over generations, we get a *very* strong and *strongly bounded* idea of culture and cultural identity. This definition of culture, though not actually genetic or biological, is often experienced as if it were a part of our biological nature because it is tied up with the sharing of the culture between members with a long and unbroken common genealogy, kinship, residence and descent.

We call this very strong, well-bounded version of cultural identity *ethnicity*. Ethnicity arises wherever shared activities and meaning systems in one place are underpinned by shared kinship and blood-ties, evidence of which can sometimes be 'read' into certain shared physical features and

characteristics of a population. Where people share not only a culture but an *ethnos*, their belongingness or binding into group and place, and their sense of cultural identity, are very strongly defined. Indeed, ethnicity is a form of cultural identity which, though historically constructed like all cultural identities, is so unified on so many levels over such a long period that it is experienced as if it were imprinted and transmitted by Nature, outside what we would call Culture or History. This is where culture appears, and is defended as, part of one's kith and kin, and cultural identity is based on ideas of 'Blood and Soil'.

Place, in short, is one of the key discourses in the systems of meaning we call culture, and it functions to help stabilize cultural patterns and fix cultural identities, as they say, 'beyond the play of history'. . . .

There is little doubt that this association with place is one, very powerful, version of our commonsense understanding about culture. It seems to persist even when it has no strong historical basis in fact. We are therefore driven to the conclusion that we think of cultures as strongly placed, not because all cultures *are* but *because that is how we imagine them.* To put the point simply, when we think of or imagine cultural identity, we tend to 'see' it in a place, in a setting, as part of an imaginary landscape or 'scene'. We *give* it a background, we put it in a frame, in order to make sense of it. Can we think of 'Englishness' without seeing, somewhere, in our mind's eye, England's 'green and pleasant land', rose-trellised thatched cottages, village green and church steeple, a 'sceptred isle', 'this precious stone set in a silver sea'? Can we think of what it is like to be typically 'Greek' or 'Italian' without a Mediterranean scenario – white beaches, azure-blue sea – flashing before us? These are stereotyped representations of *national* identities; but I am suggesting that all cultural identities tend to have . . . their 'landscapes of the mind', their 'imaginary geographies'. There is a strong tendency to 'land-scape' cultural identities, to give them an imagined place or 'home', whose characteristics echo or mirror the characteristics of the identity in question. . . .

The association of national cultures and identities with particular landscapes therefore helps to construct and to fix in place a powerful association between culture and 'home'. We think of our culture as a *home* – a place where we naturally belong, where we originally came from, which first stamped us with our identity, to which we are powerfully bonded, as we are to our families, by ties that are inherited, obligatory and unquestioning. To be among those who share the same cultural identity makes us feel, culturally, at *home*. Cultures give us a powerful sense of belongingness, of security and familiarity.

[. . .]

. . . . Who belongs [to England], who doesn't? When did the many different peoples and nations coalesce into *one* identity? Does the use of different terms, like 'English' and 'British', signal differences within the 'United Kingdom' which still persist? The Scots, the Welsh and the Northern Irish are 'British' too, but 'Englishness' seems to have a quintessential relation to 'Britishness' which the others don't quite carry. Is national identity, then, also a power relationship? And if the Romans and Normans have 'become' English, how about more recent settler populations? How long does it take to *become* 'English'? Can you be 'black *and* English'? Great Britain has been a relatively unified political entity for centuries (at least since the Act of Union), but it would be hard to prove that the 'political roof' of the British nation-state covers what were originally a single people belonging to a single culture (*ethnos*). The UK is in fact the product of a series of invasions, settlements and conquests, by different ethnic groups, belonging to different cultures, speaking different languages and worshipping different gods – it evolved only gradually into one composite nation. The job of the national culture was therefore not to reflect in its political arrangements an already unified people and culture but to *produce* a culture in which, with luck, the different elements could gradually be unified into a sense of common belongingness – a process of cultural unification which has been, at best, *only party successful*. One has only to think of the regional, cultural, class, gender, 'racial', economic and linguistic differences which still persist within its boundaries; of the tensions which now accompany this idea of a 'united' kingdom; and of the role of 'Englishness' as the hegemonic culture in relation to the other 'nations' within the kingdom – a fact which irritates many Scots, Welsh and Northern Irish people, and which fuels nationalist sentiment and aspirations in different parts of the UK. Even after

centuries, the 'unity' of the United Kingdom remains somewhat precarious. Indeed, it has become in recent years, as a result of the move toward the EU and the growth of nationalist sentiment within the UK, a source of growing public anxiety and debate.

The role of the national culture – perhaps of cultures in general – is therefore not to express the unitary feelings of belongingness which are 'always there in the culture', but to represent what are, in fact, real differences *as a unity*; to produce, through its ongoing 'narrative of the nation' (in education, literature, painting, the media, popular culture, the historical heritage, the leisure industry, advertising, marketing, etc.) an identification, a sense of belongingness which, without constant nurturing, would not be sufficient to bind the nation together across the divisions of class, region, gender, 'race' and the unevenness of economic development.

If this is true of the culture of the United Kingdom, a relatively stable nation-state which has not suffered invasion or conquest for several centuries, how much more so is it true of other Western European national cultures, many of which (like Germany and Italy) were not unified until the nineteenth century, and which within their sovereign borders include very sharp ethnic and cultural differences between regions, especially between 'North' and 'South'. Europe is the product of successive conquests, and of the subjugation of peoples, often within the framework of empires which long preceded the formation of the nation-state (e.g. the Holy Roman Empire). This is certainly not to deny that there are shared cultural features between the peoples and cultures clustered under the roof of these 'core' Western European nation-states. But it is to insist on the profound *differences* which persist and which these shared national cultures have constantly had to negotiate. This points us towards a rather different conception of culture – one where 'a culture' is never a simple, unified entity, but always has to be thought of as composed of similarities and differences, continuities and new elements, marked by ruptures and always crosscut by *difference*. Its meanings are the result of a constant, ongoing *process* of cultural negotiation which is constantly shifting and changing its contours to accommodate continuing tensions. Cultures conceptualized in this way do *not* stretch backwards, unchanged, into 'time immemorial'. They offer no fixed, single point of origin which can stabilize cultural identities forever, thereby guaranteeing that all those who ever belong to them were, have become, or are destined to remain *the same* (i.e. identical).

This suggests that what we call nationalism – the ideology of belongingness as one people to what is sometimes referred to as the 'family' of the nation and national culture – arises not (as the national story tells us) because that is what it always was in the past. Rather, it is a key element in the ongoing *process* of unifying or binding people together, creating through these discourses an idea of the nation with which they can identify, and in that way binding up differences and cementing the nation in the present, for the sake of the future. Nationalism often invokes a return to past glories or virtues (think of Mrs Thatcher's remark at the time of the Falklands War, that 'Britain can be "Great" again'). But its aim is really to produce something – a unified culture – in the future. Nationalism is always rehearsing the 'narrative of the nation' as a return to lost or forgotten origins. But its project is actually to create something which does not yet fully exist – a unified culture.

The Western European nation-states which we think of now as the 'motors' of modernity and world development are, in fact, all culturally and ethnically mixed formations – 'hybridized', to use a term we shall examine at greater length below. The situation in Central and Eastern Europe, and in what was formerly the Soviet Union, is even more complicated. Here, nation-states were often the relatively recent product of the disintegration of former empires (the Austro-Hungarian and Ottoman empires, for example); or of the expansion and forcible absorption of one people by another (the Baltic States); or of the arbitrary solutions which the 'Great Powers' reached at conference tables (the former Czechoslovakia). The framework of a national culture has proved even less durable here than it has in the West. It has proved itself incapable of unifying peoples of different cultures, languages, ethnic origins or religious persuasions, many of whom have lived for decades in 'multicultural' communities – as recent events in the former Yugoslavia tragically demonstrate.

Europe, East and West, is so culturally mixed that the effort to oblige each 'homeland' to contain only one ethnic group, to constitute nations out of one culture . . . to unify one people, one ethnos, under one 'political roof', only encounters the many minorities living in the same place, who are *not* of the same culture but persist in considering that place to be their 'home' too. . . . A unified, well-bounded, internally coherent 'Serb' culture can only at this late stage of history be carved out of this hotchpotch by violence – the sort of 'ethnic cleansing' we have seen in Bosnia-Herzegovina is its direct result.

In short, though cultures are sometimes 'placed' and we tend to imagine them as strongly unified and homogeneous, integrated by tradition in a landscape and tied to a homeland, the effort – against the complex and tortured background of modern history – to actually *make* 'culture' and 'place' correspond with one another turns out to be a hopeless, expensive and sometimes violent and dangerous illusion.

[. . .]

In some accounts, a more open and diverse conception of culture is closely linked with the process of globalization. It is often argued that the *pace* of globalization in the last few decades is particularly intense. Time and space have been globally condensed in ways which were unimaginable a few decades ago, and many of the forms which organize the latest phase of globalization are novel. However . . . globalization is not a new or recent process. It is more or less coterminous with the whole of the historical period we now call 'modernity'.

The Europe of the nation-states . . . was a relative late-comer, historically. Before that there was no such thing as Europe or a European culture or a European identity, at least as a single or unified object. Greece, the so-called cradle of European civilization, was a series of city-states, with closer connections to North Africa than to the countries bordering the North Sea. In pre-classical times, the territory we call Europe was occupied by dispersed tribes and peoples and formed no single unity of any kind. Europe took shape through a succession of larger groupings, mostly of an imperial kind, which were very extensive in spatial terms: the conquests of Alexander the Great spread all over the eastern and much of the western Mediterranean and a large part of the Near East; the Roman Empire stretched from Scotland to the

African Sahara and from Gibraltar to the Euphrates, and its trade routes reached India, Indochina and the Far East. By definition, these empires ruled over many different cultures and peoples.

[. . .]

Globalization is the process by which the relatively separate areas of the globe come to intersect in a single imaginary 'space'; when their respective histories are convened in a time-zone or time-frame dominated by the time of the West; when the sharp boundaries reinforced by space and distance are bridged by connections (travel, trade, conquest, colonization, markets, capital and the flows of labour, goods and profits) which gradually eroded the clear-cut distinction between 'inside' and 'outside'. The history of the New World, of India, Asia and Africa gradually became a subordinated part of the 'internal' history of the West. It was the beginning of that very uneven time-frame we call 'global' time.

By globalization we refer, in a long historical perspective, to a number of different processes.

- The exploration by the West of hitherto 'unknown' parts of the globe (unknown to Europe, that is).
- The expansion of world trade and the early stages of the construction of a 'world market'.
- The movements of capital investment and the transfer of profits and resources between metropolis and periphery.
- The large-scale production of raw materials, food, minerals and commodities for industries and markets elsewhere.
- The process of conquest and colonization which imposed systems of rule and other cultural norms and practices on subordinated cultures.
- The migrations which were set in motion and the settlements and colonized outposts which were established.
- The establishment, even where direct colonization was avoided, of powerful imperial spheres of cultural influence: Britain, France, the Netherlands and Portugal in the Middle East and the Far East; the British, Spanish, Portuguese and Dutch in Latin America; the colonizations by the Dutch, British and French of the Pacific; the scramble by the great powers for colonies in Africa.

In its wake, the idea of culture as a set of autonomous, self-enclosed meaning systems and practices begins to seem anachronistic. The cultures of many parts of the world had for the first time to negotiate with the colonizing cultures imposed on them through conquest, settlement, trade or direct administration and government. The colonizers exported European culture to the places they conquered. Other cultural influences – religion, language, education, legal systems, conceptions of property – followed the flag. The indigenous ways of life were often broken up and destroyed; new habits and values were implanted in their place. Movements of populations, planned and unplanned, followed the shifts in power. Cultures began to be defined, not in terms of their own indigenous values, but by their relationship (usually of power) with *other* cultures. Culture began to form one of the critical circuits through which power of different kinds – economic, political, religious, gendered, racial – circulated. The world, we might say, became, through this process, 'global' for the first time.

[. . .]

Far from some gentle movement of ideas, culture and power were intimately connected in the process of globalization from its early inception. The establishment of spheres of cultural influence, the hierarchical relations of dominance and subordination between colonizing and colonized cultures, between different racial groups, between the 'civilized' and the barbarians', the shifting relations of cultural power which followed in the wake of the successive phases of globalization, are difficult to exaggerate. Much the same process was set in train elsewhere around the globe – for example, in the Indian subcontinent, in South East Asia and in Africa – in the wake of the 'Euro-imperial adventure'.

[. . .]

. . . As time and space compression has deepened, so the cultures of more and more places become translated. As modern migration has stimulated the flows of peoples, cultures have characteristically become composed not of single but of diverse cultural traditions and patterns. There is a further argument that, whereas globalization in the earlier phases of colonial expansion tended to impose metropolitan cultural values and norms on the cultures of the periphery, the process – greatly supplemented by the global character of modern communications and consumer industries – is now complicated by the reverse flow: from the impoverished cultures of the 'margins' to the metropolitan centres. In that sense, the transculturation typical of the old colonial cities like Kingston, Mombasa, Bombay, Saigon or Hong Kong is now supplemented and expanded by the multicultural processes which are reshaping and hybridizing London, Paris, Los Angeles and New York . . . and 'third world' cities of the first world like Toronto, Marseilles or Miami.

. . . However, this is too evolutionary a way of describing what are in fact bitterly contested *alternative strategies* in the process of cultural change. It makes it sound as if societies will smoothly and unproblematically evolve from one of our models to the other. What such an account leaves out is the question of *cultural power* and *cultural resistance*; the issues of cultural politics which underpin such transitions; the different, often contradictory, currents which are at play in the modern version of 'the culture wars'; the way in which the question of cultural identity has become *the* key issue – what is ultimately at stake, being struggled over, in these cultural shifts.

For many groups, cultural survival has always been seen to depend on keeping the culture 'closed' – intact, homogeneous, unified within, and with strongly marked boundaries separating it from 'others'. Creolization, transculturation, hybridization – whichever of these terms we wish to use to describe culture under the impact of globalization – is seen as threatening the integrity of the culture and weakening the sense of cultural identity that holds the group together. From this perspective, migration weakens the bonds of belongingness, and creolization is seen as a form of dilution – even, perhaps, of cultural pollution. The only viable strategy is to hold fast to cultural traditions, to reaffirm those elements of the culture which maintain the links to one's past, to keep the connection 'pure', to resist all forms of syncretism as, in effect, a 'loss of identity', to counter this potential loss by a 'return' to one's cultural roots, to hold fast to one's founding identity; and thus in these ways to *close up the community around its foundational cultural beliefs and values*.

We may call this strategy the *revival of ethnicity* – the attempt to restore strong, closed definitions of what constitutes a culture. The surprising thing is that this response to globalization is to be found in the late twentieth century in *both*

'colonizing' and 'colonized' peoples, at *both* the 'centre' and the 'periphery', in what are usually thought of as *both* traditional and modern societies. Indeed, it has made a strong, and somewhat unexpected, return in recent years to the world stage. It cuts across the usual political alignments of 'left' and 'right' and it sometimes divides members of the same cultural community from one another.

Thus, we can see evidence of this tendency in the revival of nationalism which is to be found in resurgent forms in Central and Eastern Europe and in the former Soviet Union. But you can also see it in the 'Little England' reaction in Britain to the fear of losing sovereignty to 'Europe'; and it is very much in evidence in the growth of racism and the rebirth of racist and neofascist movements across the New Europe, focused on the threat of migrants coming into Europe from Africa, the Middle East, the Caribbean, Asia, or the fear of Europe being overrun by 'economic migrants' in search of a better life, or by the refugee problem. Forms of cultural racism have appeared in recent years, not only in the so-called 'backward' parts of Europe in the East, but right in the centre of *modern* Europe – in Britain, Germany, France, Spain and Italy. You can certainly find an equally exclusive definition of 'culture' – this time with religion rather than 'race' or ethnicity providing the focus – in the versions of Islamic fundamentalism which have gained ground in recent years in Iran, Egypt and Algeria, as well as in the success of the Hindu fundamentalist movement in India. But 'fundamentalism' is not confined to the so-called 'third world'. We can see a respectable version (though not perhaps so closely related to religion) beginning to attract support amongst those groups in both the US and the UK who want to roll back 'multiculturalism' in education and go back to much more traditional and exclusivist definitions of 'Englishness'; or among those who have reacted strongly against including the literatures of other peoples, or indeed works by women writers from other cultures, in the literary canon being taught in colleges and schools – despite the fact that British and American urban schools and campuses have been transformed in recent years by migrations from Latin America, the Caribbean and Asia.

These are all, of course, very different examples with different specific histories, and I am not trying to represent them as all the same. What I am suggesting, however, is that they do all share *one* important feature in common: their response to globalization is to turn back to more 'closed' definitions of culture, in the face of what they see as the threats to cultural identity which globalization in its late-twentieth-century forms represents. They mark the revival of an attachment to more 'local', or fixed, or placed aspects of culture. This may not be such a surprising turn in a world where globalization is increasingly transgressing boundaries, mixing up traditions, confusing 'us' and 'them', 'inside' and 'outside', and constructing identities based on less 'grounded' forms of identification.

[. . .]

. . . There are other ways of imagining communities of belongingness, which are not centred in the nation-state or the national identity 'story', which cut across and disrupt many of these boundaries and borderlines, and provide alternative resources for constructing identity and fashioning culture.

There are many ways of trying to describe this . . . transcultural response to the globalization of culture: the idea of the 'global city' is one way; the notion of 'multiculturalism' and its effects within the hitherto settled cultural frontiers of the western nation-state is another. Yet another way of framing the new relationship between culture, place and identity is to be found in the concept which many cultural critics are beginning to use . . . the idea of the *diaspora*.

The term *diaspora* can, of course, be used in a 'closed' way, to describe the attempt of peoples who have, for whatever reason, been dispersed from their 'countries of origin', but who maintain links with the past through preserving their traditions intact, and seeking eventually to return to the homeland – the true 'home' of their culture – from which they have been separated. But there is another way of thinking about diasporas. 'Diaspora' also refers to the scattering and dispersal of peoples who will *never* literally be able to return to the places from which they came; who have to make some kind of difficult settlement with the new, often oppressive, cultures with which they were forced into contact; and who have succeeded in remaking themselves and fashioning new kinds of cultural identity by, consciously or unconsciously, drawing on more than one cultural *repertoire*. These are people who, as Salman Rushdie wrote in his essay in *Imaginary Homelands*, 'having been borne across the world . . . are translated men (and women)'. They are people who belong to more than one world, speak more than one language (literally and metaphorically), inhabit more than one identity, have more than one home; who have learned to negotiate and translate *between* cultures, and who, because they are irrevocably the product of several interlocking histories and cultures, have learned to live with, and indeed to speak from, *difference*. They speak from the 'in-between' of different cultures, always unsettling the assumptions of one culture from the perspective of another, and thus finding ways of being both *the same as* and at the same time *different from* the others amongst whom they live. Of course, such people bear the marks of the particular cultures, languages, histories and traditions which 'formed' them; but they do not occupy these as if they were pure, untouched by other influences, or provide a source of fixed identities to which they could ever fully 'return'.

They represent new kinds of identities – new ways of 'being someone' – in the late-modern world. Although they are characteristic of the cultural strategies adopted by marginalized people in the latest phase of globalization, more and more people in general – not only ex-colonized or marginalized people – are beginning to think of themselves, of their identities and their relationship to culture and to place, in these more 'open' ways. It is certainly one of the greatest sources of cultural creativity today – and what much late-modern culture (novels, poems, paintings, images, films, video, etc.) seems to be about.

Used in this way, the concept of *diaspora* provides an alternative framework for thinking about 'imagined communities'. It cuts across the traditional boundaries of the nation-state, provides linkages across the borders of national communities, and highlights connections which intersect – and thus disrupt and unsettle – our hitherto settled conceptions of culture, place and identity.

Because it is spatially located, but imagined as belonging not to one but to several different places, the *diaspora* idea actively contests the way in which *place* has been traditionally inserted into the story of *culture* and *identity*. It therefore forges a new relationship between the three key terms – culture, identity and place. From the *diaspora* perspective, identity has many imagined 'homes' (and therefore no one, single, original homeland); it has many different ways of 'being at home' – since it conceives of individuals as capable of drawing on different maps of meaning, and of locating themselves in different imaginary geographies at one and the same time – but is not tied to one, particular place.

It also breaks with a certain conception of *tradition* – the thing which is supposed to link us to our origins in culture, place and time. In the 'closed' version of culture, tradition is thought of as a one-way transmission belt; an umbilical cord, which connects us to our culture of origin. Ultimately, if we keep the links pure, they will lead us back to where we belong. The 'closed' version assumes that the further you get from your origins, the more you are separated from your true culture. It is a *linear* conception of culture. In 'diaspora' conceptions of culture, the connections are not linear but circular. We should think of culture as moving, not in a line but through different circuits. Paul Gilroy argues in his book *The Black Atlantic* that, if you wanted to tell the story of black music, you wouldn't construct a story of how 'authentic' black music *started* in Africa and became diluted with each subsequent transformation – the blues, reggae, Afro-Cuban, jazz, soul and rap – all representing 'loss of tradition' the further the music gets dispersed from its *roots*. Instead, you would have to pay attention to the way black music has travelled across and around the diaspora by many, overlapping *routes*.

You would show how *different* were the many 'African' musics and rhythms which slaves originally brought with them from Africa; how much these were transformed, first by life on the slave plantations, then by the impact of different 'European' musics, and especially by the influence of religious music; how in America, in the blues – urban or rural, secular or religious – these already complex musical traditions were further modified, as blacks migrated to the cities; how, in jazz, in soul, in reggae, in rap music, one can see the musical forms being constantly reworked and transformed to produce, not a *diluted* version of 'African' music, but a variety of *new* black diaspora musics (plural). One can also see how all of these were once again transformed in the conditions of post-war migration. Examples would include the influence of American rhythm and blues on Jamaican folk music that produced ska and reggae itself (which was a new, not a traditional music); or the influence of Jamaican music, played on the great 'sound systems' of the 1970s, which was taken to the Bronx by Jamaican DJs, and fused with the tradition of DJ 'talk over' rooted in the powerful black-vernacular traditions of American soul music, and later with rhythms from the Hispanic migrant community to provide the matrix out of which contemporary rap music first emerged. Or one could follow, along a different track, the adaptation of Jamaican 'roots' reggae to British conditions, and the way it was influenced by black soul and other popular music in the UK to create the distinctive sounds of contemporary British black music.

These stories connect the different 'black' diasporas around the Atlantic, linking Africa to the settlements of blacks in the New World created by slavery and to the resettlements in the Old World formed by post-war migration. But they don't connect them in one direction only. There are also contraflows: for example, black British rap artists moving between the UK and the US markets; or black music of the diaspora imported back into Africa, and played there, influencing the growth of *new* kinds of African popular music which are both urban and modern (not 'tribal' and traditional). Each variant of black music takes some 'African' elements – rhythmic and vocal elements especially – from traditions which are continuous and common between all the diasporas. Much of these were already transformed by the culture of plantation slavery. But these musics are, nevertheless, not the *same* because each has also fused these elements with different new elements. Each has taken the shape and imprint of the national contexts in which it developed (compare Afro-Cuban with New Orleans jazz; or American rhythm and blues with British 'Northern Soul'). Thus, what connects one part of the black diaspora to another is *not a tradition which remains the same*, but a complex combination of *continuities and breaks, similarities and differences*: what Gilroy calls a conception of tradition as the 'changing same'.

[. . .]

You may like, in conclusion, to consider for yourself how far the shift which was signalled from 'roots' to 'routes' as a way of thinking about culture applies not only to the ex-colonized, ex-enslaved, marginalized peoples of the diasporas but is slowly and unevenly becoming a more general model of how culture and identity are being reconstructed everywhere in late modernity.

"National Geographic: The Rooting of Peoples and the Territorialization of National Identity among Scholars and Refugees"

from *Cultural Anthropology* 7, 1 (1992): 24–44

Liisa Malkki

Editors' introduction

As the French philosopher Simone Weil noted in the quote in the opening sentence of Liisa Malkki's 'National geographic', there exists a deeply felt human need for attachment to place. That Weil wrote this in 1942, when France (and indeed much of Western Europe) was in the midst of the turmoil of World War II, is not coincidental. Times of heightened upheaval and human displacement, occasioned by events such as wars, spotlight the apparently profound importance of rootedness for human beings. Of course, there is no proof that ties to place are in truth an essential requirement of the human soul. You may well know people who seem to thrive on constant movement – you may even consider yourself to be a modern nomad. So there is something of a debate, one that is only heightened by the fast-paced nature of globalization, about movement versus rootedness when it comes to questions of identity, community, and belonging.

Malkki does not attempt to resolve this debate. Rather, she takes on the question of refugees, in order to explore how the figure of the involuntary migrant brings to light cultural tensions surrounding place and identity. Because we tend to use botanical metaphors, of roots and trees in particular, to discuss people's connections to place, the condition of being *uprooted* is viewed as a potentially deadly one, if not in literal then in figurative terms. Displaced peoples are assumed to be unnaturally out of place, disruptive of the supposedly stable moral and political order, and as such they constitute a threat. This is particularly true when viewed in terms of the nation-state and its supposed integrity. In these global times, characterized by the ever-increasing movement of displaced peoples across national boundaries – what literary theorist Edward Said called a "generalized condition of homelessness" (see p. 357) – these questions become ever more pressing. The spatiality of asylum is explored by Jennifer Hyndman in *Managing Displacement: Refugees and the Politics of Humanitarianism* (2000). Gilles Deleuze and Félix Guattari have also considered organic metaphors in the representation of culture at length, particularly in *A Thousand Plateaus: Capitalism and Schizophrenia* (1987).

In keeping with one of the key themes of this part, Malkki questions why it is that so many of us take for granted a notion of earth's space that is carved into discrete nations, often imagined as colorful blocks of territory juxtaposed on a map or in an atlas? Nations, Malkki and others argue, are not natural units; rather they are thoroughly socially constructed entities. Thus we know that they are also deeply involved in reflecting and solidifying, as well as contesting, relations of power in society. The social construction of places is discussed at length at the scale of the nation-state by Benedict Anderson, in *Imagined Communities: Reflections on the Origin and Spread of Nationalism* (1983), and at a larger scale in *The Myth of the Continents: A Critique of Metageography*, by Martin Lewis and Karen Wigen (1997).

Liisa Malkki is an Associate Professor of Anthropology at Stanford University in California. She has written a book, titled *Purity and Exile: Violence, Memory, and National Cosmology among Hutu Refugees in Tanzania* (1995). Malkki continues to focus on questions of culture, identity, and nation. Her current research interests involve humanitarian interventions and the work of the International Committee of the Red Cross.

INTRODUCTION

"To be rooted is perhaps the most important and least recognized need of the human soul," wrote Simone Weil in wartime England in 1942. In our day, new conjunctures of theoretical enquiry in anthropology and other fields are making it possible and necessary to rethink the question of roots in relation – if not to the soul – to identity, and to the forms of its territorialization. The metaphorical concept of having roots involves intimate linkages between people and place – linkages that are increasingly recognized in anthropology as areas to be denatured and explored afresh.

. . . [N]otions of nativeness and native places become very complex as more and more people identify themselves, or are categorized, in reference to deterritorialized "homelands," "cultures," and "origins." There has emerged a new awareness of the global social fact that, now more than perhaps ever before, people are chronically mobile and routinely displaced, and invent homes and homelands in the absence of territorial, national bases – not *in situ*, but through memories of, and claims on, places that they can or will no longer corporeally inhabit.

Exile and other forms of territorial displacement are not, of course, exclusively "postmodern" phenomena. People have always moved – whether through desire or through violence. Scholars have also written about these movements for a long time and from diverse perspectives. What is interesting is that now particular theoretical shifts have arranged themselves into new conjunctures that give these phenomena greater analytic visibility than perhaps ever before. Thus, we . . . have old questions, but also something very new.

The recognition that people are increasingly "moving targets" of anthropological enquiry is associated with the placing of boundaries and borderlands at the center of our analytical frameworks, as opposed to relegating them to invisible peripheries or anomalous danger zones. Often, the concern with boundaries and their transgression reflects not so much corporeal movements of specific groups of people, but, rather, a broad concern with the "cultural displacement" of people, things, and cultural products. Thus, what [Edward] Said, for example, calls a "generalized condition of homelessness" is seen to characterize contemporary life everywhere.

In this new theoretical crossroads, examining the place of refugees in the national order of things becomes a clarifying exercise. On the one hand, trying to understand the circumstances of particular groups of refugees illuminates the complexity of the ways in which people construct, remember, and lay claim to particular places as "homelands" or "nations." On the other, examining how refugees become an object of knowledge and management suggests that the displacement of refugees is constituted differently from other kinds of deterritorialization by those states, organizations, and scholars who are concerned with refugees. Here, the contemporary category of refugees is a particularly informative one in the study of the sociopolitical construction of space and place.

The major part of this article is a schematic exploration of taken-for-granted ways of thinking about identity and territory that are reflected in ordinary language, in nationalist discourses, and in scholarly studies of nations, nationalism, and refugees. The purpose here is to draw attention to the analytical consequences of such deeply territorializing concepts of identity for those categories of people classified as "displaced" and "uprooted." These scholarly views will then be juxtaposed very briefly with two other cases. The first of these derives from ethnographic research among Hutu refugees who have lived in a refugee camp in rural Western Tanzania since fleeing the massacres of 1972 in Burundi. It will be traced how the camp refugees' narrative construction of homeland, refugee-ness, and exile challenges scholarly constructions and common sense. In the second case, the ethnography moves among those Hutu refugees in Tanzania who have lived (also since 1972) outside of a refugee camp, in and around the township of Kigoma on Lake Tanganyika. These "town refugees" present a third, different conceptual constellation of links between people, place, and displacement – one that stands in antagonistic opposition to views from the camp, and challenges from yet another direction scholarly maps of the national order of things.

MAPS AND SOILS

To begin to understand the meanings commonly attached to displacement and "uprootedness" in the contemporary national order of things, it is necessary to lay down some groundwork. This means exploring widely shared commonsense ideas about countries and roots, nations and national identities. It means asking, in other words, what it means to be rooted in a place. Such commonsense ideas of soils, roots, and territory are built into everyday language and often also into scholarly work, but their very obviousness makes them elusive as objects of study. Common sense, as [Clifford] Geertz has said, "lies so artlessly before our eyes it is almost impossible to see."

That the world should be composed of sovereign, spatially discontinuous units is a sometimes implicit, sometimes stated premise in much of the literature on nations and nationalism . . . much like any school atlas with yellow, green, pink, orange, and blue countries composing a truly global map with no vague or "fuzzy spaces" and no bleeding boundaries. The national order of things . . . usually also passes as the normal or natural order of things. For it is self evident that "real" nations are fixed in space and "recognizable" on a map. One country cannot at the same time be another country. The world of nations is thus conceived as a discrete spatial partitioning of territory; it is territorialized in the segmentary fashion of the multicolored school atlas.

The territorialization expressed in the conceptual, visual device of the map is also (and perhaps especially) evident on the level of ordinary language. The term "the nation" is commonly referred to in English (and many other languages) by such metaphoric synonyms as "the country," "the

land," and "the soil." For example, the phrase "the whole country" could denote all the citizens of the country or its entire territorial expanse. And "land" is a frequent suffix, not only in "homeland," but also in the names of countries (Thailand, Switzerland, England) and in the old colonial designations of "peoples and cultures" (Nuerland, Basutoland, Nyasaland). One dictionary definition for "land" is "the people of a country," as in "the land rose in rebellion." Similarly, soil is often "national soil." Here, the territory itself is made more human.

This naturalized identity between people and place is also reflected and created in the course of other, nondiscursive practices. It is not uncommon for a person going into exile to take along a handful of the soil (or a sapling, or seeds) from his or her country, just as it is not unheard of for a returning national hero or other politician to kiss the ground upon setting foot once again on the "national soil." Demonstrations of emotional ties to the soil act as evidence of loyalty to the nation. Likewise, the ashes or bodies of persons who have died on foreign soil are routinely transported back to their "homelands," to the land where the genealogical tree of their ancestors grows. Ashes to ashes, dust to dust: in death, too, native or national soils are important.

The powerful metaphoric practices that so commonly link people to place are also deployed to understand and act upon the categorically aberrant condition of people whose claims on, and ties to, national soils are regarded as tenuous, spurious, or nonexistent. It is in this context, perhaps, that the . . . events in Carpentras, Southern France, should be placed. On the night of 9 May 1990, 37 graves in an old Jewish cemetery were desecrated, and the body of a man newly buried was disinterred and impaled with an umbrella. One is compelled to see in this abhorrent act of violence a connection to "love of country" in the ugliest sense of the term. The old man's membership in the French nation was denied because he was of the category "Jew." He was a person in the "wrong" soil, and was therefore taken out of the soil.

ROOTS AND ARBORESCENT CULTURE

The foregoing examples already suggest that the widely held commonsense assumptions linking people to place, nation to territory, are not simply territorializing, but deeply metaphysical. To begin to understand the meaning of displacement in this order of things, however, it is necessary to explore further aspects of the metaphysic. The intent in this section is to show that the naturalizing of the links between people and place is routinely conceived in specifically botanical metaphors. That is, people are often thought of, and think of themselves, as being rooted in place and as deriving their identity from that rootedness. The roots in question here are not just any kind of roots; very often they are specifically arborescent in form.

Even a brief excursion into nationalist discourses and imagery shows them to be a particularly rich field for the exploration of such arborescent root metaphors. . . .

But more broadly, metaphors of kinship (motherland, fatherland, *Vaterland*, *patria*, *isänmaa*) and of home (homeland, *Heimat*, *kotimaa*) are also territorializing in this same sense. . . . Motherland and fatherland, aside from their other historical connotations, suggest that each nation is a grand genealogical tree, rooted in the soil that nourishes it. By implication, it is impossible to be a part of more than one tree. Such a tree evokes both temporal continuity of essence and territorial rootedness.

Thinking in terms of arborescent roots is, of course, in no way the exclusive province of nationalists. Scholars, too, often conceptualize identity and nation-ness in precisely such terms. . . . Thinking about nations and national identities may take the form of roots, trees, origins, ancestries, racial lines, autochthonism, evolutions, developments, or any number of other familiar, essentializing images; what they share is a genealogical form of thought, which . . . is peculiarly arborescent. . . .

THE NEED FOR ROOTS AND THE SPATIAL INCARCERATION OF THE NATIVE

Two kinds of connection between the concept of the nation and the anthropological concept of culture are relevant here. First, the conceptual order of the "national geographic" map is comparable to the manner in which anthropologists have often conceptualized the spatial arrangement of "peoples and cultures." This similarity has to do with the ways in which we tend to conceptualize space in general. . . .

This spatial segmentation is also built into the lens of cultural relativity that . . . made the world appear as culture gardens separated by boundary-maintaining values. . . . The conceptual practice of spatial segmentation is reflected not only in narratives of "cultural diversity," but also in the internationalist celebration of diversity in the "family of nations."

A second, related set of connections between nation and culture is more overtly metaphysical. It has to do with the fact that, like the nation, culture has for long been conceived as something existing in "soil." Terms like "native," "indigenous," and "autochthonous" have all served to root cultures in soils; and it is, of course, a well-worn observation that the term culture derives from the Latin for cultivation. . . . Here, culture and nation are kindred concepts: they are not only spatializing but territorializing; they both depend on a cultural essentialism that readily takes on arborescent forms.

A powerful means of understanding how "cultures" are territorialized can be found in . . . the ways in which anthropologists have tended to tie people to places through ascriptions of native status. . . . The spatial incarceration of the native operates . . . through the attribution not only of physical immobility, but also of a distinctly ecological immobility. Natives are thought to be ideally adapted to their environments – admirable scientists of the concrete mutely and deftly unfolding the hidden innards of their particular ecosystems, PBS-style. . . . [T]hese ways of confining people to places have deeply metaphysical and moral dimensions.

The ecological immobility of the native . . . can be considered in the context of a broader conflation of culture and people, nation and nature – a conflation that is incarcerating but also heroizing and extremely romantic. . . .

On a certain North American university campus, anthropology faculty were requested by the Rainforest Action Movement (RAM) Committee on Indigenous Peoples to announce in their classes that "October 21st through the 28th is World Rainforest Week. The Rainforest Action Movement will be kicking the week off with a candlelight vigil for Indigenous Peoples." (The flyer also lists other activities: a march through downtown, a lecture "on Indigenous Peoples," and a film.) One is, of course, sympathetic with the project of defending the rainforests and the people who live in them, in the face of tremendous threats. The intent is not to belittle or to deny the necessity of supranational political organizing around these issues. However, these activities on behalf of "The Indigenous," in the specific cultural forms that they take, raise a number of questions: Why should the rights of "Indigenous People" be seen as an "environmental" issue? Are people "rooted" in their native soil somehow more natural, their rights somehow more sacred, than those of other exploited and oppressed people? And one wonders, if an "Indigenous Person" wanted to move away, to a city, would his or her candle be extinguished? The dictates of ecological immobility weigh heavily here.

But something more is going on with the "Indigenous Peoples' Day." That people would gather in a small town in North America to hold a vigil by candlelight for other people known only by the name of "Indigenous" suggests that being indigenous, native, autochthonous, or otherwise rooted in place is, indeed, powerfully heroized. At the same time, it is hard not to see that this very heroization – fusing the faraway people with their forest – may have the effect of subtly animalizing while it spiritualizes. Like "the wildlife," the indigenous are an object of enquiry and imagination not only for the anthropologist but also for the naturalist, the environmentalist, and the tourist.

[. . .]

It is when the native is a national native that the metaphysical and moral valuation of roots in the soil becomes especially apparent. In the national order of things, the rooting of peoples is not only normal; it is also perceived as a moral and spiritual need. . . .

A SEDENTARIST METAPHYSICS

The territorializing, often arborescent conceptions of nation and culture explored here are associated with a powerful sedentarism in our thinking. . . . This is a sedentarism that is peculiarly enabling of the elaboration and consolidation of a national geography that reaffirms the segmentation of the world into prismatic, mutually exclusive units of' "world order". This is also a sedentarism that is taken for granted to such an extent that it is nearly invisible. And, finally, this is a sedentarism that is

deeply metaphysical and deeply moral, sinking "peoples" and "cultures" into "national soils," and the "family of nations" into Mother Earth. It is this transnational cultural context that makes intelligible the linkages between contemporary celebratory internationalisms and environmentalisms. . . .

UPROOTEDNESS: SOME IMPLICATIONS OF SEDENTARISM FOR CONCEPTUALIZING DISPLACEMENT

Conceiving the relationships that people have to places in the naturalizing and botanical terms described above leads, then, to a peculiar sedentarism that is reflected in language and in social practice. This sedentarism is not inert. It actively territorializes our identities, whether cultural or national. And as this section will attempt to show, it also directly enables a vision of territorial displacement as pathological. The broader intent here is to suggest that it is in confronting dis placement that the sedentarist metaphysic embedded in the national order of things is at its most visible.

That displacement is subject to botanical thought is evident from the contrast between two everyday terms for it: transplantation and uprootedness. The notion of transplantation is less specific a term than the latter, but it may be agreed that it generally evokes live, viable roots. It strongly suggests, for example, the colonial and postcolonial, usually privileged, category of "expatriates" who pick up their roots in an orderly manner from the "mother country," the originative culture-bed, and set about their "acclimatization" in the "foreign environment" or on "foreign soil" – again, in an orderly manner. Uprootedness is another matter. Even a brief overview of the literature on refugees as uprooted people shows that in uprooting, the orderliness of the transplantation disappears. Instead, broken and dangling roots predominate – roots that threaten to wither, along with the ordinary loyalties of citizenship in a homeland.

The pathologization of uprootedness in the national order of things can take several different (but often conflated) forms, among them political, medical, and moral. After the Second World War, and also in the interwar period, the loss of national homeland embodied by refugees was often defined by policymakers and scholars of the time as a politico-moral problem. For example, a prominent 1939 historical survey of refugees states, "Politically uprooted, he [the refugee] may sink into the underworld of terrorism and political crime; and in any case he is suspected of political irresponsibility that endangers national security."

It is, however, the moral axis that has proven to command the greatest longevity in the problematization of refugees. . . .

. . . . The point to be underscored here is that these refugees' loss of bodily connection to their national homelands came to be treated as *a loss of moral bearings*. Rootless, they were no longer trustworthy as "honest citizens."

The theme of moral breakdown has not disappeared from the study of exile and displacement. . . .

The more contemporary field of "refugee studies" is quite different in spirit from the postwar literature. However, it shares with earlier texts the premise that refugees are necessarily "a problem." They are not ordinary people, but represent, rather, an anomaly requiring specialized correctives and therapeutic interventions. It is striking how often the abundant literature claiming refugees as its object of study locates "the problem" not in the political conditions or processes that produce massive territorial displacements of people, but, rather, within the bodies and minds (and even souls) of people categorized as refugees.

[. . .]

The point here is obviously not to deny that displacement can be a shattering experience. It is rather this: Our sedentarist assumptions about attachment to place lead us to define displacement not as a fact about sociopolitical context, but rather as an inner, pathological condition of the displaced.

[. . .]

NATIONALS AND COSMOPOLITANS IN EXILE

. . . Based on one year of anthropological field research in rural western Tanzania among Hutu refugees who fled the genocidal massacres of 1972 in Burundi, this work explores how the lived experiences of exile shape the construction of national identity and historicity among two groups

of Hutu refugees inhabiting two very different settings in Tanzania. One group was settled in a rigorously organized, isolated refugee camp, and the other lived in the more fluid setting of Kigoma Township on Lake Tanganyika. Living outside of any camp context, these "town refugees" were dispersed in non-refugee neighborhoods. Comparison of the camp and town settings revealed radical differences in the meanings ascribed to national identity and homeland, and exile and refugee-ness.

The most striking social fact about the camp was that its inhabitants were continually engaged in an impassioned construction and reconstruction of their history as "a people." Ranging from the "autochthonous" origins of Burundi as a "nation" to the coming of the pastoral Tutsi "foreigners from the North" to the Tutsi capture of power from the autochthons by ruse to, finally, the culminating massacres of Hutu by Tutsi in 1972, which have been termed a "selective genocide," the Hutu refugees' narratives formed an overarching historical trajectory that was fundamentally also a national trajectory of the "rightful natives" of Burundi. The camp refugees saw themselves as a nation in exile, and defined exile, in turn, as a moral trajectory of trials and tribulations that would ultimately empower them to reclaim (or create anew) the "Homeland" in Burundi.

Refugee-ness had a central place in these narrative processes. Far from being a "spoiled identity," refugee status was valued and protected as a sign of the ultimate temporariness of exile and of the refusal to become naturalized, to put down roots in a place to which one did not belong. Insisting on one's liminality and displacement as a refugee was also to have a legitimate claim to the attention of "international opinion" and to international assistance. Displacement is usually defined by those who study refugees as a subversion of (national) categories, as an international problem. Here, in contrast, displacement had become a form of categorical purity. Being a refugee, a person was no longer a citizen of Burundi, and not yet an immigrant in Tanzania. One's purity as a refugee had become a way of becoming purer and more powerful as a Hutu.

The "true nation" was imagined as a "moral community" being formed centrally by the "natives" in exile. The territorial expanse named Burundi was a mere state. . . . Here, then, would seem to be a deterritorialized nation without roots sunk directly into the national soil. Indeed, the territory is not yet a national soil, because the nation has not yet been reclaimed by its "true members" and is instead governed by "impostors".... [T]he Hutu nation has reterritorialized itself precisely in displacement, in a refugee camp. The homeland here is not so much a territorial or topographic entity as a moral destination. And the collective, idealized return to the homeland is not a mere matter of traveling. The real return can come only at the culmination of the trials and tribulations in exile.

These visions of nation, identity, and displacement challenge the commonsense and scholarly views discussed in the first section of this article, not by refuting the national order of things, but, rather, by constructing an alternative, competing nationalist metaphysic. It is being claimed that state and territory are not sufficient to make a nation, and that citizenship does not amount to a true nativeness. Thus ... Burundi is an "impostor" in the "family or nations."

In contrast, the town refugees had not constructed such a categorically distinct, collective identity. Rather than defining themselves collectively as "the Hutu refugees," they tended to seek ways of assimilating and of manipulating multiple identities – identities derived or "borrowed" from the social context of the township. The town refugees were not essentially "Hutu" or "refugees" or "Tanzanians" or "Burundians," but rather just "broad persons." Theirs were creolized, rhizomatic identities – changing and situational rather than essential and moral. In the process of managing these "rootless" identities in township life, they were creating not a heroized national identity, but a lively cosmopolitanism – a worldliness that caused the camp refugees to see them as an "impure," problematic element in the "total community" of the Hutu refugees as "a people" in exile.

For many in town, returning to the homeland meant traveling to Burundi, to a spatially demarcated place. Exile was not a moral trajectory, and homeland was not a moral destination, but simply a place. Indeed, it often seemed inappropriate to think of the town refugees as being in exile at all. Many among them were unsure about whether they would ever return to Burundi, even if political changes were to permit it in the future. But more important, they had created lives that were located in the present circumstances of Kigoma, not in the past in Burundi.

The town refugees' constructions of their lived circumstances and their pasts were different from both the national metaphysic of the camp refugees and that of scholarly common sense. Indeed, they dismantled the national metaphysics by refusing a mapping and spurning origin queries altogether. They mounted instead a robust challenge to cultural and national essentialisms; they denaturalized those scholarly, touristic, and other quests for "authenticity" that imply a mass traffic in "fake" and "adulterated" identities; and, finally, they trivialized the necessity of living by radical nationalisms....

CONCLUSION

.... [T]he nation – having powerful associations with particular localities and territories – is simultaneously a supralocal, transnational cultural form. In this order of things, conceptualizations of the relations between people and place readily take on aspects of the metaphysical sedentarism described here. It is these naturalized relations that this article has tried to illuminate and decompose through the three-way comparison of sedentarist common sense, of the Hutu in the refugee camp, and of the cosmopolitan refugees in Kigoma. These ethnographic examples underscore what a troubled conceptual vehicle "identity" still is, even when the more obvious essentialisms have been leached out of it. Time and again, it reappears as a "root essence," as that "pure product" of the cultural, and of the national, soil from which it is thought to draw its nature and its sustenance. That many people (scholars included) see identity through this lens of essentialism is a cultural and political fact to be recognized. But this does not mean that our analytical tools must take this form. The two main oppositions in this article – first, that between sedentarism and displacement in general, and, second, that between "the nationals" and "the cosmopolitans" in exile in Tanzania – suggest alternative conceptualizations.

They suggest that identity is always mobile and processual, partly self-construction, partly categorization by others, partly a condition, a status, a label, a weapon, a shield, a fund of memories, et cetera. It is a creolized aggregate composed through

FIVE

bricolage. The camp refugees celebrated a categorical "purity," the town refugees a cosmopolitan "impurity." But both kinds of identity were rhizomatic, as indeed is any identity, and it would not be ethnographically accurate to study these as mere approximations or distortions of some ideal "true roots."

[...]

Observing that more and more of the world lives in a "generalized condition of homelessness" – or that there is truly an intellectual need for a new "sociology of displacement," a new "nomadology" – is not to deny the importance of place in the construction of identities. On the contrary, as this article has attempted to show . . . deterritorialization and identity are intimately linked. . . . To plot only "places of birth" and degrees of nativeness is to blind oneself to the multiplicity of attachments that people form to places through living in, remembering, and imagining them.

"Shades of Shit"

from *Wisdom sits in Places: Landscape and Language among the Western Apache* (1996)

Keith H. Basso

Editors' introduction

Keith Basso is a Professor of Anthropology at the University of New Mexico. Basso studies language and culture, and his work has involved extensive field research in Australia and the American Southwest. In this selection, he recounts a conversation that occurred when he first began working with a group of Western Apache residing in Cibecue, New Mexico. In the late 1950s Basso received a grant to map Western Apache place names, in an attempt to study and preserve their specific history encapsulated in these place names. Basso is conversing with Morley, Charles, and Jason, who are some of the Western Apache collaborators on his mapping project, about how the place called "Shades of Shit" acquired its colorful name.

This selection emphasizes the importance of storytelling among this group of Native Americans. As the book it is drawn from, *Wisdom sits in Places*, emphasizes, these place-based stories function to create a historical and moral continuity among a group that is forever in danger of losing its identity *vis-à-vis* mainstream American culture, dissolving the boundaries between this specific place and the identity of its inhabitants. Because they have their own laws and borders, Indian reservations can be understood as *states* within the larger United States. In addition, because these are people who share a deep sense of belonging together that is rooted in a common language and historical experience, they clearly fit the definition of a *nation*. Thus the Western Apache in this excerpt are a *nation-state*, one that is conjured into being and maintained through stories, and one that is facing disintegration as the lure of life off the reservation entices residents to leave.

"Shades of Shit" might just as easily have been placed in a different section of this *Reader*. One of these might have been Part Three, "landscape." Place names, or *toponyms*, provide important clues to the histories layered-in to the landscape. Naming is always a gesture of power, and place names are no exception. For example, the fact that over half of US state names are derived from Native American words tell us of the enduring linguistic and cultural importance of Native Americans to the history of the United States. Though the mainstream culture of the United States tends to overlook or devalue Native American contributions, these are in fact encoded into the very places whose names are repeated – often by those unaware of their origins – on a daily basis. In another example, note how place names on a map of Antarctica consist mostly of the names of European explorers or sovereigns. Conquerors and explorers imposed toponyms on the land, in effect laying claim to ownership of place through naming it. In each case, place names give us clues to who holds claim to place, and thus to power and identity. Though work on toponyms was more common in an earlier era, contemporary cultural geographers have done relatively little work on toponyms. An intriguing exception is Catherine Nash's work on Ireland, in "Irish Placenames: Post-colonial Locations," in *Transactions of the British Institute of Geographers* 24, 4 (1999): 457–80.

The people that Basso interviewed do not have a written history. Rather, they use stories to recall significant historical events, and to instruct the younger members of the community in proper behavior. These stories are condensed into place names, such as Shades of Shit, and the others Basso mentions in this selection. Merely by speaking the place name, an entire historical moment or moral tale unfolds, without the necessity of recounting the full story. In this selection, the place name refers to the history of how Shades of Shit acquired its name. Folded-in to this history is a moral tale of the importance of sharing. The residents of Shades of Shit refused to share their corn, an act which might make perfect sense to those used to the notion of accumulating a surplus. However, in subsistence societies like that described in the tale behind Shades of Shit, refusing to share could have deadly consequences for group members. Thus merely speaking the name "Shades of Shit" serves to remind the audience of the perils of stinginess.

Another section where this selection might have been included is Part Seven, "*Difference*." In the United States, Native Americans make up a negligible fraction of those receiving PhDs in geography. This may go some way toward explaining why there is relatively little work by geographers on indigenous peoples. However, it does not fully explain this phenomenon, which is in all likelihood more closely related to the cultural submersion and silencing of Native Americans in mainstream US culture than it is to a dearth of Native Americans with PhDs in geography. Some examples of work by geographers focusing on indigenous people include Matt Sparke's essay titled "A map that roared and an original atlas: Canada, cartography, and the narration of nation," in *Annals of the Association of American Geographers* 88, 3 (1998): 463–495; and Sarah Radcliffe and Sallie Westwood, *Remaking the Nation: Place, Identity and Politics in Latin America* (1996).

Wisdom sits in Places emphasizes the importance of the stories behind place names. At some level, because "geography" is literally *geo-graphy*, or place-writing, all geography involves storytelling. More specifically, however, narrative is an emerging focus of some cultural geographers, particularly those looking at place-making. Work in this vein asks questions like: How do places acquire their particular associations or "character"? How are contentious histories crystallized into landscapes? How are the ties between place and identity formed and maintained over time? Some recent geographically oriented work on narrative includes Belden Lane's *Landscapes of the Sacred: Geography and Narrative in Native American Spirituality* (1988); William Cronon's essay titled "A place for stories: nature, history, and narrative," in *Journal of American History* 78, 4 (1992): 1347–1376; Simon Schama's *Landscape and Memory* (1995); Patricia Price's *Dry Place: Landscapes of Belonging and Exclusion* (2004); and John Wiley's article titled "Becoming-icy: Scott and Amundsen's south polar voyages, 1910–1913," in *Cultural Geographies* 9 (2002): 249–265.

It is now mid-July and our topographic maps of the Cibecue region are getting increasingly crowded. Dozens of dots and shaded areas mark the locations of places bearing Apache names, and numbers next to these index the names themselves, which are listed in separate notebooks. Morley says admiringly that some of the maps look like they were blasted with a shotgun – and more than once. Charles, modulated as always, expresses his approval in less effusive ways. Jason, who studies the maps whenever he gets a chance, has yet to voice an opinion. I am struck by the mounting number of named localities – we have charted 109 in only five weeks – and the consistent manner in which they cluster, mainly around sources of water and past and present farmsites.

But what impresses me most of all is the rich descriptive imagery of Western Apache place-names. Lately, with ear and eye jointly enthralled, I, have stood before

Tséé Dotł'izh Ténaahijaahá (Green Rocks Side By Side Jut Down Into Water; a group of mossy boulders on the bank of a stream).
Tséé Ditłige Naaditiné (Trail Extends Across Scorched Rocks; a crossing at the bottom of a canyon).
T'iis Ts'ósé Bił Naagolgaiyé (Circular Clearing With Slender Cottonwood Trees; a meadow).
Túzhį Yaahigaiyé (Whiteness Spreads Out Descending To Water; a sandstone cliff next to a spring).

Tséé Yaaditł'ishé (Line Of Blue Below Rocks; a mineral deposit).
Yaahiłbigé (Stunted Rising Up; a small mountain).
Kaiłbáyé Bił Naagozwodé (Gray Willows Curve Around A Bend; a point on a stream).

and a number of other places whose handsomely crafted names – bold, visual, evocative – lend poetic force to the voices of the ancestors.

Just as expressive are other Apache place-names, different from these, that do not give close descriptions of the places to which they refer. Commemorative in character and linked to traditional stories, they allude instead to historical events that illuminate the causes and consequences of wrongful social conduct. And in this important capacity, as I would discover at a place named Shades of Shit (Chąą Bi Dałt'ohé), they invest the Apache landscape with a sobering moral dimension, dark but instructive, that place-makers can exploit to deeply telling effect.

The shades, or brush-covered ramadas, are no longer standing. They collapsed, Charles says, a long time ago. Yet the place where they stood, a tree-covered knoll southwest of Cibecue, is avoided to this day. "No one wants to come here," he explains, as we slowly approach a vantage point a hundred yards away. "The people who lived here had farms down below, probably next to the creek. This was long after they settled in this valley. Then they did something bad, very bad, and they came close to dying. There is a story about it I was told by my grandfather. It's short." And it is . . .

It happened here at Shades of Shit.
They had much corn, those people who lived here, and their relatives had only a little. They refused to share it. Their relatives begged them but still they refused to share it.
Then their relatives got angry and forced them to stay at home. They wouldn't let them go anywhere, not even to defecate. So they had to do it at home. Their shades filled up with it. There was more and more of it! It was very bad! Those people got sick and nearly died.
Then their relatives said, "You have brought this on yourselves. Now you live in shades of shit!" Finally, they agreed to share their corn.
It happened at Shades of Shit.

An uneasy silence settles over our group. Jason looks suddenly wan. Morley spits in disgust. A soft breeze, recalling a terrible stench it could not possibly carry, ruffles the morning air. When Charles speaks again, he says that he wonders what really happened here: it couldn't have been as simple as the story suggests. And even if it were, he adds, the story gives no sense of why events unfolded as they did or how the people involved might have reacted to them. "What were they *thinking*?" he asks rhetorically in a tone of disbelief. "How must they have *felt*?" Charles would like to know these things, he says, though he doubts he ever will. And then, speaking as if he knew them very well, he tells his grandfather's story again, fleshing it out at length and constructing for us an astonishing world as surely revealing of Apache social values as it is violently offensive to their most basic sensibilities.

It must have been late in the summer. Those people had harvested their corn and were drying it and roasting it. They must have been grateful and happy. "Now we have much to eat," they are saying.
Their relatives envied them. Their own corn had not grown well. (Sometimes it happens that way. Some fields produce a lot, and those right next to them do not. It happens that way, and no one knows why, and sometimes they talk of witchcraft prompted by revenge for something that was done to them in the past.) Their own corn was meager and small but they were not yet afraid or angry. "Our more fortunate relatives will help us," they said, speaking among themselves. "They have more than enough corn. They will want to share it with us. We have always helped each other. That is how it should be."
Then they waited for their relatives to help them. They waited in vain. Their relatives kept their corn to themselves, eating it every day and making big shits when they went off into the brush. They did nothing for their relatives, although they noticed their plight. "They have enough food, even though they harvested little corn. They probably have plenty of beans and squash. Some of them are skilled hunters. Soon they will have plenty of deer meat to eat. We will keep our corn for ourselves, so that our children will not be hungry during the winter."

Now their poor relatives are becoming scared and puzzled. "Why do they not offer to help us?" they said. "They're treating us like we don't exist, as if we are nothing to them. We will have a hard time unless they change their minds and give us some of their corn."

Then they sent someone to talk to the people who lived here. "We are your relatives," he said to them. "We must help each other. You have plenty of corn. We have seen it. But we have only a little and soon it will be gone. Soon our children will be crying because they have nothing to eat. Give us some of your corn. Give us some of your corn. We will be grateful. This is how it should be."

Then they waited again, and still their relatives did nothing for them. They talked again among themselves. "Our relatives are not going to help us," they said. "They have become greedy and stingy. They think only of themselves. They have put themselves above us, ignoring us like we don't exist. We have waited long enough. We must do something!"

Then they became angry at their own relatives. "We will make them stay at their homes. They will not go anywhere. We will make them live with their own big shits!" This is what they decided to do.

Then they came over here and surrounded their relatives' homes. They told them to stay there. They did this day and night. "We will harm you if you try to leave," they said. "You have brought this on yourselves. You can eat all you want. Only now you will shit at your homes. This is not how it should be, but we are doing it anyway," they said.

Then those people must have thought they were joking. "They don't really mean what they say," they said. "They will not harm us," they said. So they chose a man to leave his home. He was forced back by his relatives. Another man tried to leave. He was also forced back. "They mean what they say," they said. "Now we are in for trouble," they thought.

Then they started to shit in their shades. Some of them said, "This is very bad. We should share our corn and put an end to it." Others said, "No! If we give away some of our corn, they will want it all. We must not give in to them. This is their way of leaving us with nothing."

Then they ate less and less but still they fouled their shades.

There was more and more of it! It was visible everywhere! The sight and smell could not be avoided! There were swarms and swarms of flies! Huge swarms! They no longer cooked in their shades. Eating became something they detested. It was terrible!

Then they started to get sick from the sight and smell of their own filth. Some of them were constantly dizzy. Others had trouble walking straight. Their children started moaning. They themselves were moaning. "We could die from this!" they said. "We could die from our own filth."

Then a man of the people who had little corn went and talked to them. "You have brought this on yourselves," he said. "You should have shared your corn with us as soon as you knew you had more than enough. You didn't do this! You gave us nothing at all. You were greedy and stingy, thinking only of yourselves. Because of this we had to beg you to share your corn with us. Even then, you did nothing. You just kept on eating, more and more, knowing that we had little food of our own. You ignored us – your own relatives – as if we were nothing! This is not how it should be. As relatives we make each other rich because we help each other in times of need. It has been this way since the beginning. What made you forget this? What made you ignore us? Well, I don't know. But now you live in shades of shit! Now you are getting sick!"

Then he laughed at them. He laughed at them.

Then those people talked among themselves. "What he says is true," they said. "*Look what has become of us!* We were thinking only of ourselves. Our greed is responsible for our trouble. We looked down on our own relatives and gave them nothing. Look what has become of us!"

Then they shared their corn. Finally, they did this. Their relatives took the corn away, saying nothing, saying nothing. Now those people were allowed to leave their homes.

Then those people said, "We must leave here and go somewhere else to live. This is a bad place. It stinks with signs of our stinginess and greed."

"It could have happened that way," Charles says almost casually. And then, a bit sternly, "Let's move on. We've been here long enough."

"Culture sits in Places: Reflections on Globalism and Subaltern Strategies of Localization"

from *Political Geography* 20 (2001): 139–174

Arturo Escobar

Editors' introduction

In "Culture sits in places" we see a conscious reference to the title of Keith Basso's book, *Wisdom sits in Places* (see p. 283). Both Escobar and Basso focus their field research on a racialized minority group; in Arturo Escobar's case, black Colombians resident on the country's Pacific coast. For years Escobar has researched the political mobilization of this group in their efforts to maintain their richly biodiverse environment in the face of global pressures to extract resources from this region. In "Culture sits in Places" Escobar notes the importance of cross-scale interaction of this group with the Colombian government, non-government organizations (NGOs), academics, and others engaged in similar place-based struggles. He also notes the broader interaction of human and non-human elements of the ecosystem, forming various life-corridors. Together these cross-scale and nature–culture interactions form what Escobar refers to as a *region-territory*.

This selection resonates with the others in Part Five in its concern with place *vis-à-vis* the forces of globalization. However, the place-based activism of Escobar's study population is presented as progressive, not reactive. Many scholars from a variety of disciplines have struggled with the question of the maintenance of place-based distinctiveness in the face of globalization. See for example Arjun Appadurai's *Modernity at Large: Cultural Dimensions of Globalization* (1996) and David Harvey's *Spaces of Global Capitalism: A Theory of Uneven Geographic Development* (2006). Edward Casey, in his book *The Fate of Place: A Philosophical History* (1997), gives a longer-range view to place as it has been largely devalued *vis-à-vis* space by philosophers throughout history. Like Casey, Escobar argues in defense of place.

A growing number of scholars are interested in the cultural aspects of economic development, natural resources, and grass-roots political organizing. Anna Lowenhaupt Tsing's *Friction: An Ethnography of Global Connection* (2004) is a good example of this sort of work. Geographers have been particularly active in the cultural dimensions of political ecology; see Roderick Neumann's *Imposing Wilderness: Struggles over Livelihood and Nature Preservation in Africa* (2002), Karl Zimmerer and Thomas Bassett (eds.) *Political Geography: An Integrative Approach to Geography and Environment Development Studies* (2003), and Richard Peet and Michael Watts (eds.) *Liberation Ecologies* (1996).

Arturo Escobar is a Colombian national who is a Professor of Anthropology at the University of North Carolina, Chapel Hill. Escobar's academic background also includes training in engineering, biochemistry, international nutrition, and food science. Escobar has a long-standing interest in Colombia's Pacific coast black community, and the relationship between social mobilization, development, and the environment. Selected other publications include *Encountering Development: The Making and Unmaking of the Third World* (1994), *Cultures of Politics/Politics of Culture* (co-edited with Sonia Alvarez and Evelina Dagnino, 1998), and "Beyond the Third World: imperial globality, global coloniality, and anti-globalization social movements," in *Third World Quarterly* 25, 1 (2004): 207–230.

INTRODUCTION: CULTURE AND THE MARGINALIZATION OF PLACE

The question of "place" has been newly raised in recent years from a variety of perspectives – from its relation to the basic understanding of being and knowing to its fate under globalization and the extent to which it continues to be an aid or a hindrance for thinking about culture and the economy. This questioning, of course, is not coincidental; for some, placelessness has become the essential feature of the modern condition, and a very acute and painful one in many cases, such as those of exiles and refugees. Whether celebrated or decried, the sense of atopia seems to have settled in. This seems to be as true of discussions in philosophy, where place has been ignored by most thinkers; theories of globalization, that have effected a significant discursive erasure of place; or debates in anthropology, which have seen a radical questioning of place and place making. Yet the fact remains that place continues to be important in the lives of many people, perhaps most, if we understand by place the experience of a particular location with some measure of groundedness (however, unstable), sense of boundaries (however, permeable), and connection to everyday life, even if its identity is constructed, traversed by power, and never fixed. There is an "implacement" that counts for more than we want to acknowledge. . . .

To be sure, the critique of place in anthropology, geography, communications, and cultural studies of recent times has been both productive and important, and continues to be so. New spatial concepts and metaphors of mobility – deterritorialization, displacement, diaspora, migration, traveling, border-crossings, nomadology, etc. – have made us aware of the fact that the principal dynamics of culture and economy have been significantly altered by unprecedented global processes. Yet there has been a certain asymmetry in these debates. . . .[T]his asymmetry is most evident in discourses of globalization, where the global is often equated with space, capital, history and agency, and the local with place, labor, and tradition. Place has dropped out of sight in the "globalization craze" of recent years, and this erasure of place has profound consequences for our understanding of culture, knowledge, nature, and economy. It is perhaps time to reverse some of this asymmetry by focusing anew – and from the perspective afforded by the critiques of place themselves – on the continued vitality of place and place-making for culture, nature, and economy. Restoring some measure of symmetry, as we shall see, does not entail an erasure of space as a domain of resistance and alterity, since both place and space are crucial in this regard, as they are in the creation of forms of domination. It does mean, however, a questioning of the privilege accorded to space in analyses of the dynamics of culture, power, and economy.

This is, indeed, an increasingly felt need of those working at the intersection of environment, culture and development, despite the fact that the development experience has meant for most people a sundering of local life from place of greater depth than ever before. Not only are scholars and activists in environmental studies confronted with social movements that commonly maintain a strong reference to place and territory, but faced with the growing realization that any alternative course of action must take into account place-based models of nature, culture, and politics. While it is evident that "local" economies and culture are not outside the scope of capital and modernity, it also needs to be newly acknowledged that the former are not produced exclusively by the latter; this place specificity, as we shall see, enables a different reading of culture and economy, capitalism and modernity. The inquiry into place is of equal importance for renewing the critique of eurocentrism in the conceptualization of world regions, area studies, and cultural diversity. The marginalization of place in European social theory of the nineteenth and twentieth centuries has been particularly deleterious to those social formations for which place-based modes of consciousness and practices have continued to be important. This includes many contemporary societies, perhaps with the exception of those most exposed to the de-localizing, disembedding and universalizing influence of modern economy, culture and thought. The reassertion of place thus appears as an important arena for rethinking and reworking eurocentric forms of analysis.

[. . .]

CULTURE SITS IN PLACES: THE AVATARS OF PLACE IN RECENT ANTHROPOLOGICAL LITERATURE

The disregard of place in Western theory and social science has been most pointedly stated by phenomenologists. For philosopher Edward Casey, this disregard has been endemic and long-standing. Since Plato, Western philosophy – oftentimes with the help of theology and physics – has enshrined space as the absolute, unlimited and universal, while banning place to the realm of the particular, the limited, the local, and the bound. Seventeenth and eighteenth century philosophers, from Descartes to Leibniz, assumed that places are only momentary subdivisions of a universal and homogeneous space. For this to happen, space had to be dissociated from the bodies that occupy it and from the particularities that these bodies lent to the places they inhabit. Scientific knowledge welcomed this notion of the void, even if a void with extension and structure that made possible the Cartesian project of a *mathesis universalis* and the mathematization of nature. Despite the hegemony of space . . . there has always been an undercurrent of interest in, and theorizing of, place which has remained understudied. . . . This interest in place has spilled over into disciplines such as architecture, archaeology, anthropology, geography, and historical ecology. . . . Common to many of these tendencies is an anti-essentialist notion of place, an interest in finding place at work, place being

constructed, imagined and struggled over. One could say that today there is an emerging philosophy and politics of place even if it still is clearly under construction.

The disregard of place in the social and human sciences is the most puzzling since, as Casey passionately argues, it is our inevitable immersion in place, and not the absoluteness of space, that has ontological priority in the generation of life and the real. It certainly does so in the accounts and practices of most cultures, echoed in the phenomenological assertion that, given the primacy of embodied perception, we always find ourselves in places. We are, in short, placelings. . . . Place is, of course, constituted by sedimented social structures and cultural practices. Sensing and moving are not presocial; the lived body is the result of habitual cultural and social processes. It is thus imperative that we "get back into place" and reverse the long-standing disempowerment of place in both modern theory and social life. This means recognizing that place, body and environment integrate with each other; that places gather things, thoughts and memories in particular configurations; and that place, more an event that a thing, is characterized by openness rather than by a unitary self-identity. From an anthropological perspective, it is important to highlight the emplacement of all cultural practices, which stems from the fact that culture is carried into places by bodies – bodies are encultured and, conversely, enact cultural practices. . . .

This also means that people are not only "local"; we are all indissolubly linked to both local and extralocal places through what might be called networks – of which the kula ring and internet networks would be contrasting variations in terms of the ways in which they connect persons and places. Places concatenate with each other to form regions, which suggests that porosity of boundaries is essential to place, as it is to local constructions and exchange. Locality, in this way, becomes marked by the interplay between position, place and region; by the porosity of boundaries; and by the role of the lived body between enculturation and emplacement. . . . Against this view militate migration, wars, the new information and communications technologies (NICTs), speed and, of course, the abstractions of space and much of Western thought. . . .

I will argue that some social movements are taking the lead in this "getting back into place" to which Casey summons us. Not only social movements, of course, because there are multiple sources in this endeavor including, among others, feminist body politics, phenomenological biology, new form of dwelling in architecture, alternative thinking on land and community, and the like. In commenting on the economic prejudice against the small and the desecration of nature and moral communities in the United States, Wendell Berry, the poet farmer, for instance, underscores ways of being rooted in the land; this leads him to envisage the historical possibility of creating "the party of the local community," that is, of local communities becoming more aware of themselves in their opposition to a postagricultural, postnatural and posthuman world that he sees as insidiously settling in. This party has a double commitment: to the preservation of ecological diversity and integrity, and to the renewal of local economies and communities. As we shall see, this double goal of transforming ecology and economy can provide a powerful interface for the renewal of place-based theory and practice.

[. . .]

SOCIAL MOVEMENTS AND SUBALTERN STRATEGIES OF LOCALIZATION

The Pacific region of Colombia is a vast rainforest area about 900 km long and 50–180 km wide, stretching from Panama and Ecuador, and between the westernmost chain of the Andes and the Pacific Ocean. It is known as one of the "hot spots" of biological diversity in the world. Afro-Colombians, descendants from slaves brought beginning in the sixteenth century to mine gold, make up about 90 per cent of the population, with indigenous peoples from various ethnic groups accounting for about 5 per cent of the region's population of close to a million. About 60 per cent of the population still lives in rural settlements along the numerous rivers that, in the southern part, flow from the Andes towards the ocean. Although the region has never been completely isolated, two factors have brought watershed changes to it in recent years: the radical neo-liberal opening of the country to the world economy adopted by the government after 1990; and the granting of col-

lective territorial and cultural rights to the black communities in 1993 (the so-called Ley 70 or Law 70), following the implementation of a new national constitution in 1991. It was in the context of this conjuncture that the three changes with which this account is concerned need to be situated. First, the increased pace of capitalist extractivist activities, such as the rapid expansion of African palm plantations and industrial shrimp cultivation in the southern part of the region. Second, the growing concern with the destruction of biological diversity, leading to the implementation of an innovative project for its conservation, with the region's social movements as one of the project's main interlocutors. Third, the rise of important ethnic movements, particularly the social movement of black communities.

How can the production of this peculiar "rainforest" region be analyzed in terms of place? Generally speaking, the "Pacífico biogeográfico," as the region is known, is constructed through processes involving the human, biophysical non-human, and machinic worlds operating at many scales, from the microbiological to the transnational. These processes can schematically be seen as follows:

1 Historical processes of geological and biological formation. Geologists and paleo-scientists present a view of the region in terms of geological and evolutionary time in ways that account for its specificity, particularly its unusually high levels of endemism and biological diversity.

2 Historical processes constituted by the daily practices of the local black, indigenous and mestizo communities. Through their laborious daily practices of being, knowing and doing the communities have been actively constructing their socio-natural worlds for several centuries, even if in the midst of other forces.

3 Historical processes of capital accumulation at all scales, from the local to the global. Capital is doubtless one of the most powerful forces constructing this and most rainforest regions of the world. Nevertheless, the construction of the Pacific as place cannot be explained solely in terms of capital. Indeed, it could be posited that forms of non-capitalism exist and are actually being created today out of the dynamics of place-based cultural and ecological practices,

even if in the decided engagement with capital and the state.

4 Historical processes of incorporation of the region into the State, particularly through development representations and strategies. These processes have taken on great importance in the last few decades, when the government finally sought to incorporate the region fully into its development apparatus. In the early 1980s, the Colombian Pacific was for the first time represented as a "developmentalizable" region by state discourses. Capital and development constitute a two-pronged strategy for the territorialization of the Pacific as a modern space of thought and intervention.

5 The cultural-political practices of social movements. After the 1990s, black and indigenous movements became an important contender for the representation and construction of the Pacific as place and region-territory. These movements have set into motion a cultural politics which operates chiefly through a process of ethnicization of identity in close connection with ecological and alternative development strategies.

6 The discourses and practices of techno-science at all scales, from the local to the global, particularly in the areas of biodiversity conservation and sustainability. "Biodiversity" has become a powerful discourse and has originated a network of sites that embraces ever more domains of cultural, political and ecological action. Since the early 1990s, the biodiversity network has become an important element in the struggle over the Colombian Pacific as place and territory.

In a very schematic fashion, these processes can be further divided into two overall strategies. These strategies, let it be emphasized, are not bounded and discrete, but overlapping and in many ways co-produced:

1 Strategies of global localization by capital, the state, and technoscience. Capital, state and technoscience engage in a politics of scale that attempts to negotiate the production of locality in their own favor. Nevertheless, to the extent that these strategies are not place-based, they inevitably induce a delocalizing effect with respect to local places, despite their efforts at

articulating with localities. (I will not discuss here those related territorial strategies based on the violence of weapons and intimidation, which unfortunately have gained ascendancy in the region since the late 1990s, causing massive displacement of people in a number of places.)

2 Subaltern strategies of localization by communities and, particularly, social movements. These strategies are of two kinds: place-based strategies that rely on the attachment to territory and culture; and glocal strategies through meshworks that enable social movements to engage in the production of locality by enacting a politics of scale from below. Social movements engage in the politics of scale by engaging biodiversity networks, on the one hand, and through coalition making with other place-based struggles.

Activists of the Process of Black Communities (PCN) have progressively articulated a political ecology framework in their interaction with community, state, NGO and academic sectors. One of the important contributions of the Biodiversity Conservation Project (PBP) has been to initiate research and conceptualization of the "traditional production systems" of the river communities. For PBP staff and PCN activists alike, it is clear that these systems are more geared towards local consumption than to the market and accumulation; they have operated as forms of resistance, even if they have also contributed to the region's marginalization. Also commonly appreciated is that traditional practices have been sustainable to the extent that they have enabled the reproduction of the cultural and biophysical ecologies (Sanchez and Leal, 1995). This sustainability has been increasingly put into question for most communities for at least the past two decades. Activists have introduced other important conceptual innovations in this context. The first one is the definition of "biodiversity" as "territory plus culture." Closely related to it is a view of the entire Pacific rainforest region as a "region-territory of ethnic groups," that is, an ecological and cultural unit that is laboriously constructed through the daily cultural and economic practices of the communities. The region-territory is also thought about in terms of "life corridors," veritable nodes of articulation between sociocultural forms of use and the natural environment.

There are, for instance, life corridors linked to the mangrove ecosystems; to the foothills; to the middle part of the rivers, extending towards the inside of the forest; and those constructed by particular activities, such as traditional gold mining or women's shell collecting in the mangrove areas. Each of these corridors is marked by particular patterns of mobility, social relations (gender, kindred, ethnicity), use of the environment and links to other corridors; each involves a particular use and management strategy of the territory.

The region-territory is a category of inter-ethnic relations that points toward the construction of alternative life and society models. It entails an attempt to explain biological diversity from the endogenous perspective of the eco-cultural logic of the Pacific. More concretely, the territory is seen as the space of *effective appropriation* of the ecosystem, that is, as those spaces used to satisfy community needs and for social and cultural development; it is multidimensional space for the creation and recreation of the ecological, economic and cultural practices of the communities. For a given river community, this appropriation has longitudinal and transversal dimensions, sometimes encompassing several river basins. Thus defined, the territory cuts across several landscape units; more importantly, it embodies a community's life project. The region-territory, on the contrary, is conceived of as a *political construction* for the defense of the territories and their sustainability. In this way, the region-territory is a strategy of sustainability and vice versa: sustainability is a strategy for the construction and defense of the region-territory. The region-territory can thus be said to articulate the *life project* of the communities with the *political project* of the social movement. The struggle for territory is thus a cultural struggle for autonomy and self-determination. This explains why for many people of the Pacific the loss of territory would amount to a return to slavery or, worse perhaps, to becoming "common citizens."

The issue of territory is considered by PCN activists as a challenge to developing local economies and forms of governability that can support its effective defense. The strengthening and transformation of traditional production systems and local markets and economies; the need to press on with the collective titling process; and working towards organizational strengthening and the development of forms of territorial governability are all important components of an overall strategy centered on the region. Finally, it is clear that communities themselves increasingly have a sense of the loss of territory at present and what it might take to defend it. Those in river communities are prone to point at the "loss of traditional values and identity" as the most immediate source of loss of territory. Other factors seem to converge on this variable; loss of traditional production practices, irrational exploitation of resources, state development policies oriented by purely external criteria, increased pace of industrial extraction, and the existence of totally inappropriate and alienating educational models for the young people are cited as the most common factors associated with the loss of values and territory. In more substantial discussions with community leaders and social movement activists, a series of other factors linked to the loss of territory start to emerge, such as: the spread of plantations and specialization of productive activities; changes in production systems; internal conflicts in the communities; the cultural impact of national media, education and culture; out migration and the arrival of people foreign to the region espousing the ethics of capitalism and extractivism; and of course inadequate development policies, the neo-liberal opening to world markets, and the demands of the global economy.

To sum up, to the strategies of production of locality by capital (and, in different ways, techno-science), social movements oppose strategies of localization which, as we have seen, focus in the first instance on the defense of territory and culture. The idioms of biodiversity, sustainability, traditional production systems, cultural rights, and ethnic identities are all interwoven by movement activists into a discourse for the defense of place and a political ecology framework that enables them to articulate a political strategy. Social movements such as the movement of black communities of the Colombian Pacific can thus be seen as advancing a triple localizing strategy for the defense of their territories: a place-based localizing strategy for the defense of local models of nature and cultural practices; a further strategy of localization though an active and creative engagement with translocal forces, such as similar identity or environmental movements or various global coalitions against globalization and free trade; and a shifting political strategy linking identity, territory and culture at local, regional, national, and transnational levels. . . .

[. . .]

PLACE, DIFFERENCE, AND THE POLITICS OF SCALE

. . . Theoretically, it is important to learn to see place-based cultural, ecological, and economic practices as important sources of alternative visions and strategies for reconstructing local and regional worlds, no matter how produced by "the global" they might also be. Socially, it is necessary to think about the conditions that might make the defense of place – or, more precisely, of *particular constructions of place* and the *reorganization of place* this might entail – a realizable project. As I mentioned, in their triple localizing strategy, some rainforest social movements engage in what geographers call "the politics of scale", they jump from one scale to another in their political mobilization. The results occur at various scales, from the local territories to the construction of regional socio-natural worlds, such as the Pacific as a "region-territory of ethnic groups." Alternative ecological public spheres might be opened up in this way against the imperial ecologies of nature and identity of capitalist modernity.

It is true that capital and globalization achieve dramatic scaling effects. They control places through the control of space. As geographers point out . . . we are witnessing an important geographic re-scaling by capital which shifts power primarily to the global level and global forms of governance (for example, NAFTA, the EU, GATT and WTO). Most times these maneuvers are undemocratic and disempowering; they are fueled by discourses of free trade, development and the unrestricted work of markets. However, social movements and progressive NGOs often times also create networks that achieve supra-place effects that are not negligible. The various networks of indigenous peoples of the Americas are already well known in this regard, but there are transnational networks emerging around a host of issues worldwide. The anti-WTO demonstrations in Seattle in November, 1999, are a case in point. They were actually the result of networks of organizations in ascension since at least the

anti-GATT protests in India in the early 1990s. These networks propitiate the reorganization of space from below and some measure of symmetry between the local and the global. They can be seen as creating "glocalities," that is, cultural and spatial configurations that connect places with each other to create regional spaces and regional worlds. Glocality means that everything is local and global, to be sure, but not global and local in the same way. In other words, not only capital but place-based struggles reorganize space through networks, and they do so according to different parameters and concerns.

This is also to suggest that the politics of place has to be found at the intersection of the scaling effects of networks, on the one hand, and emergent identities, such as the black and indigenous identities of the Colombian Pacific, on the other. Social movements and local communities are not just trapped in places, awaiting the liberating hand of capital, technology or development to join the networks of transnational flows of commodities, images, and the like. In constructing networks and glocalities of their own, even if of course in their engagement with dominant networks, social movements might contribute to democratize social relations, contest visions of nature (such as in biodiversity debates), challenge current technoscientific hype (and in the case of transgenic agriculture and genetically modified organisms, GMOs), and even suggest that economies can be organized differently from current neo-liberal dogmas (as in the resurgence of barter and local-currency economies and the continued survival of non-capitalist practices). Social movements suggest that "the gestalt of space" needs to be approached not only from the perspective of capital's spatialization but from the side of the production of space by place-based networks. It is also vital that researchers recognize both the social production and the cultural construction of space; the scaling-up of networks has cultural effects that are often missed in conceptions of space and networks, including those that have been most enlightening and influential.

A politics of difference based on place-based practices and networks is greatly aided today by the creative use of NICTs. Information and networking have been shown to be of vital importance to the political strategies of a number of cultural rights movements, including the Zapatista and the Maya culturalist movement, women's movements, and other ethnic, environmental, and indigenous movements. It might seem paradoxical at first to use NICTs, known for their de-localizing effects at the service of capital and global media, for a defense of place-based practices. But the fact is that people rooted in local cultures are finding ways to have a stake in national and global society precisely as they engage with the conditions of transnationalism in defense of local cultures and ecologies. This is so because these networks are the location of emergent local actors and the source of promising cultural practices and possibilities. They are most effective when they rely on an ongoing tacking back and forth between cyberpolitics and place politics – that is, between political activism on the internet and other network-mediated spaces and activism in the physical location in which the networkers sit and live. Because of their historical attachment to places and the cultural and ecological difference they embody, women, environmentalists and ethnic social movements in some parts of world are particularly suited to this task of weaving the virtual and the real, and culture, gender, environment and development into an innovative cultural-political practice.

[. . .]

In sum, social movements and many progressive NGOs and scholars are finding it increasingly necessary to posit a defense of place and place-based practices against the economic and cultural avalanche of recent decades. Most times, this project does not take the form of an intransigent defense of "tradition" but rather of a creative engagement with modernity and transnationalism, oftentimes aided by NICTs. These social actors do not seek so much inclusion into the global network society but its reconfiguration in such a way that their visions of the world may find minimum conditions for their existence. Despite tensions and conflicts, they create networks and glocalities with a more decidedly plural character: glocalities in which many cultural politics and political cultures can coexist, giving new meaning to democracy. Popular glocalities might be able to establish structures of power that do not impose homogeneous conceptions of the good on all of its participants. Here we might find a new hope for a reasonable pluralism. The fact that a growing

number of people and groups demand the right to their own cultures, ecologies, and economies as part of our modern social world can no longer be denied, nor can it these demands be easily accommodated into any universalist liberal or neo-liberal doctrine. It is no longer the case, as neoliberal globalizers would have it, that one can only contest dispossession and argue for equality from the perspective of inclusion into the dominant culture and economy. In fact, the opposite is becoming the case: the position of difference and autonomy is becoming valid, if not more, for this contestation. Appeals to the moral sensibility of the powerful ceased to be effective, if they ever were. It is time to try out other strategies, like the power strategies of groups connected in networks, in order to negotiate contrasting conceptions of the good and the value of different forms of life and to re-state the long-standing predicament of difference-in-equality. It is time for thinking more openly about the potential healing effects of a politically enriched alterity.

CONCLUSION

It might seem paradoxical to assert that the identities that can been as emerging in the cultural-environmental domain today might simultaneously be attached to place and most open to what remains unimagined and unthought in biological, cultural, and economic terms. These identities engage in more complex types of mixing and dialectics than in the most recent past. The dynamic of place, networks and power at play today in many ambits suggests that this is the case. Subaltern strategies of localization still need to be seen in terms of place; places are surely connected and constructed yet those constructions entail boundaries, grounds, selective connection, interaction and positioning, and in some cases a renewal of history-making skills. Connectivity, interactivity and positionality are the correlative characteristics of the attachment to place, and they derive greatly from the modes of operation of the networks that are becoming central to the strategies of localization advanced by social movements (and, of course, by capital in different ways). Networks can be seen as apparatuses for the production of discourses and practices that connect nodes in a discontinuous space; networks are not necessarily hierarchical but can in some cases be described as self-organizing, non-linear and non-hierarchical meshworks, as some theorists of complexity think of them at present. They create flows that link sites which, operating more like fractal structures than fixed architectures, enable diverse couplings (structural, strategic, conjunctural) with other sites and networks. This is why I say that the meaning of the politics of place can be found at the intersection of the scaling effects of networks and the strategies of the emergent identities. . . .

It has been said that the ideas and practices of modernity are appropriated and re-embedded in locally-situated practices, giving rise to a plethora of modernities through the assemblage of diverse cultural elements, and that oftentimes this process results in counter-tendencies and counter-development, defined as "the process by which multiple modernities are established." The challenge for this constructive proposal is to imagine multiple modernities from multiple directions, that is, from multiple genealogies of place-based (if clearly not place-bound) practices. It is at this level that "the postdevelopment moment" is of relevance, at least in some recent reinterpretations of the concept. . . . A movement towards the defense of place might well be an element in this strategy. This defense is of course not the only source of hope and change, but an important dimension of them.

The critique of the privilege of space over place, of capitalism over non-capitalism, of global cultures and natures over local ones is not so much, or not only, a critique of our understanding of the world but of the social theories on which we rely to derive such understanding. This critique also points at the marginalization of intellectual production on globalization produced in the "peripheries" of the world. The critique, finally, is an attempt to bring social theory into line with the views of the world and political strategies of those who exist on the side of place, non-capitalism and local knowledge – and effort to which anthropologists and ecologists are usually committed. Dominance and subalternity . . . are complex social and epistemological phenomena. Those frameworks that elide the historical experience of the subaltern and that participate in the erasure of subaltern strategies of localization can also be said to participate in the prose of counter-insurgency. Conversely, if it is true that politically enriched forms of difference are always under construction, there is hope that they could get to constitute new grounds for existence and significant rearticulations of subjectivity and alterity in their economic, cultural and ecological dimensions.

In the last instance, anthropology, political geography and political ecology can contribute to re-state the critique of current hegemonies as a question of the utopian imagination: Can the world be reconceived and reconstructed from the perspective of the multiplicity of place-based practices of culture, nature and economy? Which forms of "the global" can be imagined from multiple place-based perspectives? Which counter-structures can be set into place to make them viable and productive? What notions of politics, democracy and the economy are needed to release the effectivity of the local in all of its multiplicity and contradictions? What role will various social actors – including technologies old and new – have to play in order to create the networks on which manifold forms of the local can rely in their encounter with the multiple manifestations of the global? Some of these questions will have to be given serious consideration in our efforts to give shape to the imagination of alternatives to the current order of things.

"No Place like Heimat: Images of Home(land)"

from *Spaces of Identity: Global Media, Electronic Landscapes, and Cultural Boundaries* (1995)

David Morley and Kevin Robins

Editors' introduction

All of the selections in Part Five have, in one way or another, questioned the naturalness of the nation-state in light of globalization. Yet the fact remains that today the nation-state remains one of the – if not *the* – primary sites of political identification for many people in the world. Before anything else, we are "Chinese" or "American" or "South African." But what of Europe's experiment to construct a supra-national entity? Can one reasonably say, today, that before all else one is "European," as opposed to "English" or "Belgian" or "Greek"? The question of identification with place is one of the most intriguing, and vexing, aspects of the European Union. Since 1951 the European Union has evolved into its current status as the world's largest confederation of nation-states. Yet, particularly in recent years as the EU has moved toward closer political and military, as well as economic, integration, one of its largest challenges has been for citizens of member countries to agree to forgo markers of their national heritage in exchange for symbols of *pan-European identity*. For the decision to adopt the euro in exchange for the traditional pesetas, liras, and francs has involved much more than simply a monetary transition. Conversion to the euro has involved giving up a fundamental aspect of *national identity* in exchange for a *supra-national identity* that voters of some nations – for example Sweden and Britain – have found unacceptable.

As the European Union considers membership applications from countries not uniformly agreed to be culturally a part of Europe, questions of identity and boundaries loom even larger. Does Europe necessarily consist only of nations that are both Christian and democratic? If so, should the European Union then be closed off from new member states and immigrants, to form a "fortress Europe"? Or has Europe already changed so profoundly through in-migration that, even if a defensive stance were attempted, it would not succeed? And what of nations like Germany, Spain, and Italy which, though they fall squarely within the common conceptualization of "European" nations, have relatively recent fascist pasts? In "No place like Heimat: images of (Home)land," David Morley and Kevin Robins consider these questions from the vantage point of Germany. Morley and Robins suggest that the outcome of struggles over representing Germany's past has important implications for Germany's future. Furthermore, Germany – while providing an extreme example in some respects – may well provide a test case for other European nations in the matter of the scale of belonging and identity in a global world.

In order to build their argument, Morley and Robins centralize the tension surrounding a particular reconstruction of German history depicted in the American television series titled *Holocaust*. Its broadcast into the homes of millions of West Germans in 1979 provoked intense debate about who has the right to represent a country's history and memories. In response, in 1984 German film maker Edgar Reitz produced the television series titled *Heimat*, to reclaim from the Americans the right to depict German historical memory. Films, television, music, and other media are often used to construct history, national identity, and political loyalties. In other words, there is an ideological dimension to visual media that is the subject of critical cultural analysis. The work of film theorist Laura Mulvey has provided a touchstone in this field; her essays are collected in *Visual and other Pleasures* (1989). Visual culture and its importance to cultural geography is discussed further in the selection introduction for Gillian Rose (see p. 171).

The term *Heimat* refers to a *sub-national* identity: the opposite of what is invoked by *supra-national* entities such as the European Union. As discussed at more length in Part Two, the sub-national region is a mainstay of European geography. The underlying question of "At what scale should our identities be fashioned?" involves, in part, what counts as home. The subject of home finds resonance in work by French philosopher and poet Gaston Bachelard, who discussed the issue in *The Poetics of Space*, which was translated into English in 1964, and in the work of German philosopher Martin Heidegger; in particular his consideration of dwelling in "Building dwelling thinking," an essay translated and published in English in *Poetry, Language, Thought* (1971). More recently, cultural geographers have considered home as memory and everyday domestic practice; an example is the home-themed issue of the journal *Cultural Geographies* (11, 2004) edited by Allison Blunt and Ann Varley.

David Morley is a Professor of Communications at Goldsmiths' College, University of London. Kevin Robins is a Professor of Sociology at City University, London, and a Visiting Fellow at Goldsmiths' College, University of London. Together, Morley and Robins edited the book *British Cultural Studies* (1991).

INTRODUCTION

Our concern . . . is with the questions of identity and memory in the construction of definitions of Europe and European culture. It is in this context that we address the centrality of the idea of Heimat (home/land). We take as a particular instance the debates opened up in Germany by Edgar Reitz's 1984 film, *Heimat* (and further developed in his sequel *Die Zweite Heimat* (1990)), centred around the opposition between Heimat and Fremde ('homeland' and 'foreignness'). This provides the focus for a broader discussion of the relations between European and 'Other' cultures in the post-war period, and, more particularly, of the representation of the European past as constructed through the media. Our argument is that we see played out here, in these debates over who holds the franchise on the representation of the past, an illuminating 'echo' of debates as to who has the right to determine Germany's future. This is, of course, no local matter, but is crucial to the future of Europe as a whole. We take the 'German story' to be a symbolic condensation of many of the most problematical themes of the European past and a central issue in the contemporary *Realpolitik* of Europe. . . .

If Germany, the past somehow reconciled . . . is to be united in more than name, and Europe no longer divided by the 'Iron Curtain', then the question arises, inescapably, as to where Europe ends (what is the status of *Mitteleuropa* or Eastern Europe?), and against what 'Other' (besides America) Europe and European culture are to be defined, if no longer against Communism. Our argument is that, if America continues to supply one symbolic boundary, to the 'West', there is also, implicit in much recent debate, a reworking of a rather ancient definition of Europe – as what used to be referred to as 'Christendom' – to which Islam, rather than Communism, is now seen to supply the 'Eastern' boundary. Our concern is with identifying some of the threads from which this pattern is being woven – the better, hopefully, to unravel it.

BRINGING IT ALL BACK HOME

[. . .]

It is this idea of 'home' that interests us. Home in a world of expanding horizons and dissolving boundaries . . . In pre-modern times . . . this sense of trust and security was rooted in kinship systems, in local community, in religious beliefs and in the continuity of tradition. The effect of the great dynamic forces of modernity . . . has been to 'disengage some basic forms of trust relation from the attributes of local contexts'. Places are no longer the clear supports of our identity.

If anything, this process of transformation has become accelerated, and time-space compression has come to be ever more intense. It is through the logic of globalisation that this dynamic of modernisation is most powerfully articulated. Through proliferating information and communications flows and through mass human migration, it has progressively eroded territorial frontiers and boundaries and provoked ever more immediate confrontations of culture and identity. . . . Through this intermixture and hybridisation of cultures, older certainties and foundations of identity are continuously and necessarily undermined. The continuity of identity is broken too. . . . There is a desire to be 'at home' in the new and disorientating global space.

Home, homeland, Heimat. It is around the meaning of European culture and identity in the new global context that this image – this nostalgia, this aspiration – has become polemically activated. Consider [leader of the Soviet Union from 1985 to 1991] Mikhail Gorbachev's appeal to a 'common European home':

> Europe is indeed a common home where geography and history have closely interwoven the destinies of dozens of countries and nations. Of course, each of them has its own problems, and each wants to live its own life, to follow its own traditions. Therefore, developing the metaphor, one may say: the home is common, that is true, but each family has its own apartment, and there are different entrances, too.

This notion of a single Europe, from the Atlantic to the Urals, has an obvious appeal. But what does it really amount to? What kind of community does it offer?. . . . [One possibility] is a defensive identity, a fortress identity, defined against the threat of other cultures and identities (American, Japanese, Islamic, African or whatever). This reassertion of European cultural identity amounts to a refusal to confront the reality of a fundamental population shift that is undermining 'the little white "Christian" Europe' of the nineteenth century. . . .

. . . . The European Heimat invokes the past grandeur of Europe as a bastion against future uncertainties. This is a Europe that divides those who are of the Community from those who are *extracommunitari* and, effectively, extraterrestrial.

There are those, however, who are less committed to this particular vision of a European home. They are, to appropriate Gorbachev's metaphor, more interested in the different apartments than in the common home. For them, a faceless Europeanism is inimical to the rich diversity of national cultures and identities that are, supposedly, the basis of a more authentic sense of belonging; they feel that it is only in the sense of nationhood that one can feel truly 'at home'. Throughout Europe, we can now see the rekindling of national and nationalist sentiments. It is more apparent in Central and Eastern Europe, where national aspirations of sixty and seventy years ago are currently being reactivated through the reassertion of ethnic, religious and cultural differences. But also in Western Europe, particularly in the context of German reunification (*Deutschland, einig Vaterland*), national allegiance is asserting itself as a powerful way of belonging. . . .

As an alternative to continental Europeanism and to nation statism, there is yet another kind of 'homely' belonging. This is the identity rooted in the Heimat of regions and small nations. . . . the rich pluralism of regional traditions, languages, dialects – and cultures as the true basis for authentic identities. . . . This 'small is beautiful' ideal of a Europe of the regions clearly seems to offer a richer and more radical way to belong. There is a romantic utopianism in this celebration of small nationalism and regionalism, a utopianism of the underdog. . . .

Yet Heimat is an ominous utopia. Whether 'home' is imagined as the community of Europe or of the nation state or of the region, it is steeped in the longing for wholeness, unity, integrity. It is about community centred around shared traditions and memories. As the German film-director Edgar Reitz puts it:

> The word is always linked to strong feelings, mostly remembrances and longing. Heimat always evokes in me the feeling of something lost or very far away, something which one cannot easily find or find again. . . . It seems to

me that one has a more precise idea of Heimat the further one is away from it.

Heimat is a mythical bond rooted in a lost past, a past that has already disintegrated. . . . It is about conserving the 'fundamentals' of culture and identity. And, as such, it is about sustaining cultural boundaries and boundedness. To belong in this way is to protect exclusive, and therefore excluding, identities against those who are seen as aliens and foreigners. The 'Other' is always and continuously a threat to the security and integrity of those who share a common home. Xenophobia and fundamentalism are opposite sides of the same coin. For, indeed, Heimat-seeking is a form of fundamentalism. . . . In contemporary European culture, the longing for home is not an innocent utopia.

COMMUNICATIONS, MEMORY AND IDENTITY

These questions of identity, memory and nostalgia are inextricably interlinked with patterns and flows of communication. The 'memory banks' of our times are in some part built out of the materials supplied by the film and television industries. It is to the role of these industries in the construction of memory and identity that we now turn. . . .

One of the first questions concerns how we are to understand the 'national', and what the role of media institutions is in the construction of national identities. . . . The discourses of 'art', 'culture' and 'quality' have . . . been mobilised against Hollywood and used to justify various nationally specific economic systems of support and protection for indigenous film-making.

The role of the state is crucial in this respect, in so far as government policies have often determined the parameters and possibilities of various national cinemas. . . . This is, necessarily, a contentious business. Definitions of national cinema always involve the construction of an imaginary homogeneity of identity and culture, apparently shared by all national subjects; this involves mechanisms of inclusion and exclusion whereby one definition of 'the nation' is centralised and others are marginalised . . . [in] a process of 'internal cultural colonialism'.

It is a question of recognising the role of the stories we tell ourselves about our past in constructing our identities in the present. One key issue concerns the power of the idea of the nation to involve people in a common sense of identity and its capacity to work as an inclusive symbol which provides integration and meaning as it constructs and conscripts public images and interpretations of the past 'to re-enchant a disenchanted everyday life'. In this fashion . . . the idea of the national past is constantly reworked and represented within the historical experience of a particular nation state. Identity is a question of memory, and memories of 'home' in particular.

Film and television media play a powerful role in the construction of collective memories and identities. It is in this context that we address the centrality of the idea of Heimat, principally with reference to the debates opened up in the mid-1980s in the Federal Republic of Germany by Edgar Reitz's film/television series of that name. The Heimat film is, of course, a well established genre in Germany. One obvious question concerns whether one can work within this traditionally reactionary genre and yet give the material new and different meanings. Reitz's attempts to do just this have to be seen in the context of the political revitalisation of the rural 'Heimat' tradition in West Germany in the 1970s – as an attempt by a coalition of ecological and anti-nuclear groupings to 'reclaim' these traditions for the left, by means of the rediscovery and revaluation of regional and folk traditions, dialect poetry and so on, in an anti-centralist (and anti-urban) political movement. This turn to ecology represents an important shift, and, in this context . . . in the face of the steady destruction of the environment, 'homeland' ceased to be a dirty word. . . . Heimat is a place no one has yet attained, but for which everyone yearns. Reitz notes that 'Heimat, the place where you were born, is for every person the centre of the world'; the idea, or ideal, is not simply territorial, but rather invokes a 'memory of origin' and involves the notion of an 'impossible return' to roots or origins.

When the American-produced television series *Holocaust* was shown in West Germany in 1979 it was watched by more than twenty million Germans, who were confronted with this version of their own history in their own living rooms. When *Heimat* was shown in the autumn of 1984, it was much more than a television series: it provided the focus and stimulus for a wide-ranging debate on German identity and history.... Both these series acquired the status of television events; it was absolutely necessary for people to watch them if they were to be able to participate effectively in the public debates that were generated in daily conversation.

This raises the question of who has the power to structure discourse in the 'instant public sphere' (an issue that was again raised in early 1994 by the release of Spielberg's film, *Schindler's List*).... Edgar Reitz, of course, explicitly conceived *Heimat* as the German 'answer' to this American series. For Reitz, *Holocaust* was a 'glaring example [of an] international aesthetics of commercialism [for which] the misery produced by the Nazis is nothing but a welcome background spectacle for a sentimental family story'. He was concerned that German filmmakers should establish the 'rights' to their own history, reclaiming them from the Americans. For Reitz the real scandal was 'German history – Made in Hollywood': hence the subtitle to *Heimat*, 'Made in Germany'. With *Holocaust*, he believed,

> the Americans had stolen our history . . . taken narrative possession of our past. . . . I watched the crocodile tears of our nation and I saw how it was all taken seriously and how the question of guilt in German history was being discussed by all the great German intellectuals on the basis of this travesty.

It is worth noting that when *Heimat* was shown in the United States, many critics responded negatively, deeming the series to be a dangerous whitewash of German history. Clearly, the history of a world war does not belong to any single nation. In these debates over the politics of representing the German past, what is at issue is who has the right to determine Germany's future.

A number of useful parallels can be drawn between the debates surrounding *Heimat* and the filmic representation of 'Vietnam' in the United States. Here again we see the pertinence of the argument that the representation of the past is very much a question of active processes in the present – as the Vietnam War continues to be waged symbolically on television, in bookshops and at a cinema near you. The historical Vietnam War, a specific set of conflictual events, policies and conditions, has

been transformed into a symbolic 'Vietnam', just as with the German (and thus the European) past in *Holocaust* and *Heimat*. In the case of both *Heimat* and the Vietnam films, we have the questions not only of loss and mourning, but also, and more problematically, the cultural blockage created by questions of guilt, and how *that* is to be represented. In both cases, we also have the question of whether it is possible to undertake a 'progressive' reappropriation of patriotic sentiment, along with the further issue of the potential usurpation of the role of victim by the perpetrators of the initial violence. And then, of course, we have the question of the silences in these discourses: on the one hand, the marginalisation of the Holocaust itself in *Heimat*'s sixteen hours; on the other, the almost total absence of anything other than caricature representations of the Vietnamese themselves in Hollywood's Vietnam films.

[. . .]

HOW EUROPEAN IS IT?

. . . The debates around the concept of 'home' and 'homeland' occasioned by *Heimat* have now, of course, also to be seen in the transformed context of Gorbachev's call for the construction of a 'common European home' to transcend the Cold War division of Europe, which found its most dramatic expression in the division of Germany. As we have already suggested, the debates over who should hold the franchise rights on the story of the German past have many parallels in the debates as to who should have the right to determine Germany's future. Current debates concerning the reunification of the country have a necessary centrality to our argument, not least in so far as, in the context of *perestroika* and *glasnost*, the very concept of 'Europe' now becomes geographically less distinct.

Questions of religion and race are also lurking in the definition of Europe and European culture. As the Cold War order crumbles, we are seeing the reassertion of religion as both a buttress of cultural identity and a token of membership of the 'civilised' world. In this context, the debates generated by Turkey's application to join the European Community offer a number of interesting insights into the issues at stake.

At one level, the issue is simple. On the one hand Turkey, on account of its membership to NATO, its possession of a small but important triangle of land on the European side of the Bosphorus and the modern secular framework of institutions bequeathed by Kemal Ataturk, has a strong *prima facie* case for membership of the Community. On the other hand, there is a complex set of questions concerning trade barriers, the potential impact of cheap Turkish agricultural (and increasingly, electrical) products on existing member countries, and, of course, there is the continuing question of Turkey's record on human rights. However, we suggest that, at base, something far more fundamental is at stake: the question of whether in contemporary debates 'Europe' is being defined as co-extensive with what used to be called Christendom. Or, to put it the other way round, can an Islamic (albeit secularised) state be fully accepted as part of Europe? Consider that historically the Ottoman Empire provided an image of difference and threat (and, indeed, dread), against which Europe defined itself. Consider, too, that today's European Community was founded by Christian bureaucrats (indeed, Catholics) across Europe.

Certainly, in recent years there has been a marked increase in the anxiety and suspicion with which many Europeans view the Islamic world. Across Europe we can see an emerging pattern of racial hostility towards Muslims – dramatised in complex ways by the [Salman] Rushdie affair in Britain, and by violence and hostility to Turkish immigrant workers in Germany and to North African immigrants in France and in Italy. One could argue that the oil crisis of the 1970s, images of PLO terrorists and Lebanese hostage-takers and the image of Islamic fundamentalism throughout the Middle East, have all been aggregated in the popular media to produce a greater sense of 'Islamic threat' to Europe than at any time since the seventeenth century. The French mass-circulation news magazine *Le Point* headlined a story about Islamic fundamentalism in Algeria, 'The Holy War at our Gates', a story full of references to the Muslim 'danger' and its 'threat' to French national identity. Jean-Marie Le Pen, leader of the French National Front, claims Joan of Arc as his inspiration. The director of the Turkish Foreign Policy Institute in Ankara puts it quite simply: 'In Europe, many people see us as a new version of the Ottoman empire, attacking this time in the form of guest workers and terrorists.' It can be argued that Islam is now the primary form in which the Third

World presents itself to Europe, and that the North–South divide, in the European context, has been largely inscribed onto a pre-existing Christian–Muslim division. . . .

However, there is more to it than that, in so far as the relation between these two terms, or rather, the significance of this relation, has itself been shifted by the current transformation of East–West relations. . . . [T]he deep-seated anxieties about European identity (and the centrality of Christianity to that definition) were driven underground by the Cold War, during which Stalin's empire provided Europe with a *de facto* eastern frontier. During this period, whatever was not 'Communist' was 'Western' (that is, European). In this context, as a member of NATO, and a strategically crucial one at that, the European credentials of Turkey were accepted without much question. Certainly many Turks regard their membership of NATO as proof of their Western status. But with the collapse of the Soviet bloc, all this is now called into question. Central and Eastern Europe is reasserting its identity in large part as a Christian one. Europe is suddenly feeling the need to re-establish its psychic boundaries anew. And, as it redefines itself, the question of who is to be excluded – that is to say, in contradistinction to whom or to what 'European' identity is to be defined – is being refocused. Turkey suddenly finds itself in a different context, one in which its European credentials have been dramatically devalued.

There is, it seems, no place like home – and apparently no place in that home for some who wish to dwell there. Our common European home remains to be built: but the stories we tell ourselves about our common (and uncommon) past are already shaping our understanding of how it should be constructed, how many floors it should have (a basement for the servants?), which way it should face and who should have the keys to the door.

'THE BORDER RUNS RIGHT THROUGH MY TONGUE'

Our discussion has, at various points, focused on Germany because of its particular strategic and symbolic importance in the contemporary transformation of Europe. Germany, once again the question mark of Europe. Germany has been divided against itself, and this divide has also marked the separation of the eastern and western halves of Europe. Now the dividing wall has been deconstructed: what was protectively solid has apparently evaporated into air. . . .

Now the two components have come into direct contact. What compound mixture is being distilled in the process? If Germany had until recently been seen as a kind of 'post-national' society, questions of national culture and identity are once again on the political agenda. What does it mean to be German today, after forty years of division? What is 'German' now? The border ran right through German identity and now it has been dissolved and Germany re-encounters itself, across space and also across time. . . . [There is] a kind of historical 'doppelganger' effect: West Germans must now see their past, their history, reflected back at them; and East Germans have the dislocating and disorientating experience of confronting their future. Who now are 'we the people'?

The tragedy will be if reunification provokes a defensive and exclusivist form of nationalism. The defeat will be if German identity is refounded in terms of a closed community, with boundaries drawn between those who belong and those who do not. 'Germany is one' and 'we are one people' were the slogans chanted outside the Berlin opera house in Karl Marx Square. One people. One homeland. . . . [It has been suggested that] nationalist sentiments are akin to infantile attachments to the family. The nation . . . is both mother and father. . . . This complex allegiance, this 'matripatriotism', expresses itself . . . in a strong sense of rootedness, of belonging to a home and a homeland. . . . One people, one family, one homeland: belonging together, with common origins. 'We the people' defined against the 'Others' who do not belong, and have different origins.

The question of a German home, as we have argued at length, has been a central motif in recent cultural debates in the Federal Republic. At the heart of the New German Cinema the problem of identity and the quest for origins has centred around the theme of the family, the damaged relation to the (absent) father and the fixation on the mother figure. For many, this has been about trying to find a way home; it has been about becoming reconciled to German culture and identity. The romantic utopia of Heimat, with all its connotations

of remembrance and longing, has been about reconnecting with a national heritage and history. For others, however, the issue is far more complex. National integrity is a vain ideal; one people, a false utopia. The cinema of Wim Wenders, particularly, has been about the state of homelessness that seems to be a necessary expression of the condition of modernity. . . . In his films there is no easy recourse to the security of origins, rootedness and authenticity. . . . Wenders is concerned with journeys, with crossing borders, with exile, with the relation between inside and outside. What he seeks to explore, particularly through his relationship with 'America', are the realities of difference, Otherness and estrangement. For Wenders, there is no utopia of home and homeland:

> The idea is that, not being at home [my heroes] are nevertheless at home with themselves. In other words, not being at home means being more at home than anywhere else. . . . Maybe the idea of being more oneself when one's away is a very personal idea. . . . Identity means not having to have a home. Awareness, for me, has something to do with not being at home. Awareness of anything.

Being away, not being at home, is what Wenders aspires to.

Not being at home is, of course, the permanent destiny of so many people and peoples ('involuntary cosmopolitans') in the modem world. It is the condition of those millions of so-called *Ausländer* or *Gastarbeiter* who live precarious and unsettled lives in the German homeland itself. . . . *Uberfremdung* (overforeignisation) has been perceived as a threat to national integrity and culture. Now it is the 1.5 million Turks living in Germany who have become the salient and disturbing 'Other'. 'We the people' are now defined, in Germany, against the 'Islamic Other'. The question is whether Germany can come to terms with this 'Islam within', or whether the new nation will be imagined on the basis of an exclusive and excluding racism. It is also a question of whether Germany can understand that it is not one, can never be one, because it is multiple, because it contains many peoples, Germans of different ethnicities.

What must be recognised is that, if Germany is a home for some, then it is at the same time exile for others. What must be understood is the relation between Heimat and Fremde. If Heimat is about security and belonging, Fremde evokes feelings of isolation and alienation. . . . Germany – the real, rather than the imaginary, Germany – is at once Heimat and Fremde. Is it possible to come to terms with this relational truth, rather than taking refuge in the comforting absolute of Heimat? Is it possible to live with this complexity and ambivalence? In his poem 'Doppelmann', Zafer Senocak writes of his Germany:

I carry two worlds within me
  but neither one whole
they're constantly bleeding
  the border runs
right through my tongue.

It is this experience that is fundamental to questions of German – and also European – culture and identity today. And it is out of this tension – between homelessness and home – that we might begin to construct more meaningful, more complex, identities. . . .

Our discussion has been about images of home and homeland, and it has arrived at the reality of homelessness. It has focused particularly on the idea of a German home to illuminate the powerful appeal of Heimat throughout a changing Europe. Whether it is in terms of a national home, a regional home or a common European home, the motivating force is a felt need for a rooted, bounded, whole and authentic identity. And yet Heimat is a mirage, a delusion. . . . It is a dangerous delusion. Heimat is rooted in that intolerance of difference, that fear of the 'Other', which is at the heart of racism and xenophobia.

The crucial issue that now confronts European culture, we would argue, is whether it can be open to the condition and experience of homelessness. The questions posed by Wim Wenders are at the heart of the matter. Can we imagine an identity, an awareness, grounded in the experience of not having a home, or of not having to have a home? Can we see home as a necessarily provisional, always relative, truth?. . . It is this experience of transit that is fundamental to the culture. . . .

There can be no recovery of an authentic cultural homeland. In a world that is increasingly characterised by exile, migration and diaspora, with all the consequences of unsettling and hybridisation, there can be no place for such absolutism of the pure and authentic. In this world, there is no longer any place like Heimat. More significant, for European cultures and identities now, is the experience of displacement and transition. . . . What is most important is to live and work with this disjuncture and ambivalence. Identity must live out of this tension. Our feet must learn to walk on both banks of the river at the same time.

PART SIX

Home and Away

Courtesy of Alex Dorfsman

INTRODUCTION TO PART SIX

Mobility has become a kind of leitmotif of the modern era. It has become a banal observation to say that we live in increasingly 'fast paced' societies. As infrastructures of travel have allowed for faster and more commonplace movement, it is easier and easier to conduct our lives "on the road." Tourism is now one of the largest industries in the world, and the chief source of income for many developing countries. Labor migrants wrap the globe in their networks of travel between home and workplace, from the Philippines to Singapore, India to Kuwait, Turkey to Germany, and Mexico to the United States. In China, over 150 million migrants have left their rural homes to labor in the cities, fueling industrialization and urbanization at an unprecedented scale. The cities are full of strangers, and it is this fact perhaps more than any other which has captured the imagination of scholars of modernity. Collectives of the uprooted, cities are places of mobility. This has also given the city a liberating quality for many. When everyone's a stranger, the conventions of traditional communities don't apply. More than simply uprooting populations and moving them from one place to another, then, mobility is related to more fundamental social changes.

To say this, however, implies a binary between mobility–modernity on the one hand, and dwelling–tradition on the other. And as Part Two of the *Reader* made clear, the work of many past geographers was organized around exactly this kind of division: modern, mobile, industrial cities versus traditional, fixed, rural villages. As was pointed out in various selections in Part Two, much early work in geography concentrated on the rural, the folk-cultural, the pre-industrial. Culture was seen as "growing," organically, out of the soil, like a plant. The term, after all, has etymological roots in the tillage and cultivation (of plants, crops). For geographers like Ratzel (see p. 83) or Vidal de la Blache (see p. 90) culture dwelled in a fixed place, was planted in the soil, and this organic link with the land is what produced regional distinctions, or Vidal's *genres de vie*. The processes that uprooted people from these organic culture regions could not help but be viewed, in this context, as disruptive.

The idea of nations and societies being similarly rooted in the land was a logical extension of this kind of thinking. And when geographers sought to describe the true cultural identity of a nation, they not surprisingly looked toward the undisturbed rural folk for the last vestiges of a pure culture. Again, mobility was viewed as an aberration in the definition of a society or a nation. An important exception to this, of course, can be found in much of the rhetoric about the American character, which has typically been thought of in much more mobile terms (more on this below).

In addition to cultures, nations, and societies being conceived in fixed, place-bound, territorial terms, *home* itself has been typically conceived as a rooted and fixed place. And perhaps because of this, these ideas of home, territory, and nation have often been conflated. The "homeland" or *Heimat* represents a linking of identity with the idea of fixed place, a territorial space with clear boundaries, inhabited by people with a common culture that has itself "grown" out of the soil of that place (see Morley and Robins, p. 296). It is no mistake that homeland is now explicitly linked with security in the post-9/11 United States. And while the notion of *Heimat* itself has a troubling geopolitical history – it was a term used to articulate the expansionist designs of Germany under the Nazis – having roots in a home place is certainly a deeply felt "need of the human soul" (see Malkki, p. 275). Dwelling in one's home is, for philosophers such as Gaston Bachalard (*The Poetics of Space*, 1964) and Martin Heidegger ("Building, Dwelling, Thinking," 1971), the most basic form of being that there is (see also the introduction to Part Five).

Yet, while dwelling in place is therefore thought of as the *normal* state of things, there have always been travelers, nomads, wanderers, and strangers who have never fit this norm. In one respect, such outsiders have always been crucial to helping insiders view themselves as normal. Outside others, as Edward Said (p. 357) has pointed out, play a role in helping insiders see themselves as sharing a common cultural identity (see also Sibley, p. 380). But it should also be remembered that the exile, nomad, and wanderer occupies a central place in the canon of Western literature, myth, and folklore. This is reflected at the most basic level, for instance, in the exile of Adam and Eve from the Garden of Eden, Jesus Christ's journey into the desert wilderness, the many exiles of the Jews, or Abraham's wanderings. It is reflected in the observation that a surprising number of classics in Western literature were written by writers living in exile.[1] It is reflected, as claimed by Simmel in this part's first selection, by the central role of "the stranger" in the constitution of modern society. And it is also reflected in Said's claim that the modern condition – with its masses of refugees and other uprooted people – conveys more than anything a "generalized condition of homelessness" (see p. 357).

It is perhaps odd, then, that modern social science should be dominated by what Liisa Malkki (p. 275) refers to as a "sedentarist metaphysics" in which dwelling in a fixed, bounded place is regarded as the norm. While the selection by Jon May in this part makes clear that one should not doubt the basic human need for a "home place" that is in some way fixed and reliable, the home–dwelling/ homeless–mobility binary makes it difficult to think of mobility in any way other than as a dysfunction or aberration. Dwelling and mobility need not be viewed as opposite and distinct ways of being, one normal and one not. Instead, dwelling and mobility must be regarded as necessarily linked and, like all dualisms, mutually constitutive. Even phenomenologists of dwelling recognize that while dwelling might be the basic condition of being, mobility is also a fundamental part of dwelling. As Edward Casey argues (*Getting back into Place*, 1993), at the most basic level it's the habitual movements of the body, our regularized circuits of micro-travel, the spaces of movement that we eke out for ourselves – these are what make a home-place. And then there's the general observation that one has to leave home and return before one can appreciate it. In China, the idea of native-place identity did not emerge until enough uprooted sojourners could find each other in the distant cities and establish native-place associations.

So we may dwell by being both home and away, by moving in place. Of course now it is easier to see an opposite problem. If a traditional "sedentarist metaphysics" represented the norm that social science must move beyond, an opposite problem is raised by Crang in a commentary on mobility: "What does it mean to dwell in a mobile society?"[2] If mobility is increasingly the norm, the challenge may in fact become one of understanding how dwelling can still be possible. And this is the problem that confronts Mathias Woo in *A Very Good City*, as discussed in the Introduction to the *Reader*. Does home travel? Does place travel? In one of the following sections, James Clifford argues that cultures are not fixed in place; instead, cultures travel. But clearly, if homes travel, they do so with great difficulty for most people. And with the world's population of refugees only growing, Clifford's celebratory view of traveling cultures – while technically accurate – may be an insult to those who have been forced to leave their homelands.

Nevertheless, it is clear, as John Urry argues in *Sociology beyond Societies* (2000), that in our conceptions of culture and society, metaphors of network, flow, and travel are replacing metaphors of region and home. Urry believes that "People dwell in and through being both at home and away, through the dialectic of roots and routes or what Clifford terms 'dwelling in travel'" (pp. 132–133).

As Part Two made clear, cultural geographers have had much to say about organic homelands, fixed places of rooted culture. Until recently they have had much less to say, however, about mobility. The selections in Part Six offer a range of responses to how mobility is being addressed within the context of cultural geography, and how ideas of home have also been changed as a result.

NOTES

1 M. Bradbury and J. McFarlane, (eds.) *Modernism, 1890–1930* (1976).
2 Mike Crang, "Commentary – between places: producing hubs, flows, and networks," *Environment and Planning A* 34 (2002): 569–574.

S
I
X

"The Stranger"

from *On Individuality and Social Forms: Selected Writings*, ed. and trans. D. Levine (1971)

Georg Simmel

Editors' introduction

Much early work in cultural geography, as is clear from most of the selections in Part Two of the *Reader*, tended to focus on rural places and peasant folk cultures as exemplars of the "organic" connection between people and their environment. Indeed, some of this early geography (see Sauer, p. 96; Hoskins, p. 105) betrayed a distinct suspicion of industrial and urban places and landscapes as modern destroyers of traditional culture. The culture of cultural geography was a culture firmly rooted to the soil. Ratzel (p. 83) had claimed that it was like a tree. He was fond of recalling the etymology of the term: to cultivate, to tend, as in agri*culture*. In fact, Ratzel went further than this, noting that the most advanced societies are those that have been firmly rooted for ages. He compared the development of advanced civilization to "hoarding" – a kind of cumulative warehousing of all the best of the previous generations. Such storage of course requires fixity in place.

It's not surprising, then, that cultural geographers didn't have much to say about people who were *not* fixed in place. Clearly such people were marginal to the kind of "organic" communities that geographers like Vidal de la Blache (p. 90) concerned themselves with. Uprooted people and wanderers were people who left such communities behind to join other wanderers in the growing cities. Urban, industrial places were full of strangers. And for the German sociologist and philosopher Georg Simmel (1858–1918), the stranger was an emblematic figure of modernity and urbanity. Of course, industrialization did not *create* strangers; societies have *always* had their strangers. If many geographers, particularly in Europe, saw an organic connection between people and places as the foundation of their science, then the stranger clearly disrupted this organic connection. It therefore took a sociologist to provide the pioneering treatise on a topic – mobility – that has much later become quite central to the practice of cultural geography. Of course, geographers had in many ways concerned themselves with mobility, examining the diffusion of cultures and so on. But Simmel was perhaps the first to consider mobility in the context of *social structure*. Simmel was interested in the role that strangers played within a larger social and cultural system, and, more specifically, the stranger as vehicle for understanding the social insecurities of modernity.

Simmel's brief essay traverses a great deal of conceptual ground in only a few pages. He begins by considering the stranger as an historical figure: the trader who settles down in some foreign place (for one is only a stranger once he tries to join the organic community), but who must earn a living without access to land. Ultimately, however, Simmel is interested in the *idea* of the stranger, an idea that relates to our understanding of *objectivity* (the point of view of the dispassionate outsider – that is, the stranger), and an idea conveying something that is *near and remote* at the same time. This strangeness, Simmel argues, is a general characteristic of all relationships, even the most intimate. All human relations have an element of strangeness to them. But at the social level, the stranger exemplifies the characteristic of 'the Other' – something that is completely different from the self, yet contains repressed or hidden elements of the self.

Georg Simmel is recognized as one of the most important early figures in German sociology. Despite his prolific writing, his associations with other established social theorists such as Max Weber and Ferdinand Tönnies, and his general brilliance, Simmel was something of a stranger himself in the German academy. He was not awarded a professorship until 1914, four years before his death. "The Stranger" was first published in 1908.

Simmel's best known works are probably *The Philosophy of Money* (1900) and the essay "The Metropolis and Mental Life" (1903). Other major monographs include *On Social Differentiation* (1890) and *Sociology: Investigations on the Forms of Sociation* (1908). While his work has been less central to the development of cultural geography, his pioneering studies of the spatial transformations associated with modernity have found strong echoes in contemporary geography as seen, for example, in David Harvey's *The Condition of Postmodernity* (1989).

If wandering, considered as a state of detachment from every given point in space, is the conceptual opposite of attachment to any point, then the sociological form of "the stranger" presents the synthesis, as it were, of both of these properties. (This is another indication that spatial relations not only are determining conditions of relationships among men, but are also symbolic of those relationships.) The stranger will thus not be considered here in the usual sense of the term, as the wanderer who comes today and goes tomorrow, but rather as the man who comes today and stays tomorrow – the potential wanderer, so to speak, who, although he has gone no further, has not quite got over the freedom of coming and going. He is fixed within a certain spatial circle – or within a group whose boundaries are analogous to spatial boundaries – but his position within it is fundamentally affected by the fact that he does not belong in it initially and that he brings qualities into it that are not, and cannot be, indigenous to it.

In the case of the stranger, the union of closeness and remoteness involved in every human relationship is patterned in a way that may be succinctly formulated as follows: the distance within this relation indicates that one who is close by is remote, but his strangeness indicates that one who is remote is *near*. The state of being a stranger is of course a completely positive relation: it is a specific form of interaction. The inhabitants of Sirius are not exactly strangers to us, at least not in the sociological sense of the word as we are considering it. In that sense they do not exist for us at all; they are beyond being far and near. The stranger is an element of the group itself, not unlike the poor and sundry "inner enemies" – an element whose membership within the group involves both being outside it and confronting it.

The following statements about the stranger are intended to suggest how factors of repulsion and distance work to create a form of being together, a form of union based on interaction.

In the whole history of economic activity the stranger makes his appearance everywhere as a trader, and the trader makes his as a stranger. As long as production for one's own needs is the general rule, or products are exchanged within a relatively small circle, there is no need for a middleman within the group. A trader is required only for goods produced outside the group. Unless there are people who wander out into foreign lands to buy these necessities, in which case they are themselves "strange" merchants in this other region, the trader *must* be a stranger; there is no opportunity for anyone else to make a living at it.

This position of the stranger stands out more sharply if, instead of leaving the place of his activity, he settles down there. In innumerable cases even this is possible only if he can live by trade as a middleman. Any closed economic group where land and handicrafts have been apportioned in a way that satisfies local demands will still support a livelihood for the trader. For trade alone makes possible unlimited combinations, and through it intelligence is constantly extended and applied in new areas, something that is much harder for the primary producer with his more limited mobility and his dependence on a circle of customers that can be expanded only very slowly. Trade can always absorb more men than can primary production. It is therefore the most suitable activity for the stranger, who intrudes as a supernumerary, so to speak, into a group in which all the economic positions are already occupied. The classic example of this is the history of European Jews. The stranger is by his very nature no owner of land – land not only in the physical sense but also metaphorically as a vital substance which is fixed, if not in space, then at least in an ideal position within the social environment.

Although in the sphere of intimate personal relations the stranger may be attractive and meaningful in many ways, so long as he is regarded as a stranger he is no "landowner" in the eyes of the other. Restriction to intermediary trade and often (as though sublimated from it) to pure finance gives the stranger the specific character of *mobility*. The appearance of this mobility within a bounded group occasions that synthesis of nearness and remoteness which constitutes the formal position of the stranger. The purely mobile person comes incidentally into contact with *every* single element but is not bound up organically, through established ties of kinship, locality, or occupation, with any single one.

Another expression of this constellation is to be found in the objectivity of the stranger. Because he is not bound by roots to the particular constituents and partisan dispositions of the group, he confronts all of these with a distinctly "objective" attitude, an attitude that does not signify mere detachment and nonparticipation, but is a distinct structure composed of remoteness and nearness, indifference and involvement. I refer to my analysis of the dominating positions gained by aliens, in the discussion of superordination and subordination, typified by the practice in certain Italian cities of recruiting their judges from outside, because no native was free from entanglement in family interests and factionalism.

Connected with the characteristic of objectivity is a phenomenon that is found chiefly, though not exclusively, in the stranger who moves on. This is that he often receives the most surprising revelations and confidences, at times reminiscent of a confessional, about matters which are kept carefully hidden from everybody with whom one is close. Objectivity is by no means nonparticipation, a condition that is altogether outside the distinction between subjective and objective orientations. It is rather a positive and definite kind of participation, in the same way that the objectivity of a theoretical observation clearly does not mean that the mind is a passive tabula rasa on which things inscribe their qualities, but rather signifies the full activity of a mind working according to its own laws, under conditions that exclude accidental distortions and emphases whose individual and subjective differences would produce quite different pictures of the same object.

Objectivity can also be defined as freedom. The objective man is not bound by ties which could prejudice his perception, his understanding, and his assessment of data. This freedom, which permits the stranger to experience and treat even his close relationships as though from a bird's-eye view, contains many dangerous possibilities. From earliest times, in uprisings of all sorts the attacked party has claimed that there has been incitement from the outside, by foreign emissaries and agitators. Insofar as this has happened, it represents an exaggeration of the specific role of the stranger: he is the freer man, practically and theoretically; he examines conditions with less prejudice; he assesses them against standards that are more general and more objective; and his actions are not confined by custom, piety, or precedent.

Finally, the proportion of nearness and remoteness which gives the stranger the character of objectivity also finds practical expression in the more *abstract* nature of the relation to him. That is, with the stranger one has only certain *more general* qualities in common, whereas the relation with organically connected persons is based on the similarity of just those specific traits which differentiate them from the merely universal. In fact, all personal relations whatsoever can be analyzed in terms of this scheme. They are not determined only by the existence of certain common characteristics which the individuals share in addition to their individual differences, which either influence the relationship or remain outside of it. Rather, the kind of effect which that commonality has on the relation essentially depends on whether it exists only among the participants themselves, and thus, although general within the relation, is specific and incomparable with respect to all those on the outside, or whether the participants feel that what they have in common is so only because it is common to a group, a type, or mankind in general. In the latter case, the effect of the common features becomes attenuated in proportion to the size of the group bearing the same characteristics. The commonality provides a basis for unifying the members, to be sure; but it does not specifically direct *these* particular persons to one another. A similarity so widely shared could just as easily unite each person with every possible other. This, too, is evidently a way in which a relationship includes both nearness and remoteness simultaneously. To the extent to which the similarities assume a universal nature, the warmth of the connection based on them will acquire an element of coolness, a sense of the contingent nature of precisely *this* relation – the connecting forces have lost their specific, centripetal character.

In relation to the stranger, it seems to me, this constellation assumes an extraordinary preponderance in principle over the individual elements peculiar to the relation in question. The stranger is close to us insofar as we feel between him and ourselves similarities of nationality or social position, of occupation or of general human nature. He is far from us insofar as these similarities extend beyond him and us, and connect us only because they connect a great many people.

A trace of strangeness in this sense easily enters even the most intimate relationships. In the stage of first passion, erotic relations strongly reject any thought of generalization. A love such as this has never existed before; there is nothing to compare either with the person one loves or with our feelings for that person. An estrangement is wont to set in (whether as cause or effect is hard to decide) at the moment when this feeling of uniqueness disappears from the relationship. A skepticism regarding the intrinsic value of the relationship and its value for us adheres to the very thought that in this relation, after all, one is only fulfilling a general human destiny, that one has had an experience that has occurred a thousand times before, and that, if one had not accidentally met this precise person, someone else would have acquired the same meaning for us.

Something of this feeling is probably not absent in any relation, be it ever so close, because that which is common to two is perhaps never common *only* to them but belongs to a general conception which includes much else besides, many *possibilities* of similarities. No matter how few of these possibilities are realized and how often we may forget about them, here and there, nevertheless, they crowd in like shadows between men, like a mist eluding every designation, which must congeal into solid corporeality for it to be called jealousy. Perhaps this is in many cases a more general, at least more insurmountable, strangeness than that due to differences and obscurities. It is strangeness caused by the fact that similarity, harmony, and closeness are accompanied by the feeling that they are actually not the exclusive property of this particular relation, but stem from a more general one – a relation that potentially includes us and an indeterminate number of others, and therefore prevents that relation which alone was experienced from having an inner and exclusive necessity.

On the other hand, there is a sort of "strangeness" in which this very connection on the basis of a general quality embracing the parties is precluded. The relation of the Greeks to the barbarians is a typical example; so are all the cases in which the general characteristics one takes as peculiarly and merely human are disallowed to the other. But here the expression "the stranger" no longer has any positive meaning. The relation with him is a non-relation; he is not what we have been discussing here: the stranger as a member of the group itself.

As such, the stranger is near and far at the *same time*, as in any relationship based on merely universal human similarities. Between these two factors of nearness and distance, however, a peculiar tension arises, since the consciousness of having only the absolutely general in common has exactly the effect of putting a special emphasis on that which is not common. For a stranger to the country, the city, the race, and so on, what is stressed is again nothing individual, but alien origin, a quality which he has, or could have, in common with many other strangers. For this reason strangers are not really perceived as individuals, but as strangers of a certain type. Their remoteness is no less general than their nearness.

This form appears, for example, in so special a case as the tax levied on Jews in Frankfurt and elsewhere during the Middle Ages. Whereas the tax paid by Christian citizens varied according to their wealth at any given time, for every single Jew the tax was fixed once and for all. This amount was fixed because the Jew had his social position as a *Jew*, not as the bearer of certain objective contents. With respect to taxes every other citizen was regarded as possessor of a certain amount of wealth, and his tax could follow the fluctuations of his fortunes. But the Jew as taxpayer was first of all a Jew, and thus his fiscal position contained an invariable element. This appears most forcefully, of course, once the differing circumstances of individual Jews are no longer considered, limited though this consideration is by fixed assessments, and all strangers pay exactly the same head tax.

Despite his being inorganically appended to it, the stranger is still an organic member of the group. Its unified life includes the specific conditioning of this element. Only we do not know how to designate the characteristic unity of this position otherwise than by saying that it is put together of certain amounts of nearness and of remoteness. Although both these qualities are found to some extent in all relationships, a special proportion and reciprocal tension between them produce the specific form of the relation to the "stranger."

"Traveling Cultures"

from: *Routes: Travel and Translation in the late Twentieth Century* (1997)

James Clifford

Editors' introduction

If the majority of early work by cultural geographers emphasized rural, pre-industrial, folk-cultural landscapes, they were certainly not alone in the broader academic context. Anthropologists, for example, had also tended to study in rural, pre-industrial, even exotic places throughout what is now identified as the "Third World" or the "developing world." Because of this focus, ideas about the concept of culture were in many ways linked to empirical studies of very small-scale communities – villages, for the most part – described in vivid ethnographic accounts that often made such sites appear to be worlds unto themselves. Traditional ethnographies tended to describe communities as if they had little or no connection to the outside world. This occurred in part because the outside world was often regarded as a modern, industrial, and urban world that was threatening the traditional world of small-scale communities.

Perhaps the most iconic figure associated with ethnographies of isolated traditional communities is Bronislaw Malinowski, whose description of the Trobriand Islanders in *Argonauts of the Western Pacific* (1922) set the standard for what is now thought of as traditional ethnography. A photograph near the beginning of *Argonauts* shows 'the Ethnographer's tent' among the Trobriand dwellings. Malinowski's research, at the time, was a radical departure from typical research styles in the late nineteenth and early twentieth centuries. He rejected the superficial way in which anthropologists studied culture groups at the time ("calling up 'informants' to talk culture in an encampment or on a verandah," in the words of James Clifford), and instead argued that one must live "with the natives" for an expended period of time, learn their language, and participate in their daily lives. This intensive focus, however, had the effect, particularly in the ethnographic account later written about the fieldwork, of viewing the community as if it existed in a bubble, cut off from the outside world.

The image in *Argonauts* of Malinowski's tent serves as a jumping-off point for James Clifford to interrogate the inadequacies of this style of ethnographic research. The tent, after all, is a dwelling built for travel, and Malinowski was himself a traveler, a "stranger" in Simmel's sense of the term (see p. 311). In "Traveling cultures" Clifford thus begins by arguing that traditional ethnography has hidden the travel that occurs in the construction of accounts like Malinowski's *Argonauts*. "I've been arguing," Clifford writes, "that ethnography (in the normative practices of twentieth-century anthropology) has privileged relations of dwelling over relations of travel." This has meant, for example, that the "field" of fieldwork has been a place traveled to rather than a space of travel itself. In fact, Clifford argues, not only is ethnography really a practice of travel, but ethnographers typically depend on informants who are themselves "travelers," people who have already traveled beyond their home communities and are therefore prepared to translate their cultural practices in ways that outsiders can understand.

After suggesting that we think about hotels as an alternative kind of space in which culture also happens, Clifford argues for a broad rethinking of the spaces of culture and a consideration of what such rethinking does to our understanding of culture itself. (There are distinct parallels here with the arguments advanced by Gupta and Ferguson, p. 60.) This means, he suggests, thinking comparatively "about the distinct routes/roots of tribes, barrios, favelas, immigrant neighborhoods" in which ideas of "home" and "homeland" undergo constant debate, revision, and reinvention. It means looking at new sites of research such as "borders," and "circuits" rather than well defined or enclosed communities. Ultimately, then, Clifford calls for the uprooting of culture from Ratzel's tree metaphor (see p. 83).

Trained as an historian, James Clifford has been a Professor of History of Consciousness at the University of California, Santa Cruz, since 1978. With books such as *Writing Culture: the Poetics and Politics of Ethnography* (1986), which he co-edited with George Marcus, *The Predicament of Culture: Twentieth Century Ethnography, Literature, and Art* (1988), and *Routes: Travel and Translation in the Late Twentieth Century* (1997), he has been an influential critic of ethnography and culture. His work has had a broad impact on cultural studies beyond anthropology, and his arguments about rethinking the spaces in which we conceptualize culture have direct bearing on cultural geography. In particular, his work has foreshadowed the development in cultural geography and sociology of mobility studies (see Cresswell, p. 325). Clifford's call for the "opening" of the closed places of traditional ethnography echoes Doreen Massey's call for a "progressive sense of place" (see p. 257). His focus on travel, however, has also been criticized by geographers for overlooking the continuing power of place-based attachments (see, again, Cresswell, p. 325).

Remarks at a conference entitled "Cultural Studies, Now and in the Future," Champaign–Urbana, Illinois, April 6, 1990

To begin, a quotation from C.L.R. James in *Beyond a Boundary:* "Time would pass, old empires would fall and new ones take their place. The relations of classes had to change before I discovered that it's not quality of goods and utility that matter, but movement, not where you are or what you have, but where you come from, where you are going and the rate at which you are getting there."

Or begin again with hotels. Joseph Conrad, in the first pages of *Victory:* "The age in which we are encamped like bewildered travelers in a garish, unrestful hotel." In *Tristes Tropiques*, Lévi-Strauss evokes an out-of-scale concrete cube sitting in the midst of the new Brazilian city of Goiania in 1937. It's his symbol of civilization's barbarity, "a place of transit, not of residence." The hotel as station, airport terminal, hospital: a place you pass through, where the encounters are fleeting, arbitrary.

A more recent avatar: the hotel as figure of the postmodern in the new Los Angeles "downtown" – John Portman's Bonaventure Hotel, evoked by

Fredric Jameson in an influential essay "Postmodernism, or the Cultural Logic of Late Capitalism." The Bonaventure's glass cliffs refuse to interact, reflecting back their surroundings; there's no opening, no main entrance. Inside, a confusing maze of levels frustrates continuity, hinders the narrative stroll of a modernist *flâneur*.

Or begin with June Jordan's "Report from the Bahamas" – her stay in something called the Sheraton British Colonial Hotel. A black woman from the United States on vacation . . . confronting her privilege and wealth, uncomfortable encounters with people who make the beds and serve food in the hotel . . . reflections on conditions for human connection, alliances cutting across class, race, gender, and national locations.

Begin again with a London boardinghouse. The setting for V. S. Naipaul's *Mimic Men* – a different place of inauthenticity, exile, transience, rootlessness.

Or the Parisian hotels, homes away from home for the Surrealists, launching points for strange and wonderful urban voyages: *Nadia, Paysan de Paris*. Places of collection, juxtaposition, passionate encounter – "l'Hôtel des Grands Hommes."

Begin again with the hotel stationery and restaurant menus lining (with star charts) Joseph Cornell's magical boxes. Untitled: Hotel du Midi, Hotel du Sud, Hotel de I'Etoile, English Hotel, Grand Hotel de I'Univers. Enclosed beauty of chance encounters – a feather, ball bearings, Lauren Bacall. Hotel/*autel*, reminiscent of, but *not the same as* – no equal sign – marvelous-real altars improvised from collected objects in Latin American popular religions, or the home "altars," constructed by contemporary Chicano artists. A local/global fault line opening in Cornell's basement, filled with souvenirs of Paris, the place he never visited. Paris, the Universe, basement of an ordinary house in Queens, New York, 3708 Utopia Parkway.

This, as we often say, is "work in progress," work *entering* a very large domain of comparative cultural studies: diverse, interconnected histories of travel and displacement in the late twentieth century. This entry is marked, empowered and constrained, by previous work – my own, among others. And so I'll be working, today *out of* my historical research on ethnographic practice in its twentieth-century exoticist, anthropological forms. But the work I'm *going toward* does not so much build on my previous work as locate and displace it.

Perhaps I could start with a travel conjuncture that has, to my thinking at least, come to occupy a paradigmatic place. Call it the "Squanto effect." Squanto was the Indian who greeted the pilgrims in 1620 in Plymouth, Massachusetts, who helped them through a hard winter, and who spoke good English. To imagine the lull effect of this meeting, you have to remember what the "New World" was like in 1620: you could smell the pines fifty miles out to sea. Think of coming into a new place like that and having the uncanny experience of running into a Patuxet just back from Europe.

A disconcertingly hybrid "native" met at the ends of the earth – strangely familiar, and different precisely in that unprocessed familiarity. The trope is increasingly common in travel writing: it virtually organizes "postmodern" reports like Pico Iyer's *Video Night in Kathmandu*. And it reminds me of my own historical research into specifically anthropological encounters, in which I'm always running up against a problematic figure, the "informant." A great many of these interlocutors, complex individuals routinely made to speak for "cultural" knowledge, turn out to have their own "ethno-graphic" proclivities and interesting histories of travel. Insider-outsiders, good translators and explicators, they've been around. The people studied by anthropologists have seldom been homebodies. Some of them, at least, have been travelers: workers, pilgrims, explorers, religious converts, or other traditional "long-distance specialists." In the history of twentieth-century anthropology "informants" first appear as natives; they emerge as travelers. In fact, as I will suggest, they are specific mixtures of the two roles.

[. . .]

Localizations of the anthropologist's objects of study in terms of a "field" tend to marginalize or erase several blurred boundary areas, historical realities that slip out of the ethnographic frame. Here is a partial list. (1) The means of transport is largely erased – the boat, the land rover, the mission airplane. These technologies suggest systematic prior and ongoing contacts and commerce with exterior places and forces which are not part of the field/object. The discourse of ethnography ("being there") is separated from that of travel ("getting there"). (2) The capital city, the national context, is erased. This is what Georges Condominas has called the *préterrain,* all those places you have to go through and be in relation with just to get to your village or to that place of work you will call your field. (3) Also erased: the university home of the researcher. Especially now that one can travel more easily to even the most remote sites and now that all sorts of places in the "First World" can be fields (churches, labs, offices, schools, shopping malls), movement in and out of the field by both natives and anthropologists may be very frequent. (4) The sites and relations of *translation* are minimized. When the field is a dwelling, a home away from home where one speaks the language and has a kind of vernacular competence, the cosmopolitan intermediaries – and complex, often political, negotiations involved – tend to disappear. We are left with participant-observation, a kind of hermeneutic freedom to circle inside and outside social situations.

Generally speaking, what's hidden is the wider global world of intercultural import–export in which the ethnographic encounter is always already enmeshed. But, as we shall see, things are changing. Moreover, in various critiques of anthropology – which are responses in part to anticolonial upheavals – we see the emergence of the informant as a complex, historical subject, neither a cultural *type* nor a unique *individual*, My own work, to take only one among many examples, has questioned the oral-to-literate narrative hidden in the very word "informant." The native speaks; the anthropologist writes. The writing/inscribing practices of indigenous collaborators are erased. My own attempt to multiply the hands and discourses involved in "writing culture" aims not to assert a naive democracy of plural authorship, but to loosen at least somewhat the monological control of the executive writer/anthropologist and to open for discussion ethnography's hierarchy and negotiation of discourses in power-charged, unequal situations.

If thinking of the so-called informant as writer/inscriber shakes things up a bit, so does thinking of her or him as *traveler*. Arjun Appadurai challenges anthropological strategies for localizing non-Western people as "natives." He writes of their "confinement," even "imprisonment," through a process of representational essentializing that he calls "metonymic freezing," a process in which one part or aspect of peoples' lives come to epitomize them as a whole, constituting their theoretical niche in an anthropological taxonomy. India equals hierarchy, Melanesia equals exchange, and so forth. "Natives, people confined to and by the places to which they belong, groups unsullied by contact with a larger world, have probably never existed."

In much traditional ethnography, the ethnographer has localized what is actually a regional/national/global nexus, relegating to the margins the external relations and displacements of a "culture." This practice is now increasingly questioned. The title of Greg Dening's superb ethnographic history of the Marquesas is indicative: *Islands and Beaches* (1980). Beaches, sites of travel interaction, are half the story. Eric Wolf's *Europe and the People without History* (1982), though it may tip the local/global cultural dialectic a little too strongly toward "external" (global) determinations, is a dramatic and influential step away from an ethnographic focus on separate, integral cultures. "Rather than thinking of social alignments as self-determining," Wolf writes, "we need – from the start of our inquiries – to visualize them in their multiple external connections." Or, in another current anthropological

vein, consider a sentence from the opening of James Boon's intricate work of ethnological "crisscrossing," *Affinities and Extremes* (1990): "What has come to be called Balinese culture is a multiply authored invention, a historical formation, an enactment, a political construct, a shifting paradox, an ongoing translation, an emblem, a trademark, a nonconsensual negotiation of contrastive identity and more." Anthropological "culture" is not what it used to be. And once the representational challenge is seen to be the portrayal and understanding of local/global historical encounters, co-productions, dominations, and resistances, one needs to focus on hybrid, cosmopolitan experiences as much as on rooted, native ones. In my current problematic, the goal is not to *replace* the cultural figure "native" with the intercultural figure "traveler." Rather, the task is to focus on concrete mediations of the two, in specific cases of historical tension and relationship. In varying degrees, both are constitutive of what will count as cultural experience. I am recommending not that we make the margin a new center ("we" are all travelers) but that specific dynamics of dwelling/traveling be understood comparatively.

In tipping the balance toward traveling, as I am doing here, the "chronotope" of culture (a setting or scene organizing time and space in representable whole form) comes to resemble as much a site of travel encounters as of residence; it is less like a tent in a village or a controlled laboratory or a site of initiation and inhabitation, and more like a hotel lobby, urban café, ship, or bus. If we rethink culture and its science, anthropology, in terms of travel, then the organic, naturalizing bias of the term culture – seen as a rooted body that grows, lives, dies, and so on – is questioned. Constructed and disputed *historicities,* sites of displacement, interference, and interaction, come more sharply into view.

To press the point: Why not focus on any culture's farthest range of travel while *also* looking at its centers, its villages, its intensive fieldsites? How do groups negotiate themselves in external relationships, and how is a culture also a site of travel for others? How are spaces traversed from outside? To what extent is one group's core another's periphery? If we looked at the matter in this way, there would be no question of relegating to the margins a long list of actors: missionaries, converts, literate or educated informants, people of mixed blood, translators, government officers, police, merchants, explorers, prospectors, tourists, travelers, ethnographers, pilgrims, servants, entertainers, migrant laborers, recent immigrants. New representational strategies are needed, and are, under pressure, emerging. Let me evoke quickly several examples – notes for ways of looking at culture (along with tradition and identity) in terms of travel relations.

Ex-centric natives. The most extreme case I know of traveling "indigenous" culture-makers is a story I learned about through Bob Brosman, a musician and nonacademic historian of music, who for some years has been bringing traditional Hawaiian music into the continental United States. Brosman became very involved with the Moe (pronounced "Moay") family, a group of veteran performers who play Hawaiian guitar, sing, and dance. Their work represents the most authentic version of early twentieth-century Hawaiian slide guitar and vocal styles. But to approach "traditional" Hawaiian music through the Moes brings some unexpected results, because their experience has been one of almost uninterrupted travel. For various reasons, the Moes spent something like fifty-six years on the road, almost never going back to Hawaii. They played Hawaiian music in "exoticist" shows all over the Far East, South Asia, the Middle East, North Africa, eastern and western Europe, and the United States. And they performed, too, the gamut of hotel-circuit pop music. Now in their eighties, the Moes have recently returned to Hawaii, where, encouraged by revivalists like Brosman, they are making "authentic" music from the teens and twenties.

Bob Brosman is working on a film about the Moes which promises to be quite remarkable, in part because Tal Moe made his own home movies everywhere he went. Thus, the film can present a traveling Hawaiian view of the world, while posing the question of how the Moe family maintained a sense of identity in Calcutta, Istanbul, Alexandria, Bucharest, Berlin, Paris, Hong Kong. How did they compartmentalize their Hawaiianness in constant interaction with different cultures, musics, and dance traditions – influences they worked into their act, as needed? How, for fifty-six years in transient, hybrid environments, did they preserve and invent a sense of Hawaiian "home"? And how currently, is their music being recycled in the continuing invention of Hawaiian authenticity? This story of dwelling-in-travel is an extreme case, no doubt. But the Moes' experience is strangely resonant. (By the way, I also learned from Brosman's research that the National Steel Guitar, an instrument popular across the United States in the twenties and thirties and often called the "Hawaiian Guitar," was actually invented by a Czech immigrant living in California.)

Several more glimpses of an emergent culture-as-travel-relations ethnography *Joe Leahy's Neighbors,* a film by Bob Connolly and Robin Anderson, is a good example. (Its better-known predecessor, *First Contact*, is set in early twentieth-century New Guinea.) Joe Leahy, a mixed-blood colonial product, is a successful entrepreneur – kids in Australian schools, satellite dish behind his house in the New Guinea highlands. Connolly and Anderson include Leahy's own travels to Port Moresby and to Australia, while focusing on his ambiguous relations with the highland locals, his relatives. The entrepreneur seems to be exploiting his "neighbors," who resent his wealth. Sometimes he appears as an uncontrolled individualist impervious to their demands; on other occasions he distributes gifts, acting as a "big man" within a traditional economy. Joe Leahy seems to move in and out of a recognizably Melanesian culture. . . . Here, not only is the "native" a traveler in the world system, but the focus is on an atypical character, a person out of place but not entirely – a person *in history*. Joe Leahy is the sort of figure who turns up in travel books, though not in traditional ethnographies. Yet he is not simply an eccentric or acculturated individual. Watching Connolly and Anderson's film, we remain uncertain whether Joe Leahy is a Melanesian capitalist or a capitalist Melanesian – a new kind of big man, still bound in complex ways to his jealous, more traditional neighbors. He is and is not of the local culture.

[. . .]

Traveling cultures. One could cite many more examples, opening up an intricate comparative field. So far, I have been talking about the ways people leave home and return, enacting differently centered worlds, interconnected cosmopolitanisms. To this I should add: sites traversed – by tourists, by oil pipelines, by Western commodities, by radio and television signals. For example, Hugh

Brody's ethnography *Maps and Dreams* (1982) focuses on conflicting spatial practices – ways of occupying, moving through, using, mapping – by Athapascan hunters and the oil companies that are driving pipelines across their lands. But here a certain normative concept and history built into the word "travel" begins to weigh heavily. (Can I, without serious hesitations, translate Athapascan hunting as travel? With what violence and what loss of specificity?)

The anthropologist Christina Turner has pressed me on this point. Squanto as emerging norm? Ethnographic informants as travelers? But informants are not all travelers, and they're not natives either. Many people choose to limit their mobility, and even more are kept "in their place" by repressive forces. Turner did ethnographic work among female Japanese factory workers, women who have not "traveled," by any standard definition. They do watch TV; they do have a local/global sense; they do contradict the anthropologist's typifications; and they don't simply enact a culture. But it's a mistake, she told me, to insist on literal "travel." This begs too many questions and overly restricts the important issue of how subjects are culturally "located." It would be better to stress different modalities of inside–outside connection, recalling that the travel, or displacement, can involve forces that pass powerfully through – television, radio, tourists, commodities, armies.

Turner's point leads me to my last ethnographic example, Smadar Lavie's *The Poetics of Military Occupation* (1990). Lavie's ethnography of Bedouins is set in the southern Sinai, a land long traversed by all sorts of people, most recently by an Israeli occupation immediately followed by Egyptian occupation. The ethnography shows Bedouins in their tents telling stories, joking, making fun of tourists, complaining about military rule, praying, and doing all sorts of "traditional" things . . . but with the radio on, the BBC World Service (Arabic version). In Lavie's ethnography, you hear the crackle of that radio.

"Shgetef, could you pour some tea?" the Galid nonchalantly requests the local Fool. Shgetef enters the mag'ad and for the umpteenth time pours us yet more cups of hot sweet tea.

"So what did the news say?" the Galid asks the man with his ear glued to the transistor radio, but doesn't wait for an answer. "I'll tell you," he says with a half-bemused, half-serious expression. "No one will solve the problems between Russia and America. Only the Chinese will ever figure a way out. And when the day comes that they conquer the Sinai, that will be the end of that."

It's a good pun – the Arabic for "Sinai" is *Sina*, for "Chinese" is *Sini* – and we laugh heartily. But Shgetef, perhaps betraying his deep fool's wisdom, stares at us with eyes wide open.

The Galid continues, "The Greeks were here and left behind the Monastery [Santa Katarina], the Turks were here and left behind the Castle [in Nuweb'at Tarabin], and the British drew up maps, and the Egyptians brought the Russian army (and a few oil wells), and the Israelis brought the Americans who made the mountains into movies, and tourists from France and Japan, and scuba divers from Sweden and Australia, and, trust Allah to save you from the devil, we Mzeina are nothing but pawns in the hands of them all. We are like pebbles and the droppings of the shiza."

Everyone but Shgetef again roars with laughter. The Coordinator points to me with his long index finger, saying in a commanding voice, "Write it all down, The One Who writes Us!" (*Di Illi Tuktubna – one* of my two Mzeini nicknames).

[. . .]

Begin again with that odd invocation of hotels. I wrote it in the course of returning to an earlier essay on Surrealism and the Paris of the 1920s and 1930s. I was struck by how many of the Surrealists lived in hotels, or hotel-like transient digs, and were moving in and out of Paris. I was beginning to see that the movement was not necessarily *centered* in Paris, or even in Europe. . . . It all depended on how (and where) one saw the historical *outcomes* of the modernist moment. . . . I began to imagine rewriting Paris of the twenties and thirties as travel encounters – including New World detours through the old – a place of departures, arrivals, transits. The great urban centers could be understood as specific, powerful sites of dwelling/traveling.

[. . .]

In my invocation of different hotels, the relevant sites of cultural encounter and imagination began to slip away from metropolitan centers such as Paris.

At the same time, levels of ambivalence appeared in the hotel chronotope. At first I saw my task as finding a frame for negative and positive visions of travel: travel, negatively viewed as transience, superficiality, tourism, exile, and rootlessness (Lévi-Strauss's invocation of Goiania's ugly structure, Naipaul's London boarding house); travel positively conceived as exploration, research, escape, transforming encounter (Breton's Hôtel des Grands Hommes, June Jordan's tourist epiphany). The exercise also pointed toward the broader agenda I've been getting at here: to rethink cultures as sites of dwelling *and* travel, to take travel knowledges seriously. Thus, the ambivalent setting of the hotel suggested itself as a supplement to the field (the tent and the village). It framed, at least, encounters between people to some degree away from home.

[. . .]

As recycled in this talk, then, the hotel epitomizes a specific *way into* complex histories of traveling cultures (and cultures of travel) in the late twentieth century As I've said, it has become seriously problematic, in several major ways involving class, gender, race, cultural/historical location and privilege. The hotel image suggests an older form of gentlemanly occidental travel, when home and abroad, city and country, East and West, metropole and antipodes, were more clearly fixed. Indeed, the marking of "travel" by gender, class, race, and culture is all too clear.

"Good travel" (heroic, educational, scientific, adventurous, ennobling) is something men (should) do. Women are impeded from serious travel. Some of them go to distant places, but largely as companions or as "exceptions" – figures like Mary Kingsley, Freya Stark, or Flora Tristan, women now rediscovered in volumes with titles like *The Blessings of a Good Thick Skirt*, or *Victorian Lady Travellers*. "Lady" travelers (bourgeois, white) are unusual, marked as special in the dominant discourses and practices. Although recent research is showing that they were more common than formerly recognized, women travelers were forced to conform, masquerade, or rebel discreetly within a set of normatively male definitions and experiences. One thinks of George Sand's famous account of dressing as a man in order to move freely in the city, to experience the gendered freedom of the *flâneur*. Or Lady Mary Montague's envy of the anonymous mobility of veiled women in Istanbul. And what forms of displacement, closely associated with women's lives, do not count as proper "travel"? Visiting? Pilgrimage? We need to know a great deal more about how women have traveled and currently travel, in different traditions and histories. This is a very large comparative topic that's only beginning to be opened up: for example, in the work of Sara Mills, Caren Kaplan, and Mary Louise Pratt. The discursive/imaginary topographies of Western travel are being revealed as systematically gendered: symbolic stagings of self and other that are powerfully institutionalized, from scientific research work to transnational tourism. Although there are certainly exceptions, particularly in the area of pilgrimage, a wide predominance of male experiences in the institutions and discourses of "travel" is clear – in the West and, to differing degrees, elsewhere.

But it is hard to generalize with much confidence, since the serious, *cross-cultural* study of travel is not well developed. What I'm proposing here are research questions, not conclusions. I might note, in passing, two good sources: *Ulysses' Sail* (1988), by Mary Helms, a broadly comparative study of the cultural uses of geographic distance and the power/knowledge gained in travel (a study focused on male experiences); and *Muslim Travelers* (1990), edited by Dale Eickelman and James Piscatori, an interdisciplinary collection designed to bring out the complexity and diversity of religious/economic spatial practices.

Another problem with the hotel image: its nostalgic inclination. For in those *parts* of contemporary society that we can legitimately call postmodern (I do not think, *pace* Jameson, that postmodernism is yet a cultural dominant, even in the "First World"), the *motel* would surely offer a better chronotope. The motel has no real lobby, and it's tied into a highway network – a relay or node rather than a site of encounter between coherent cultural subjects. Meaghan Morris has used the motel chronotope effectively to organize her essay "At Henry Parkes Motel." I can't do justice to its suggestive discussions of nationality gender, spaces, and possible narratives. I cite it here as a displacement of the hotel chronotope of travel, for, as Morris says, "Motels, unlike hotels, demolish sense regimes of place, locale, and history. They memorialize only movement, speed, and perpetual circulation."

Other major ways in which the hotel chronotope – and with it the whole travel metaphor – becomes problematic have to do with class, race, and sociocultural "location." What about all the travel that largely avoids the hotel, or motel, circuits? The travel encounters of someone moving from rural Guatemala or Mexico across the United States border are of a quite different order; and a West African can get to a Paris *banlieu* without ever staying in a hotel. What are the settings that could realistically configure the cultural relations of these "travelers"? As I abandon the bourgeois hotel setting for travel encounters, sites of intercultural knowledge, I struggle, never quite successfully, to free the related term "travel" from a history of European, literary, male, bourgeois, scientific, heroic, recreational meanings and practices.

Victorian bourgeois travelers, men and women, were usually accompanied by servants, many of whom were people of color. These individuals have never achieved the status of "travelers." Their experiences, the cross-cultural links they made, their different access to the societies visited – such encounters seldom find serious representation in the literature of travel. Racism certainly has a great deal to do with this. For in the dominant discourses of travel, a nonwhite person cannot figure as a heroic explorer, aesthetic interpreter, or scientific authority A good example is provided by the long struggle to bring Matthew Henson, the black American explorer who reached the North Pole with Robert Peary, equally into the story of this famous feat of discovery – as it was constructed by Peary, a host of historians, newspaper writers, statesmen, bureaucrats, and interested institutions such as *National Geographic* magazine. And this is still to say nothing of the Eskimo travelers who made the trip possible! A host of servants, helpers, companions, guides, and bearers, etc. have been excluded from the role of proper travelers because of their race and class, and because theirs seemed to be a dependent status in relation to the supposed independence of the individualist, bourgeois voyager. The independence was, in varying degrees, a myth. As Europeans moved through unfamiliar places, their relative comfort and safety were ensured by a well-developed infrastructure of guides, assistants, suppliers, translators, and carriers.

Does the labor of these people count as "travel"? Clearly a comparative cultural studies account would want to include them and their specific cosmopolitan viewpoints. But in order to do so, it would have to thoroughly transform travel as a discourse and genre. Obviously, many different kinds of people travel, acquiring complex knowledges, stories, political and intercultural understandings, without producing "travel writing." Some accounts of these experiences have found their way to publication in Western languages – for example, the nineteenth-century travel journals of the Rarotongan missionary Ta'unga, or the fourteenth-century records of Ibn Battouta. But they are tips of lost icebergs. . . .

[We risk] downplaying the extent to which the mobility is coerced, organized within regimes of dependent, highly disciplined labor. In a contemporary register, to think of cosmopolitan workers, and especially migrant labor, in metaphors of "travel" raises a complex set of problems. The political disciplines and economic pressures that control migrant-labor regimes pull very strongly against any overly sanguine view of the mobility of poor, usually nonwhite, people who *must* leave home in order to survive. The traveler, by definition, is someone who has the security and privilege to move about in relatively unconstrained ways. This, at any rate, is the travel myth. In fact, as studies like those of Mary Louise Pratt are showing, most bourgeois, scientific, commercial, aesthetic, travelers moved within highly determined circuits. But even if these bourgeois travelers can be "located" on specific itineraries dictated by political, economic, and intercultural global relations (often colonial, postcolonial, or neocolonial in nature), such constraints do not offer any simple equivalence with other immigrant and migrant laborers. Alexander von Humboldt obviously did not arrive on the Orinoco coast for the same reasons as an Asian indentured laborer.

. . . [T]ravelers move about under strong cultural, political, and economic compulsions and . . . certain travelers are materially privileged, others oppressed. These specific circumstances are crucial determinations of the travel at issue – movements in specific colonial, neocolonial, and postcolonial circuits, different diasporas, borderlands, exiles, detours, and returns. Travel, in this view, denotes a range of material, spatial practices that produce knowledges, stories, traditions, comportments, musics, books, diaries, and other cultural expressions.

Even the harshest conditions of travel, the most exploitative regimes, do not entirely quell resistance or the emergence of diasporic and migrant cultures. The history of transatlantic enslavement, to mention only a particularly violent example, an experience including deportation, uprooting, marronnage, transplantation, and revival, has resulted in a range of interconnected black cultures: African American, Afro-Caribbean, British, and South American.

We need a better comparative awareness of these and a growing number of other "diaspora cultures." As Stuart Hall has argued in a provocative series of articles, diasporic conjunctures invite a reconception – both theoretical and political – of familiar notions of ethnicity and identity; unresolved historical dialogues between continuity and disruption, essence and positionality, homogeneity and differences (cross-cutting "us" and "them") characterize diasporic articulations. Such cultures of displacement and transplantation are inseparable from specific, often violent, histories of economic, political, and cultural interaction – histories that generate what might be called *discrepant cosmopolitanisms*. In this emphasis we avoid, at least, the excessive localism of particularist cultural relativism, as well as the overly global vision of a capitalist or technocratic monoculture. And in this perspective the notion that certain classes of people are cosmopolitan (travelers) while the rest are local (natives) appears as the ideology of one (very powerful) traveling culture. My point, again, is not simply to invert the strategies of cultural localization, the making of "natives," which I criticized at the outset. I'm not saying there are no locales or homes, that everyone is – or should be

– traveling, or cosmopolitan, or deterritorialized. This is not nomadology. Rather, what is at stake is a comparative cultural studies approach to specific histories, tactics, everyday practices of dwelling and traveling: traveling-in-dwelling, dwelling-in-traveling.

[. . .]

I hang on to "travel" as a term of cultural comparison precisely because of its historical taintedness, its associations with gendered, racial bodies, class privilege, specific means of conveyance, beaten paths, agents, frontiers, documents, and the like. I prefer it to more apparently neutral, and "theoretical," terms, such as "displacement," which can make the drawing of equivalences across different historical experiences too easy. (The postcolonial/postmodern equation, for example.) And I prefer it to terms such as "nomadism," often generalized without apparent resistance from non-Western experiences. (Nomadology: a form of postmodern primitivism?) "Pilgrimage" seems to me a more interesting comparative term to work with. It includes a broad range of Western and non-Western experiences and is less class- and gender-biased than "travel." Moreover, it has a nice way of subverting the constitutive modern opposition between traveler and tourist. But its "sacred" meanings tend to predominate – even though people go on pilgrimages for secular as well as religious reasons. And in the end, for whatever reasons of cultural bias, I find it harder to make "pilgrimage" stretch to include "travel" than to do the reverse. (The same is true of other terms such as "migration.") There are, in any event, no neutral, uncontaminated terms or concepts. A comparative cultural studies needs to work, self-critically, with compromised, historically encumbered tools. . . .

"The Production of Mobilities"

from *New Formations* 43 (2001): 11–25

Tim Cresswell

Editors' introduction

The theme of an inherent antagonism between nomadic pastoral societies and settled-agricultural ones has a history in geography dating all the way back to Ratzel, who wrote in *Völkerkunde* (1885–1888), "... Just as the soil of the Old World is marked by the great line of a band of plateau, extending from the Atlantic to the Pacific, bordered on either side by fertile mountains and lowlands, so there runs through all its history the struggle between nomad and settled, between herdsmen and tillers of the soil..." (see p. 83). And, like Ratzel, many geographers – and probably most social scientists too – have built their understanding of the world around an assumption that being fixed in place is the norm, and mobility represents a kind of deviation from that norm. For Ratzel, fixity in place was a necessary condition of advanced culture and civilization.

Another way of putting this preference for dwelling is to say that the power to define which societies were advanced and which were not (for example, the power of imperialism and colonialism) is also a power to define dwelling in place as a social norm, and mobility as an aberration, or a dysfunction. It is often in the interest of territorial states, for example, to minimize the physical mobility of its citizens (this was certainly the case in both the Soviet Union and China under Mao Zedong). Likewise mobility was theorized by many social scientists as the result of some kind of social "problem" (such as income inequality between cities and countryside); mobility was something that happened to "fix" things when the normal condition of dwelling did not function properly. And, continuing this theme, the arrival of immigrants, Simmel's "strangers" (p. 311), is typically viewed as a problem requiring various forms of social intervention.

It is this association of dwelling with power that has led many theorists to conceive of mobility as a subversive phenomenon. "Nomads" disturbed the norm, and offered a new model for thinking about resistance to dominant social formations (empires, colonies, nations, patriarchy). This impulse, for instance, can be seen in James Clifford's "Traveling cultures" (see p. 316). Tim Cresswell, however, seeks to formulate an idea of mobility that does not romanticize movement as inherently subversive, but rather views it as "socially produced." Mobility, for him, is movement made meaningful in any given social context.

Cresswell argues that, in geography, the normative valuing of dwelling over mobility took on the mantle of an objective, abstract science during the 1960s and 1970s 'spatial science' turn in the discipline. Movement, he notes, was viewed merely as the need to get from place A to place B, the need to change locations. Cresswell even argues that spatial science was *suspicious* of movement. Optimal conditions, in which human activities were at their most efficient, would reduce the need for movement, and therefore movement itself marked a kind of social dysfunction. For Cresswell, this making movement "dysfunctional" also made it *meaningful* within a specific social context: in this case, the context of modernization theory and positivist science.

Cresswell's idea of mobility, then, seeks to explore the many ways movement has been socially produced and made meaningful within specific historical and geographical contexts. He argues that this approach avoids an necessary linking of mobility with resistance to power, because he points out that mobility can just as easily be the privilege of the powerful. Indeed, this is precisely the sort of criticism that has been leveled against Clifford's "Traveling cultures": that rethinking culture through travel risks losing the groundedness of location, where social contexts laden with power relations and structured in difference still condition behavior. For Cresswell, "socially produced" mobility keeps a firm eye on such power relations and structures while not necessarily privileging resistance to them.

Tim Cresswell studied with Yi-Fu Tuan at the University of Wisconsin, and is now Professor of Human Geography at Royal Holloway, University of London. Along with numerous essays, articles, and edited collections, he is the author of *In Place/Out of Place: Geography, Ideology and Transgression* (1996), *The Tramp in America* (2001), and *On the Move: Mobility in the Modern Western World* (2006). He is also a member of the editorial board for the journal *Mobilities*. Cresswell's work on mobility stems from a long-term interest in geographies of exclusion and inclusion and how such geographies are given expression through ideas of place, space, and mobility. Geography, Cresswell argues, is a constitutive force in social life, not simply a reflection of other forces. Thus, the geographical imaginations – for example, thinking of dwelling in place as "normal" – have very real influences on people's lives (as seen, in the selection below, in the arrest of Fred Edwards for assisting in the migration of his brother-in-law to California in 1939).

Related work on similar themes of mobility include John Urry's *Sociology beyond Societies* (2000), Rosler's *In the Place of the Public* (1998), and Gottdiener's *Life in the Air* (2001).

In December 1939 Fred Edwards left his home in Marysville, California, and drove to Spur, Texas, with the intention of bringing back his wife's brother in law, Frank Duncan, a citizen of the United States and resident of Texas. They left Spur on New Year's Day in an old jalopy. Duncan who had been unemployed for some time had twenty dollars with him. They entered California on 3 January reaching Marysville on 5 January. By the time they arrived Duncan had spent his twenty dollars. He remained unemployed for ten days before getting relief from the Farm Security Administration. The movements of Edwards and Duncan were far from exceptional. Migration into California from Texas, Arkansas, Oklahoma and other states to the east had been the subject of varying degrees of moral panic since the late 1920s. Migrants known as Okies and Arkies had moved to California in order to get promised work in the new agribusiness centres of the California valleys following the dust storms of the Great Plains. By the time Duncan entered California, migrants were mostly looking for work in the defense industry. It was in the defense industry that Duncan was finally employed – in a chemical plant in Pittsburg, California. What made Duncan's trip noteworthy was that it led to his story, and that of his brother in law, being told in the Supreme Court. On 17 February 1940, Fred Edwards was convicted in the Justices Court of Marysville township, county of Yuba, of a violation of section 2615 of the Welfare and Institutions Code of the State of California. The section read: 'Every person, firm or corporation, or officer or agent thereof that brings or assists in bringing into the State any indigent person who is not a resident of the State, knowing him to be an indigent person, is guilty of a misdemeanor.' The case was taken to the Superior Court of the State of California on 26 June 1940 where the judgment was upheld. Edwards then petitioned for an appeal at the Supreme Court of the United States where Edwards was represented by Samuel Slaff of the American Civil Liberties Union. Oral arguments were made on 28 and 29 April 1941 and again on 21 October 1941. The court gave its judgment on 24 November. The question before the Court was whether section 2615 of the Welfare and Institutions Code of California violated the Federal Constitution: 'The Court was asked to answer the question whether, in a nation which protects the free movement across State lines of the products of its fields, factories, and mines, an employable citizen of that nation did not enjoy the same freedom of movement accorded to articles of commerce.'

The Supreme Court rejected California's statute with three different opinions that nicely illustrate the always-differentiated politics of mobility. Justice Byrnes argued that the statute was an unconstitutional burden on interstate commerce. Byrnes delivered the opinion of the court noting that similar statutes had been in effect in California since 1860. His judgment was based on Article 1, section 8 of the Constitution which delegates to Congress the authority to regulate interstate commerce. Byrnes believed that the transportation of persons across state borders constituted commerce and that the California statute constituted an unconstitutional barrier on interstate commerce By creating and enforcing a law that protected California from an alleged threat to health, morals and finance the state had effectively tried to remove itself from problems common to all states by placing restraints on movement. Justice Byrnes' legal judgment, therefore, was that Duncan was protected because his mobility was no different from any other mobility that might constitute commerce – indigents were no different from oranges, farm machinery or capital.

Mr Justice Douglas, concurring, wanted to widen the terms on which the statute was unconstitutional. 'I am of the opinion,' he wrote, 'that the right of persons to move freely from State to State occupies a more protected position in our constitutional system than does the movement of cattle, fruit, steel and coal across state lines.' The right to move, he argued, was protected not by interstate commerce law but by the privileges and immunities clause of the 14th Amendment to the Constitution. Douglas referred to a previous case (*Crandall* v. *Nevada* 1867) in which a Nevada tax on people leaving the state by common carrier was struck down because of the judgment that the right to move freely between states was a national citizenship issue. The right to move from place to place according to inclination was, in the view of Douglas, an attribute of personal liberty protected by the 14th amendment. Placing an impediment on personal mobility would therefore result in a dilution of the rights of national citizenship and an impairment of the principles of equality.

Although Justices Byrnes and Douglas came to the same conclusion, they used very different rationales to reach that conclusion. Byrnes defined the mobility of Duncan as no different for the purposes of law in this case than the mobility of any article of commerce. In this sense Duncan's mobility was the same as that of a piece of machinery. Douglas made his case on the broader grounds of national citizenship – a specifically human attribute that couldn't be shared by oranges and machinery.

The final concurring opinion came from Justice Jackson who further differentiated mobilities in his judgment. He not only found a different logic for defending Duncan's mobility but he explicitly denied the legitimacy of Byrnes' commerce defense. 'The migrations of a human being, of whom it is charged that he possesses nothing that can be sold and has no wherewithal to buy,' he argued, 'do not fit easily into my notions as to what is commerce. To hold that the measure of his rights is the commerce clause is likely to result eventually either in distorting the commercial law or in denaturing human rights.' Jackson's argument instead looked to the fact of Duncan's citizenship of the United States – a fact that made it impossible for states to abridge his immunities and privileges and which included mobility. Jackson did not stop there though. He argued, contra Douglas, that the right to mobility that constituted part of citizenship was in fact limited. He pointed out, for instance, that states were able to prevent movement of fugitives from justice and people likely to cause contagion. The crux of the issue, for Jackson, was whether or not there was something characteristic of being an indigent that could provide a legal basis for curtailing interstate mobility. His conclusion was that '"Indigence" in itself is neither a source of rights nor a basis for denying them. The mere state of being without funds is a neutral fact – constitutionally an irrelevance, like race, creed, or color.'

To summarise, in the case of *Edwards* v. *California* the Supreme Court justices made decisions based on differences and similarities between forms of mobility. In the course of three judgments the journey of Duncan and Edwards was compared, in legal terms, to that of oranges, a bus full of people leaving Nevada and unnamed fugitives from justice. While Justice Byrnes ruled against California based on perceived similarities between commerce of goods and movement of people, others concurred by stating that the mobility of people was uniquely protected by the 14th amendment and was not analogous to commerce. It was further argued that, unlike criminality or disease, indigence was not a human characteristic that could be used to prevent mobility. In each case mobility was being reconfigured in relation to other mobilities.

I recount this story for three reasons. First, it self-evidently concerns the issue of movement and mobility. It involves the displacement of people from A to B. Second it points towards the fact that human movement is made meaningful in social and cultural context. In this case in a court of law. Third it indicates the crucial nature of perceived and actual differences between forms of mobility. It illustrates the politics of mobility. I want to take this opportunity to reflect on all of these points with the help of a number of other stories.

My aim is to explore the possibilities for an account of the production of mobilities. I say possibilities because I do not believe that such a thing has been developed explicitly and thoroughly. By account I mean an interpretative framework for thinking about human movement – I want to sketch a way of thinking about mobilities that allows for the specificity of particular types of movement – a framework that has difference built into it but can account for that difference. My starting point is to paraphrase Henri Lefebvre's tautology [in *The Production of Space* (1991)] about the production of space and say that (social) mobility is a (social) product. What am I talking about when I talk about mobility?

STARTING POINTS

To think of mobility as produced indicates the social nature of movement. I will come to this later. Let us start, instead, with movement. Movement stripped of its social meaning is an abstraction. Movement describes the idea of an act of displacement that allows objects, people, ideas – things – to get between locations (usually given as point A and point B in abstract and positivist discussions of migration). Movement is the general fact of displacement before the type, strategies and

social implications of that movement are considered. This notion of desocialised movement can be illustrated by the work of geographers in the 1970s on migration. It is helpful to think of geography here, not as a subject or discipline, but as a cultural artifact that needs interpretation. Just as law, as a cultural system, was busy interpreting the mobility of Duncan and Edwards, so geography has been busy trying to understand, describe and interpret mobility.

Movement and spatial science

Geography textbooks of the 1970s are littered with references to movement, mobility and migration. Indeed the phenomenon known as spatial interaction was instrumental to the development of the theoretical approaches of spatial science. Spatial interaction was defined as all forms of movement between two or more places. Central to spatial interaction models were questions concerning reasons for movement between one location and another. For example, the textbook called *The Geography of Movement* [Lowe and Moryadas, 1975] explains that: 'Movement occurs to the extent that people have the ability to satisfy their desires with respect to goods, services information. or experience at some location other than their present one, and to the extent that these other locations are capable of satisfying such desires'. At the very core of these models of movement was not movement itself but the relative merits of locations between which movement of people, goods and information occurs. Another textbook suggests that:

> We can think of each migrant assigning one value to his present location and other values to places where he could be. He compares his present status with potential status elsewhere. Then he weights the different alternatives according to their distances and how risky he thinks each of them is. Finally he picks a strategy he thinks will be best for him.

The causes for movement lie in locations that might be left and others which may be moved to. Notions such as place utility and disutility, complementarity (specific match of needs and supply), and transferability (the ability of something to move) were given as causes for movement. Studies of residential mobility asked why people move, how they choose where to move to, and the constraints and limits on movement. When the act of moving itself was the focus it was only as a map of routes noting the density of movement along such routes. Some differentiation among types of movement was also attempted. Thus, the point to point movement of an individual person or thing was differentiated from the diffusion of an idea or general population which did not involve an original departure from point A to get to point B.

On the whole the aim of these 'scientific' considerations of movement was to exclude those very factors which differentiated mobilities and thus provide a general model of movement. In the words of Lowe and Moryadas:

> we have tried to provide an exposition of the underlying regularities and processes that may lead to an understanding of human movement . . . Movement in both Asia and North America can be adequately treated in the same framework. Our examples are taken mostly from the American context, but alternatives can be rather easily substituted. In any event, this point may be irrelevant since examples are meant to support generalizations: hence whether a trip for medical attention. for example. is to a witch doctor or to a medical complex is totally immaterial.

In a similar vein, Abler, Adams and Gould [*Spatial Organization: The Geographer's View of the World*, 1972] give some fascinating descriptions of what they call 'ideal movements'. One type of ideal movement is that from an area to a line. This type of movement describes diverse things such as water flowing from a roof to a gutter, commuters moving from a suburb to a highway and sheet erosion of soil into a ditch. This is explained by the principle of 'least net effort' which is said to regulate movement and accounts for observed patterns in nature and in social life. Interestingly, movement is said to be, by some within spatial science, 'dysfunctional', in so far as spatial structures are supposed to be organised in such a way as to minimise the need for movement. Thus the arrangement of rooms in houses and the hierarchy of streams in a drainage basement are both spatial alignments created to reduce the necessity for movement and thus the degree of dysfunctionality. This label of dysfunctionality points towards the wider suspicion of all things mobile. A suspicion that has been reversed and celebrated more recently in social and cultural theory.

Mobility as resistance

It is relatively easy to critique the abstractions of spatial science but there are other, more seductive, ways in which mobility can be made to appear curiously isomorphic. Many of the heroes and some of the heroines of modern critical thought have argued implicitly and explicitly that mobility has a privileged relationship to resistance, just as the production of spaces and boundaries is often associated with domination. Mobility seems to have a furtive and transgressive character to it, crossing boundaries, breaking definitions of the proper. Indeed a number of theorists, ranging from Bakhtin to de Certeau to Deleuze and Guatarri. take mobility as a central trope for anti-systemic movements of one kind or another. Bakhtin's development of the grotesque and carnivalesque [in *Rabelais and his World*, 1984] frequently posits the laughter and movement of the everyday profane body as an antidote to the monumental seriousness of fixity and stability. His descriptions of carnival are full of references to fluidity and movement set against rigidity and propriety. The formal finished world is static while everyday life is mobile. Michel de Certeau's discussions of the strategy and the tactic repeat this [*The Practice of Everyday Life*, 1984]. The strategy is the weapon of the strong and depends for its power on the fixing of boundaries and the definition of a proper place from which to operate. The tactic as the weapon of the weak is mobile and furtive, marked by words such as cunning and sabotage. Tactics work for the moment and are never fixed. The powerful make spaces but the weak use them. Deleuze and Guatarri [*Nomadology: The War Machine*, 1986] mobilise the figure of the nomad as an emblem of smooth and mobile space unmarked by the striations of arboreal state organisation. More empirically James Scott [*Seeing like a State*, 1998] in his discussions of power and resistance has often remarked that the state and formal organised power inscribe themselves by way of order and formality, which are more often than not used to control and prevent mobile people. Finally feminist theory has also had its fair share of nomads. For example Rosi Braidotti in her book, *Nomadic Subjects* (1994), has celebrated and, against her best intentions no doubt, romanticised the mobility of the nomad as a liberating identity which flits between cultures, places and languages in a constant state of becoming.

Three cheers for mobility then. Mobility for its own sake or as a way of life is often seen as a threat to normality. The fact of property relations, of the split between home, work and leisure and of place-based law and regulation makes it somewhat inevitable that certain forms of mobility are going to upset the people who have invested in those fixities. Tramps in America in the 1880s, Gypsies in Europe and Jews across the world have been treated as objects of loathing due to their actual or alleged mobility. They are seen as mobile and rootless. Nomads. itinerants, runaway slaves, travelling salespeople and new age travellers have all, at one time or another, upset residents in their rooted existence. The agents of the state have constantly sought to make society legible by sedentarising the rootless or regulating and channelling mobility into acceptable pathways. According to Liisa Malkki [see p. 275] there is a tendency in the modern world to locate people and identities in particular spaces and within particular boundaries. The corollary of this is to think of mobile people in wholly negative ways. Connected to this, she continues, are ways of thinking which are also rooted and bounded. She refers to this as a sedentary metaphysics, It is just such a metaphysics that Braidotti, Deleuze and others are seeking to relegate to history.

[. . .]

Nomad thought in a mobile world

Edward Said has commented [in *Culture and Imperialism*, 1994] that mobility and migration are the markers of our time. He carefully links the characteristic experience of the exile – a forced migrant – with the mobile thoughts of postmodern theorists. On the one hand he argues there are the modernist and reactionary forces of 'confinement' – education, nationalism, hospitals, asylums, the military. On the other are the transgressive and mobile forces of the refugees, boat people, itinerants, the homeless, guest workers and exiles:

> For surely it is one of the unhappiest characteristics of the age to have produced more refugees, migrants, displaced persons, and exiles than ever before in history, most of them as an accompaniment to and, ironically enough, as afterthoughts of great post-colonial and imperial conflicts. As the struggle for independence produced new states and new boundaries, it also produced homeless wanderers, nomads, vagrants, unassimilated to the emerging structures of institutional power, rejected by the established order for their intransigence and obdurate rebelliousness.

In addition to mass migrations of mobile people (either forced or voluntary) the postmodern world includes the experiences of communication and transportation on a scale and speed hitherto unknown – the phenomenon David Harvey calls 'time–space compression' [in *The Condition of Postmodernity*, 1989]. This increased mobility and interconnectedness results in characteristic landscapes of mobility – landscapes which include such sites as motorway service stops, bus stations, motorways and airports. To a geographer influenced by phenomenology such as Edward Relph these sites would have been condemned as 'placeless' – lacking in roots and authenticity. A completely 'other-directed' place full of people from elsewhere, going elsewhere . . .

[. . .]

But the celebration of the mobile does not help us to diagnose difference. It shares with spatial science and the idea of abstract movement an inability to examine human mobility ranging from the mundane trip to the supermarket to the most celebrated of grand voyages. It replaces a longstanding distaste for, and suspicion of, mobility with an overly general celebration and romanticisation. . . . Janet Wolff has made similar comments noting how discourses of mobility tend to ignore the gendering of motion. She insists that the actual practices of mobility have tended to exclude women and that this exclusion is carried over into theoretical travel: '. . . the problem with terms like "nomad", "maps" and "travel" is that they are not usually located and hence (and purposely) they suggest ungrounded and unbounded movement – since the whole point is to resist selves/viewers/subjects. But the consequent suggestion of free and equal mobility is itself a deception, since we don't all have the same access to the road'. It is for this reason that an analysis of the production of mobility becomes both useful and necessary.

THE PRODUCTION OF MOBILITIES

[. . .]

Think of the way more bounded notions such as space and location have been taken out of the realm of the given and the absolute and placed firmly in the meaningful and power-laden world of ideology. Space and place are not pre-existent givens but social productions. They are not *just* productions but constitutive moments in the production of the social, the cultural and political. Mobility, like social space and place, is produced. Mobility is to movement what place is to location. It is produced and given meaning within relations of power. There is, then, no mobility outside of power. Mobility, unlike movement, is contextualized. It is a word for produced movement. Seeing as it is produced and in common with produced space, it is not inherently implicated in any form of domination or resistance. . .

To think of mobility as produced, I insist, is to think of it as differentiated. Some mobilities are acts of freedom, transgression and resistance in the face of state power which seeks to limit movement, police boundaries and inscribe order in space. It would be a mistake however to think of mobilities as in any way essentially transgressive. Other mobilities are produced to support the state, to support patriarchy or to support the power of multi-national corporations in the globalised world of flexible capitalism. Manuel Castells has recently taken a rather different tack by suggesting that the 'spaces of flows' are where power often resides in a Network society.

[. . .]

The mobility of some can immobilise others

Legal geographer Nick Blomley provides a story [in *Law, Space and the Geographies of Power*, 1994] which illustrates the differential mobilities which are often involved in political struggle. In the 1984–1985 British miners' strike there was a clear distinction between the miners of Yorkshire, South Wales and Kenton the one hand and the miners of Nottinghamshire on the other. While the former were more or less firmly behind the national strike, the Nottinghamshire miners continued to work and eventually formed a breakaway non-striking union. This resulted in a stand-off between the two groups with the more militant miners of the National Union of Miners (NUM) sending flying pickets to Nottinghamshire in order to stop work at the Nottinghamshire pits. Flying pickets were made illegal by the Conservative government and the police from counties across England were mobilized in order to stop the movement of flying pickets. Nottinghamshire was more or less completely surrounded by police road blocks for months on end and anyone who looked like they might be a radical miner was stopped and turned back. Blomley explores the language that was used by both sides of the legal debate that ensued. The government rested much of its case on what it called the right to work. As Blomley makes clear, what they meant was not the right to a job but the right to go to a place of work. As the Attorney-General put it:

> . . . the fundamental proposition of our law is that each of us has the right to go about his daily work free from interference from anyone else. Each one of us is free, as an individual, to come and go as he pleases to his place of work . . . People have no right to link arms or otherwise prevent access to the place that they are picketing.

A clear link is made between work, mobility and rights. The action against the NUM pickets is thus defendable as a way of protecting two rights – to work and to move. Part of the power of this line of discourse comes from the fact that mobility as a right is deeply entrenched in western liberal thought. The idea of liberty and the idea of mobility have long been intertwined. As long ago as 1765 one commentator wrote that the personal liberty of individuals 'consists in the power of locomotion, of changing situation or removing one's person to whatsoever place one's own inclination may direct; without imprisonment or restraint, unless by due course of law'. Mobility as a right for some is enshrined in the Magna Carta. The Attorney-General and others were appealing to long standing ideals when they linked work and mobility. The irony of course, and the important point for what we are discussing, is that this language was used to support the road blocks that led to the stopping of flying pickets who were practising an entirely different form of movement. Blomley remarks that the difference between the two mobilities was that they had different relationships to authority and control. One kind of movement – the journey to work – was seen as properly structured, while the other, that of the pickets, was seen as a threat to a form of social power which relied upon ordered and bounded spaces. According to Blomley: 'The "heroic" mobility of the strike breaker reinscribes the liberal universe of individually contracting employees. Consequently, given the assumed threat to such action by the picket line, the mobility of secondary pickets is not only suspect but unprotected.'

[. . .]

Just as some mobilities are dependent on the immobilities of others so it is the case that one mobility may be symbiotically related to other mobilities with entirely different cultural and social characteristics. Changi Airport in Singapore is a case in point. The airport lounge of the postmodern *flâneur* – the archetypical non-place described by French anthropologist of supermodernity Marc Auge – is indeed the space of the privileged business traveller: the western adventure traveller heading for somewhere cheaper or the puzzled academic ruminating on the increased mobility of the joined-up world. However, Changi Airport is also the space of immigrant labour from the Indian subcontinent brought in to build the new terminal and then asked to leave. Further it is the space of the people who work there – the people who staff the check-in desks and the people who clean the toilets and empty the bins who come in from the city on a daily commuting cycle. The already differentiated traveller, the immigrant workers and the airport workers are all mobile. While their mobilities were enabled by the construction of a node in a network, each is brimming over with different forms of significance. The general observation that the world as a more mobile place does not do justice to this richness.

Changi is a major transfer point for western tourists visiting Asia. As Cynthia Enloe has argued [in *Bananas, Beaches, and Bases: Making Feminist Sense*

of International Politics, 1989], '[t]ourism is as much ideology as physical movement. It is a package of ideas about industrial, bureaucratic life. It is a set of presumptions about manhood, education and pleasure.' Tourism, she argues, is rooted in a political history that includes business trips, exploration and military duty. This political history is obviously one that men and women have experienced differently. This difference of experience has spilled over into contemporary international tourism.

Take the political economy of tourism for instance. As Enloe points out, the tourist experience is one of being served by flight attendants and chambermaids. It is women who hold the low paid, menial tasks in the tourism industry. In the most extreme case – sex tourism – it is literally women who are the destination. In Bangkok there are half a million more women than men, yet male tourists outnumber female tourists 3 to 1. Most of the women who sell sex to men in Bangkok are people who themselves experience mobility in a quite different way. The sex industry in Bangkok is fed by female migration from the rural margins where pay and conditions are even worse. Many Thai prostitutes move cyclically from country to city and back again providing their families with extra income. One source indicates that out of 50 prostitutes interviewed, 46 sent as much as half their earnings back to rural families. In addition it has been reported that Bangkok has become a site of international migration by women (and some men) for jobs in the sex trade. Myanmar, Russia and the Sichuan province of China all provide female labour.

[. . .]

It is clear then, that looking at the tourist opens up a whole host of questions about meaning and power. The movement of people is never just velocity – getting from A to B – it is imbued with an interrelated set of power-relations and meanings. The mobility of the tourist to South-East Asia is very clearly different from that of rural–urban migrants to Bangkok or the movements of Filipino 'entertainment workers' to Japan.

. . . If we return to the story of Fred Edwards and Frank Duncan we have a snapshot of the mobilities being produced. The courtroom is one particularly powerful site where the raw fact of motion gets given meaning and becomes embedded in relations of power. This is not to say that

that particular journey was not already produced by other contexts – particularly the context of the political economy of the United States at the time when the defence industry of California had become a major magnet for migrant labour. There is no pure movement to start with. But the arguments about oranges, Nevada buses, convicts and disease vectors were arguments about sameness and difference which enabled particular forms of mobility to continue and, at least for a while, made other types of mobility impossible or dangerous. The question of how mobilities get produced – both materially and in terms of 'ideas' of mobility – means asking: Who moves? How do they move? How do particular forms of mobility become meaningful? What other movements are enabled or constrained in the process? Who benefits from this movement? Questions such as these should get us beyond either an ignorance of mobility on the one hand or sweeping generalisations on the other.

"Of Nomads and Vagrants: Single Homelessness and Narratives of Home as Place"

from *Environment and Planning D: Society and Space* 18 (2000): 737–759

Jon May

Editors' introduction

It would be an understatement to say that there has been a tendency in Western literature and art to romanticize life "on the road." After all, Western cultural tradition is replete with myths and stories of pilgrimage, exile, diaspora, and nomadism. From Adam and Eve's expulsion from the Garden, Abraham's wandering, and Homer's Odysseus to Jack Kerouac's Dean and Sal, and Ridley Scott's Thelma and Louise – mobility has long been part of our myths of sin, healing, redemption, shame, and freedom. And while it is not necessarily surprising that cultural geographers traditionally emphasized not mobility but place-based organic attachment of people to the land, it is nevertheless curious that such a central element as mobility of Western culture has not had a more prominent place in the traditions of the discipline. That, of course, has been changing, as this part of the *Reader* makes clear. But, in the enthusiastic embrace of mobility currently found in Western theory (discussed, for example, in Cresswell, p. 325), we are cautioned to avoid the kind of romanticism that drove Thelma and Louise off a cliff in a blaze of suicidal glory.

It is caution for this kind of romanticism – for the cleansing of a pilgrimage, for the freedom of the "open road" – that informs Jon May's study of the life histories of homeless men in southern England. May's research suggests that lest we romanticize being on the road, the desire to establish a home and something resembling an organic sense of place remains very powerful. In making this argument, May seeks to also broaden the meaning of homelessness, from a narrow focus on being without a residence to a broader sense of being out of place. As he puts it, "we might usefully extend a consideration of homelessness as the absence of home as residence to take account of feelings of the absence of home as place." May finds that the homeless men he interviews are "relatively immobile," that they do not typically think of their lives as mobile as much as "displaced" and constantly searching for a new place in which to settle.

This finding raises some broader questions that are touched upon in various ways throughout this part of the *Reader*. Does place travel? Can one be "at home" on the road? While May's research suggests not, the questions remain on the theoretical agenda of much work in mobility studies. For example, John Urry notes in *Sociology beyond Societies* (2000) that there are a variety of ways of dwelling and "almost all involve complex relationships between belongingness *and* travelling, within and beyond the boundaries of national societies. People can indeed be said to dwell in various mobilities; bell hooks (p. 373) writes: 'home is no longer one place. It is locations'". James Clifford (p. 316) refers to the need to conceive of "dwelling in travel." And, in a more empirical example that raises many further questions about power, Yan Hairong ("Neoliberal governmentality and neohumanism: organizing suzhi/value flow through labor recruitment networks", *Cultural*

Anthropology 18, 2003) has observed that the city in China – where millions of rural migrants travel to find work – is regarded by the state "as a 'comprehensive social university' (*shehui zonghe daxue*) in which millions of peasants can go to develop their suzhi ["quality"] at no cost to the state, requiring no investment!"

Jon May is Professor of Geography at Queen Mary, University of London, where he conducts research on the social and cultural geographies of cities, migrant labor in global cities, and homelessness. The co-editor of several books, including *Cultural Geography in Practice* (2003) and *TimeSpace: Geographies of Temporality* (2001), May has also written many articles on homelessness, globalization, and place identity. He is one of the leading scholars on the issue of homelessness, migrant worker identities, and the provision of services to the homeless and migrant workers.

INTRODUCTION

The homeless are forced in to constant motion not because they are going somewhere, but because they have nowhere to go. Going nowhere is simultaneously being nowhere: homelessness is not only being without home, but more generally without place. Unlike movement from place to place of travel or migration, the itinerant movement of the homeless is a mode of movement peculiar to the condition of placelessness. (Samira Kawash, *The Homeless Body*, 1998)

Within a rapidly developing literature on homelessness attention has begun to focus upon homeless people's own experiences of homelessness and, by extension, their understandings of home. Whilst noting the diverse ways in which those who are currently 'homeless' articulate a sense of 'home', and the varying circumstances that shape their understandings of 'home', such work has in the main been limited to a consideration of homeless people's constructions of what might be termed 'home as residence'.

Yet, as Kawash reminds us, 'homelessness is not only being without home, but more generally without place'. Whether considering the micro-geographies of the urban homeless, the wider migrations of those in search of work or accommodation, the nomadic cycles of those moving around a 'hostels circuit', or the movements of those seeking simply to escape the wider circumstances that precipitated the loss of their home, it is clear that the experience of homelessness cannot be considered apart from the experience of movement – of varying kinds and at a variety of scales. Such movements clearly impact upon a person's day-to-day experiences of homelessness in different ways. But as the homeless move within, between, and through places – sometimes by necessity, sometimes by 'choice' – these movements must also impact upon any subsequent articulations of a wider sense of home not only as residence but as place.

Drawing upon a reconstructive life-history approach, I explore the notion of 'home as place' articulated by men living in night shelter and hostel accommodation in a large town on the south coast of England. Building upon work that has explored the movements of these men in some detail, I set the understandings of 'home as place' of each respondent within the broader context of the mobility that has characterised his homeless career.

Rather than the pattern or experience of that mobility, the primary focus of this paper is an examination of the manner in which those for whom the experience of homelessness has included differing degrees of movement between places articulate this wider sense of home. Examining these men's experiences four narratives of home as place are outlined, relating to the experiences of the '(dis)placed', the 'homesick', of those articulating a 'spectral geography', and of the 'new nomads'.

[. . .]

OF (DIS)PLACEMENT

In contrast to those who would position the homeless as Other than the housed population by virtue of a life of extended and extensive movement, the biographical sequences recounted by the men interviewed here would suggest that a significant proportion of those using night shelters and hostels for the homeless are in fact relatively immobile. This is not to suggest that single homelessness should be considered a 'local phenomenon'. Rather, although the majority of respondents had in fact moved to the town in which the study was conducted having first become (visibly) homeless elsewhere, for many the move was one of only a few times, if indeed not the first time, they had moved away from the places in which they had previously spent the main part of their lives. Far from extensively or unusually mobile, these men's biographies therefore revealed very similar levels of mobility to those found in studies of the long-term unemployed, a group considerably less mobile over the life-course than the professional or managerial classes.

Those biographies also revealed most to be neither only recently nor truly long-term homeless. Rather, with housing and employment histories dominated by intermittent or long-term unemployment and the use of private rented accommodation, many had experienced numerous though short-lived periods of homelessness over the years. It is this history of 'episodic' homelessness coupled with (and to some extent driven by) the continued reliance upon often insecure private rented housing that most obviously sets these men's experiences apart from the experiences of other poor and unemployed people but which also complicates any easy or neat distinctions between a 'visibly' homeless population and those experiencing some form of 'hidden' homelessness. Indeed, on becoming homeless these men had in the past more often turned to friends or relatives (in their home towns) in times of crisis than they had to a formal network of emergency services, only occasionally staying in night shelters or hostels – most usually following a period of sleeping rough.

Given such histories it is not difficult to understand why on becoming homeless (again) a person might (finally) move . . . [E]ven though a number of respondents suggested that moving in such circumstances was to leave little behind, for others the decision finally to leave was clearly more complicated, as Peter, recounting the period following the separation from his wife, explains:

There were people I could have stayed with in [place name] I guess [pause] you know, if things had been different. But I don't really fancy going back there, having to rely on people there . . . not yet, like . . . even though my kiddies are there and that . . . What the situation is, you've gone downhill sort of thing and you don't really want people to know. Because. some people would have a laugh on it and some people would be upset over it. So, er, you come somewhere like this – to get out of the picture sort of thing." (Peter, age 44, 24 December 1997)

Peter's experiences suggest that the decision finally to move away from the place in which one becomes homeless is as often shaped by the desire to leave a place as to go somewhere else and, even when not explicitly related to the need to escape the type of situations described by those tracing the movements of younger homeless people, that that decision may primarily be shaped by the powerful sense of shame associated for many of the respondents with the experience of street homelessness. In such a situation the decision of where to move to may in turn be shaped not by the desire to access informal networks of support but the opposite – to go to a place where one's homelessness is rendered less 'visible'.

At the same time, it would appear that even in those cases where a person had moved in an attempt to find support from friends and relatives, such support was frequently unavailable – as people often found their friends in similar situations as themselves, unable to accommodate them for long if at all as either lack of space or pressure from landlords made it difficult to offer help. For the majority of respondents, unable to find accommodation with landlords unwilling to take those on benefits and unable to secure a bed in night shelters or hostels that are almost always full, arriving homeless in a (strange) town and attempting to find one's way around the emergency-services network was therefore deeply traumatic. especially for those with no previous experience of sleeping rough . . . Nor . . . did this sense of isolation and confusion diminish over time as the men eventually found a space in one of the town's night shelters or hostels. Instead, with its constant turnover of residents. the overwhelming feature of hostel life for most of the respondents was its loneliness, as Albert's description makes plain. When asked about the other people at the hostel, he replied,

Well, I've never had any trouble with anyone. I mean, where I am – at the top here – there's [name], he's a nice fella. They're alright up here. I don't know what it's like down there of course [the floor below]. I don't even know anybody down there. I keep myself to myself, to be honest . . . I've always found that the best way. Mind my own business. I don't bother no one, no one bothers me. I mean, there are people I talk to and they talk to me. But er [pause] I don't go visiting anybody in their room or anything like that. (Albert, age 56, 4 December 1997)

Nor was this sense of isolation restricted to life at the hostel. Rather, it extended to the respondents' experience of the other spaces that shaped their day-to-day lives (the local day centres and soup runs, for example) and to the town as a whole, rendering what could be termed a 'hollowed-out' experience of place:

[The thing is] everyone you come across is in the same sort of position . . . you meet up with a few guys, but they're just passing through, sort of thing. There's nothing definite with them . . . no one would really say that this place is their home. because it isn't . . . you've got no ties . . . no commitments . . . no one's fixed here . . . It's like all my friends are homeless too, you know. so they could all change in the next few days . . . They're just acquaintances really. (Peter)

Ironically, then, even as a number of these men had moved in the hope of building a home for themselves, having lost their homes elsewhere, the result of that move was simply to extend a sense of homelessness as lack of residence to homelessness as lack of place. That sense of homelessness is best expressed by Michael, whose experiences perhaps suggest not so much the simple feeling of placelessness (the absence of a sense of home as place) than one of (dis)placement:

Interviewer. Why leave London?
Michael. I dunno. I just wanted some place new. Get out of it. I was fed up with it [long pause]. You know, erm, I probably didn't think about it, to be honest . . . I wasn't going to stay where I was [in a hostel following the

separation from his wife] so I said, well, 'Anywhere has got to be better than this.'
Interviewer. So moving wasn't like a wrench for you?
Michael. No, I wouldn't say that . . . [It wasn't as if] I was leaving anything behind, you know? I mean, you've just got to pack your bags and you're off.
Interviewer. And does this feel like your town now?
Michael. [Ironic laughter.] I don't know. [Pause.] Yeah, I suppose it is – at the moment . . . But at the end of the day you always talk about 'going home' don't you? . . . I mean, I've made friends with a few people, [pause] you know, go for a drink and that. Fine, but . . . . no, I wouldn't call this place my home . . . where you come from's home isn't it?" (Michael, age 33, 4 December 1997)

THE HOMESICK

Although articulating an experience of loss, by definition the notion of (dis)placement presupposes an earlier experience of home – one reaching beyond the boundaries of residence to include that wider sense of belonging more usually described as a 'sense of place', in this sense, though apparently similar, the experience of (dis)placement is in fact quite different from simple homesickness, described by Bauman [*Life in Fragments*, 1995] not as the *nostalgic* yearning for home but the 'dream of belonging': a dream that situates it firmly within the 'future' tense. Even though experiencing that same sense of isolation and confusion described by Albert, Peter, and the others, it is this notion of homesickness rather than (dis)placement that better describes the experiences of a second set of respondents. These experiences have their roots in (and give shape to) a quite different pattern of movement and quite different experiences of homelessness, mobility, and home as place.

Asked what it is he wants, David's description of the sense of home he seeks, for example, is at first depressingly familiar with regard to the kind of accommodation that has so far characterised his own life and the lives of most of the men interviewed here and with regard to the simple desire for anything better being so quickly dismissed as an unrealisable dream:

> *Interviewer.* What do you want?
> *David.* I want a flat, but I can't get one. I want my own, my own independence?
> *Interviewer.* And what would make it a home?
> *David.* Um, decent people for starters. Somewhere where you haven't got all your drunks, people starting fights all the time . . . a decent place [not one where] there are holes in the wall, where the carpet smells, that's full of drunks and druggies . . . a decent job . . . and a bird?
> *Interviewer.* And does it matter *where* it is?
> *David.* No. Anywhere, really. It doesn't matter to me. (David, age 22, 10 December 1997)

Yet as he continues it becomes apparent that there is also something else lacking from David's account. Although able to articulate the sense of home as residence he so badly desires (even if he has little previous experience on which to ground that desire) there is at the same time a peculiar placelessness to David's description of his future home. That sense of placelessness has its roots in David's past and it projects a sense not of (dis)placement but of homesickness into his future.

Growing up in an industrial port city a little further along the coast from the town in which he was interviewed. David left 'home' at the age of sixteen years to move within the same town into a flat provided by the local social services. His memories of early family life were in fact confused, as he found it difficult, for example, to remember which of his siblings had already been placed in care by the time he had left home. Much clearer, though, were his memories of the area in which he grew up and the powerful sense of exclusion that dominated those memories:

> I was born in [place name], one of the worst areas of [place name]. And what it was, was like gangs. All the youngsters round there from age anything, from about 14 up to about eighteen, nineteen, hung around in gangs. If you didn't join a gang [very quietly] you got picked on, basically.

Never part of a gang and frequently bullied (partly because of his size), having left home David also quickly left the town, moving no fewer than six times over the next six years. A number of these moves were, ostensibly at least, made in search of work, others as he took up the kind of training schemes offering only low-paid, temporary employment that are a common feature in the lives of the young homeless and unemployed. Each time he moved David also found himself homeless, sleeping rough for a few nights as he reached a new town before finding a bed in a night shelter or hostel and moving from there to a bedsit or room in a shared house. Significantly, failing to find either a (decent) place to live or a job, between each new move David returned to his home town to find accommodation in hostels or bedsits located in or close to his old neighbourhood. Though unrelated to the geographies of the hostels circuit (insofar as a number of these moves took him to places where no hostel accommodation is to be found) David's movements have therefore been entirely contained within that wider network of day centres and hostels, multiple occupancy houses, and cheap rented bedsits which forms the contemporary 'skid row' network, as he has moved in a cyclic pattern between similar such neighbourhoods across the country.

For David, and the other (mainly younger) respondents with similar biographies, the sense of homelessness he describes might usefully be understood as emergent out of a sense of dislocation not only from 'mainstream' society but mainstream (consumerist) youth culture . . . Yet, if as Bauman argues, homesickness is the 'urge to feel at home, to recognise one's surroundings and belong there . . . the [dream] of being, for once, *of* the place, not merely *in*', there is another way of understanding the sense of homelessless David articulates. Here, it is the continued search for a feeling of home as place, rather than any simple search for work or even for home (as residence) that drives David's movements. Continually frustrated in that search each time he moves elsewhere, the search itself (and setting him in contrast to the first set of respondents) has its roots in the original absence of such feelings for his 'home' town (to which he continually returns). Recalling his most recent return 'home', for example, when he stayed for a little over 18 months, moving through a variety of hostels and bedsits, David remembers that:

I made one or two friends, but apart from that – not very many . . . they still come down there, you see, the [place name] lot. [The thing is] I love the town . . . [pause] I just don't like the people there. They're just, I don't know how to explain it. They're just – not so friendly . . . if they aren't causing problems, they're ignoring you.

Perhaps the most tragic aspect of David's account is that, denied this more familiar sense of home, so too he remains unable to access any alternative constructions. As he talked with a certain sense of pride of his knowledge of the hostels circuit (claims unsupported by the pattern of his movements) and of the 'dodges' he had learnt to bypass the more mundane but nonetheless crucial problems associated with, for example, transferring one's benefits each time one moves, it might be suggested that such claims to knowledge speak of David's desire to belong at least within those Other spaces that have shaped his life over the past six years. Yet each time he arrived in a strange place having to find a safe place to sleep and negotiate an unfamiliar scene, the feelings of isolation and disorientation described by the first set of respondents are replicated, rendering a sense of homelessness for David here too:

[If you want to find a hostel, what you have to do is] see someone looking a bit scruffy . . . and ask them . . . But the thing is, you don't know if they're going to turn round and tell you to flick off, or beat you up. So you have to take your chances?

SPECTRAL GEOGRAPHIES

It is precisely the sense of home to be found in these Other spaces that apparently characterises the experiences of a third set of respondents. These are the 'men of the road' for whom a sense of home is predicated upon that very mobility that renders the experience of home so difficult for David to achieve. As they work their way around a 'hostels circuit', the concept of home supposedly assumes an alternative shape: provided by the occasional rather than sustained encounter with others following similar routes. Though yet to be examined

in such a manner, these men's experiences might reasonably be set within the world of the nomad, as described by Deleuze and Guattari [in *Nomadology: The War Machine*, 1986] for whom the nomad's sense of identity and of home is to be located not within the movement between places or with a lingering in place but within the network of waystations and rest-points that shape their travels. Predictable and cyclic in their movements, returning on a regular basis to a familiar set of night shelters and hostels, such men may therefore (potentially at least) be seen as reordering a traditional geography of movement, space, home, and place.

The difficulty with such arguments is that, if not entirely abstract, the accounts on which they are based have rarely examined the movements of these men themselves in any detail. Nor have they provided any context within which their movements, and any subsequent understandings of home, may be set – largely ignoring the broader life histories of those who have taken to a 'life on the road' or working unduly from secondary sources.

A heavy drinker in his late forties with long hair and unkempt beard, amongst the men interviewed here it was Don who most obviously fitted the description of a 'travelling man'. Already identified as such by other residents (with whom he rarely mixed) as well as by the hostel's staff, Don himself seemed keen to foster this identification, frequently recounting tales of the numerous night shelters and hostels in which he had stayed over the years. Moreover, though Don also talked of the possibility of 'settling down' now he was 'getting a bit older', perhaps even giving a bedsit 'a go', more often his talk was of his future travel plans. By the time he was formally 'interviewed Don's travels had therefore already assumed something of a mythical air, with his understandings of home too seemingly in fitting with the kind of understandings often articulated by those with a long history of hostel use or rough sleeping if not also considerable mobility:

Interviewer. So. when you said earlier you consider yourself homeless that's less to do with whether you're sleeping rough or in a hostel or in your own flat. It's actually whether you feel at home?

Don. At home. Yeah, that's right . . . the physical side is not important . . . Whether I have
a roof over my head or I'm sleeping on the
grass, it's immaterial. It's an . . . emotional
thing. Do you understand what I mean?
(Don, 5 December 1997)

But to assume from this that for Don a sense of
home is to be found on the road or in his encounters with the residents of those hostels to which
he regularly returns would be premature. As he
expands upon his earlier remarks, for example, it
is clear that not only have these hostels rarely if
ever provided Don with a sense of home but also
that he does not hold any romantic illusions about
life on the road:

I wouldn't regard this as a home . . . I wouldn't
regard any of these places as home . . . Home
is where you're emotionally and physically at
home. Satisfied. Home is satisfaction . . . You
understand? Emotional, sort of rest. Where you
don't want to go anywhere. Where there is
'where I want to spend the rest of my life here'.
That to me is home . . . It's that emotional side.
It's – people who move around, they're desperately unhappy, you know? My old man knew that,
you know, and, well . . . so that's that, more or less.
We've got that bit. What do you want now?

[. . .]

Constantly returning to Birmingham in a frustrated
attempt to reconnect with an earlier life, and to
London in an effort to find again those men who
(for a while at least) provided a sense of belonging, if the experiences of the first set of respondents
described a sense of (dis)placement, and the
movements of David are best viewed as driven by
a feeling of homesickness, Don's movements
might therefore be understood as articulating what
we can call a 'spectral geography' as he continually returns to places that were once meaningful
but which now contain only the ghosts of previous
relationships.

THE NEW NOMADS

Though offering a peculiarly powerful illustration
of such a geography, Don was by no means the only

person to outline these kinds of experiences. In the
case of another of the long-term homeless respondents, for example, time since leaving the Merchant
Navy had been almost entirely spent retracing the
geographies of an earlier life – moving around
Britain's coastal ports, picking up occasional work
on the docks. For another, several years sleeping
rough had been spent moving between a resort town
a little further along the coast (in which he had been
born and in which his children still lived) and a
nearby market town to which his wife had moved
following their divorce. Though occasionally using
night shelters and hostels, the 'hostels circuit' was
therefore of only limited importance to these
men's lives, the pattern of which was set by a rather
different geography.

[. . .]

Such experiences are best illustrated by Martin,
whose biography reads like a condensed history of
the New Age traveller movement. Leaving care at
the age of sixteen and having spent time in a number of hostels for young homeless people, Martin
spent the next four years of his life moving
between a series of squats in the north London area.
Supporting himself through a combination of what
Carlen [in *Jigsaw*, 1996] refers to as 'survivalist
crimes', begging, small-time drug dealing, and the
use of day centres and soup runs, during his time
in London Martin made contact with a wider network of New Age travellers and squatters extending across the country.

When continued police pressure led to the
eventual dispersal of the London squatting scene
Martin and a number of his friends moved to the
town where he was later interviewed – attracted in
part by the area's well-established network of day
centres and soup runs of which a number had
prior experience. On arriving in the town, Martin
and his friends quickly joined up with a group who
had begun to squat the seafront and pier in protest
at the area's redevelopment, linking up with
protesters from a nearby travellers' site located a
few miles outside of the town. When this too was
broken up following pressure from local businesses, Martin moved first into a small seafront
bedsit until, growing depressed, he bought a car
and began moving between a number of New Age
traveller sites throughout northern and eastern
England and Wales – supporting himself between
times by drawing upon his knowledge of day

centres and soup runs to be found in different parts of the country. Eventually, driving without tax or insurance, Martin was arrested. Sent first to a bail hostel in the north of England, Martin was subsequently moved three times under warrant, the last such move bringing him to the hostel in which he was interviewed.

Tying this history together is a complex geography of informational networks connecting the worlds of the homeless with those of the New Age traveller and the urban squatter, ensuring access to resources for members of a community stretched across space. Yet, rather than to provide for a life of continual movement, the purpose of such networks is to provide for the possibility of (re)establishing a sense of home for those who (like Martin) are liable always to be seen as 'out of place' and who must therefore regularly move if they are to find that sense of home they seek. Thus ... Martin's decision to live in his car, and his movement around these overlapping networks, should not necessarily be understood as articulating the desire for a mobile lifestyle. Rather, for Martin at least, the decision to 'take to the road' came only as his attempts to establish an alternative space of home in place was continually frustrated – first in London and later on at the south coast:

I've never had a family, you know? ... I've lived in care ... in hostels ... The hostel was OK. But there were so many restrictions ... These are not places where you are entitled to your freedom ... So squatting was a pleasure ... I felt at home. People were very friendly and they made you feel welcome. If you wanted anything, they'd help you. You moved in and, in a week, a day, it was 'We've got this, we know our way around, we'll show you' ... We formed a sort of community, we helped each other out ... Myself and others, we went out dealing, stealing, picking up food from Sainsbury's, from ... We'd put it all in a big box and we'd cook a meal for everyone ... otherwise it was soup runs, day centres ... it was pretty easy ... But in the end, the longest you could stay anywhere was a couple of weeks. More than that and the police were in there ... they'd come in, evict you, take your stuff, arrest you ... [With the] new regulations there was a different atmosphere and it was impossible to squat anymore.

[The same thing happened down here] when the hotels started to complain ... They got the council and blocked up the front of the chalets so we couldn't get in ... Then they pulled out all the cables so we couldn't use any heating or do any cooking ... Then they called the police ... [and] arrested [anyone they found] ... [so in the end] I was living out of the back of my car ... Always driving somewhere else. To day centres, soup runs. That was my life ... and they took it off me. Made me feel like a piece of shit. (Martin, age 23, 15 December 1997)

Where for each of the other respondents a sense of homelessness emerged as they either lost, were unable to find, or could not reconnect with a sense of home (as place) the most profound sense of homelessness is perhaps reserved for Martin and others like him who, in response to the continual frustration of their attempts to build an alternative space of home in place, end up transgressing normative constructions of home as both residence and place. ...

CONCLUSIONS

... The biographical sequences which form the context for the discussion here suggest that, though single homelessness cannot be considered a 'local phenomenon', the majority of those using night shelters and hostels for the homeless are in fact relatively immobile. Yet this is not the same as suggesting that homelessness can be understood without reference to questions of mobility. Rather, although the majority of the men interviewed had moved only infrequently throughout their lives, when they did at last move away from the place in which they (most recently) lost their homes such movement had a profound impact upon the experience both of their homelessness and upon any subsequent ability to regain that sense of home they had lost. The narratives of people such as Simon, Albert, Peter, and Michael suggest that, although the experience of moving to a new place once homeless may be extremely traumatic (especially for those forced to sleep rough), in the longer term it is the feelings of disorientation and isolation that continue even once a person has found shelter that are often harder to cope with. Such feelings contribute to the experience of what has been called a 'hollowed-out' sense of place and, for those who continue to feel that their home lies elsewhere, may lead to a powerful sense of (dis)placement.

In contrast to the experiences of the (dis)placed are those whose lives are now shaped by the contours of what has been called a 'spectral geography'. Previous studies have suggested that these men, part of a (declining) population of (mainly older) single homeless men, may through a 'life on the road' articulate an alternative sense of home constructed within and between the spaces of a 'hostels circuit'. Without denying there may be those who have chosen to travel rather than to settle and who find within this more mobile lifestyle a different sense of identity and of home, the narratives of Don and the other long-term homeless respondents interviewed here whose biographies revealed similar homeless careers suggest that such accounts are at worst overly romantic, at best now nostalgic. Indeed, though highly mobile, the movements of such men would in any case rarely seem to be shaped by such a circuit, and in their descriptions of the night shelters, hostels, and bed and breakfast hotels that provide the setting for their encounters with the 'acquaintances' they now share their life with it is difficult to find anything equating to a sense of home. Rather, in their returns to those places the more obvious impression is of a sense of unavoidable and unending repetition, as they move in continual search of a sense of home and of place not merely lost but (perhaps) irretrievable . . .

"The Tourist at Home"

from *On the Beaten Track* (1999)

Lucy Lippard

Editors' introduction

John Urry (*The Tourist Gaze*, 1990) once wrote that, more and more, we are all tourists in our daily lives. He meant that on a daily basis we increasingly inhabit and interact with spaces and landscapes that are themed. In many of these landscapes we consume not just food, clothing, and other retail items, but *experiences*. In part, this is because consumers are more likely to spend their money if their purchase has some additional experience or meaning attached to it. Some products are attached to environmental or social causes, such as saving the rain forests or supporting indigenous groups. Some products are simply sold in a themed environment: a rain forest café, or a shopping mall in a heritage site such as a converted factory. Indeed, the distinction between theme parks like Disneyland and themed shopping malls like the West Edmonton Mall or the Mall of America by now escapes many of us. More than this, however, entire neighborhoods and towns are often subject to themed development, sometimes as heritage sites, sometimes based on a local specialty product, and sometimes for no apparent reason at all, other than garnering some attention from outsiders. In Shanghai, new housing developments springing up around the city have been built according to a variety to themes. There is a "Thames Town" with an English country village theme, along with other towns with "Canadian," "traditional Chinese," "Swedish," "German," "Italian," and "Spanish" themes. Such themed property developments can be found also in Japan and many other places around the world.

But there's more to being "a tourist in your home" than this. Urry sought to make a point about changing practices of consumption, and about the way post-industrial capitalism was increasingly producing images and experiences for sale, rather than simply products. His claim about being a tourist every day was a claim about the changing nature of society. In the selection below, feminist cultural critic Lucy Lippard takes a somewhat different perspective on "the tourist at home." While she certainly would agree with Urry that our experience of daily life is increasingly mediated by spectacle landscapes, advertising images, and contrived themes, she would also agree with Simmel (see p. 311), that there is a certain kind of "objectivity" to be gained by looking at a place through the eyes of a "stranger." For Urry, one is a tourist at home because post-industrial capitalism has made it increasingly impossible *not* to be. "We are tourists," he quips, "whether we want to be or not." For Lippard, being a tourist at home involves a conscious effort to "take care" of one's home place, to step outside the taken-for-granted routines of one's daily life and see place anew. "Travel is the only context in which some people ever look around," she argues. "If we spent half the energy looking at our own neighborhoods, we'd probably learn twice as much." Lippard notes the particularly important role that local artists and performers play in helping residents achieve this strangers' gaze, and her essay describes some of the projects that artists have created for this purpose. One notable group (though not mentioned in Lippard's essay) is Wrights & Sites, in Exeter, England. They have produced a "guidebook" for tourism at home: *An Exeter Mis-guide* (2003), and have since extended their project to a much broader idea of "home" in *A Mis-guide to Anywhere* (2006). In many cases, these projects encourage "tourists" to encounter the hidden sides of their homes, the marginalized, the "dangerous." Such encounters might result in people taking better care of their home places and the many different kinds of people who live there.

Seeing our homes through the eyes of a stranger reminds us, also, of Simmel's basic point. "The Stranger" reminds us that the relationships and interactions among people that go into making a home-place are expressions of a self–other binary through which we create our world. We inhabit our homes as both insiders and outsiders; we know our most intimate partners both as strangers and lovers. To be a tourist at home, then, is to remind ourselves of this binary, to understand that it is both mobility and dwelling the creates a place.

Lucy Lippard has been writing about art, politics, feminism, and place since the 1960s. Her most well known book among geographers is probably *The Lure of the Local* (1997), which explored from many angles the relationship between art and place-making. *On the Beaten Track* (1999) is an extended exploration of that relationship in the context of tourism. Lippard claims that "It's more about staying home than traveling." In its focus on "how much people move around trying to find their place," Lippard's book echoes the homeless men of Jon May's study (see p. 334). Recent work in geography on the relationship between home and travel, place and mobility, includes the collections by Minca and Oakes, *Travels in Paradox: Remapping Tourism* (2006), Cartier and Lew's *Seductions of Place* (2005), and Crang and Coleman's *Tourism: Between Place and Performance* (2002).

▪ ▪ ▪ ▪ ▪ ▪

This village loves this village because its river banks are full of iguanas sunning themselves and its fishes love to bite. (*Santiago Chub*)

"What's here?" asked some friends from Maine as I walked them through the New Mexican village I live in. They had seen the place written up in a guidebook as "picturesque." "Nothing," I said with a certain mendacious pleasure, thinking how opaque the village's surface is.

"Is there anything over there?" asked a couple I met on the bridge; they were staying at the local inn. "Depends on what you're looking for," I replied, secure in the knowledge that there was nothing over there they would see.

Yet when I give my own walking tours through the rutted dirt streets (and few of my visitors escape them), it seems to me that everything is here: culture, nature, history, art, food, progress, and irony. There is the old village itself and its vestigial claims to "authenticity"; the church (relatively new as Southwestern churches go, having replaced an older one in 1884); the eighteen-year-old upscale development to the west for contrast (and for an architectural tour of another nature; it's a good survey of imagined "Santa Fe style"); the movie set in the distance; the curandero's "office" with its skull on a pole; what used to be here and there (scattered adobe ruins); the quite new community center and the brand new firehouse (partially built by community work parties); yard art; an extensive petroglyph site; the cloud shows and encompassing light on ranch-lands and mountains; the (diminishing) biological diversity of the creek and bosque; the mouth-watering tamales at the Tienda Anaya; and, of course, the people. We have it all, but for an outsider, it's hard to find.

The next question is, should it be easier? What's in it for a town like this, with few local businesses? Who would profit from a higher profile? Will signs begin to proliferate along the highway? Will local artists lend themselves to making this place a "destination" rather than a fly-through? Will a proposed café/gallery and/or restaurant change our identity? We may soon have to answer these questions, as the state and county tourism bureaus look farther and farther afield for attractive "authenticity." Dean MacCannell has said that the concept of the authentic is a potential "stake driven into the heart of local cultures"

The local is defined by its unfamiliar counterparts. A peculiar tension exists between around here and out there, regional and national, home and others' homes, present and past, outsiders and insiders. This tension is particularly familiar in a multicentered society like ours, where so many of us have arrived relatively recently in the places we call home, and have a different (though not lesser) responsibility to our places than those who have been living in the area for generations. Jody Burland has remarked on the "peculiar reciprocity of longing" at the heart of tourism which binds outsiders to insiders. Tourists may long for warmth, beauty, exoticism, whereas locals may long for escape, progress, and an improved economy: "Between us there can be a moment of strange, perhaps misleading comprehension." Local residents both possess and become a "natural resource which produces more pleasure, and tourists are necessary to its conversion to wealth." Smiles and solicitude are part of the negotiations. The exchange contains the contradictions that define a multicentered society.

Tourism is the apotheosis of looking around, which is the root of regional arts as well as how we know where we are. Travel is the only context in which some people ever look around. If we spent half the energy looking at our own neighborhoods, we'd probably learn twice as much. When we are tourists elsewhere seeing the sights, how often do we stop and wonder who chose the sights we are seeing and how they have been constructed for us? We do often wonder about the sights we're not seeing – houses and gardens glimpsed behind the walls, historic sites and natural wonders sequestered on private property or closed on Tuesdays.

The tourist experience is a kind of art form if it is, as Alexander Wilson says, its own way of organizing the landscape and our sense of it. "We tour the disparate surfaces of everyday life as a way of reintegrating a fragmented world." It is an art form best practiced domestically, challenging artists to work in the interstices between the art scene and local audiences. This can mean demythologizing local legends and constructing antimyths that will arm residents against those who would transform their places in ways that counter local meaning (which in itself is unstable). So the resident who accepts the role of tourist at home becomes responsible not only for the way the place is seen

but for how it is *used*. Jim Kent, a sociologist based in the legendary Colorado ski town notes, "So many people complain about the people who bought Aspen. What about the people who sold Aspen?"

Being here and being there, being home and being away, are more alike than we often think. Even as we learn them, our places change, because no place is static, and no resident remains the same as s/he lives and changes with the experiences life and place provide. People visit, they like the place, they retreat or retire there, becoming what have been called "amenity migrants." Then, prey to the "drawbridge syndrome," they begin to complain about the tourists and other newcomers. In Aspen, they say "If you've been here a year, you remember the good old days." A former county commissioner observes, "What defines you as a local, in my mind, is whether you give more than you take." Yet locals can be takers too, from a littering habit that pervades the rural United States to more permanently destructive behavior. It was a local, mad at his girlfriend, who poisoned and brought down the great historic tree called the Austin Treaty Oak, in Texas. At Higgins Beach, near Portland, Maine, drunken partygoers deliberately stomped on the nests and eggs of endangered terns. All over the West, local people target-shoot at ancient rock art. Vandalism, not necessarily by "foreign" tourists, recently destroyed an arch in Canyonlands. The examples are chillingly ubiquitous.

Many towns are not so much potential destinations as service stops along the way to more desirable places. Considered negligible, they are unseen, recalling tourism in its innocence, when travelers were the strangers, providing entertainment for locals, when the passing tourists looked out upon views that were the same before they came and after they left. But all too soon came the deluge. Opposing tourism in the West, if only theoretically, has suddenly become "like being against ranching, or Christianity," writes Donald Snow in a bitter elegy for Montana titled "Selling Out the Last Best Place":

We're getting the endless strips of motels, junk food restaurants, and self-serve gas depots out along the interstates that make our towns look like every other greasy little burg everywhere else in Walt Disney's Amerika. We've got increasingly egregious pollution problems now, here in the paradise of the northern plains, and we have seriously outstripped the abilities of local government to handle even modest levels of new home development. Recent news in my hometown paper is that a new hydrologic study of Missoula County has found significant levels of septic contamination in every single well . . . including one well drilled 220 feet down to bedrock. . . . If all that isn't stupid enough, we spend what paltry money we raise from a tourism tax right back on more tourism.

Over a period of years, John Gregory Peck and Alice Shear Lepie have studied three North Carolina communities and charted the effects of rapid growth, slow growth, and "transient development" (weekend and special event tourist trade) on three criteria of central importance to local people: *power* (land ownership, sources of financing, local input, and the relationship of local traditions to development projects); *payoff* (benefits and potential upward mobility for how many residents); and *tradeoffs* (the social impact on communities). Under the best of conditions, balance seems achievable. Yet when tourism becomes the only option for economic survival, our labor force becomes a nation of service workers, dolled up to look like our ancestors as we rewrite the past to serve the present. Although this situation might provide a chance for retrospection, the romantics, the generalizers, and the simulators usually get there first. Towns can wither on the vine as they preserve the obsolete out of stubbornness or impotence, or they can inform their residents' current lives. Past places and events can be used to support what is happening in the present, or they can be separated from the present in a hyped-up, idealized no-place, or pseudo-utopia, that no longer belongs to the people who belong there.

In recent years, a lot of cities around the country have come up with PR campaigns called Be a Tourist in Your Own Town. It's an interesting idea if it's taken way past the overtly commercial motives that inspire it. Instead of discounted trips to restaurants and museums offered in order to stimulate local markets, this could be a time to focus on latent questions about our own places – areas we've never walked through, people we've never met, history we don't know, issues we aren't well-informed about, political agendas written on the landscape. It is a task taken seriously by the innovative Center for Land Use Interpretation and its publications and tours.

John Stilgoe's studies of "locally popular" places such as the Blue Springs Café, just off the interstate at Highland, Indiana, and the Ice House Café in Sheridan, Arkansas, or even the ubiquitous Wal-Marts, suggest that people are less interested in the visual impact, in the architectural containers, than in "something different" from the corporate style and, above all, in "high quality product and service within the container." Thus the tourist looking for the locally validated, the truly "authentic," is unlikely to stumble into it because *from the outside* it looks like nothing special. Most locals, perhaps even some proprietors, would like to keep it that way. In tourist towns, at least, residents feel displaced. They need their own refuges, which are always endangered – potential tourist spots, if the secret gets out. MacCannell has pointed out that in San Francisco, "everything that eventually became an attraction certainly did not start out as one. There was a time when . . . Fisherman's Wharf was just a fisherman's wharf, when Chinatown was just a neighborhood settled by the Chinese."

Years ago, theater innovator Richard Schechner got a job as a tour guide to prepare for an article on tourist performances. Creative Time's 1995 *Manhattan Passport – 7 Two Token Tours* encouraged New Yorkers to rediscover their borough by "re-contextualizing typical tourist attractions with areas of the island other than those in which you might live and work." The seven cleverly titled excursions included "Take the A Train" (Harlem), "More Than Harlem on My Mind" (Washington Heights and Inwood), the "Melting Pot Tour" (the U.N. and Roosevelt Island), and the "Cultural Cornucopia Tour" (Lower East Side and East Village). The latter stopped at the Henry Street Settlement, Gus's Pickle Stand, Liz Christie Gardens, the 6th Street Indian restaurant row, Nuyorican Poets Café, St. Marks on the Bowery, the Russian and Turkish Baths, and four yoga dens. This is cultural tourism at its liveliest, though it is unclear how actively it addressed the problematics of gentrification, homelessness, redlining and other pressing social ills in relation to cultural issues.

A provocative community exercise in being a tourist in our own towns would be to ask people what local existing sites or buildings, artifacts, places they'd like to see preserved, and *why*. Times Square might not have been on my list during the forty years I lived in New York, but now that it's too late, it is suddenly on everyone's list. As early as 1914 the area was described as "a little bit of the underworld, a soupçon of the half-world – there you have the modern synthesis of New York as revealed in the neighborhood of Forty-Second Street." First-run movies showed in the odorous and slightly dangerous theaters of my youth, now gone, as is Grant's, where New Yorkers could buy a hot dog and be goggle-eyed tourists on the seamier side of their home town – not to mention the squalid porn shops, hustlers of every stripe, adult movies billed as "XXXstasy," and the sometimes violent street life. All gone now, replaced by Disneyfication, to make Times Square a safe place for mallrats and anathema for locals.

What is at stake in Times Square, according to literary scholar Andreas Huyssen, is "the transformation of a fabled place of popular culture in an age in which global entertainment conglomerates are rediscovering the value of the city and its millions of tourists for its marketing strategies." And where will Times Square's marginalized population (which made the place what it was for better and for worse) go now? Wherever they turn up, it's unlikely that a new place will ever achieve the historical and populist grandeur of its predecessor. Or can Times Square be reincarnated elsewhere? Maybe Disney will take that on too, if we don't. As architect Michael Sorkin concludes sadly, "of course, it's terribly true that the demise of Times Square, its conversion to another version of the recursion of Vegas (which has now built its own Times Square, even more pared down and distilled than the vanishing 'original'), must be blamed squarely not on the energetic advocates of sanitized fun but on our own failures to propose a better idea."

San Francisco has often been celebrated by its multifaceted artist and writer population. In 1984, a group of "activist punks" organized street theater action tours of corporations involved in nuclear energy and military intervention (modeled on the "Hall of Shame" tours of nuclear corporations in 1981 and preceding the "War Chest Tours" at the 1984 Democratic Convention). These enabled the anarchist Left, wrote David Solnit, to "collectively take their politics out of the underground shows and

into public spaces." Fourteen years later, his sister Rebecca Solnit lauds San Francisco's scale and its street life, which "still embodies the powerful idea of the city as a place of unmediated encounters," unlike other Western cities which are "merely enlarged suburbs, scrupulously controlled and segregated."

The city has been the site of several artists' tours, such as Jo Hanson's $5 tour of "Illegal Sights/ sites" in the early 1980s. Conceived as part of her "Art that's Sweeping the City," the environmental tours to ten sites were guided by community activists, exploring "the living city under the tourist attractions . . . focusing on the web of urban issues/ relationships through litter and dumping." The selected sites included Chinatown ("where you will see more of the alleys and markets than the tourist shops"), "Bay View and Hunter's Point, the Shadow of Candlestick Park, the victimization of unique Black communities by illegal dumping from outside," "Ocean Beach and its devastation," and "Twin Peaks, the breathtaking grand view strewn with litter down its steep slopes."

In 1994, artist Bernie Lubell and writers/professors/activists/artists Dean MacCannell and Juliet Flower MacCannell led a carefully considered bus tour of "unconventional sites" in San Francisco. Before they began they asked their passengers to make "metaphors of the city," handing out blank notebooks, a blank postcard, and a sheet of "suggestive words." They hoped to reconnect "the tourist quest to the fundamental human desire to see and know something else, resisting the conventional forms that have grown up around that desire, over-organizing and killing it." Aiming to produce "a common narrative of the city," the tour guides were convinced that "only emptiness can fill a space with possibility."

The tour ended up at the Palace of the Legion of Honor, where a new wing was being hastily built over a century-old paupers' graveyard despite protests from archaeologists. By involving their tourists in a current, unresolved controversy, Lubell and the MacCannells forced an intimate relationship with the place – -and with death, and perhaps with the imposed stasis, which is tourism itself.

A year later the three artists met again to discuss the tour, revealing some of their own motives. Lubell's "biggest surprise was the contribution of the tourists on the bus," the bits of information and insights garnered from them. Dean MacCannell saw the "tour itself as the missing key: the buttonholing, imperious, insistent sharing of the overlooked." Juliet MacCannell wanted to end the tour by "getting lost" and calling attention to the fog, "so they wouldn't be able to 'see' in any usual sense," so that San Francisco would become for them "a conspicuously imaginary object." Imagined places are, after all, one result of conventional tourism.

In 1996, Capp Street Project sponsored the four members of the Chicago collective called Haha (Richard House, Wendy Jacob, Laurie Palmer, and John Ploof) who took turns living with four local residents and shadowing them as they went about their daily lives. In the gallery they exhibited the resulting audio tours of the city sites visited, providing a curious inversion in which a local viewer was made privy to an outsider's experiences of nearby familiar sites, literally "shadows" of real experience.

Susan Schwartzenberg's *Cento: A Market Street Journal* was commissioned by the San Francisco Art Commission as part of a series intended to animate and illuminate the Market Street area for those passing through it. This compact, densely illustrated 116-page artist's book is "a combination walking tour guide, personal journal, and map," which was offered free to visitors from June 1996 to January 1997. The artist has lived in the city for some twenty years and makes no attempt to oversimplify her experience for rapid consumption. She describes Market Street as "not a place, but a sequence of places strung together where all manner of life experiences are acted out." After researching a variety of tour guides and historical archives, Schwartzenberg took to the streets with camera and tape recorder and interviewed anyone who was ready to talk. "Sometimes we talked about San Francisco and Market Street, but more often we talked about work, success and failure – life and its uncertainties."

Beautifully designed, crammed with images, quotations, interviews, and pockets of unexpected history, the Market Street journal is more experimental art than visitors' guide, but what a rich compendium of the kind of miscellany that turns out to be significant when a place means enough to enough people. For example, the hotels that historically harbored merchant seamen and

immigrants – they now cater to the elderly, the transient, the homeless, and to single (often Filipino) men – have been the first to fall to the wrecking ball. Development decrees that soon someone will build new hotels aimed at an entirely different clientele. Schwartzenberg's book also seems to be intended more for resident tourists than for those from elsewhere. As "a collage of voices and impressions," it replicates the random screen of daily encounters more accurately than the ordered view that is demanded, and needed, if the average tourist wants to make sense out of superficial experience. At the same time, the tourist's goal is supposedly to go backstage; if s/he succeeds, s/he is more likely to fall into the Market Street montage – "a confusing string of events and encounters we try endlessly to decipher." Schartzenberg quotes a private investigator: "It's a kind of psychological archeology. Everyone leaves traces – it's a matter of looking for them." Perhaps the ultimate in guidebooks is *The Visitor's Key to Iceland*, described by poet Eliot Weinberger as following "every road in the country step by step, as though one were walking with the Keeper of Memories. Iceland has few notable buildings, museums or monuments. What it has are hills and rivers and rocks, and each has a story the book recalls. Here was a stone bridge that collapsed behind an escaping convicted murderer, proving his innocence . . . This farm refused shelter to a pregnant woman, and was buried in a landslide that night . . . Here lived a popular postman in the eighteenth century . . . What other modern society so fully inhabits the landscape it lives in? Where else does the middle class still remember?"

Tourism has long-term effects on our places, given its connection to development, traceable back to the 1820s when fashionable tours wended north from New York, transforming Saratoga Springs, the Erie Canal abuilding, and Niagara Falls. Now that tourism is the last economic straw to be grasped (as it often was in the mid-nineteenth century too) the damnedest places are deemed tour-worthy. If your town hasn't been naturally endowed, if it's too new or too flat or too modernized to be intriguing, then attractions must be created from scratch – a theme park, an amusement park, a marina, a spa, a museum, or just a vast shopping opportunity. (On the other hand, I have heard of one town that posted a sign on its outskirts reading: WE AIN'T QUAINT.)

Theme parks and proliferating bed and breakfasts are not isolated phenomena of individual entrepreneurship. Resource exploitation and tourist development often go hand in hand. Corporations whitewash clear-cutting and strip-mining by masking their devastations with "parklands" as a "gift" to a gullible public, while government, too, is hardly above disguising its own agendas. In 1995, for instance, the New Mexico Department of Tourism ads featured Carlsbad Caverns, with no mention of the pending Waste Intensive Pilot Plant (WIPP), a national nuclear waste dump about to open at a nearby site that will imperil the same scenic routes through the "Land of Enchantment" that tourists traverse to get to the Caverns. For those aware of the federally funded WIPP, the glowing encrustaions pictured in the ad bear an eerie subtext, predicting one effect of the nuclear waste to be dumped there.

From turn-of-the-century boosterism when the hoopla was aimed at attracting railroads, businesses, and permanent settlers to the age of rapid transience when money spent is what counts, any place can be marketed, developed, and drastically changed in the process. Whether all the residents will like the transformation is another story. Some may be reluctant to make their homes a zoo but can be persuaded by the promise of jobs. Some would rather run their own business or farm their own land than build roads, clean toilets, make beds, and provide valet parking for the more fortunate. Some are forced on to the highway not by choice but by the need to follow an elusive seasonal job market. In the process, the dangers of tourism have escalated from being cheated to being murdered.

Viewed from the perspective of the places "visited," even in those disaster areas that are victims of downsizing and deindustrialization exacerbated by NAFTA and GATT, tourism is a mixed blessing – sometimes economically positive, usually culturally negative, and always resource-depleting (as measured by the "demo-flush" figures in which toilet use becomes an indicator of success for summer and weekend resort towns). Tourism leads to summer people leads to year-round newcomers leads to dispossession and a kind of internal colonialism. As an increasing amount of

the world's acreage is "opened up," the search for the "unspoiled" intensifies, exposing the most inaccessible places to commercial amenities and barbarities, from vandalism to jet-skis.

Tourists come to the American West, for instance, looking for places destroyed by shifting economies: Indian ruins, ghost towns, abandoned farms, deserted mines, and nineteenth-century spaces frozen in the governmentally managed wildernesses. For years now, Oregon, Colorado, and other states whose tourists have tended to come back and stay, have engendered a "bumper sticker jingoism": WELCOME TO OREGON, NOW GO HOME or, more brutally, GUT SHOOT 'EM AT THE BORDER. A popular Cape Cod T-shirt reads: I CHEAT DRUNKS AND TOURISTS. In Maine, some "natives" put out signs at the southern end of the turnpike: NEXT TIME JUST SEND THE MONEY, and the state has spent a good deal of money on campaigns begging the locals to be nice to tourists. One motive is plain old territorialism. Those same tourists at home may have similar attitudes about Mainers invading their own turf.

Sometimes we are tourists, sometimes we are toured. Even those who hate to travel, even those who live in out-of-the-way spots, have been exposed to tourists either in passing or as a sight to be seen. As we live what we perceive as ordinary lives, we are under surveillance – if not by the government then by the citizenry. I remember how startled I was when my picture was taken by some Japanese tourists leaning out of a bus as I schlepped my laundry through SoHo. Disheveled and purposeful in my black jeans, I was obviously a native. Turnabout is fair play, though as a tourist I'm more given to sidelong glances than stares and lenses. In Lower Manhattan, tourists are merely a nuisance. But in poorer "destinations," the divide is far greater, as Jamaica Kinkaid writes:

> That the native does not like the tourist is not hard to explain. . . . Every native would like to find a way out, every native would like a rest, every native would like a tour. But some natives – most natives in the world – cannot go anywhere. They are too poor. They are too poor to go anywhere. They are too poor to escape the reality of their lives; and they are too poor to live properly in the place where they live, which is the very place you, the tourist, want to go – so when the natives see you, the tourist, they

envy you, they envy your ability to leave your own banality and boredom, they envy your ability to turn their own banality and boredom into a source of pleasure for yourself.

If appalling disparities between classes are taken for granted, bypassed, and forgotten in metropolitan and international tourism ("It's none of my business; I can't do anything about it: it's their country"), they remain glaringly obvious when one is a tourist at home. Negotiating the contradictions demands a sensibility finely tuned to local politics and the global forces that drive it. Places presented to tourists as false unities are then broken down into thousands of fragments, since no place is seen exactly the same way by several people – let alone by people from different backgrounds, temperaments, and needs. As novelist John Nichols has said of his hometown:

> To make it palatable to visitors, our living culture in Taos is embalmed, sanitized, and presented much like a diorama in a museum: picturesque and safe. Tourists would rather not know that in many respects life here approximates the way four fifths of the globe survives . . . When commerce and social interaction take place solely with transients, culture and responsibility die. The town itself develops a transient soul.

One of the obvious contradictions in tourism concerns what is being escaped from and to. Absence (sometimes) makes the heart grow fonder. If we live away from native ground and then go home to visit, we can see the place anew, with fresh eyes. Some return to their hometowns to find the mines and factories they escaped now glorified as museums. The sad tale of Flint Auto World, a theme park simulating the immediate past in the corporate-abandoned town of Flint, Michigan, was told with tragicomic wit in Michael Moore's film, *Roger and Me*. Long popular in socialist countries, industrial tourism is catching on again in the United States. An edifying example is the themification of the history of tumultuous labor relations in Lawrence, Massachusetts, and in Lowell, where the first national park devoted to industry has caught on. Are the visitors simply the curious, the history buffs? Are they those who worked in factories or remember their parents' and grandparents' experiences? Are they lefties looking for landmarks of rebellion? The ways in which places and their histories are hidden, veiled, preserved, displayed, and perceived provide acute measures of the social unconscious. Yet their relationships to broad economic issues seldom surface overtly in daily lives. We live in a state of denial officially fostered by State denial.

Such grand-scale abdication from the present does not bode well for the future. One can only wonder what our hometowns will look like when the fad passes. Will the ghosts of fake ghost towns haunt the twenty-first century? Or will our places be ghosts smothered in new bodies we would never recognize as home? Tourism is a greased pig, a slippery target, like its offspring – sprawl. The new populations that spring up around tourist sites become a necessary evil, as purported insurance in the event that the tourist boom falls slack. But "sustainable tourism" may be an oxymoron along the lines of "military intelligence." Those of us at home in towns, counties, and states soon to be converted for display value are unprepared for these changes. We wake up only when it's too late to channel or control them. Soon my private village walking tours may run into other, more public ones. There may even be "sights" to see. Two centuries ago, William Blake wrote, "You never know what is enough, unless you know what is more than enough."

PART SEVEN

Difference

Courtesy of Alex Dorfsman

INTRODUCTION TO PART SEVEN

What is invisible, unspoken, and excluded from the world around us speaks every bit as loudly about what, and who, is important in society. Race, ethnicity, physical (dis)ability, sexuality, wealth, weight, age, and gender are only some of the ways in which societies distinguish, and oftentimes discriminate, amongst their members. Many cultural geographers today consider their work to be *critical* in some way. In other words, it is attentive to injustice and how injustice is spatially manifest, reinforced, and contested. Critical scholarship may also offer strategies for scholars, policymakers, or activists to help address these injustices. Because perceptions of cultural difference so often translate into a variety of types of injustice, critical cultural geography spans the gamut of possible research topics. Thus many of the selections in other parts of this *Reader* deal with difference, too. We debated the merits of including a separate part on difference at all, given that so many contemporary cultural geographers have incorporated in their work attentiveness to difference in its myriad forms. Ultimately, we determined to include a separate part on difference. Though the day when it is no longer necessary to underscore the importance of attending to difference may be near, there are still some who are unsure of the real value of focusing on, say, gender or sexuality or race. It is in fact useful to remind ourselves of how recently it was that work in these areas was considered to be marginal, often going unfunded and unpublished, and even viewed as threatening.

Difference in its multiple, and often intersecting, forms is closely linked to issues of *access*. As cultural geographers, we might think of access in at least two related ways. First, access can be taken literally, referring to the physical ease of, or impediments to, moving about in space. Doubtlessly you are aware – either through first-hand experience or through the experiences of friends or family members – that to be above-average weight, or openly gay, or poor, or female, very old or very young, or non-white, or making use of a wheelchair, limits where you can go in a very real spatial sense. You may not fit comfortably in coach class airline seats. You may have insults such as "fag" or "dike" directed at you in public spaces, which make you so uncomfortable or afraid that you choose to leave. You may not be able to afford the literal price of admission to privatized spaces, such as clubs; or the figurative "price" of admission to spaces that seem public but really are not – like shopping malls – if you are, or simply appear to be, poor. If you are black, you might well be followed by security guards in that mall; if you appear to be Arab, you may be stopped and searched every time you fly; if you appear to be Latino, it may be assumed that you are an immigrant and asked to document the legality of your presence time and again: all of this detainment and searching based on assumptions regarding your physiognomy, your accent, or your name. You, or your mother or sister, may have experienced catcalls from men on the street, groping on the subway, or even sexual attack. Perhaps you've used crutches after a fall. The need to plan elaborate travel trajectories around out-of-the-way ramps and out-of-service elevators brings home the hardships faced by those whose mobility is restricted every day because of a physical challenge. In all of these cases, choices about where and when to move about in space are often shaped by these very real barriers, which make certain people feel literally out of place.

Closely related to the question of physical access is access to the figurative, but no less important, benefits of cultural belonging. These benefits may include jobs, promotions, and raises; the likelihood of being elected to public office or even the right to vote; or a choice of who you may date, or if you may adopt children or marry. Indeed, there is a whole host of social, cultural, political, and economic "goods" that is placed off-limits to certain categories of people who are considered different, marginal, and thus not deserving of the benefits of cultural belonging. In the United States, the struggles of many lesbians and gays to extend work-related health benefits to their domestic partners is emblematic of these exclusions; while the controversy over headscarves for Muslim women in French public schools – a controversy that in great measure turns on the issue of allowing visible differences in public places (discussed in the Introduction to the *Reader*) – also speaks to issues of difference and exclusion.

Why is it that most, if not all, societies make such sharp distinctions between who is considered normal and who is considered different? All of the pieces in this section deal with this question; in particular, however, the selections by Edward Said and David Sibley address the question in the most encompassing ways. Said develops the notion of *Orientalism*, which he defines as the West's long-standing practice of defining the East (or Orient, though Said concentrates particularly on Muslim Southwest Asia/North Africa) as radically unlike the Christian West in all meaningful ways. Sibley uses *object relations theory*, derived from psychoanalytic theory, to suggest that as individuals and as societies we establish our identity through a process of exclusion, through specifying and then spatially marginalizing those we determine to be different from ourselves. Though they take different intellectual routes to do so, both Said and Sibley are describing *the Other*. The notion of the Other is used together with its counterpart, *the Self*, to arrive at the dyad of Self and Other which, for many scholars across the social sciences and humanities, is the basis of identity formation. Central to establishing a firm sense of Self is what may seem at first glance the rather backward process of establishing a firm sense of what the Self *is not*: in other words, specifying who the Other *is*. These notions are central to *postcolonial studies*, which draw largely on foundational work by scholars from formerly colonized places – such as Edward Said – who explore how Western identity formation has often violently turned on the establishment, and subsequent marginalization, of the Other.

For geographers, this has important spatial dimensions. Take for instance the practice of defining, policing, and defending a country's borders. Why is it so important to do this? Why are countries so defensive when these borders are challenged by outsiders, either through military aggression or immigration? Cultural geographers have addressed these questions in a variety of ways (see, for example, Part Five). With regard to the question of difference, national borders establish a spatial demarcation between "us" (citizens who belong) and "them" (outsiders, or barbarians, who are different from us). Indeed, designating an Other "over there" is a fundamental step toward constructing an "us" that exists here, and by contrast to our marginalized and excluded other. When political borders are challenged, it is a challenge to national integrity, which hinges on the maintenance of sameness though assimilating outsiders, or preventing them from entering in the first place. As several of the selections illustrate, the scale of analysis can be shifted closer-in to examine Self/Other boundaries in cities, schools, and even to conceptualize the body itself as a sort of boundary.

It is only relatively recently that human geography as a discipline has become concerned with difference. The 1970s, for example, saw an interest on the part of Marxist geographers in socio-economic class as an important, but often overlooked, aspect of societal inclusion and exclusion. This scholarship was often *radical* in spirit, meaning that revolutionary social change was posited as the best way to redress class-based inequality. In the early 1980s (with some isolated work in the 1970s, as well) some human geographers began to explore difference through gender. Some of the earliest of this work utilized a *socialist-feminist* framework, arguing that the oppression of women was similar in some respects to the oppression of the working class, but that a focus on the private sphere of the home, in addition to the public sphere of the workplace, was needed to fully understand the subordination of women in capitalist society. The notion of *patriarchy*, or the systematic male dominance of women, became an important analytical lens for cultural geographers studying gender difference. Monk and Hanson's essay, included here, made the case that the omission of the experiences and perspectives of half of humanity – women – had resulted in an impoverished stock of geographic knowledge. Though they posit that the inclusion of women practitioners and women's perspectives will provide one way to gradually redress the *masculinist* (male-oriented) bias of mainstream geography – a *liberal* perspective – they are also quite aware that, given the depth of gender discrimination in geography, more radical transformation would also have to occur.

Since these early forays into the terrain of difference by geographers, critical cultural geography has on the whole become much more nuanced in its approach. With respect to gender, contemporary cultural geographers have by and large moved beyond a totalizing notion of patriarchy and binary male–female and public–private distinctions, focusing instead on a rich variety of historically and place-specific topics ranging from gender as a performance, gender and the nation-state, gendered constructions of home, embodiment, and masculinities. Feminist geographers have by and large shifted away from a focus on material inequalities between men and women – though there is widespread recognition that documentation of these continues to be a crucial endeavor – and toward theoretically informed under-standings of identity, embodiment, and sexuality. So-called *Third World feminisms* have emphasized the importance of place, power, and history to geographers' understandings of the negotiation, performance, and resistance of gendered subjectivities (see Abu-Lughod, p. 50). These, and other, scholars – for example those working on border studies, Africana and New World studies, American studies, and legal studies – have brought into the conversation considerations of racialized difference. *Postmodernism*, with its insistence on the instability of language, and its focus on symbol and representation, has also influenced cultural geographies of difference. Finally, cultural geographers working on issues of differ-ence have always had a keen awareness, if not active involvement themselves, in advocating social change to redress the injustices they document. Thus there has been a significant expansion and sophistication of gender studies in cultural geography, as well as a flowering of studies of difference by cultural geo-graphers that do not have gender as their emphasis. As these contemporary cultural geographies of difference diverge from the subfield's early and enduring focus on gender, a delicate ground must be tread between apprehending the social world in all its interrelated complexity versus emphasizing one aspect of difference above all others in order to make an intellectual or political point. As Robyn Longhurst put it, "While it is difficult, and not always politically strategic, to examine simultaneously multiple axes of subjectivity, in some instances it may prove enlightening."[1]

In some key ways, vast institutional, as well as conceptual, strides have been made over the past twenty-five years, changes that work toward realizing Monk and Hanson's vision of a more-inclusive human geography. Non-traditional students, particularly women, have become better represented among the ranks of geographers, and some have gone on to achieve prominent positions in academia and national geographic associations. Many established scholars have reconsidered and broadened their research agendas to be more attentive to difference. Methodologies that are closely associated with exploring aspects of difference in other disciplines (particularly anthropology) have become more widespread among cultural geographers, particularly qualitative methods such as interviews, focus groups, journaling, and photo elicitation. That said, the discipline of geography still has a long way to go in becoming a truly inclusive endeavor. Today, this is particularly true with regard to the inclusion of non-white scholars among the ranks of practicing geographers who publish, teach, and represent the discipline to society at large.

NOTE

1   R. Longhurst, 'Introduction: Subjectivities, Spaces and Places,' in K. Anderson, M. Domosh, S. Pile, and N. Thrift (eds.) *Handbook of Cultural Geography* (2003), p. 286.

"Imaginative Geography and its Representations: Orientalizing the Oriental"

from *Orientalism* (1978)

Edward Said

Editors' introduction

In *Orientalism*, literary critic Edward Said advanced the thesis that specialists on non-Western areas of the world have constructed and perpetuated a collective fantasy about the Orient. In even the most supposedly objective scholarship, the Orient is depicted as all the West is not: feminized, where the West is masculine; weak, where the West is strong; corrupt, where the West is righteous; inscrutable, where the West is rational; tradition-bound, where the West is progressive. These oppositions, according to Said, serve to construct a positive image of the West by contrasting it to a negative mirror image: the Orient. In other words, the Orient is the West's *Other*. This has enabled the political, economic, cultural, and social domination of the West not just during colonial times, but also in the present. One has only to examine the popular press's generally negative portrayal of Muslim South West Asia (or of Muslims in Europe and the United States, for that matter) to appreciate how relevant Said's point still is today.

Orientalism's influence reverberated throughout the academic world, forming a cornerstone of *post-colonial studies*. However, *Orientalism* has had special resonance for geographers, given its inherently spatial argument. The notion that human societies typically form place-based identities wherein "others" are "over there," while "we" are "here," along with the understanding of places and knowledge about them as ideological constructions, rather than straightforward facts, underlies a great deal of contemporary critical geographic scholarship. Cultural geographers working on issues of postcolonialism, travel writing, and economic development owe a great intellectual debt to Said. Examples of such work include Catherine A. Lutz and Jane L. Collins's *Reading National Geographic* (1993), which examines the social construction of Western understandings of the so-called Third World in the magazine's photography and text that invariably exoticizes unfamiliar people and places; while Derek Gregory, in *The Colonial Present: Afghanistan, Palestine, Iraq* (2004), focuses a geographer's eye on post-September 11 understandings of terrorism, US–Middle East relations, and the colonial roots of contemporary military conflicts.

Edward Wadie Said (1935–2003) was born in Jerusalem. His father, a Christian Palestinian who had become an American citizen, was a businessman; while his mother, a Christian of Lebanese and Palestinian ancestry, instilled in Edward and his siblings a love of literature and music. The wealthy family traveled between Cairo, Egypt, and Jerusalem (then politically a part of Palestine), and spent their summers in Lebanon. Edward Said attended elite British colonial-style schools in Cairo and Jerusalem until 1948, when the Arab–Israeli war saw the family home in Jerusalem annexed by the newly created state of Israel. At fifteen years of age, Said was sent to a private school in Massachusetts, attended college at Princeton, and graduate school at Harvard.

His autobiography, *Out of Place: A Memoir* (1999) details Said's early years, and emphasizes his enduring sense of being an outsider. Yet he was able to draw productively on his in-between status as a Christian steeped in Muslim Arab culture, a wealthy man dispossessed of his Palestinian homeland, and academic royalty who nevertheless took a critical stance toward the academy.

Edward Said's supporters recognized the subtle humanism inflecting his trenchant critique of imperialism, his tireless advocacy on behalf of the Palestinians, and his many intellectual and artistic gifts. For example, Said was renowned for his musical talent as a pianist. Politically, however, as an advocate of a single Jewish–Arab state, Said found himself in a difficult position. On the one hand, Said's pro-Palestinian detractors criticized him for his concessions to Zionism while, on the other hand, his pro-Israeli foes were angered at his denouncement of human rights violations by Israel, and his criticism of U.S. foreign policy in the Middle East. Intellectually, too, some saw Said's criticism of colonialism to be hypocritical, given that Said himself had benefited handsomely from his family's wealth, his privileged education, and the rewards he reaped as an academic luminary.

A Professor of English and Comparative Literature at Columbia University, as well as an outspoken advocate for the Palestinian cause, Said was the consummate public intellectual. A prolific writer, Said's most acclaimed publications are *Orientalism*, which is excerpted here, and *Culture and Imperialism* (1993). Many of his scholarly publications, interviews, and essays emphasized Said's conviction that the duty of a public intellectual is to "speak the truth to power"; see for example *Representations of the Intellectual* (1994). Said also published extensively on the Palestinian question; see for example *The Question of Palestine* (1979) and *The End of the Peace Process: and After* (2001).

Strictly speaking, Orientalism is a field of learned study. In the Christian West, Orientalism is considered to have commenced its formal existence with the decision of the Church Council of Vienne in 1312 to establish a series of chairs in "Arabic, Greek, Hebrew, and Syriac at Paris, Oxford, Bologna, Avignon, and Salamanca." Yet any account of Orientalism would have to consider not only the professional Orientalist and his work but also the very notion of a field of study based on a geographical, cultural, linguistic, and ethnic unit called the Orient. Fields, of course, are made. They acquire coherence and integrity in time because scholars devote themselves in different ways to what seems to be a commonly agreed-upon subject matter. Yet it goes without saying that a field of study is rarely as simply defined as even its most committed partisans – usually scholars, professors, experts, and the like – claim it is. Besides, a field can change so entirely, in even the most traditional disciplines like philology, history, or theology, as to make an all-purpose definition of subject matter almost impossible. This is certainly true of Orientalism, for some interesting reasons.

To speak of scholarly specialization as a geographical "field" is, in the case of Orientalism, fairly revealing since no one is likely to imagine a field symmetrical to it called Occidentalism. Already the special, perhaps even eccentric attitude of Orientalism becomes apparent. For although many learned disciplines imply a position taken towards, say, *human* material (a historian deals with the human past from a special vantage point in the present), there is no real analogy for taking a fixed, more or less total geographical position towards a wide variety of social, linguistic, political, and historical realities. A classicist, a Romance specialist, even an Americanist focuses on a relatively modest portion of the world, not on a full half of it. But Orientalism is a field with considerable geographical ambition. And since Orientalists have traditionally occupied themselves with things Oriental (a specialist in Islamic law, no less than an expert in Chinese dialects or in Indian religions, is considered an Orientalist by people who call themselves Orientalists), we must learn to accept enormous, indiscriminate size plus an almost infinite capacity for subdivision as one of the chief characteristics of Orientalism – one that is evidenced in its confusing amalgam of imperial vagueness and precise detail.

All of this describes Orientalism as an academic discipline. The "ism" in Orientalism serves to insist on the distinction of this discipline from every other kind. The rule in its historical development as an academic discipline has been its increasing scope, not its greater selectiveness. Renaissance Orientalists like Erpenius and Guillaume Postel were primarily specialists in the languages of the Biblical provinces, although Postel boasted that he could get across Asia as far as China without needing an interpreter. By and large, until the mid-eighteenth century Orientalists were Biblical scholars, students of the Semitic languages, Islamic specialists, or, because the Jesuits had opened up the new study of China, Sinologists. The whole middle expanse of Asia was not academically conquered for Orientalism until, during the later eighteenth century, Anquetil-Duperron and Sir William Jones were able intelligibly to reveal the extraordinary riches of Avestan and Sanskrit. By the middle of the nineteenth century Orientalism was as vast a treasure-house of learning as one could imagine. There are two excellent indices of this new, triumphant eclecticism. One is the encyclopedic description of Orientalism roughly from 1765 to 1850 given by Raymond Schwab in his *La Renaissance orientale*. Quite aside from the scientific discoveries of things Oriental made by learned professionals during this period in Europe, there was the virtual epidemic of Orientalia affecting every major poet, essayist, and philosopher of the period. Schwab's notion is that "Oriental" identifies an amateur or professional enthusiasm for everything Asiatic, which was wonderfully synonymous with the exotic, the mysterious, the profound, the seminal. ... A nineteenth-century Orientalist was therefore either a scholar (a Sinologist, an Islamicist, an Indo-Europeanist) or a gifted enthusiast (Hugo in *Les Orientales*, Goethe in the *Westostlicher Diwan*), or both (Richard Burton, Edward Lane, Friedrich Schlegel).

The second index of how inclusive Orientalism had become since the Council of Vienne is to be found in nineteenth-century chronicles of the field itself. The most thorough of its kind is Jules Mohl's *Vingt-sept ans d'histoire des études orientales*, a two-volume logbook of everything of note that took place in Orientalism between 1840 and 1867. Mohl was the secretary of the Société asiatique in Paris, and for something more than the first half of the nineteenth century Paris was the capital of the Orientalist world. ... There is scarcely anything done by a European scholar touching Asia during those twenty-seven years that Mohl does not enter under "études orientales." His entries of course concern publications, but the range of published material of interest to Orientalist scholars is awesome. Arabic, innumerable Indian dialects, Hebrew, Pehlevi, Assyrian, Babylonian, Mongolian, Chinese, Burmese, Mesopotamian, Javanese: the list of philological works considered Orientalist is almost uncountable. Moreover, Orientalist studies apparently cover everything from the editing and translation of texts to numismatic, anthropological, archaeological, sociological, economic, historical, literary, and cultural studies in every known Asiatic and North African civilization, ancient and modern. ...

Such eclecticism as this had its blind spots, nevertheless. Academic Orientalists for the most part were interested in the classical period of whatever language or society it was that they studied. Not until quite late in the century ... was much attention given to the academic study of the modern, or actual, Orient. Moreover, the Orient studied was a textual universe by and large; the impact of the Orient was made through books and manuscripts. ... Even the rapport between an Orientalist and the Orient was textual, so much so that it is reported of some of the early nineteenth-century German Orientalists that their first view of an eight-armed Indian statue cured them completely of their Orientalist taste. When a learned Orientalist traveled in the country of his specialization, it was always with unshakable abstract maxims about the "civilization" he had studied; rarely were Orientalists interested in anything except proving the validity of these musty "truths" by applying them, without great success, to uncomprehending, hence degenerate, natives. Finally, the very power and scope of Orientalism produced not only a fair amount of exact positive knowledge about the Orient but also a kind of second-order knowledge – lurking in such places as the "Oriental" tale, the mythology of the mysterious East, notions of Asian inscrutability – with a life of its own, what V.G. Kiernan has aptly called "Europe's collective daydream of the Orient." One happy result of this is that an estimable number of important writers during the nineteenth century were Oriental enthusiasts: It is perfectly correct, I think, to speak of a genre of Orientalist writing as exemplified in the works of Hugo, Goethe, Nerval, Flaubert, Fitzgerald, and the like. What inevitably goes with such work, however, is a kind of free-floating mythology of the Orient, an Orient that derives not only from contemporary attitudes and popular prejudices but also from what Vico called the conceit of nations and of scholars. . . .

Today an Orientalist is less likely to call himself an Orientalist than he was almost any time up to World War II. Yet the designation is still useful, as when universities maintain programs or departments in Oriental languages or Oriental civilizations. There is an Oriental "faculty" at Oxford, and a department of Oriental studies at Princeton. As recently as 1959, the British government empowered a commission "to review developments in the Universities in the fields of Oriental, Slavonic, East European and African studies . . . and to consider, and advise on, proposals for future development." The Hayter Report, as it was called when it appeared in 1961, seemed untroubled by the broad designation of the word Oriental, which it found serviceably employed in American universities as well. For even the greatest name in modern Anglo-American Islamic studies, H.A.R. Gibb, preferred to call himself an Orientalist rather than an Arabist. Gibb himself, classicist that he was, could use the ugly neologism "area study" for Orientalism as a way of showing that area studies and Orientalism after all were interchangeable geographical titles. But this, I think, ingenuously belies a much more interesting relationship between knowledge and geography. I should like to consider that relationship briefly.

Despite the distraction of a great many vague desires, impulses, and images, the mind seems persistently to formulate what [the anthropologist] Claude Lévi-Strauss has called a science of the concrete. A primitive tribe, for example, assigns a definite place, function, and significance to every leafy species in its immediate environment. Many of these grasses and flowers have no practical use; but the point Lévi-Strauss makes is that mind requires order, and order is achieved by discriminating and taking note of everything, placing everything of which the mind is aware in a secure, refindable place, therefore giving things some role to play in the economy of objects and identities that make up an environment. This kind of rudimentary classification has a logic to it, but the rules of the

logic by which a green fern in one society is a symbol of grace and in another is considered maleficent are neither predictably rational nor universal. There is always a measure of the purely arbitrary in the way the distinctions between things are seen. And with these distinctions go values whose history, if one could unearth it completely, would probably show the same measure of arbitrariness. This is evident enough in the case of fashion. Why do wigs, lace collars, and high buckled shoes appear, then disappear, over a period of decades? Some of the answer has to do with utility and some with the inherent beauty of the fashion. But if we agree that all things in history, like history itself, are made by men, then we will appreciate how possible it is for many objects or places or times to be assigned roles and given meanings that acquire objective validity only *after* the assignments are made. This is especially true of relatively uncommon things, like foreigners, mutants, or "abnormal" behavior.

It is perfectly possible to argue that some distinctive objects are made by the mind, and that these objects, while appearing to exist objectively, have only a fictional reality. A group of people living on a few acres of land will set up boundaries between their land and its immediate surroundings and the territory beyond, which they call "the land of the barbarians." In other words, this universal practice of designating in one's mind a familiar space which is "ours" and an unfamiliar space beyond "ours" which is "theirs" is a way of making geographical distinctions that *can be* entirely arbitrary. I use the word "arbitrary" here because imaginative geography of the "our land – barbarian land" variety does not require that the barbarians acknowledge the distinction. It is enough for "us" to set up these boundaries in our own minds; "they" become "they" accordingly, and both their territory and their mentality are designated as different from "ours." To a certain extent modern and primitive societies seem thus to derive a sense of their identities negatively. A fifth-century Athenian was very likely to feel himself to be non-barbarian as much as he positively felt himself to be Athenian. The geographic boundaries accompany the social, ethnic, and cultural ones in expected ways. Yet often the sense in which someone feels himself to be not-foreign is based on a very unrigorous idea of what is "out there," beyond one's own territory. All kinds of suppositions, associations, and fictions appear to crowd the unfamiliar space outside one's own.

The French philosopher Gaston Bachelard once wrote an analysis of what he called the poetics of space. The inside of a house, he said, acquires a sense of intimacy, secrecy, security, real or imagined, because of the experiences that come to seem appropriate for it. The objective space of a house – its corners, corridors, cellar, rooms – is far less important than what poetically it is endowed with, which is usually a quality with an imaginative or figurative value we can name and feel: thus a house may be haunted, or home-like, or prison-like, or magical. So space acquires emotional and even rational sense by a kind of poetic process, whereby the vacant or anonymous reaches of distance are converted into meaning for us here. The same process occurs when we deal with time. Much of what we associate with or even know about such periods as "long ago" or "the beginning" or "at the end of time" is poetic – made up. For a historian of Middle Kingdom Egypt, "long ago" will have a very clear sort of meaning, but even this meaning does not totally dissipate the imaginative, quasi-fictional quality one senses lurking in a time very different and distant from our own. For there is no doubt that imaginative geography and history help the mind to intensify its own sense of itself by dramatizing the distance and difference between what is close to it and what is far away. This is no less true of the feelings we often have that we would have been more "at home" in the sixteenth century or in Tahiti.

Yet there is no use in pretending that all we know about time and space, or rather history and geography, is more than anything else imaginative. There are such things as positive history and positive geography which in Europe and the United States have impressive achievements to point to. Scholars now do know more about the world, its past and present, than they did. . . . Yet this is not to say that they know all there is to know, nor, more important, is it to say that what they know has effectively dispelled the imaginative geographical and historical knowledge I have been considering. We need not decide here whether this kind of imaginative knowledge infuses history and geography, or whether in some way it overrides them. Let us just say for the time being that it is there as something

more than what appears to be merely positive knowledge.

Almost from earliest times in Europe the Orient was something more than what was empirically known about it. At least until the early eighteenth century . . . European understanding of one kind of Oriental culture, the Islamic, was ignorant but complex. For certain associations with the East – not quite ignorant, not quite informed – always seem to have gathered around the notion of an Orient. Consider first the demarcation between Orient and West. It already seems bold by the time of the *Iliad*.

[. . .]

The two aspects of the Orient that set it off from the West . . . will remain essential motifs of European imaginative geography. A line is drawn between two continents. Europe is powerful and articulate; Asia is defeated and distant. . . . It is Europe that articulates the Orient; this articulation is the prerogative, not of a puppet master, but of a genuine creator, whose life-giving power represents, animates, constitutes the otherwise silent and dangerous space beyond familiar boundaries. . . . Secondly, there is the motif of the Orient as insinuating danger. Rationality is undermined by Eastern excesses, those mysteriously attractive opposites to what seem to be normal values. . . . Hereafter Oriental mysteries will be taken seriously, not least because they challenge the rational Western mind to new exercises of its enduring ambition and power.

But one big division, as between West and Orient, leads to other smaller ones, especially as the normal enterprises of civilization provoke such outgoing activities as travel, conquest, new experiences. In classical Greece and Rome geographers, historians, public figures like Caesar, orators, and poets added to the fund of taxonomic lore separating races, regions, nations, and minds from each other; much of that was self-serving, and existed to prove that Romans and Greeks were superior to other kinds of people. But concern with the Orient had its own tradition of classification and hierarchy. From at least the second century B.C. on, it was lost on no traveler or eastward-looking and ambitious Western potentate that Herodotus – historian, traveler, inexhaustibly curious chronicler – and Alexander – king warrior, scientific conqueror – had been in the Orient before. The Orient was therefore subdivided into realms previously known, visited, conquered, by Herodotus and Alexander as well as their epigones, and those realms not previously known, visited, conquered. Christianity completed the setting up of main intra-Oriental spheres: there was a Near Orient and a Far Orient, a familiar Orient . . . and a novel Orient. The Orient therefore alternated in the mind's geography between being an Old World to which one returned, as to Eden or Paradise, there to set up a new version of the old, and being a wholly new place to which one came as Columbus came to America, in order to set up a New World (although, ironically, Columbus himself thought that he had discovered a new part of the Old World). Certainly neither of these Orients was purely one thing or the other: it is their vacillations, their tempting suggestiveness, their capacity for entertaining and confusing the mind, that are interesting.

Consider how the Orient, and in particular the Near Orient, became known in the West as its great complementary opposite since antiquity. There were the Bible and the rise of Christianity; there were travelers like Marco Polo who charted the trade routes and patterned a regulated system of commercial exchange, and after him Lodovico di Varthema and Pietro della Valle; there were fabulists like Mandeville; there were the redoubtable conquering Eastern movements, principally Islam, of course; there were the militant pilgrims, chiefly the Crusaders. Altogether an internally structured archive is built up from the literature that belongs to these experiences. Out of this comes a restricted number of typical encapsulations: the journey, the history, the fable, the stereotype, the polemical confrontation. These are the lenses through which the Orient is experienced, and they shape the language, perception, and form of the encounter between East and West. What gives the immense number of encounters some unity, however, is the vacillation I was speaking about earlier. Something patently foreign and distant acquires, for one reason or another, a status more rather than less familiar. One tends to stop judging things either as completely novel or as completely well known; a new median category emerges, a category that allows one to see new things, things seen for the first time, as versions of a previously known thing. In essence such a category is not so much a way of receiving new information as it is a method of controlling what seems to be a threat to some established view of things. If the mind must suddenly deal with what it takes to be a radically new form of life – as Islam appeared to Europe in the early Middle Ages – the response on the whole is conservative and defensive. Islam is judged to be a fraudulent new version of some previous experience, in this case Christianity. The threat is muted, familiar values impose themselves, and in the end the mind reduces the pressure upon it by accommodating things to itself as either "original" or "repetitious." Islam thereafter is "handled": its novelty and its suggestiveness are brought under control so that relatively nuanced discriminations are now made that would have been impossible had the raw novelty of Islam been left unattended. The Orient at large, therefore, vacillates between the West's contempt for what is familiar and its shivers of delight in – or fear of – novelty.

Yet where Islam was concerned, European fear, if not always respect, was in order. After Mohammed's death in 632, the military and later the cultural and religious hegemony of Islam grew enormously. First Persia, Syria, and Egypt, then Turkey, then North Africa fell to the Muslim armies; in the eighth and ninth centuries Spain, Sicily, and parts of France were conquered. By the thirteenth and fourteenth centuries Islam ruled as far east as India, Indonesia, and China. And to this extraordinary assault Europe could respond with very little except fear and a kind of awe. Christian authors witnessing the Islamic conquests had scant interest in the learning, high culture, and frequent magnificence of the Muslims. . . . What Christians typically felt about the Eastern armies was that they had "all the appearance of a swarm of bees, but with a heavy hand . . . they devastated everything": so wrote Erchembert, a cleric in Monte Cassino in the eleventh century.

Not for nothing did Islam come to symbolize terror, devastation, the demonic, hordes of hated barbarians. For Europe, Islam was a lasting trauma. Until the end of the seventeenth century the "Ottoman peril" lurked alongside Europe to represent for the whole of Christian civilization a constant danger, and in time European civilization incorporated that peril and its lore, its great events, figures, virtues, and vices, as something woven into the fabric of life. . . . The point is that what remained current about Islam was some necessarily diminished version of those great dangerous forces that it symbolized for Europe. . . . [T]he European representation of the Muslim, Ottoman, or Arab was always a way of controlling the redoubtable Orient, and to a certain extent the same is true of the methods of contemporary learned Orientalists, whose subject is not so much the East itself as the East made known, and therefore less fearsome, to the Western reading public.

There is nothing especially controversial or reprehensible about such domestications of the exotic; they take place between all cultures, certainly, and between all men. My point, however, is to emphasize the truth that the Orientalist, as much as anyone in the European West who thought about or experienced the Orient, performed this kind of mental operation. But what is more important still is the limited vocabulary and imagery that impose themselves as a consequence. The reception of Islam in the West is a perfect case in point. . . . One constraint acting upon Christian thinkers who tried to understand Islam was an analogical one; since Christ is the basis of Christian faith, it was assumed – quite incorrectly – that Mohammed was to Islam as Christ was to Christianity. Hence the polemic name "Mohammedanism" given to Islam, and the automatic epithet "imposter" applied to Mohammed. . . . Islam became an image . . . whose function was not so much to represent Islam in itself as to represent it for the medieval Christian.

[. . .]

Our initial description of Orientalism as a learned field now acquires a new concreteness. A field is often an enclosed space. The idea of representation is a theatrical one: the Orient is the stage on which the whole East is confined. On this stage will appear figures whose role it is to represent the larger whole from which they emanate. The Orient then seems to be, not an unlimited extension beyond the familiar European world, but rather a closed field, a theatrical stage affixed to Europe. An Orientalist is but the particular specialist in knowledge for which Europe at large is responsible, in the way that an audience is historically and culturally responsible for (and responsive to) dramas technically put together by the dramatist. In the depths of this Oriental stage stands a prodigious cultural repertoire whose individual items evoke a fabulously rich world: the Sphinx, Cleopatra, Eden, Troy,

Sodom and Gomorrah, Astarte, Isis and Osiris, Sheba, Babylon, the Genii, the Magi, Nineveh, Prester John, Mahomet, and dozens more; settings, in some cases names only, half-imagined, half-known; monsters, devils, heroes; terrors, pleasures, desires. The European imagination was nourished extensively from this repertoire: between the Middle Ages and the eighteenth century such major authors as Ariosto, Milton, Marlowe, Tasso, Shakespeare, Cervantes, and the authors of the *Chanson de Roland* and the *Poema del Cid* drew on the Orient's riches for their productions, in ways that sharpened the outlines of imagery, ideas, and figures populating it. In addition, a great deal of what was considered learned Orientalist scholarship in Europe pressed ideological myths into service, even as knowledge seemed genuinely to be advancing.

[. . .]

This whole didactic process is neither difficult to understand nor difficult to explain. One ought again to remember that all cultures impose corrections upon raw reality, changing it from free-floating objects into units of knowledge. The problem is not that conversion takes place. It is perfectly natural for the human mind to resist the assault on it of untreated strangeness; therefore cultures have always been inclined to impose complete transformations on other cultures, receiving these other cultures not as they are but as, for the benefit of the receiver, they ought to be. To the Westerner, however, the Oriental was always *like* some aspect of the West; to some of the German Romantics, for example, Indian religion was essentially an Oriental version of Germano-Christian pantheism. Yet the Orientalist makes it his work to be always converting the Orient from something into something else: he does this for himself, for the sake of his culture, in some cases for what he believes is the sake of the Oriental. This process of conversion is a disciplined one: it is taught, it has its own societies, periodicals, traditions, vocabulary, rhetoric, all in basic ways connected to and supplied by the prevailing cultural and political norms of the West. . . .

[. . .]

Imaginative geography . . . legitimates a vocabulary, a universe of representative discourse peculiar to the discussion and understanding of Islam and of the Orient. What this discourse considers to be a fact – that Mohammed is an imposter, for example – is a component of the discourse, a statement the discourse compels one to make whenever the name Mohammed occurs. Underlying all the different units of Orientalist discourse – by which I mean simply the vocabulary employed whenever the Orient is spoken or written about – is a set of representative figures, or tropes. These figures are to the actual Orient – or Islam, which is my main concern here – as stylized costumes are to characters in a play. . . . In other words, we need not look for correspondence between the language used to depict the Orient and the Orient itself, not so much because the language is inaccurate but because it is not even trying to be accurate. What it is trying to do . . . is at one and the same time to characterize the Orient as alien and to incorporate it schematically on a theatrical stage whose audience, manager, and actors are *for* Europe, and only for Europe. Hence the vacillation between the familiar and the alien; Mohammed is always the imposter (familiar, because he pretends to be like the Jesus we know) and always the Oriental (alien, because although he is in some ways "like" Jesus, he is after all not like him).

Rather than listing all the figures of speech associated with the Orient – its strangeness, its difference, its exotic sensuousness, and so forth – we can generalize about them as they were handed down through the Renaissance. They are all declarative and self-evident; the tense they employ is the timeless eternal; they convey an impression of repetition and strength; they are always symmetrical to, and yet diametrically inferior to, a European equivalent, which is sometimes specified, sometimes not. For all these functions it is frequently enough to use the simple copula *is*. Thus,

Mohammed *is* an imposter. . . . No background need be given; the evidence necessary to convict Mohammed is contained in the "is." One does not qualify the phrase, neither does it seem necessary to say that Mohammed *was* an imposter, nor need one consider for a moment that it may not be necessary to repeat the statement. It *is* repeated, he *is* an imposter, and each time one says it, he becomes more of an imposter and the author of the statement gains a little more authority in having declared it. Thus Humphrey Prideaux's famous seventeenth-century biography of Mohammed is subtitled *The True Nature of Imposture*. Finally, of course, such categories as imposter (or Oriental, for that matter) imply, indeed require, an opposite that is neither fraudulently something else nor endlessly in need of explicit identification. And that opposite is "Occidental," or in Mohammed's case, Jesus.

Philosophically, then, the kind of language, thought, and vision that I have been calling Orientalism very generally is a form of radical realism; anyone employing Orientalism, which is the habit for dealing with questions, objects, qualities, and regions deemed Oriental, will designate, name, point to, fix what he is talking or thinking about with a word or phrase, which then is considered either to have acquired, or more simply to be, reality. Rhetorically speaking, Orientalism is absolutely anatomical and enumerative: to use its vocabulary is to engage in the particularizing and dividing of things Oriental into manageable parts. Psychologically, Orientalism is a form of paranoia, knowledge of another kind, say, from ordinary historical knowledge. These are a few of the results, I think, of imaginative geography and of the dramatic boundaries it draws. . . .

"On not Excluding Half of the Human in Human Geography"

from *The Professional Geographer*
34 (1982): 11–23

Janice Monk and Susan Hanson

Editors' introduction

It is hard to believe that merely three decades ago, gender was not on the radar screen of cultural geography. Gender was neither a focus of geographic theory or research, nor did it constitute much of a concern regarding the demographics of the discipline's practitioners. On the latter, in 1973 Wilbur Zelinsky (see also p. 113) published two linked articles in *The Professional Geographer* (25, 2): "The Strange Case of the Missing Female Geographer" (pp. 101–105); and "Women in Geography: a brief factual account" (pp. 151–165). The first article begins with these ominous words: "I bear evil tidings. By every objective measure that can be mustered, the lot of the female geographer is, and has been, a discouraging one; and there is little assurance of substantial improvement during the foreseeable future" (p. 101). Zelinsky noted the institutional biases that discouraged women from attaining doctorates, from working outside the home, or from continuing with their work once they married or had children. If women persisted despite these cultural barriers, disciplinary gender biases practically assured that professional women geographers would publish less, earn less, and fail to be promoted. Today, the statistics are more encouraging than in Zelinsky's time, particularly with regard to the slightly higher number of women than men who earn bachelor's degrees in geography. Yet when women's progress through the professional ranks of academia is examined, there is less to be enthusiastic about. For example, in the United States in 1970 – the year Zelinsky took as his base – women held 6.2 per cent of assistant professorships in geography (the lowest tenure-stream rank in the US academic system) and a mere 2.9 per cent of full professorships (the highest rank in the US academic system); by 1998 women assistant professors' numbers had risen to 28.3 per cent of total assistant professorships, yet women full professors held a mere 8 per cent of all full professorships.[1] The continued gender imbalance in geography's professional ranks makes the discipline's demographic profile appear more like that of the natural sciences than the social or human sciences, which – at least on the surface – seem to be friendlier toward female scholars.

Increasing the number of women geographers as a strategy for overcoming gender bias in geography exemplifies a liberal feminist approach: the so-called "add women and stir" solution. Yet the mere addition of women to the ranks of geography professionals is no assurance that explicitly feminist theoretical approaches and topics related to women (or gender more generally) will become more visible or accepted. In addition, not all women are feminists; indeed, some noted feminist geographers are in fact men (Zelinsky is a case in point here). By the 1980s, it had become increasingly clear that the gender bias in geography ran deeper than many had thought. As Monk and Hanson elaborate in this selection, "On not Excluding Half of the Human in Human Geography," the gender bias in geography has roots in the dearth of professional women geographers; but it also arises, and is reinforced, by sexist assumptions that underpin theory, research design, methodological practices, and the interpretation of findings, that are biased against the full inclusion of women. Near the end of this piece, Monk and Hanson ask a key question: "Is the purpose of geographic research to accumulate facts and knowledge in order to improve our understanding of current events . . . or is the purpose to go beyond asking why things are the way they are to consider the shapes of possible futures?" Thus they suggest going beyond liberal feminist approaches that would be able to better account for women's experiences to a transformative feminist geography that would help shape a non-sexist future.

Since the publication of their article in 1982, human geography has seen an efflorescence of broadly gender-aware, and explicitly feminist, research. Today, one can find a substantial body of feminist research conducted by economic, social, political, medical, population, development, and cultural geographers, to name just some of human geography's many sub-disciplines. Feminist approaches to various aspects of cultural geography are included in this *Reader*: see for example selections by Abu-Lughod (p. 50), Rose (p. 171), Massey (p. 257), and McDowell and Court (p. 457). Since the 1980s, feminist geographers have broadened their focus from women to include masculinities as a vital component of gender. Recently, the stability of 'gender' as an analytical category has itself come under scrutiny, following on the heels of the foundational theoretical work by Judith Butler in *Gender Trouble: Feminism and the Subversion of Identity* (1990). Several useful overviews of gender and human geography include Linda McDowell, *Gender, Identity and Place: Understanding Feminist Geographies* (1999); Mona Domosh and Joni Seager, *Putting Women in Place: Feminist Geographers Make Sense of the World* (2001); and Lise Nelson and Joni Seager (eds.) *A Companion to Feminist Geography* (2005).

Janice Monk is a Professor in the Department of Geography and Regional Development at the University of Arizona. She also directed the Southwest Institute for Research on Women (SIROW) for over two decades. SIROW is an interdisciplinary center that focuses on projects related to women's employment, education, health, and culture, and that fosters collaborations among women on both sides of the US–Mexico border. Monk is a feminist social and cultural geographer who has published extensively on the history of women in geography, feminist perspectives on landscape, and geographic careers and higher education. Her publications include "Women, gender, and the histories of American geography," in *Annals of the American Association of Geographers* 94, 1 (2004): 1–22; "Many roads: the personal and professional lives of women geographers," pp. 167–187 in Pamela Moss (ed.) *Placing Autobiography in Geography* (2001); and (with Vera Norwood) *The Desert is no Lady: Southwestern Landscapes in Women's Writing and Art* (1996).

Susan Hanson is a Professor in the School of Geography at Clark University in Worcester, Massachusetts. Hanson is an urban geographer, and has published extensively on gender and labor markets, transportation, and sustainability. Her publications include (with Genevieve Giuliano, eds.) *The Geography of Urban Transportation*, third edition (2004); "Who are 'we'? An important question for Geography's future," in *Annals of the American Association of Geographers* 94, 4 (2004): 715–722; and *Ten Geographic Ideas that Changed the World* (1997).

Both Monk and Hanson are past Presidents of the Association of American Geographers.

NOTE

1 Data from J. Winkler, "Faculty reappointment, tenure, and promotion: barriers for women," *Professional Geographer* 52, 4 (2000): 737–750.

Recent challenges to the acceptability of traditional gender roles for men and women have been called the most profound and powerful source of social change in this century, and feminism is the "ism" often held accountable for instigating this societal transformation. One expression of feminism is the conduct of academic research that recognizes and explores the reasons for and implications of the fact that women's lives are qualitatively different from men's lives. Yet the degree to which geography remains untouched by feminism is remarkable, and the dearth of attention to women's issues, explicit or implicit, plagues all branches of human geography.

Our purpose here is to identify some sexist biases in geographic research and to consider the implications of these for the discipline as a whole. We do not accuse geographers of having been actively or even consciously sexist in the conduct of their research, but we would argue that, through omission of any consideration of women, most geographic research has in effect been passively, often inadvertently, sexist. It is not our primary purpose to castigate certain researchers or their traditions, but rather to provoke lively debate and constructive criticism on the ways in which a feminist perspective might be incorporated into geography.

There appear to us to be two alternative paths to this goal of feminizing the discipline. One is to develop a strong feminist strand of research that would become one thread among many in the thick braid of geographic tradition. We support such research as necessary, but not sufficient. The second approach, which we favor, is to encourage a feminist perspective within all streams of human geography. In this way, issues concerning women (some of which are discussed later in this paper) would become incorporated in all geographic research endeavors. Only in this way, we believe, can geography realize the promise of the profound social change that would be wrought by eliminating sexism. In this paper we first briefly consider the reasons for the meager impact of feminism on the field to date, and review the nature of feminist scholarship in other social sciences and the humanities. We then examine the nature of sexist bias in geographic research, and, through examples of this, demonstrate ways in which a nonsexist geography might evolve.

WHY THE NEGLECT OF WOMEN'S ISSUES?

Why has geography for the most part assiduously avoided research questions that embrace half of the human race? We believe the answer lies very simply in the fact that knowledge is a social creation. The kind of knowledge that emerges from a discipline depends very much upon who produces that knowledge, what methods are used to produce knowledge, and what purposes knowledge is acquired for. The number of women involved in generating knowledge in a given discipline appears to be important in determining the degree to which feminism is absorbed in that discipline's research tradition. Although the number of women researchers in geography is growing, women still constitute only 9.6 percent of the college and university faculty who are members of the Association of American Geographers. The characteristics of researchers influence the kinds of issues a discipline focuses upon. Geographers have, for instance, been more concerned with studying the spatial dimensions of social class than of social roles, such as gender roles. Yet for many individuals and groups, especially women, social roles are likely to have a greater impact than social class on spatial behavior.

Geography's devotion to strict logical positivism in recent years can also help to account for the lack of attention to women's issues. . . . [P]ositivism has not been particularly concerned with social relevance or with social change. It is a method that tends to preserve the status quo. The separation of facts from values and of subject from object are elements of positivism that would prevent positivist research from ever guiding, much less leading, social change. Researchers in the positivist tradition have tended to ask normative questions that have little to do with defining optimal social conditions. . . .

Although strict logical positivism no longer has a life-threatening grip on the discipline, alternative paradigms have done little to incorporate a feminist perspective. Marxists have championed social change but, with a few exceptions, they have not explored the effects of capitalism on women. Phenomenologists have promised a more humanistic geography, a geography that would increase self-knowledge and would focus on the full range of human experience, but even this research stream has produced few insights into the lives of women.

Finally, the purpose of much geographic research has been to provide a rational basis for informed decision making. Insofar as planners are committed to maintaining the status quo, and insofar as both researcher and decision maker were, especially in the past, likely to belong to the male power establishment, a focus on women, or even a recognition of women, was unlikely. In sum, most academic geographers have been men, and they have structured research problems according to their values, their concerns, and their goals, all of which reflect their experience. Women have not been creatures of power or status, and the research interests of those in power have reflected this fact.

[. . .]

SOME EXAMPLES OF SEXIST BIAS IN GEOGRAPHIC RESEARCH

. . . [W]e consider sexist biases in the content, method, and purpose of geographic research. We do not imply that all human geography is sexist, but aim to demonstrate the pervasive nature of the problem by drawing illustrative examples from many areas of geographic endeavor. Neither the examples given nor the topic areas covered are intended as an exhaustive expose of the problems we address. . . . Our purpose here is merely to suggest the dimensions and sketch out the character of sexist bias in geographic research.

Content

Perhaps the most numerous examples of sexist bias in geographic research concern content. Problems relating to content include inadequate specification of the research problem, construction of gender-blind theory, the assumption that a population adheres to traditional gender roles, avoidance of research themes that directly address women's lives, and denial of the significance of gender or of women's activities.

Inadequate Specification of Research Problems. Many geographic research questions apply to both men and women, but are analyzed in terms of male experiences only. We see this in two recent historical studies involving immigration of families from Europe to North America. . . . Study of the women's lives might have supported or weakened . . .

conclusions. As it stands, generalizations about communities were drawn from data on men only.

The omission of women's experience from . . . [a] text on suburbanization is more surprising than are similar omissions from the historical studies, because women might be assumed to spend more of their lives in suburbia than do men. Yet . . . [the] section on the social organization of contemporary suburbia and its human consequences fails to address women's lives directly . . . . Are women only passive followers to the suburbs? There is research suggesting that women are ambivalent about suburban life, and that husbands and wives evaluate residential choices differently.

Inadequate specification can involve male as well as female exclusion when neither type of misspecification seems warranted. Studies of shopping behavior, for example, have assumed a female consumer and have analyzed data collected for samples of women only. A problem that seems to be related to the researcher's perception of shoppers as female is the assumption, implicit in models of consumer store choice, that all shopping trips originate at home, rather than, say, being chained to the journey to work. Hence such models employ a home-to-store distance variable rather than some other, possibly more important, variable such as workplace-to-store.

Gender-blind Theory. A concern stemming from inadequate problem specification is the emergence of gender-blind theory. Such theory may be dangerously impoverished if gender is an important explanatory variable and is omitted. Geographers interested in theories of development have drawn extensively on work outside the discipline. Nevertheless, these writers have not cited the significant quantity of literature on women and development that followed the publication of [Danish economist Ester] Boserup's *Women's Role in Economic Development* [1970]. Thus geographers address the political economy of the international division of labor, but ignore the theoretical implications of the sexual division of labor. Study of the literature on women would extend the range of development issues worth considering. For example, is development enhanced if women have access to wage incomes or only if they are increasingly involved in decision making with regard to income allocation? Should theories focus on production or give more attention than previously to family maintenance activities?

Geographic theories aimed at problems in industrialized countries also suffer when they are gender-blind. Attempts to build theories of urban travel demand have largely overlooked the importance of gender roles in determining travel patterns, but recent work suggests the seriousness of this omission. Theories of the residential location-decision process have likewise failed to take gender roles into account. . . .

Gender-blind theory is also emerging in research on issues of social well-being and equity. Although sexual discrimination receives passing mention, few of the welfare indicators refer specifically to women, nor are data disaggregated by gender. Yet . . . there are marked differences in the spatial patterns of relative versus absolute well-being of males and females in the United States. . . . The result of the general omission of gender in welfare and equity research is that race, class, and the political economy dominate explanations, while the contributions of gender and the patriarchal organization of society to the creation of disadvantage remain invisible. So long as gender remains a variable that is essential to understanding geographic processes and spatial form and to outlining alternative futures, explanations that omit gender are in many cases destined to be ineffective. Clearly, theoretical work along diverse lines of inquiry could benefit from becoming gender-sighted rather than remaining gender-blind.

The Assumption of Traditional Gender Roles. Explicit geographic writing on women, though rare, is likely to assume traditional gender (social) or sexual (biological) roles. [Carl] Sauer's hypothesis about women's role in the origins of sedentary settlement and social life relies on his concept of the "nature of women," the "maternal bond," and associated assumed restrictions on spatial mobility. The assumption that women universally (and perhaps historically) are primarily engaged in home and child care may reflect stereotypes of Western culture in the recent past, but can lead to inaccurate generalizations. . . . [Scholars have] ignored women's central roles in agriculture in much of Africa and in many Asian countries, their provision of fuel and water and their extensive roles in marketing and petty trading. . . .

Traditional urban land use theory, assuming as it does that each household has only one wage earner and therefore need be concerned with only one journey to work, seems also to be founded upon traditional gender roles. As we have pointed out elsewhere, models and theories that simply assume that all households are "traditional" nuclear families are not particularly useful for understanding changing urban spatial structure as a function of fundamental demographic or social changes. An additional example of gender stereotyping is the practice . . . of identifying women's participation in the paid labor force as part of an index of urbanization or familism. Work outside the paid labor force is not recognized, and within the labor force is not broken down by type of occupation as it is for the male head of household on whom the social status index is therefore based. The implications appear to be that non-urban women do not work and that knowing simply that a woman works outside the home is more important than knowing how she is employed. Neither seems conceptually sound.

Review of such examples highlights the need for rethinking the concepts of work and labor force if research is to treat women accurately. Normally such concepts are used to refer to the formal sector of the economy traditionally connected with male activity. Yet women also work in the informal sector (for example, in marketing food and crafts or as baby sitters or domestic servants), in home production for the market (food processing, sewing), in subsistence production (keeping domestic animals, raising gardens), and in unpaid service work (housework, child care, community volunteer work). Among partial solutions proposed for incorporating women's work are a Japanese indicator, "net national welfare," which includes the contributions of housework (at female wage rates), and estimates of work in terms of time or energy expended. Certainly more attention to this problem is warranted.

Avoidance of Research Themes that Directly Address Women's Lives. Women are generally invisible in geographic research, reflecting the concentration on male activity and on public spaces and landscapes. Work in recent issues of the *Journal of Cultural Geography* (1980, 1981), for example, deals with farm silos, farmsteads, housing exteriors, gasoline stations, a commercial strip, and country music (identified as a male WASP form). The massive *Man's Role in Changing the Face of the Earth* [edited by William J. Thomas, 1956] is aptly named. Women make only cameo appearances in three papers in the entire volume. A sampling of research on regional cultural landscapes and historical landscape perception, such as studies of the Mormon landscape and the Great Plains, discloses a preoccupation almost entirely with public spaces and men's perception. . . . [The] Great Plains country town has streets, businesses and businessmen, railroad depots, and men marketing livestock and making the trip to the elevator. We see little of the churches, schools, homes, and other social settings where women passed their lives.

Not surprisingly, the only mention of women's lives in the Great Plains studies reviewed is by a woman historian. She described not only the hardships that space brought to men, but the loneliness and isolation of women separated from kin and friends, the oppression of emptiness, and women's terror of injury, disease, and childbirth remote from doctors. She also compared barriers to social interaction for ranch and farm wives. Such insights suggest how research on women, the family, and social spheres would enrich our understanding of place. Beginning research on domestic interiors and symbolic uses of space similarly indicates how the horizons of cultural geography might be extended by attention to places closer to women's lives.

In the urban realm, geographic research could profit from assessing the effects of the availability of such facilities as shopping areas, day care, medical services, recreation, and transportation on female labor-force participation and on labor in the home. Take, for example, the provision of child care, a topic practically untouched by geographic researchers yet one of great consequence in the lives of women. Compare the trickle of research on this issue with the virtual torrent of material produced in the past few years on the provision of mental health care, an area that touches the lives of fewer people. Pursuing research themes that directly address the lives of women will do more than merely flesh out a bony research agenda: such research should also provide needed insights on the diversity of women's experiences and needs.

Dismissing the Significance of Gender or Women's Activities. Preconceived notions of significance lead some authors to dismiss women's activities or to overlook gender as a variable, despite evidence to the contrary. . . .

[An] . . . interesting example comes from incomplete interpretations of the findings . . . that Atlanta's unemployed are mainly black female heads of families. . . . [I]n drawing conclusions from [this study scholars have] focused on racial or "racial and other" discrimination. [Scholars] missed the double bind of gender and race.

[. . .]

Method

Sexist bias can afflict geographic research in the methods used as well as in content. A number of specific methodological concerns enter into empirical research design and execution regardless of the general approach (e.g., positivist or humanist) of the researcher. Here we address a few of these concerns and the ways in which they are susceptible to sexist bias.

Variable Selection. We have identified several inappropriate or inadequate practices in the selection and interpretation of variables in studies in which women are or should be included. One problem is the use of data on husbands to describe wives. For example, two of eight variables included . . . in a study of housewives' perceptions of neighborhoods in Cambridge, England, were "location of husband's work" and "husband's occupation". . . . Such use of husband's occupation as a surrogate for social class is problematic. Its appropriateness and the identification of alternatives is a concern of feminist sociologists as well as geographers insofar as geographers use measures of social class in their own research.

The assumption that data on males adequately describes the entire population is also suspect. . . . [W]e know there are gender differences in educational access and attainment, and that this varies spatially.

The diversity among women and the range of women's needs often goes unrecognized in variable selection. Male occupational categories are invariably differentiated, but women are recorded only by "female labor force participation" or "female activity rate." Social welfare studies would better reflect women's condition if indicators were included on such topics as women's legal situation, rape rates, or the provision of services such as day care.

Lack of awareness of women is also evident in variable interpretation and factor naming. For example . . . "old age" [was chosen] as the salient feature to name a factor that had high loadings on female divorce rate, illegitimate birth rate, high proportions of persons over sixty, low proportions in younger age groups, small households and shared dwellings. Without denying the significance of the elderly, the factor could be identified more comprehensively as "female-headed households." Such gender-blind naming of factors has theoretical and policy implications.

Respondent Selection. There is a need to rethink the unit of observation in survey research. Frequently data are collected on one individual yet reported as representative of the household; in particular, researchers like to rely upon responses from the "head of household." This practice presents several problems. First, it assumes one person represents the household, which is questionable. Second, aggregation by head of household may mask important gender differences, given that there are substantial and increasing numbers of female-headed households throughout much of the world. Third, cultural custom may lead to an assumption of male headship, even when the male does not have principal responsibilities for household support. Collection of data on individuals (or appropriately varying combinations of individuals) would help to avoid this male bias in data. Problems also arise when authors indicate that the sampling unit was the head of household but do not indicate whether or not other household members were surveyed, or when the sex composition of the sample is not given despite the clear theoretical importance of considering gender differences in that research context. Clear, complete reporting of research methodology and disaggregating samples by gender would alleviate these problems.

Interviewing Practices. Research results can be colored by interviewing practices such as having other members of a household present when one member is being interviewed. Interpretation of survey responses may raise problems, particularly on topics relating to women's role in family support or decision making. Either subjects or interviewers may discount or underestimate the importance of women's involvement. . . . [R]ural Mexican women described themselves as "helping" the family, rather than working for its support, despite substantial activity in planting, harvesting, animal care, and food processing. . . . [In another study] New Hebridean women offered passive reasons for moves, described as largely directed by parents or husbands. This may be, but we might question whether his interpretation reflected the cultural expectations of a foreign male researcher or of the women themselves.

Inadequate Secondary Data Sources. Convenience or the nature of secondary data sources can contribute to the omission of women from research. [Certain m]igration studies . . . demonstrate this problem. They drew . . . on electoral registrations (women could not be traced because of name changes) and male apprenticeship registrations. The US Census definition of household head prior to the 1980 census makes difficult the use of census data for investigating certain research questions related to women.

Purpose

One purpose of geographic research has been to provide a basis for informed policy and decision making. Yet policy-oriented research that ignores women cannot help to form or guide policy that will improve women's conditions. In fact, there are numerous examples of the results of policies that have overlooked or have minimized the needs of women. One is the urban transportation system that is organized to expedite the journey to work for the full-time worker but not travel for other purposes.

Is the purpose of geographic research to accumulate facts and knowledge in order to improve our understanding of current events or to formulate policy within the context of the status quo, or is the purpose to go beyond asking why things are the way they are to consider the shapes of possible futures? Feminist scholars emphasize the need for research to define alternative structures in which the lot of women is improved.

A geography that avoids or dismisses women and their activities, that is gender-blind, or that assumes traditional gender roles can never contribute to the equitable society feminists envision. For such purposes we need a cultural and historical geography that would permit women to develop the sense of self-worth and identity that flows from awareness of heritage and relationship to place and a social and economic geography that goes beyond describing the status quo. . . .

TOWARD A MORE FULLY HUMAN GEOGRAPHY

A more sensitive handling of women's issues is essential to developing a non-sexist, if not a feminist, human geography. Moreover, we believe that eliminating sex biases would create a more policy-relevant geography. As long as gender roles significantly define the lives of women and men, it will be fruitful to include gender as a potentially important variable in many research contexts. Through examples of sexist bias in the content, method, and purpose of geographic research, we have attempted to indicate some of the ways in which women's issues can be included in research designs. Many of the problems we have identified are problems that are easily solved (e.g., the need to disaggregate samples by gender), but others, such as the need for nonsexist measures of social class, are more challenging. Although we encourage an awareness of gender differences and of women's issues throughout the discipline now (so that the geography of women does not become "ghetto-ized"), we would like to see gender blurred and then erased as a line defining inequality.

"Representing Whiteness in the Black Imagination"
From *Cultural Studies* (1992)

bell hooks

Editors' introduction

As you read bell hooks' essay, "Representing whiteness in the black imagination," you might ask why this essay was not included instead in Part Six. Indeed, it could have just as easily been placed there. In this essay, hooks draws upon her early memories as a child in the segregated Southern US through to her international travels as an acclaimed public intellectual, to discuss the ways that her mobility has been impaired and continuously infused with fear. Hooks discusses the practice of gathering knowledge about whites and white-dominated society as a survival strategy of sorts, one that dates from slavery but that she still observes in her classroom discussions with students. Bell hooks has been both criticized and celebrated for voicing an explicitly "black" perspective on these issues. In this essay, for example, hooks calls whites "terrorists," openly discussing her fear, distrust, and unease in moving through white society. Some have claimed that hooks' perspective merely turns white racism on its head to become its mirror image: black racism. Others, however, note that in most scholarship, the enveloping racism of white-dominant societies is simply unspoken. While white dominance may thus *seem* natural and normal, it pervades the production of knowledge. Thus hooks' contribution is to expose the unnaturalness of racism.

Hooks can be considered a *native anthropologist*; in other words, she examines the very society of which she is a part. Anthropology and other disciplines – including cultural geography – that utilize ethnographic techniques such as participant observation, typically *study down*; that is, relatively privileged scholars tend to study poor and minority populations. Beginning in the 1980s, a wave of so-called 'Third World' scholars began to reverse this practice, studying majority populations, the wealthy, and utilizing ethnographic techniques to observe dominant cultures (see also Abu-Lughod, p. 50 and Said, p. 357). According to native anthropologists, the standpoint of the observer is crucial for determining the sort of knowledge that is produced. In hooks' case, her standpoint as a black woman from a working-class background growing up in the white-dominant United States affords her a perspective that mainstream scholars cannot glimpse. Other key work in this vein includes Gayatri Spivak's essay, "Can the subaltern speak?" in Cary Nelson and Lawrence Grossberg (eds.) *Marxism and the Interpretation of Culture* (1988), pp. 271–316; Trinh T. Minh-Ha, *Woman, Native, Other: Writing Postcoloniality and Feminism* (1989); and Chela Sandoval, *Methodology of the Oppressed* (2000).

Cultural geographers have long written about race as part of a subfield known as *ethnic geography*. Most of this work, however, does not utilize a critical approach toward understanding racism, nor does it view race as a social construction. Today, critical theoretical as well as applied cultural geographies of race are flourishing. This work shares much with postcolonial studies, particularly when the Other in question is not elsewhere, rather located within: within national borders, within communities, within Selves. Some examples include Audrey Kobayashi and Linda Peake, "Racism out of place: thoughts on whiteness and an antiracist geography in the new millennium," in *Annals of the Association of American Geographers* 90, 2 (2000): 392–403; David Delaney, "The space that race makes," *Professional Geographer* 54, 1 (2002): 6–14; and Alastair Bonnett, "Geography, 'race' and whiteness: invisible traditions and current challenges," in *Area* 29, 3 (1997): 193–199. Together, these and other cultural geographers who take a critical approach to understanding race argue both that space is an important way that race is socially constructed, and that spatial strategies of resistance can provide one avenue toward an anti-racist society.

Bell hooks (who does not capitalize the first letters of her names) is a pen name. Born Gloria Jean Watson into a working-class Kentucky family, she received her PhD from the University of California at Santa Cruz. Hooks has taught at Yale University, Oberlin College, the City University of New York, and most recently at Berea College in Kentucky. An author, film maker, and an at times outspoken public speaker, hooks is a public intellectual in the sense described by Edward Said (see p. 357). Considered a key figure in American cultural studies (see also the entry for Stuart Hall, p. 264), hooks has long examined the complex interplay of race, class, sexuality, and gender in books such as *Ain't I a Woman? Black Women and Feminism* (1981), which constituted the first of her many feminist works that considered racial and other differences among women; *Black Looks: Race and Representation* (1992), in which hooks considers the problematic relationship between black culture and its commodification by the white culture industry; and *We Real Cool: Black Men and Masculinity* (2004), in which hooks examines the so-called "crisis of the black male" from a feminist perspective. *Bone Black: Memories of Girlhood* (1996) is an autobiographical account of her childhood.

■ ■ ■ ■ ■ ■ ■

Although there has never been any official body of black people in the United States who have gathered as anthropologists and/or ethnographers whose central critical project is the study of whiteness, black folks have, from slavery on, shared with one another in conversations "special" knowledge of whiteness gleaned from close scrutiny of white people. Deemed special because it was not a way of knowing that has been recorded fully in written material, its purpose was to help black folks cope and survive in a white supremacist society. For years black domestic servants, working in white homes, acted as informants who brought knowledge back to segregated communities – details, facts, observations, psychoanalytic readings of the white "Other."

Sharing, in a similar way, the fascination with difference and the different that white people have collectively expressed openly (and at times vulgarly) as they have traveled around the world in pursuit of the other and otherness, black people, especially those living during the historical period of racial apartheid and legal segregation, have maintained steadfast and ongoing curiosity about the "ghosts," "the barbarians," these strange apparitions they were forced to serve. . . .

I . . . am in search of the debris of history, am wiping the dust from past conversations, to remember some of what was shared in the old days, when black folks had little intimate contact with whites, when we were much more open about the way we connected whiteness with the mysterious, the strange, the terrible. Of course, everything has changed. Now many black people live in the "bush of ghosts" and do not know themselves separate from whiteness, do not know this thing we call "difference." Though systems of domination, imperialism, colonialism, racism, actively coerce black folks to internalize negative perceptions of blackness, to be self-hating, and many of us succumb, blacks who imitate whites (adopting their values, speech, habits of being, etc.) continue to regard whiteness with suspicion, fear, and even hatred. This contradictory longing to possess the reality of the Other, even though that reality is one that wounds and negates, is expressive of the desire to understand the mystery, to know intimately through imitation, as though such knowing worn like an amulet, a mask, will ward away the evil, the terror.

Searching the critical work of postcolonial critics, I found much writing that bespeaks the continued fascination with the way white minds, particularly the colonial imperialist traveler, perceive blackness, and very little expressed interest in representations of whiteness in the black imagination. Black cultural and social critics allude to such representations in their writing, yet only a few have dared to make explicit those perceptions of whiteness that they think will discomfort or antagonize readers. James Baldwin's collection of essays *Notes of a Native Son* (1955) explores these issues with a clarity and frankness that is no longer fashionable in a world where evocations of pluralism and diversity act to obscure differences arbitrarily imposed and maintained by white racist domination. Writing about being the first black person to visit a Swiss village with only white inhabitants, who had a yearly ritual of painting individuals black who were then positioned as slaves and bought, so that the villagers could celebrate their concern with converting the souls of the "natives," Baldwin responded:

> I thought of white men arriving for the first time in an African village, strangers there, as I am a stranger here, and tried to imagine the astounded populace touching their hair and marveling at the color of their skin. But there is a great difference between being the first white man to be seen by Africans and being the first black man to be seen by whites. The white man takes the astonishment as tribute, for he arrives to conquer and to convert the natives, whose inferiority in relation to himself is not even to be questioned, whereas I, without a thought of conquest, find myself among a people whose culture controls me, has even in a sense, created me, people who have cost me more in anguish and rage than they will ever know, who yet do not even know of my existence. The astonishment with which I might have greeted them, should they have stumbled into my African village a few hundred years ago, might have rejoiced their hearts. But the astonishment with which they greet me today can only poison mine. ("Stranger in the Village")

Addressing the way in which whiteness exists without knowledge of blackness even as it collectively asserts control, Baldwin links issues of recognition to the practice of imperialist racial domination.

My thinking about representations of whiteness in the black imagination has been stimulated by classroom discussions about the way in which the absence of recognition is a strategy that facilitates making a group "the Other." In these classrooms there have been heated debates among students when white students respond with disbelief, shock, and rage, as they listen to black students talk about whiteness, when they are compelled to hear observations, stereotypes, etc., that are offered as "data" gleaned from close scrutiny and study. Usually, white students respond with naive amazement that black people critically assess white people from a standpoint where "whiteness" is the privileged signifier. Their amazement that black people watch white people with a critical "ethnographic" gaze is itself an expression of racism. Often their rage erupts because they believe that all ways of looking that highlight difference subvert the liberal conviction that it is the assertion of universal subjectivity (we are all just people) that will make racism disappear. They have a deep emotional investment in the myth of "sameness" even as their actions reflect the primacy of whiteness as a sign informing who they are and how they think. Many of them are shocked that black people think critically about whiteness because racist thinking perpetuates the fantasy that the Other who is subjugated, who is subhuman, lacks the ability to comprehend, to understand, to see the working of the powerful. Even though the majority of these students politically consider themselves liberals, who are anti-racist, they too unwittingly invest in the sense of whiteness as mystery.

In white supremacist society, white people can "safely" imagine that they are invisible to black people since the power they have historically asserted, and even now collectively assert, over black people accorded them the right to control the black gaze. As fantastic as it may seem, racist white people find it easy to imagine that black people cannot see them if within their desire they do not want to be seen by the dark Other. One mark of oppression was that black folks were compelled to assume the mantle of invisibility, to erase all traces of their subjectivity during slavery and the long years of racial apartheid, so that they could be better – less threatening – servants. An effective strategy of white supremacist terror and dehumanization during slavery centered around white control of the black gaze. Black slaves, and later manumitted servants, could be brutally punished for looking, for appearing to observe the whites they were serving as only a subject can observe, or see. To be fully an object then was to lack the capacity to see or recognize reality. These looking relations were reinforced as whites cultivated the practice of denying the subjectivity of blacks (the better to dehumanize and oppress), of relegating them to the realm of the invisible. . . . Reduced to the machinery of bodily physical labor, black people learned to appear before whites as though they were zombies, cultivating the habit of casting the gaze downward so as not to appear uppity. To look directly was an assertion of subjectivity, equality. Safety resided in the pretense of invisibility.

Even though legal racial apartheid no longer is a norm in the United States, the habits of being cultivated to uphold and maintain institutionalized white supremacy linger. Since most white people do not have to "see" black people (constantly appearing on billboards, television, movies, in magazines, etc.) and they do not need to be ever on guard, observing black people, to be "safe," they can live as though black people are invisible and can imagine that they are also invisible to blacks. Some white people may even imagine there is no representation of whiteness in the black imagination, especially one that is based on concrete observation or mythic conjecture; they think they are seen by black folks only as they want to appear. . . . Socialized to believe the fantasy that whiteness represents goodness and all that is benign and non-threatening, many white people assume this is the way black people conceptualize whiteness. They do not imagine that the way whiteness makes its presence felt in black life, most often as terrorizing imposition, a power that wounds, hurts, tortures, is a reality that disrupts the fantasy of whiteness as representing goodness.

Collectively, black people remain rather silent about representations of whiteness in the black imagination. As in the old days of racial segregation where black folks learned to "wear the mask," many of us pretend to be comfortable in the face of whiteness only to turn our backs and give expression to intense levels of discomfort. Especially talked about is the representation of whiteness as terrorizing. Without evoking a simplistic, essentialist "us and them" dichotomy that suggests

black folks merely invert stereotypical racist interpretations, so that black becomes synonymous with goodness and white with evil, I want to focus on that representation of whiteness that is not formed in reaction to stereotypes but emerges as a response to the traumatic pain and anguish that remains a consequence of white racist domination, a psychic state that informs and shapes the way black folks "see" whiteness. Stereotypes black folks maintain about white folks are not the only representations of whiteness in the black imagination. They emerge primarily as responses to white stereotypes of blackness. . . .

Stereotypes, however inaccurate, are one form of representation. Like fictions, they are created to serve as substitutions, standing in for what is real. They are there not to tell it like it is but to invite and encourage pretense. They are a fantasy, a projection on to the Other that makes them less threatening. Stereotypes abound when there is distance. They are an invention, a pretense that one knows when the steps that would make real knowing possible cannot be taken – are not allowed.

Looking past stereotypes to consider various representations of whiteness in the black imagination, I appeal to memory, to my earliest recollections of ways these issues were raised in black life. Returning to memories of growing up in the social circumstances created by racial apartheid, to all-black spaces on the edges of town, I re-inhabit a location where black folks associated whiteness with the terrible, the terrifying, the terrorizing. White people were regarded as terrorists, especially those who dared to enter that segregated space of blackness. As a child I did not know any white people. They were strangers, rarely seen in our neighborhoods. The "official" white men who came across the tracks were there to sell products, Bibles, insurance. They terrorized by economic exploitation. What did I see in the gazes of those white men who crossed our thresholds that made me afraid, that made black children unable to speak? Did they understand at all how strange their whiteness appeared in our living rooms, how threatening? Did they journey across the tracks with the same "adventurous" spirit that other white men carried to Africa, Asia, to those mysterious places they would one day call the third world? Did they come to our houses to meet the Other face to face and enact the colonizer role, dominating us on our own turf? Their presence terrified me. Whatever their mission they looked too much like the unofficial white men who came to enact rituals of terror and torture. As a child, I did not know how to tell them apart, how to ask the "real white people to please stand up." The terror that I felt is one black people have shared. Whites learn about it secondhand. . . .

[. . .]

To name that whiteness in the black imagination is often a representation of terror: one must face a palimpsest of written histories that erase and deny, that reinvent the past to make the present vision of racial harmony and pluralism more plausible. To bear the burden of memory one must willingly journey to places long uninhabited, searching the debris of history for traces of the unforgettable, all knowledge of which has been suppressed. . . . Theorizing black experience, we seek to uncover, restore, as well as to deconstruct, so that new paths, different journeys are possible. . . .

. . . . There is then only the fantasy of escape, or the promise that what is lost will be found, redis-covered, returned. For black folks, reconstructing an archaeology of memory makes return possible, the journey to a place we can never call home even as we reinhabit it to make sense of present loca-tions. Such journeying cannot be fully encom-passed by conventional notions of travel.

. . . . Reading . . . about theory and travel, I appreciated . . . efforts to expand the travel/the-oretical frontier so that it might be more inclusive, even as I considered that to answer the questions . . . is to propose a deconstruction of the conven-tional sense of travel, and put alongside it or in its place a theory of the journey that would expose the extent to which holding on to the concept of "travel" as we know it is also a way to hold on to imperialism. For some individuals, clinging to the conventional sense of travel allows them to remain fascinated with imperialism, to write about it seductively, evoking . . . "imperialist nostalgia." . . . Theories of travel produced outside conventional borders might want the Journey to become the rubric within which travel as a starting point for dis-course is associated with different headings – rites of passage, immigration, enforced migration, reloca-tion, enslavement, homelessness. Travel is not a word that can be easily evoked to talk about the Middle Passage, the Trail of Tears, the landing of Chinese immigrants at Ellis Island, the forced relocation of Japanese-Americans, the plight of the homeless. Theorizing diverse journeying is crucial to our understanding of any politics of location. . . .

. . . . I felt [that a "playful"] . . . evocation [of travel] would always make it difficult for there to be recognition of an experience of travel that is not about play but is an encounter with terrorism. And it is crucial that we recognize that the hegemony of one experience of travel can make it impossible to articulate another experience and be heard. From certain standpoints, to travel is to encounter the terrorizing force of white supremacy. To tell my "travel" stories, I must name the movement from a racially segregated southern community, from a rural black Baptist origin, to prestigious white uni-versity settings, etc. I must be able to speak about what it is like to be leaving Italy after I have given a talk on racism and feminism, hosted by the par-liament, only to stand for hours while I am inter-rogated by white officials who do not have to respond when I inquire as to why the questions they ask me are different from those asked the white people in line before me. Thinking only that I must endure this public questioning, the stares of those around me, because my skin is black, I am startled when I am asked if I speak Arabic, when I am told that women like me receive presents from men with-out knowing what those presents are. Reminded of another time when I was strip-searched by French officials, who were stopping black people to make sure we were not illegal immigrants and/or ter-rorists, I think that one fantasy of whiteness is that the threatening Other is always a terrorist. This projection enables many white people to imagine there is no representation of whiteness as terror, as terrorizing. Yet it is this representation of white-ness in the black imagination, first learned in the narrow confines of the poor black rural community, that is sustained by my travels to many different locations.

To travel, I must always move through fear, confront terror. It helps to be able to link this indi-vidual experience to the collective journeying of black people, to the Middle Passage, to the mass migration of southern black folks to northern cities in the early part of the twentieth century. . . . It is useful when theorizing black experience to exam-ine the way the concept of "terror" is linked to representations of whiteness.

In the absence of the reality of whiteness, I learned as a child that to be "safe" it was important to recognize the power of whiteness, even to fear it, and to avoid encountering it. There was nothing terrifying about the sharing of this knowledge as survival strategy; the terror was made real only when I journeyed from the black side of town to a predominately white area near my grandmother's house. I had to pass through this area to reach her place. Describing these journeys "across town" in the essay "Homeplace: A Site of Resistance" [an essay published in *Yearning: Race, Gender, and Cultural Politics*, 1990] I remembered:

> It was a movement away from the segregated blackness of our community into a poor white neighborhood. I remember the fear, being scared to walk to Baba's, our grandmother's house, because we would have to pass that terrifying whiteness – those white faces on the porches staring us down with hate. Even when empty or vacant those porches seemed to say *danger*, you do not belong here, you are not safe.
>
> Oh! that feeling of safety, of arrival, of homecoming when we finally reached the edges of her yard, when we could see the soot black face of our grandfather, Daddy Gus, sitting in his chair on the porch, smell his cigar, and rest on his lap. Such a contrast, that feeling of arrival, of homecoming – this sweetness and the bitterness of that journey, that constant reminder of white power and control.

Even though it was a long time ago that I made this journey, associations of whiteness with terror and the terrorizing remain. Even though I live and move in spaces where I am surrounded by whiteness, surrounded, there is no comfort that makes the terrorism disappear. All black people in the United States, irrespective of their class status or politics, live with the possibility that they will be terrorized by whiteness.

[. . .]

In contemporary society, white and black people alike believe that racism no longer exists. This erasure, however mythic, diffuses the representation of whiteness as terror in the black imagination. It allows for assimilation and forgetfulness. The eagerness with which contemporary society does away with racism, replacing this recognition with evocations of pluralism and diversity that further mask reality, is a response to the terror, but it has also become a way to perpetuate the terror by providing a cover, a hiding place. Black people still feel the terror, still associate it with whiteness, but are rarely able to articulate the varied ways we are terrorized because it is easy to silence by accusations of reverse racism or by suggesting that black folks who talk about the ways we are terrorized by whites are merely evoking victimization to demand special treatment.

Attending a recent conference on cultural studies, I was reminded of the way in which the discourse of race is increasingly divorced from any recognition of the politics of racism. I went there because I was confident that I would be in the company of likeminded, progressive, "aware" intellectuals; instead, I was disturbed when the usual arrangements of white supremacist hierarchy were mirrored both in terms of who was speaking, of how bodies were arranged on the stage, of who was in the audience, of what voices were deemed worthy to speak and be heard. As the conference progressed I began to feel afraid. If progressive people, most of whom were white, could so blindly reproduce a version of the status quo and not "see" it, the thought of how racial politics would be played out "outside" this arena was horrifying. That feeling of terror that I had known so intimately in my childhood surfaced. Without even considering whether the audience was able to shift from the prevailing standpoint and hear another perspective, I talked openly about that sense of terror. Later, I heard stories of white women joking about how ludicrous it was for me (in their eyes I suppose I represent the "bad" tough black woman) to say I felt terrorized. Their inability to conceive that my terror . . . is a response to the legacy of white domination and the contemporary expressions of white supremacy is an indication of how little this culture really understands the profound psychological impact of white racist domination.

At this same conference I bonded with a progressive black woman and white man who, like me, were troubled by the extent to which folks chose to ignore the way white supremacy was informing the structure of the conference. Talking with the black woman, I asked her: "What do you do, when you are tired of confronting white racism, tired of the day-to-day incidental acts of racial terrorism?

I mean, how do you deal with coming home to a white person?" Laughing, she said, "Oh, you mean when I am suffering from White People Fatigue Syndrome. He gets that more than I do." After we finished our laughter, we talked about the way white people who shift locations, as her companion has done, begin to see the world differently. Understanding how racism works, he can see the way in which whiteness acts to terrorize without seeing himself as bad, or all white people as bad, and black people as good. Repudiating "us and them" dichotomies does not mean that we should never speak the ways observing the world from the standpoint of "whiteness" may indeed distort perception, impede understanding of the way racism works both in the larger world as well as the world of our intimate interactions. . . . [P]rogressive white people who are anti-racist might be able to understand the way in which their cultural practice reinscribes white supremacy without promoting paralyzing guilt or denial. Without the capacity to inspire terror, whiteness no longer signifies the right to dominate. It truly becomes a benevolent absence. . . . Critically examining the association of whiteness as terror in the black imagination, deconstructing it, we both name racism's impact and help to break its hold. We decolonize our minds and our imaginations.

"Mapping the Pure and the Defiled"

from *Geographies of Difference: Society and Difference in the West* (1995)

David Sibley

Editors' introduction

In "Mapping the pure and the defiled," David Sibley illustrates how the deeply felt need to separate "us" from "them" in a psychological sense is translated into spatial terms. Using a variety of examples, ranging from ancient maps to television commercials, Sibley notes that demarcating the boundary between the civilized Self and the uncivilized Other is an enduring practice. The analysis is applied at the global scale of colonialism (as well as postcolonialism), which relied on the definition of some peoples and the places they inhabited as uncivilized as a rationale for conquering them; the geography of cities, which has long sought to segregate inhabitants by class and race; and the micro-geography of the home understood to be a bastion against the polluting influences of the outside world, particularly with regard to women and children. Sibley makes the important point that people's *feelings* about others are translated into spatial *practices* which map exclusionary sentiments on to real spaces.

Central to Sibley's ideas of social inclusion and exclusion is an approach known as *object relations theory*. Object relations theory posits that individuals as well as groups form positive identities of themselves through a process of excluding other individuals and groups thought to be deviant. Through establishing physical, psychological, and social boundaries, the polluting Other is kept at bay, and the Self is constructed as whole and pure. Transgression of these boundaries is thus more than just border-crossing; it threatens to destabilize social order altogether. Foundational work in object relations theory includes anthropologist Mary Douglas's *Purity and Danger: An Analysis of Concepts of Pollution and Taboo* (1966), which explores religious, sexual, and social taboos; philosopher, literary critic, and feminist Julia Kristeva's *Powers of Horror: An Essay on Abjection* (1982 [1980]), which develops the notion of the *abject*, or that which is abhorrent and as such rejected at a visceral level; and psychoanalyst Melanie Klein's work on children and identity formation as a tug-of-war between positive "Eros" and destructive "Thanatos"; see for example, *The Psychoanalysis of Children* (1984 [1932]).

Cultural geographers have productively used ideas associated with object relations theory, and psychoanalysis more generally. In "*Le Pratique sauvage*" (p. 241), for example, Glen Elder, Jennifer Wolch, and Jody Emel discuss the shifting boundaries between human and animal in terms that owe a debt to the work of object relations theorists. An intriguing subfield of cultural geography, psychogeography, is exemplified by the selection by Sibley reprinted here, as well as by fellow cultural geographers Liz Bondi (with Judith Fewell) in " 'Unlocking the cage door': the spatiality of counseling," in *Social and Cultural Geography* 4, 4 (2004): 527–547;

Heidi Nast, "Mapping the 'unconscious': racism and the Oedipal family," in *Annals of the Association of American Geographers* 90, 2 (2000): 215–255; and recent work by Gillian Rose (see p. 171).

Embedded in the psychoanalytic work from which these geographers draw is a focus on childhood and family dynamics. David Sibley's discussion on geographies of the life course, including particularly childhood, is central to *Geographies of Difference*, and comprises an important contribution to the emerging arena of children's geographies. Two examples are Stuart Aitken, an editor of the journal *Children's Geographies*, who has authored *Geographies of Young People: The Morally Contested Spaces of Identity* (2001), and Cindi Katz, who is the author of *Growing up Global: Economic Restructuring and Children's Everyday Lives* (2004).

David Sibley is a Professor in the Department of Geography at the University of Leeds, in West Yorkshire, United Kingdom. His early research interest in Romany people, or European Gypsies, was published in *Outsiders in an Urban Society* (1981). This work formed the basis for Sibley's long-standing fascination with the shifting social boundaries of inside and outside, exemplified in the excerpt below from *Geographies of Difference*. Sibley (with Peter Jackson, David Atkinson, and Neil Washbourne) is an editor of a reference work titled *Cultural Geography: A Critical Dictionary of Key Ideas* (2005).

There is a history of imaginary geographies which cast minorities, 'imperfect' people, and a list of others who are seen to pose a threat to the dominant group in society as polluting bodies or folk devils who are then located 'elsewhere'. This 'elsewhere' might be nowhere, as when genocide or the moral transformation of a minority like prostitutes are advocated, or it might be some spatial periphery, like the edge of the world or the edge of the city. . . . Thus, values associated with conformity or authoritarianism are expressed in maps which relegate others to places distant from the locales of the dominant majority. Images of others in the mind, in literature and other media may, however, inform practice such as the isolation of Gypsies on local authority sites in Britain or the exclusion of children from adult spaces. There may be important connections between these fantasies and the exercise of power. I will trace some of these ideas about the constitution of social space according to which some groups or peoples are deemed not to belong over a long historical time period in order to demonstrate their persistence. Portrayals of minorities as defiling and threatening have for long been used to order society internally and to demarcate the boundaries of society, beyond which lie those who do not belong. To demonstrate this point, I will make references both to political discourse in a number of historical periods and to some fictional narratives which mirror social practice. One informs the other.

The expansion of European empires and the development of the capitalist world economy required fitting dependent territories and dependent peoples into the cosmic order of the dominant powers. Beyond the spatial limits of civilization, there were untamed people and untamed nature to be incorporated into the imperial system. Attitudes to people on these peripheries were ambivalent, however. While they were regarded with disgust or fear if they violated the space of the colonizers, they were also idealized and romanticized. Thus . . . the ancient Greeks and Romans, like mediaeval European powers, saw themselves at the centre of the civilized world, and in their ordering of cultures and societies, the farther a group was from the centre the greater was its 'vice'. Some cultural difference may have been tolerated but if . . . another people's culture were considered to be too discrepant, it would be considered deviant, a 'vice', and generally judged in negative terms. Thus, on a global scale, a spatial and cultural boundary was drawn between civilization and various uncivilized, deviant 'others'. . . . Aristotle's conception of the mean or average was effectively a moral judgment about levels of civilization. Being close to it was a mark of virtue but departure from the mean signalled vice. Thus, deviation in the statistical sense was also moral deviance and a device for conceptualizing the boundaries of society.

This conception of civil society was echoed in mediaeval and early modern European cosmographies, which borrowed heavily from classical Greek sources. Thus, the 'edge' of civilization was marked by the presence of grotesque peoples. . . . These people are not entirely different from the messengers of civilization in physical appearance, but they are 'imperfect' – physically deformed and/or black and at one with nature, in other words, not quite human by civilized, white European standards. This sort of characterization . . . betrayed fears of being less than perfect on the part of the civilized. . . . [T]hose threatening people beyond the boundary represent the features of human existence from which the civilized have distanced themselves – close contact with nature, dirt, excrement, overt sexuality – but these same characteristics are exaggerated in portrayals of the uncivilized, which employ negative images of smell, colour and physical form. The world map, with civilization in the centre and the grotesque adorning the periphery, then expressed this desire for a literal distancing from the 'other'.

[. . .]

Although this kind of differentiation is dependent on disgust, the very features which are reviled are also desired because they represent those features of the civilized self which are repressed. Defiled peoples and places offer excitement.

Thus, in the early period of European exploration and the emergence of capitalist economies, there was an evident fascination with non-European cultures, but there were both moral and economic arguments for representing these cultures as less than human, a part of nature, or monstrous.

[. . .]

The economic argument for monstrous representations, opposed to the perfection of white Europeans, was to ease the way for genocide in newly discovered territories, where . . . physical resources like gold were valued above a sustainable supply of labour by the colonizing powers during the early phase of capitalist development. It could be argued that elements of the monstrous tradition have continued into the twentieth century in capitalist states, for example, myths about cannibalism among colonized peoples. . . .

In mediaeval Europe, there is some evidence that the socio-spatial structure of the city also expressed a wish to erect boundaries to protect civil society from the defiled. . . . [T]he bourgeoisie were scandalized by the behaviour of deviant groups and

attempted to control their distribution. In particular, prostitution, although legal, was spatially regulated and 'red light' districts were contested spaces, frequently objected to by respectable citizens.

[. . .]

. . . [S]tereotypes of people and place were not as clearly articulated as they were in the capitalist city of the nineteenth and twentieth centuries partly because distancing, in a physical sense, was not easily accomplished in the compact and crowded mediaeval city. This was not for want of trying. . . .

In the modern period in Europe, the language of defilement is more readily identifiable, as are the spaces to which are assigned those who belong and those who are excluded. By the eighteenth century, socio-spatial separation was becoming characteristic of large cities, like London, Dublin or Philadelphia, and boundary maintenance became a concern of the rich, who were anxious to protect themselves from disease and moral pollution. This is suggested somewhat obliquely in [Jonathan] Swift's *Gulliver's Travels* [1726], which can be read as a critique of western European society in the eighteenth century, using metaphors of purity and defilement. Swift creates a series of landscapes in which Gulliver is either polluting or is trying to protect himself from the threat of pollution. . . . Thus, Lilliput is a highly ordered society with strong rules of exclusion where Gulliver, differing not only in size but also in behaviour, is polluting. Because he is a source of defilement here, Gulliver is consigned to a polluted space, the Temple:

> At the place where the carriage stopt, there stood an ancient Temple esteemed to be the largest in the whole Kingdom, which having been polluted some years before by an unnatural Murder, was according to the zeal of these People, looked upon as Prophane, and therefore had to be applied to common Uses and all Ornaments and Furniture carried away.

Notwithstanding the defilement of this space, because the pollution taboos in Lilliput were so strong, Gulliver was unclean and anomalous even here.

[. . .]

Consciousness of pollution in Lilliput is heightened by the geometry of the landscape. In particular, the metropolis, Mildendo, had a highly ordered design with strong internal boundaries and the populace was excluded from the centre – a sacred space, the home of the emperor. As in the European Baroque city on which Mildendo was probably modelled, geometry expresses power: the representation of the masses as polluting is a means of exercising control.

In Gulliver's second voyage to Brobdingnag, there is a reversal. Attitudes to social mixing are very relaxed and pollution taboos are not in evidence.... [T]his is symptomatic of the Brobdingnagians' integrated rather than segmented view of society. One telling feature of their socio-spatial organization is that, rather than maintaining hospitals for the incarceration of the old and diseased and others who are marginal or residual in Brobdingnagian society, 'They are willing to grant their beggars the liberty to roam freely through the streets of Lorbrulgrud.' The reversal in world-view represented by Brobdingnag in relation to Lilliput has interesting consequences for Gulliver. In Lilliput, he is polluting because he is unable to conform, but in Brobdingnag, where, understandably, he fears for his survival because of his diminutive stature, Gulliver becomes preoccupied with boundaries.... Thus, he insists on the separation of basic social categories – male and female, healthy and diseased, rich and poor – because mixing and non-conformity, like expressions of sexuality outside conventional bounds, create anxiety.

[...]

There is an interesting parallel in this with the small group on the margins of industrialized societies, like Gypsy communities, for whom pollution taboos and a concern with boundaries relate to the problem of cultural survival. *Gulliver's Travels*, however, can be seen more generally as a commentary on social tensions and power relations in a developing urban society as they are expressed in the language of defilement.

The poor as a source of pollution and moral danger were clearly identified in contemporary accounts of the nineteenth-century capitalist city. As socio-spatial segregation became yet more pronounced, the distance between the affluent and the poor ensured the persistence of stereotyped conceptions of the other. Social and spatial distancing contributed to the labelling of areas of poverty as deviant and threatening, a lack of knowledge

being reflected in myths about working-class living conditions and behaviour.... [The] expression of the class divide in terms of topography and health was crucial. The poor, down there on the swampy clays, were living in their own excrement and were subject to contagious diseases like cholera. The middle classes, up there on the suburban heights, were free from disease and uncontaminated by sewage, but threatened by the poor and their diseases. In one sense, [this] ... identifies a serious public health problem which reflected rapid urbanization without provision of adequate services. In another sense, however, it is a comment on different standards of morality. The poor were not only living in appalling physical circumstances but were, from a bourgeois perspective, depraved.

[...]

The significance of excrement ... is that its stands for residual people and residual places. The middle classes have been able to distance themselves from their own residues, but in the poor they see bodily residues, animals closely associated with residual matter, and residual places coming together and threatening their own categorical scheme under which the pure and the defiled are distinguished. The separations which the middle classes have achieved in the suburb contrast with the mixing of people and polluting matter in the slum. This then becomes a judgment on the poor. The class boundary marked out in residential segregation echoes the recurrent theme: 'Evil ... is embodied in excrement.'

[...]

.... Physical cleaning, separating the poor from their residues, was to be accompanied by 'moral cleansing' or purification because 'moral filth was as much a concern as physical'....

Nineteenth-century schemes to reshape the city could thus be seen as a process of purification, designed to exclude groups variously identified as polluting – the poor in general, the residual working class, racial minorities, prostitutes, and so on. This was particularly true of grand designs like that prepared by Haussmann for Paris. One of Haussmann's objectives ... was to make central Paris fit for the bourgeoisie by creating elegant spaces which distanced them from the poor and enhanced property values....

These particular mappings of nineteenth-century urban society are not solely imaginary. There were chronic problems of sanitation, waste disposal and associated illnesses which urban reformers were intent on solving and, as progress made in methods of waste disposal weakened the association between the poor and excrement, so the bourgeois metaphors seemed less appropriate. . . . Once the bourgeoisie developed a sense of self which excluded bodily residues, they could recognize their difference from the smelly working class. . . . The abhorrence of excrement became an abhorrence of the poor, who represented what the bourgeoisie had left behind. Public health policies dealt with the problem of the putrid masses and cleaning up the poor would also help to instil ideas of discipline and order amongst them. Public health schemes brought with them regulations and were thus a means of social control. . . . However, once the indigenous poor had been sanitized . . . the same notions of dirt and disease could be used to construct images of immigrants, so defilement entered the language of racism.

I doubt whether there is a neat historical sequence here, but modern social geographies, that is, media and other popular representations of place rather than academic geographies, do suggest that spatial categories like 'the inner city' and some social categories, like Gypsies, are represented in similar language to that used to exclude the poor from bourgeois space in the nineteenth century. One crucial difference is that large sections of the working class are now more conscious of their own purified identity. In modern western societies, defilement is usually suggested in more muted tones than was the case in the nineteenth century. Material improvements in housing, water supply and sewage disposal have literally cleaned up the city, but . . . places associated with ethnic and racial minorities, like the inner city, are still tainted and perceived as polluting in racist discourse, and place-related phobias are similarly evident in response to other minorities, like gays and the homeless.

Class-based geographies of defilement were still evident in the first half of the twentieth century, however, particularly where working people threatened the sanctity of middle-class preserves. Thus, middle-class commentaries on the countryside in Britain until the 1950s expressed concern about disorder, litter, advertising, and so on, associated with developments which were perceived to

be catering for the working class, like roadside cafés and some housing developments. Plotlands, as working-class creations, were considered particularly abhorrent. . . . Establishment figures like the Cambridge geographer, J.A. Steers, made very strong statements about working-class housing in the countryside, making it clear that it was considered to be a form of pollution. Steers described Canvey Island, on the south Essex coast, as 'an abomination . . . a town of shacks and rubbish . . . *It caters for a particular class of people* and, short of total destruction and a new start, little if anything can be done.' Vocal objectors to spontaneous housing development in the countryside . . . had an important influence in shaping the legislation which formed the basis of town and country planning in England and Wales after 1945, and the power given to local authorities to control or eradicate 'disorderly development' under the 1948 Town and Country Planning Act contributed to the exclusion of working people from middle-class space, particularly in areas of extensive plotland development, like Sussex and Essex. The rhetoric had an important bearing on practice, although the language of pollution was translated into less emotive terms, like non-conforming use.

[. . .]

MODERN MEDIA REPRESENTATIONS

Urban society, as it is currently projected in literature, film and television commercials, provides further visions of purity and pollution where the polluting are more likely to be social, and often spatially marginal minorities, like the gays, prostitutes and homeless. . . . Media representations are mostly fictional, imaginary constructions, but they draw on the same stereotyped images of people and places which surface in social conflicts involving mainstream communities and 'deviant' minorities. The media, particularly television, are also important because they comprise a major source of images for the representation of others, remotely consumed and requiring no engagement with the people they characterize as different. They are thus more likely to be received uncritically.

One graphic and probably not grossly exaggerated depiction of the pure and the defiled in the city is Martin Scorsese's *Taxi Driver*. This is a stark cinematic portrayal of prostitution in New York City, expressed largely in metaphors of defilement. The main character, Travis Bickle, expresses strong feelings of disgust and desire in relation to women. Thus, he is fascinated by pornography, but as he cruises the streets in the red-light districts in his taxi, he sees only 'filth'. His commentary on the city is all about dirt and the need to purify the spaces populated by prostitutes and the sexually deviant.

Travis writes in his diary:

> May 10th ... Thank God for the rain which has helped to wash away the garbage and trash from the sidewalks ... All the animals come out at night – whores, skunk pussies, buggers, queens, fairies, dopers, junkies; sick, venal; some day, a real rain will come and wash all this scum off the streets.

And, similarly, when asked by a presidential hopeful, Palantine, for his view on what is wrong with the country, Travis volunteers this about New York City:

> You should clean up this city here because this city here is like an open sewer, it's full of filth and scum and sometimes I can hardly take it. Whoever becomes the President should just really clean it up, you know what I mean. Sometimes I go out and I smell it. I get headaches, it's so bad, you know, they just like never go away, you know. It seems like the President should just clean up the whole mess here, should just flush it down the fucking toilet.

Against this background of defilement, Betsy, the woman Travis idolizes, personifies purity: 'She was wearing a white dress; she appeared like an angel out of this filthy mess; she is alone, they cannot touch her.' His own anxieties about dirt dominate the film, so when Iris, a child prostitute he hopes to save from the streets, suggests that she might go to a commune in Vermont, Travis feels uncomfortable. He says that places like that are dirty and he couldn't go to a place like that. His final purifying act is to destroy the pimps with extreme violence and this act of purification, notwithstanding the violence, makes him a local hero. While *Taxi Driver* is a film about personal obsession, it could also be seen as a moral geography which has a wider

currency. ... Similarly, in a letter to the *New York Post*, a resident of the East Village laments the decline of her neighbourhood:

> Our cars and apartments are being burglarized by the street peddlers who sell their stolen bounty on Second Avenue and the adjoining, streets; St. Mark's Place has become a haven for pimps, prostitutes, drug dealers, head shops selling drug paraphernalia and assault weapons, illegal immigrants and a sundry collection of other undesirables. A neighborhood that was once the hub of multi-culturalism and neighborly pride has, over the past 20 years, become a cesspool of vermin. (*New York Post*, 21 September 1994)

Travis, the young fascist and the resident of a deteriorating East Village, express in strong terms attitudes towards people and place which are deeply embedded in western societies, although in liberal discourse they are conveyed with greater subtlety. These vivid social maps of the city are also used for navigational purposes by banks, insurance companies, the police and the social services, but their spatial demarcations become visible only through practices such as the withdrawal of financial services from 'high-risk' localities.

Subtlety is evident in some modern advertising which, while presenting more restrained comment on the 'other' than *Taxi Driver*, still presents urban society in oppositional terms, stressing the virtues of the pure by setting it against images of pollution. ... [T]elevision commercials and the modern media generally have used either city landscapes or urban sub-cultures to make distinctions between a positively valued inside and a threatening exterior world.

Some car commercials have made particular use of images of threat and danger to convey the idea of the car as a protective capsule which insulates the owner from the hazards of an outside world populated by various 'others'. One example is a Volkswagen commercial which made use of a young child to symbolize purity in the defiled environment of New York City. The commercial implies that the car will transport her securely through the city, to the safety of the suburbs or a commisionaired apartment building. The city's street people – homeless, mentally ill, drug addicts – are represented as remote but threatening, part of another world viewed from the safety of the Volkswagen. . . .

Other geographies have been suggested in detergent commercials where, predictably, purification through cleaning, attaining a state of whiteness and virtue, is a continuing theme. . . . Thus, in a . . . concentrated Persil [laundry soap] commercial, children are depicted as a part of 'the wild', untamed nature, which is their natural habitat but one which renders them uncivilized. Mother wonders how the children get so dirty at school. The children are then shown in an imaginary sequence tearing through the wilderness, but with the boys doing more adventurous things than the girl. Their place in nature is confirmed by dirty clothes, face paint and headdresses. The suggestion of an American Indian stereotype is interesting because this also locates the minority in nature rather than as a part of society. The children, however, are returned to society, cleaned and cared for by mother, with the help of Persil. . . . The children are portrayed as 'naturally' wild but it is clear that Persil is a civilizing influence, a necessary commodity in the suburban home, contributing to the creation of a purified environment in which children behave according to standards set by adults. The family home is the setting for a struggle against dirt and natural wildness. Consumption is encouraged by suggesting the undesirability of the soiled and polluted.

It is interesting to compare the representation of the child in the city in the Volkswagen commercial with these images of childhood conveyed by the Persil commercial. In the first, the child is pure and the city, or rather some of the stereotyped inhabitants of the city, constitute a threat to this purity. In the second, the children are defiled through their association with nature and purified by the civilizing influences of mother, home and detergent. Children can be simultaneously pure and defiled. Nature, likewise, can provide images of purity, often in contrast to the defiled city, or, as wilderness, it can be associated with people – children, indigenous minorities, and so on – in order to represent them as less than civilized and in need of purification. These shifts in the use of images demonstrate the contradictions and ambiguities which characterize stereotypes and the complex associations of people and places which are used to map the spaces of the same and the other.

[. . .]

Finally, I want to suggest how such imaginary geographies translate into practice. The kinds of representations described here in literature and the visual media confirm stereotypes of people and places and inform attitudes to others. These attitudes assume significance in community conflicts and in the day-to-day routines of control. This is evident, for example, in the case of European Gypsies, to whom opposition is expressed in a consistent and highly predictable form. Here, the problem is that Gypsies' dependence on the residues of the dominant society, scrap metal in particular, and their need to occupy marginal spaces, like derelict land in cities, in order to avoid the control agencies and retain some degree of autonomy, confirm a popular association between Gypsies and dirt. The fact that Gypsies have strong pollution taboos and high standards of cleanliness, where there are adequate facilities for keeping their trailers [caravans] or houses clean, is irrelevant. Because of their frequent association with residues and residual spaces, the perception of many *gaujes* (non-Gypsies) is that Gypsies are dirty. Consequently, the fear of 'polluting Gypsies' leads to attempts by the dominant society to consign them to residual spaces where the stereotypical associations are confirmed. . . . The representation of social categories either side of a boundary defined by notions of purity and defilement and the mapping of this boundary onto particular places are not solely a question of fantasy. They translate into exclusionary practice.

CONCLUSION

The idea of society assumes some cohesion and conformity which create, and are threatened by, difference, although what constitutes a threatening difference has varied considerably over time and space. Nation-states may or may not claim to accommodate diversity but at the local level social and cultural mixing is frequently resisted. . . . [T]here are enduring images of 'other' people and 'other' places which are combined in the construction of geographies of belonging and exclusion, from the global to the local. Historically, at least within European capitalist societies, it is evident that the boundary of 'society' has shifted, embracing more of the population, with the class divide in particular becoming more elusive as a boundary marker. The imagery of defilement, which locates people on the margins or in residual spaces and social categories, is now more likely to be applied to 'imperfect people' . . . a list of 'others' including the mentally disabled, the homeless, prostitutes, and some racialized minorities. Clearly, the labels which signal rejection are challenged and there is always the hope that, through political action, the humanity of the rejected will be recognized and the images of defilement discarded. There is no clear picture of progress, however. Feelings of insecurity about territory, status and power where material rewards are unevenly distributed and continually shifting over space encourage boundary erection and the rejection of threatening difference. The nature of that difference varies, but the imagery employed in the construction of geographies of exclusion is remarkably constant.

"Some Thoughts on Close(t) Spaces"

from *Bodies: Exploring Fluid Boundaries* (2001)

Robyn Longhurst

Editors' introduction

In her poem "Notes Towards a Politics of Location" (1984), the American poet Adrienne Rich wrote of the body as "the geography closest in." Cultural geographers, particularly those concerned about the body and its place in – or absence from – mainstream scholarship in geography frequently cite this line. The title of this selection, "Some thoughts on close(t) spaces," invokes a play on words, between "close" and "closet" (cupboard). Robyn Longhurst notes that while abstract notions of "the body" are becoming more common in geographic scholarship, real bodies – the fleshy, leaky, unstable bodies that we all inhabit – generally remain a taboo subject of geographic research and publication. Longhurst argues that consideration of bodily processes and the places where they occur is viewed as disruptive.

This is so in a practical sense, regarding the boundaries of what is considered appropriate subject matter by editors, reviewers, and funding agencies: the "codes of respectability" that Longhurst refers to. All of these individuals and agencies act as gatekeepers deciding what sort of research will be financially supported and published. Thus actually researching, writing, and teaching such subjects is generally not actively encouraged; rather it is "closeted" in mainstream geography in the same way that homosexuality is closeted in mainstream society.

This is also true conceptually. For to research and write of topics such as toileting activities is to blur the boundaries between inside and outside, to "mess up" the neat, clean spaces of mainstream geography. The topics discussed in this selection are on a scale close-in: uteruses, bathrooms, boardrooms. Longhurst calls for more work in cultural geography on the micro-politics of such spaces, noting that "the body is as 'political' as the nation-state." For instance, pregnancy is a topic that is seldom seriously explored by geographers. Yet the spatiality of the pregnant body – the shifting contours of the growing body; the shifting boundaries of where it is considered appropriate for this body to be, or to wear, or to do; and the disappearing envelope of personal space that plagues pregnant women whose bellies are routinely touched by strangers – would certainly seem to provide a rich topic for cultural geographers.

This is not to say that there is no work at all by geographers on topics related to the body in the fleshy, leaky, and disruptive sense that Longhurst calls for. In particular, feminist and "queer" cultural geographers have been increasingly active in exploring the sorts of topics discussed in this selection. Indeed, the body as a general theme is one of the most important dimensions of contemporary critical cultural geography. On bathrooms (toilets) in particular, see Sally Munt, *Heroic Desire: Lesbian Identity and Cultural Space* (1998), for a discussion of the disruptive effect of "butch" lesbians using women's public bathrooms/toilets; and Kath

Brown, "Genderism and the bathroom problem: (re)materializing sexed sites, (re)creating sexed bodies," in *Gender, Place and Culture* 11, 3 (2004): 331–346.

Robyn Longhurst is a Professor of Geography at Waikato University in Aotearoa/New Zealand. She identifies as a feminist geographer whose research and teaching interests encompass maternity, "fat" bodies, masculinity studies, qualitative methods, animal tourism in New Zealand, and domestic gardens. Some of Longhurst's publications include "Plots, plants, and paradoxes: contemporary domestic gardens in Aotearoa/New Zealand," in *Social and Cultural Geography* 7, 4 (2006): 581–593; "'Man breasts': spaces of sexual difference, fluidity and abjection," pp. 165–178 in Bettina Van Hoven and Kathrin Hoerschelmann (eds.) *Spaces of Masculinities* (2005); and (with Lawrence Berg) "Placing masculinities and geographies," in *Gender, Place and Geography* 10, 4 (2003): 351–360.

... [Here] I draw together some thoughts on pregnant bodies in public places, men's bodies in toilets/bathrooms, and managers' bodies in workplaces in CBDs [central business districts, or downtowns] in order to illustrate that bodies and spaces are neither clearly separable nor stable. I attempt to destabilise notions of self/other and subject/object in relation to these spaces. I slip between talking about the body as a space (for example, the interuterine space of the pregnant body) and the intimate spaces that the body inhabits (for example, domestic toilets/bathrooms). The spaces of the body and its environs become close, intimate, merged and indeterminable as they make each other in fluid and complex ways. The interuterine spaces of pregnant bodies, defecating men, and managers whose bodies attempt, but inevitably fail, to be respectable – conjure up images of close(t) spaces. They are close spaces in that they are familiar, near and intimate. They are also close*t* spaces in that they are often socially constructed as too familiar, near, intimate and threatening to be disclosed publicly.

As close*t* spaces they function as sites of oppression and resistance. Homosexual practices are often closeted, so too are a range of other bodily practices. Women are sometimes closeted about being pregnant – 'coming out' as pregnant can be both exciting and traumatic. . . .

I do not mean to imply that a binary division ought to be drawn between 'in (closet) spaces' and 'out spaces', close spaces and 'far away' spaces, the body and the nation, the local and the global, the micro and the macro scale, views from above and below. . . .

I focus on close(t) spaces not out of a sense of voyeurism but because they are as 'political' as any other ('far away' or 'out') spaces. The instability of boundaries, whether they be the bodily boundaries of individuals or the collective boundaries of nation-states, causes anxiety and a threat to order. To ignore close(t) spaces is to ignore that which is coded as intimate, 'queer', feminine, banal and Other. Such a strategic absence allows masculinism to retain its hegemony in the discipline. Close(t) spaces need an opportunity to come out in geography. There are many censoring and discriminatory practices that operate to keep particular sights/sites in the closet. . . . [A]rticles are pulled from library collections. Secretaries sometimes refuse to type or copy certain material. There are whispers and silences from colleagues and negative press from the media. Editors have been known to refuse to publish material in geographical journals because it is 'inappropriate' (read: they are repelled by and fearful of the material). An editor of a well-known geographical journal once told me that the pregnant body is an 'inappropriate' subject for geographers to consider.

[. . .]

. . . [Despite a growing literature on 'the body' in geography] it is still difficult to speak of close(t) spaces, liminal zones, abject bodily sights/sites in the discipline. These spaces threaten to spill, soil and mess up clean, hard, masculinist geography. Codes of respectability place limits on what we can say in geography. We may be able to discuss discursive constructions of embodiment but we still cannot talk easily about the weighty materiality of flesh, or the fluids that cross bodily boundaries in daily life.

The close(t) spaces of the pregnant woman/ uterus, of toilets/bathrooms, and of supposedly respectable bodies and workplaces are both real and imaginary. They are spaces of tears/blood/sweat and spaces of discourse and representation. The pregnant woman is both self and Other, mother and fetus, one and two, subject and object. The defecating man is also both subject and object. His excrement is both of him and distant from him. Likewise the manager who attempts to remain respectable at all times at work inevitably gives way to belching, burping or farting. S/he is both a respectable self and a loathsome Other. It is worth pursuing each of these ideas in turn.

INTERUTERINE SPACES

During pregnancy the zone or space around the body changes. The zone around the pregnant stomach becomes considerably thinner and may even disappear altogether in some instances. Interpersonal relations are situated within the multiple discourses that surround pregnancy and come into play to create a new spatiality for pregnant women and for those who interact with them.

This new spatiality helps to make sense of the public touching of pregnant women's stomachs. It can also be understood in terms of the pregnant woman herself who at times is no longer sure where her body begins and ends in relationship to the geographical space that she occupies. This can lead to a sense of uneasiness, surprise and disjuncture between the image and the materiality of the body for pregnant women. The pregnant woman sometimes finds her body in places where she does not expect. For example, she may try and squeeze through a gap only to find that her stomach protrudes further than she thought. Consequently she is unable to 'pass'. The pregnant subject's anatomical, material body can grow rapidly and it often takes some time before her body image catches up. As the pregnancy proceeds the borders of the body image [shift]. . . .

This point becomes particularly evident when examining various understandings of the placenta. 'Facts' relating to the structure of the uterus, especially the placenta, have changed radically since the 1960s. . . .

Although there is an implied link between mother and fetus, they were largely understood to be almost entirely separate entities. The 'wall' of the uterus functioned to protect the fetus. . . . In the first half of the 1960s it was thought that the womb insulated the fetus from the mother. . . . This understanding of the placenta allowed for a sharp distinction or separation to be drawn between mother and fetus. The pregnant woman's body was seen as 'outer space' with the fetus being located in a sealed or walled interior – 'space capsule' – known as the womb.

This model of the placenta-as-barrier or wall meant that not only was the fetus thought to be independent from the mother, but also that the mother was understood to be independent from the fetus. This obviated some (but not necessarily all) of the present-day pressures on pregnant women to monitor or circumscribe their activities in order to safeguard their baby. Pregnant women were not expected to give up smoking, employment, sport, to take plenty of rest, to curtail activities outside the home – at least not on the grounds that these activities might threaten the fetus in some way. Pregnant women may have been pressured to give up these activities on other grounds, for example, it was not seemly or ladylike for pregnant women to engage fully in public life, but not on account of the well-being of their fetus.

It was not until a 'discovery' early in 1965 that the placenta facilitated rather than blocked communication between mother and the fetus that a new discourse leading to an increase in the surveillance of pregnant women (both self-surveillance as well as surveillance by others) began to emerge. A July 1967 column in *McCall's* noted that 'the old idea that the womb is the safest human habitat has been sharply disproven in recent years' and cautioned that 'infants' were now being 'attacked in the womb'. . . . [W]riters for women's magazines and medical professionals interpreted the new view of the placenta to mean that mothers could communicate infectious or harmful substances to fetuses. As a result, those authors cautioned mothers to practise constant self-surveillance. This self-surveillance involved getting plenty of rest, giving up employment, reducing activities outside the home and relying on housework to stay physically fit. This idea of pregnant women's behaviour directly affecting the fetus still reverberates today. . . .

[. . .]

Recognising this inter uterine space as a close(t) space may offer a way of reconceptualising the relationship between mother and fetus but also of reconceptualising space more generally. All of us have occupied interuterine space (it is perhaps the closest of all spaces) and yet it is seldom discussed in geographical discourse (it has long been closeted). It has been closeted because the maternal, fluid, indeterminate geography of the uterus is likely to mess up a masculinist knowledge based on claims to truth, objectivity and rationality.

At a broader level it is useful to recognise that all spaces (not only interuterine spaces) are a 'bleeding' of materiality, fluidity, the imaginary, the socially constructed and the psychoanalytic. Bodies as/and spaces are always already soiled – they are never self-contained but can only exist in a complex relational nexus with other bodies/spaces.

EXCREMENTAL SPACES

Like the pregnant body, the defecating or excreting body that both constitutes and is constituted by domestic toilets/bathrooms is often understood to be a potentially dangerous body. In the usually private space of toilets/bathrooms people not only excrete but also check their appearance in the mirror, check their weight, pass wind, wash and comb their hair, shower, bath, clean teeth, shave, squeeze pimples, masturbate and rub on creams. Toilets/bathrooms are also a site where subjects might attend to the bodily needs of others (for example, assisting children and/or the elderly with toileting). Such activities are not discussed widely by geographers.

There is something about toilets/bathrooms and the (excremental) bodies that they house that constructs them as an 'unspeakable space' – a close(t) space – in geographical discourse. Toilets/bathrooms tend to be considered too material, too squeamish, too uncomfortable, too unacceptable (or just plain too banal) to discuss. Toilets/bathrooms have not been considered as a material or discursive space in which bodies are imbued and inscribed by cultural practices. They have not been on geographers' research agenda.

Kayrn Kee, a librarian at the University of Waikato, searched a variety of [online] data bases. . . . Key words such as 'bathroom', 'toilet', 'lavatory' and 'water closet' were used but yielded very few 'hits'. Using various search engines on the World Wide Web also yielded very little information on domestic toilets/bathrooms. There are some sites on the design of bathrooms, planning ideas, demolition and remodelling, ceramic tiles, vinyl flooring, cabinets, marble countertops, showers and toilets but their aim is to promote bathroom products.

A great deal more geographical research has been carried out on public (rather than domestic) toilets/bathrooms. Examples of this work on public bathrooms . . . include . . . research on public toilets as a possible 'beat' for gay men. . . . Urinals as a site of cottaging for gay men. . . . [An account of] being a 'butch' in the 'Ladies' toilet' and the boundary disruptions this causes. . . . A historiography (from the mid-nineteenth to the mid-twentieth century) of public toilets in Dunedin, New Zealand. . . . [And an analysis of] the provision of public toilets for women beginning with the question 'Why is it that women invariably have to queue for the toilet in public places whereas men do not?'

Interestingly, in these examples the focus tends to rest on gay men or women, rather than on heterosexual men. Few geographers have focused explicitly on heterosexual men. . . . By and large . . . there is no geographical discussion of heterosexual men in the private spaces of toilets/bathrooms.

The lack of attention paid to domestic toilets/bathrooms is all the more interesting in the light of a substantive geographical literature on the home. Over the last two decades feminist geographers have problematised the dichotomy between public space (exterior and open) and private space (interior and closed) and the ways in which public space has been valued over private space. As a result of this critique the home has increasingly become a legitimate topic for consideration. . . .

Within this literature there is some mention of specific rooms, for example, the kitchen is mentioned as a site of women's labour. Laundries, bedrooms and 'family' rooms are also mentioned but toilets/bathrooms rarely feature in this feminist geographical literature on the home. . . . It would appear that the toilet/bathroom is even more closeted and removed from the scrutiny of others than the more public areas of homes such as lounges and kitchens.

[...]

... [T]hough toilets/bathrooms have not been completely excluded they have most certainly been largely ignored – treated as banal and unimportant. It is likely that they are considered commonplace by many geographers – unworthy of attention – but they are also sites/sights of threat. They are one of geography's abject and illegitimate sites that have been deemed (perhaps unconsciously) inappropriate and improper by the hegemons in the discipline. ... [T]oilets [are] 'liminal zones'. It is difficult to speak of liminal zones. These zones, and their articulation in language, may cause us to feel uncertain, uncomfortable, confused and/or maybe repulsed. Liminal zones are often unspeakable.

One of the reasons for a dis-ease and discomfort over what are considered to be abject sights/sites in geography is the privileging of the mind over the body in geographical work. The body tends to be Othered in geographical discourse. Questions, therefore, about the omission of particular sites/sights in geography (for example, toilets/bathrooms) slip into questions about rights (women's rights, rights to be 'queer', etc.). This Othering of toilets/bathrooms serves to marginalise certain individuals and groups (such as women, the disabled and so on who are thought to be 'tied to their bodies' and, therefore, incapable of reason). It is a specific notion of knowing as disembodied that marginalises Others in the production of geographical knowledge.

The mind/body dualism plays a vital role in determining what counts as legitimate knowledge in geography. Geographies of toilets/bathrooms run the risk of being ghettoised, feminised geographies robbed of their legitimacy. So long as the mind is privileged over the body, hegemons in geography will continue to edit out that which they consider to be 'dirty', preferring instead the clean, the clinical, the quantitative, the heroic, the solid, the straight and the scientific. What constitutes appropriate issues and legitimate topics to teach and research in geography comes to be defined in terms of reason, rationality and transcendent visions, as though these can be separated out from passion, irrationality and embodied sensation. Unlike

boardrooms, toilets/bathrooms are not considered to be the 'real' or 'serious' stuff of geography.

Toilets/bathrooms are (abject) zones or sites where bodily boundaries are broken. The insides of bodies make their way to the outside (for example, urination, excretion, vomiting, squeezing pimples) and what is outside the body may make its way to the inside (for example, the naked body may feel vulnerable to penetration). The emotions that can accompany these acts remain sealed within the privacy of toilets/bathrooms. A danger lurks here. ...

Focusing on heterosexual 'white' men's experiences of toilets/bathrooms ... is a way of rewriting male corporeality. In the spaces of toilets/bathrooms men cannot pretend to make any easy separation between subject and object, self and Other, desire and repulsion, solid and fluid, mind and body. Cartesian ontology is 'messed up'. The fluidity and permeability of heterosexual, 'white' men's bodies is exposed. Focusing on the toilet/bathroom forces heterosexual men to come out of the (water) closet.

MANAGING SPACES

Unlike toilets/bathrooms, 'professional' workplaces, especially in CBDs, are constructed as spaces in which bodies must not transgress their boundaries. Liminal zones where the insides and outsides of bodies sometimes become indeterminable – noses, vaginas, penises, eyes, sores – must be carefully monitored and kept under control at all times in workplaces. It can require enormous vigilance to construct the proper, professional and respectable body – to present a 'public face' – at work. This is one of the functions of the business suit.

The firm and straight lines of the business suit give the appearance of a body that is impervious to outside penetration. They also give the appearance of a body that is impervious to the dangers of matter that is inside the body making its way to the outside. The suit closets the body in respectability. However, although the suit helps to create an illusion of a hard, or at least a firm and respectable, body that is autonomous and in control, bodily boundaries can never continually remain intact. While business attire may reduce potential embarrassment caused by any kind of leakage it can never completely secure a body. Given that women function as the 'marked category' and that their bodies are socially constructed as 'modes of seepage' business attire takes on an added significance and importance for women.

While some workplaces are constructed as rational and cerebral (for example, banks, computer consulting companies or insurance companies) others are constructed as pleasurable and fun (for example, cafés, casinos and health clubs). Regardless, there is an assumed respectability in all these environments.... Managers, staff and even customers who do not conform to the unwritten rules that govern respectable behaviour in these spaces are treated with suspicion and caution. There are continual attempts (by managers but also by cleaners and security guards) to keep professional workplaces clean, tidy and 'nice' regardless of whether they be a clothing store, bank, restaurant or travel agency. Clean, tidy, 'nice' sites/sights help to make clean, tidy, 'nice' bodies and vice versa. Together they ensure profits for a wealthy middle class.

The boundaries separating the 'nice' from the shabby (bodies and places) are insecure and continually contested.... [Poor] bodies, like queer bodies, disabled bodies and black bodies, engender feelings of abjection for respectable bodies. Respectable bodies, however, need an Other upon which to found their (insecure) identity.... [T]here is an internal relationship between power, desire and disgust....

CONCLUSION

In all three case studies – pregnant women in public spaces, heterosexual 'white' men in domestic toilets/bathrooms, and managers in CBDs – ideas about sexed bodies, body boundaries, body fluids, abjection and (im)pure spaces have proven useful. In concluding I want to make a series of brief points....

The first point is that it is not enough to examine only broad and wide-sweeping maps of power and meaning. The micro-level politics that imbue bodies and spaces also need to be held up to scrutiny. The body is as 'political' as the nation-state. As a generalisation, geographers have been effective at looking at the broader picture but this has sometimes been at the expense of finer detail – the close(t) geographies. In order to understand the relationships between people and places it is necessary to address a range of geographical scales. To date, the body has received less attention than more macro-level analyses.

The second point I want to make concerns the production and politics of academic knowledge.... I have talked with and listened to people while at the same time trying to understand myself (including my academic practice). During the years of writing I have been prompted to think about the production and politics of academic knowledges. I have no doubt that some readers will attempt to dismiss, suppress and/or neglect the ideas contained with in this book as illegitimate.... The sites of knowledge production also matter.

I work as an academic geographer 'down under' at a small, largely unknown university in New Zealand (the 'bottom' of the world). I often find that theoretical frameworks, usually produced in the Northern hemisphere, do not travel easily and unproblematically into specific Antipodean locales. This [work], then, may be read as a kind of hybrid, impure text that subverts Euro-Anglo theory with Antipodean narratives – a coming together of 'higher' and 'lower' knowledges. To put it rather badly – it has been written from the 'bottom', about the 'bottom'. My hope is that it will prompt readers to question further some of the grounds on which geographical knowledges rests and to consider the directions and constitutions of new geographies.

The third and final point that I want to make is that the construction of the body as Other appears to have changed throughout the 1990s as geographers began to examine more explicitly the politics of embodiment and spatiality. However, often the body being examined is a kind of 'disembodied body'.... I think that in the discipline of geography over the last decade a new linguistic territory has been created. It is the territory of the body but ironically it is a fleshless territory, a territory constituted of little more than a chain of polite signifiers. It is a body that does not have specific genitalia or that breaks its boundaries. Such a body is a masculinist illusion.

Focusing on a body that has no specified materiality will not further feminist agendas.... Denying the weighty materiality of flesh and fluid will help enable masculinism (the unmarked norm) to retain its hegemonic position. In this regard, the epistemology and ontology of geographical knowledge is unlikely to change dramatically.

In examining pregnant bodies in public spaces, men's bodies in toilets/bathrooms, and managers' bodies in CBDs I have illustrated that what all these bodies share is their fluidity, volatility, and abject materiality. Although some bodies are commonly represented as more abject than others (such as pregnant bodies), in fact all bodies (including the bodies of heterosexual, 'white' men) are unstable. In addition to this point I hope to have shown that this fluid, volatile, abject corporeality cannot be plucked from the spaces it constitutes and is constituted by.

The close(t) geographies of the body challenge some of the dominant constructions of knowledge in geography. Specificity seeps into generality, a politics of fluidity seeps into a politics of solidity, and a lived messy materiality seeps into cerebral knowledge. Perhaps thinking, writing and talking about bodily fluids, abjection, orifices, and the surfaces/depths of specific bodies can offer a way of prompting different understandings of power, knowledge and social relationships between people and places.

"Contested Terrain: Teenagers in Public Space"

from *Public Space and the Culture of Childhood* (2004)

Gill Valentine

Editors' introduction

As with all of the contributions to this section on difference, Gill Valentine's work is concerned with social belonging and exclusion, and how this is expressed, reinforced, and contested spatially. In "Contested Terrain: Teenagers in Public Space," Valentine considers the special status of teenagers and the many ways that they are, for the most part, excluded from public spaces. As with many of the selections in this *Reader*, Valentine uses ethnographic methods – here, interviews – to inform her work. Conversations held with adolescents, parents, and police officers are included in the text, to support Valentine's claims that teenagers themselves feel, and are treated by adults, as *out of place*. Teenagers are under practically constant surveillance in their homes, at school, and in public as well as private spaces. They typically have little say over what activities they will engage in or their schedules; they have limited financial freedom, and limited mobility. Their presence, particularly in spaces such as parks, streets, and shopping malls, is often seen by authorities as threatening. Harkening to David Sibley's notions of purifying space (see p. 380), teenagers are often viewed as polluting influences to be got rid of (see also Cosgrove, p. 176).

The life stage known as adolescence is a fairly recent cultural invention. In the early 1960s French historian Philippe Aries established a convincing argument that the very concept of childhood, let alone adolescence, was not widespread in Europe until the late nineteenth century. Rather, young people were seen as miniature adults. They entered directly into adult society as apprentices, field and factory laborers, and the like at what seems today the shockingly young age of seven or so; see *Centuries of Childhood: A Social History of Family Life* (1962). Indeed, in some cultures today, relatively young people are expected to enter into adult roles with little or no transitional period of adolescence, through early marriage or entry into the work world. However, one might question the quite adult roles, themes, and situations – particularly with respect to sexuality and violence – that are marketed to today's American and European young people via popular films, music, video games, and other cultural products.

Though it is a relatively recent sub-field of cultural geography, the geographies of youth are a fast developing line of research. Some examples include Tracey Skelton, who has co-authored publications with Gill Valentine on the exclusion and inclusion of teenagers (see below). Madeleine Leonard, in "Teens and territory in contested spaces: negotiating sectarian interfaces in Northern Ireland," in *Children's Geographies* 4, 2 (2006): 225–238, uses stories, maps, and interviews to examine the ties of fourteen and fifteen-year-olds to place and religious identification. Peter Hopkins has written about young people in Scotland and the transition to university; see "Youth transitions and going to university: the perceptions of students attending a

Geography summer school access programme," in *Area* 38, 3 (2006): 240–247. Finally, Michael Leyshon has written about doing research with this group, in "On being 'in the field': practice, progress and problems in researching young people in rural areas," in *Journal of Rural Studies* 18 (2002): 179–191.

Gill Valentine is a Professor of Human Geography at the University of Leeds, where she also directs the Leeds Social Science Institute. Valentine has multiple research interests, encompassing her current research on children and families, deaf young people's social identities and exclusion, the culture of drinking establishments, and innovative qualitative methodologies such as photo diaries and Web logs ('blogs'). Among her many publications are (with Tracey Skelton) "Living on the edge: the marginalization and resistance of D/deaf youth," in *Environment and Planning A* 35 (2003): 301–321; "Boundary crossings: transitions from childhood to adulthood," in *Children's Geographies* 1 (2003): 37–52; (with co-editors Phil Hubbard and Rob Kitchin) *Key Thinkers on Space and Place* (2004); and (with Tracey Skelton) *Cool Places: Geographies of Youth Cultures* (1998).

■ ■ ■ ■ ■ ■

[. . .]

NOWHERE TO GO, NOTHING TO DO: TEENAGERS' EXPERIENCES OF PUBLIC SPACE

Children and teenagers have little privacy relative to adults. At home or school they are subject to the gaze of teachers, siblings and relatives who often try to channel them into organised activities . . . that conflict with their own agendas. Home, in particular, is a space that is constituted through a complex range of familial rules and regulations and, as such, boundary disputes with parents are commonplace. In particular . . . domestic tensions around home rules and the use of different rooms within the family home represent a conflict between adults' desire to establish order, regularity and strong domestic boundaries, and young people's preferences for disorder and weak boundaries. Like the home, the school is also a highly regulated institution with its clearly delimited boundaries and moral geographies. This girl describes her lack of freedom:

> *Girl.* Cos like your parents don't let you have a lot of freedom and like school's always nagging at you to do things, like when they're talking to you they talk to you as a little child not an adult. ('Working class', metropolitan area, Yorkshire)

[T]here is little public (as opposed to private) provision of facilities for young people in UK towns and cities, let alone the countryside. Moreover, teenagers commonly want to participate in adult activities rather than be corralled with young children in specialist environments. Public space is therefore an important arena for young people wanting to escape adult surveillance and define their own identities and ways of being. However, efforts to revitalise or 'aestheticise' public space as part of attempts to revive (symbolically and economically) cities . . . are increasingly resulting in the replacement of 'public' spaces with surrogate 'private' spaces such as shopping malls and festival market places. The development of these new privatised spaces of consumption, and broader processes of gentrification, are serving to homogenise and domesticate public spaces by reducing and controlling diversity in order to make these environments safe for the middle classes. . . . Among the undesirable 'others' being priced out, or driven out, of these commercial social, retail and leisure complexes by the private security industries (including guards and closed circuit television – CCTV – surveillance) are teenagers. . . . Such processes hide the extent to which the public realm is being privatised and commodified and reinforce the importance of the street for contemporary young people.

Notably, the space of the neighbourhood or city street, particularly after dark, when many adults have retreated to the sanctuary of the home, is often the only autonomous space many teenagers are able to carve out for themselves, and is therefore an important social arena where young people can be together. With nothing particular to do, young people often roam the streets looking for excitement because the street can be a place where special things happen. Moreover, the very act of doing nothing is . . . doing something because it is a time when children have the freedom and privacy from adult supervision to be themselves. This is increasingly important to young people given the fact that doing nothing is becoming less and less possible for children in the culture of contemporary parenting in which children are ferried from one institutional activity to another. . . .

[. . .]

Hanging around on street corners, in parks, under-age drinking, petty vandalism and larking about and other forms of non-adherence to order on the street become (deliberately and unconsciously) a form of resistance to adult power. . . . Groups or gangs of young people – finely delineated by age – often colonise and contest control of particular spaces, such as bus shelters or parks, as their own. They stake a claim on these spaces both by their physical presence, and by marking their territory or ownership with graffiti or other markers like rubbish. Different groups of young people use these places to play out identity struggles, excluding each other from 'their territory' through name calling, bullying and general antagonism and intimidation. Girls' single sex friendship groups are especially marginalised through such tactics by older mixed groups.

Girl 1. We hang around outside with our friends.
Girl 2. You feel safer hanging around with a gang of people.
Girl 4. More people you know.
Girl 2. Yeah in your own area.
Interviewer. Do you ever have any frightening experiences?
Girl 3. Yeah sometimes.
Girl 1. When they [other groups of teenagers] come up and start swearing and . . .
Girl 3. Bullying ya, and saying things to ya. (Group discussion, 'working class', metropolitan area, Yorkshire)

Over the recent past the public realm, rather than being a social order of civility, sociability and tolerance, has increasingly become one of apprehension and insecurity. Encounters with 'difference' are being read not as pleasurable and part of the vitality of the streets but rather as potentially threatening and dangerous. In this context, young people's nonconformity and disorderly behaviour is often read as a threat to the personal safety of other children and the elderly and as threat to the peace and order of the street. The fear that their children may come to harm at the hands of other violent children was particularly, though not exclusively, expressed by interviewees in relation to boys and was the justification some parents gave for restricting their sons' use of space. These parents explain their concerns:

Mother. [T]hat's not something I particularly fear of, not abduction. I think I'd be more frightened, more concerned that he might may be caught up by a group of lads who would really beat him up and really hurt him, not just come home crying but you know head butt him and all the rest of it. ('Working class', metropolitan area, Greater Manchester)

[. . .]

Young people are not only considered 'out of place' on urban streets, they are an equal cause of concern in rural areas, as these quotes suggest:

Mother. There is a problem with teenage children hanging around. They are hanging around at the moment, well they've been there for a while now at the bottom of the school drive, in cars . . . It is really off putting because the cubs meet there and, um, they were frightened you know. Even I was a bit, when I used to walk down and meet Robbie when it was dark, you know, just walking past these cars, it was a bit spooky. ('Middle class', rural area, Derbyshire)

[. . .]

As some of the quotes above indicate, the threat posed by 'dangerous children' is also experienced by many mothers for their own safety. Women described feeling intimidated by groups of teenagers on the streets; some even adopt precautionary measures such as crossing the road, or changing their route in order to avoid putting themselves in places where they feel at risk of teenager violence. There appears to be a different geography of women's fear in each of the neighbourhoods where the research was carried out because the gangs of young people colonise different spaces in each area. For example, in one village they were associated with the bus stop on the high street, in a city neighbourhood they were perceived to have appropriated an area near a derelict pub and in one of the non-metropolitan areas they congregated on a parade of shops, whereas in another similar area, a park was identified as their territory. These women describe their concerns:

> *Mother.* I've seen a um, standing with cans of lager outside the shop. Now I'm not saying that we didn't sort of buy a bottle of Strongbow cider or something but we wouldn't ever do it blatantly in front of adults like that outside a shop . . . they're a lot less frightened of authority than we were, we'd never have done certain things that they do, I'm sure of that . . . I tend to keep away from them because I find them threatening myself when they're all in a big group and like I say they don't have much respect for authority or older people. ('Middle class', non-metropolitan area, Cheshire)

[. . .]

Yet most research suggests that teenagers hanging around on the street do not deliberately set out to intimidate women, the elderly and other children in public space, nor do they intend to cause trouble. Rather posturing and larking around sometimes leads to laws being broken or to children disrupting adults' worlds but that this is not usually premeditated but is a by-product of natural flows of activities. Rather young people criticise the lack of public space available to them and unreasonable intervention of adults into their social worlds. These teenagers from an urban metropolitan area in Greater Manchester commented:

> If the youth club were open there wouldn't be no big gangs outside cos we'd be in the youth club. They wouldn't have to moan at us about smokin', drinkin' and shopliftin', or whatever, if the youth club were open. There's nothing else for us to do.

We're not going round smashing things up. It's cos we've got nothing to do that we hang round here . . . Every teenager does it.

Despite young people's innocent intentions, their very presence in public places is often considered not only frightening but also a potential threat to public order. . . . The suburbs, in particular, have a certain moral order based on an overwhelmingly powerful and widely understood pattern of restraint and non-confrontation. Residents have established 'norms' or appropriate ways of behaving towards each other and often have little contact with 'other' groups who are regarded as unpredictable and threatening. As such, adults regard teenagers as a menace to the moral order of neighbourhoods because of the way they are perceived to threaten property through acts of vandalism but more importantly adults' peace and tranquillity. . . . These quotes illustrate some of the way adults regard young people as bringing disorder to the streets:

> *Mother.* . . . they congregate round the Western shop area, round there, then they move along for so many weeks and they were down on Thornton Square near the shops down there, they just seem to congregate. We had a spate where they were just coming up and ripping up 'For Sale' signs just for sheer devilment . . . At the moment they've got a new game which as a crowd they're running up and down here, knocking on people's doors, ringing the bells and generally being a nuisance. ('Middle class', non-metropolitan area, Cheshire)

> *Father.* We have trouble at the back, with teenagers on the back there, only kids, vandalism more than anything. I've been out to try and stop them but it doesn't make any difference. ('Middle class', metropolitan area, Greater Manchester)

While several interviewees recalled that they too spent their own childhoods hanging around the streets in gangs, tormenting adults in their neighbourhoods, these actions were painted as innocent pranks. In contrast they interpret similar behaviour by contemporary young people as signifying that teenagers are aggressive, intimidating and out of control in public space (even though there appears to be little, aside from anecdotal, evidence of any actual increase in teenage violence). . . .

The notion of a 'moral panic' – 'invented' by the sociologist Stanley Cohen to explain the public outcry caused by the clashes between 'mods' and 'rockers' in England in the mid 1960s – is a useful concept for thinking about how various youth cultures and young people's behaviour in general is often viewed by adultist society as 'criminal' or 'deviant'. . . .

The media play a pivotal role in moral panics by representing a deviant group or event and their effects in an exaggerated way. They begin with warnings of an approaching social catastrophe. When an appropriate event happens which symbolises that this catastrophe has occurred, the media paint what is often a sensational and distorted picture of what has happened in which certain details are given symbolic meanings. In turn the media then provide a forum for the reaction to, and interpretation of, what has taken place. The public then become more sensitive to the issue raised, which means that similar 'deviations', which may otherwise have passed unnoticed, also receive a lot of publicity. This spiral of anxiety can eventually lead to punitive action being taken against the 'deviant' group or event by relevant authorities. . . . [M]oral panics are about instilling fear into people. Fear either to encourage them to turn away from complex social problems or more commonly fear in order to orchestrate consent for 'something to be done' by the dominant social order.

Moral panics are related to conflicts of interest and discourses of power and are often associated with particular 'symbolic locations' such as the street. These panics are frequently mobilised in relation to particular groups of young people, such as mods and rockers, when they appear to be taking over the streets or threatening the moral order of the suburbs. . . . [M]oral panics are increasingly less about social control and more about a fear of being out of control and an attempt to discipline the young. This process often involves nostalgia for a mythical 'golden age' where social stability and strong moral discipline were a deterrent to disorder and delinquency. . . . Such moral panics have been evident in both the US and UK at the end of

the twentieth and beginning of the twenty-first centuries.

Across the US, young people are currently the subject of popular suspicion and anxiety. One of the key triggers of this contemporary concern is the perceived omnipresence of youth gangs on the street, in which 'gang' has become a code-word for 'race' and a symbol too of drugs, guns, graffiti, gangsta rap and violence. The media have played a key role here in exaggerating the number of gangs and in distorting their activities through racialised constructions of youth violence. . . . [M]oral panics about youth gangs are indicative of more general processes of social polarisation. As a result the issue has become increasingly politicised, resulting in the inclusion of measures to prosecute gangs members and treat juveniles as adults. . . .

[. . .]

Whereas in the US this moral panic about young people is focused on gangs and is explicitly racialised, in the UK there has been a more general anxiety about what a British television documentary dubbed 'the end of childhood'. It began in the mid 1990s with the murder of a two year-old, Jamie Bulger, by two ten-year-old boys. Although extremely unusual, this murder was not completely unprecedented and quickly became a reference point for other cases of violence committed by children. Other evidence, from statistics on bullying, joy-riding and teenage crime have been mobilised by the media to fuel popular anxieties about the unruliness of young people. . . .

[. . .]

. . . [T]he blame for their behaviour has been laid at the door of parents, schooling and the State. All three stand accused of having made children ungovernable by eroding the hierarchical relationship between adults and children. It is argued that parents have traditionally had 'natural' authority over their offspring as a result of their superior size, strength, age and command of material resources. . . . However, at end of the twentieth and beginning of the twenty-first centuries understandings about what it means to be a parent are alleged to have changed, with adults voluntarily giving up some of their 'natural' authority in favour of closer and more equal relationships with their offspring. . . . [T]he balance of obligations has shifted so that the responsibility is no longer on the child to be a dutiful son or daughter but on the parent to provide for their children in particular (mainly material) ways. Indeed, in an individualised culture the stress is on children to be socialised into independence. As such . . . parent/child relationships are problematic because parental ideas of setting boundaries as moral codes or guidelines can conflict with young people's demands for more autonomy.

This social change has also been accompanied by a general shift in both legal and popular attitudes to young people, away from an adults-know-best approach towards an emphasis on the personhood of the child and children's rights. As a result of which some commentators argue that there has been a decay in childhood as a separate category and that the distinction between children and adults is increasingly becoming blurred. The growth in lone parent households has also meant that an increasing number of children are living in situations where they share emotional and financial responsibilities with a parent.

[. . .]

As a result a number of studies have concluded that parents feel that they no longer have any moral or psychological resources to exercise authority over young people. This mother and police officer describe their fears about the ungovernablity of contemporary young people:

Mother. I just think there's a whole different, you know, they're a lot less frightened of authority than we were, we'd have never done certain things that they do now, I'm sure of that . . . I don't think that they do have much respect for authority or older people now at all . . . But like I say when . . . they're all hanging around I find it really intimidating and to be, to feel like that about kids really, it's not nice to think that, not nice. ('Middle class' non-metropolitan area, Cheshire)

Police Officer. I've been in the Police twelve years and I've certainly found a difference in their [teenagers'] attitude to the police and to people. I mean I can I know it sounds a bit corny or whatever but I remember when I was a lad I would never dream of shouting at some person walking down the street. But now they don't bother and that they do all sorts. (Cheshire)

As these quotes suggest many parents hark back to the 'golden age' of their own childhood when children had respect for adults. . . . Now, parents argue that children's rights campaigners, such as EPOCH (End Physical Punishment of Children) have undermined the foundations on which this respect was built. Firstly, because they have encouraged a shift in both legal and popular attitudes to children, away from an 'adults know best' approach, towards an emphasis on the personhood of children (for example, if it is wrong to hit a person it must also be wrong to hit a child). Secondly, and perhaps most significantly in the eyes of the parents, teachers and police interviewed, the State has increasingly adopted a liberal line towards the physical punishment of children (for example, childminders are not permitted to smack children, and State schools are no longer able to use corporal punishment to discipline pupils) and has therefore removed the ultimate tool adults (parents, those who act *in loco parentis*, such as teachers, and the police) had at their disposal to enforce their authority and superiority in both private and public space. . . .

[. . .]

Father. They don't discipline them any more. You know, they're [schools and police] not allowed to touch them any more, so they can get away with a lot more. There's no corporal punishment and I think that's wrong. They're making their own decisions in life, that's what it is – The parents are not making the decisions for children. We were brought up to respect the elders – honestly, I mean I sound old-fashioned but that is the way we were brought up at school. Like now they just do what they want to do and like no one can change what they do. ('Middle class', metropolitan area, Greater Manchester)

Mother. I think it's getting to the stage where you can't control children. You know you daren't smack 'em, you daren't reprimand them in front of somebody else, otherwise they've only got to utter a word in school and you've got the Social Services knocking on the door . . . So I think that's why children are like they are, because the schools can't give 'em the cane any more; you daren't smack 'em any more, so they know that they're going to be able to get away with it, so they just do it because they know there's nothing you can do to stop it. ('Working class', metropolitan area, Greater Manchester)

Father. I know if I ever got into trouble, I mean you hear it in the papers now a copper's clipped a lad's ear and he's in court. I know when I were a lad, if a copper clipped my ear and I'd told me mother she'd have given me another one. You know, it's all sort of changed now. ('Working class', rural area, Derbyshire)

[...]

SEVEN

"The Geography Club"
from *Geography Club* (2003)

Brent Hartinger

Editors' introduction

Most of this *Reader*'s contents have been drawn from academic work, published either in book form or as articles in scholarly journals. *Geography Club*, however, is a novel written for young adults. In *Geography Club*, writer Brent Hartinger describes the lives of gay and lesbian high school students who attend the fictional Robert L. Goodkind High School. Far from being good or kind, the environment at the school for gay and lesbian teens is hostile. The school's queer students are particularly anxious to find a meeting place that would allow them to socialize. Yet in every place they have congregated – the library, the pizza parlor, the park – they have come close to being caught and "outed" by fellow students.

So the students decide to form a club – "a club that's so boring, nobody would ever in a million years join it!" – the Geography Club. The Geography Club provides a cover and a designated meeting place in a school classroom. But to their surprise, the Geography Club holds some appeal to the other students, and the group must decide whether or not to allow unaware straight students to join. Eventually, the students form a Gay–Straight–Bisexual Alliance, the high school moves a step closer to accepting difference, and the novel's characters mature in sometimes painful ways.

Why include an excerpt from a teen novel in a cultural geography *Reader*? Though *Geography Club* is not about geography, strictly speaking, Hartinger draws on spatial language to describe the exclusionary social terrain of Goodkind High School. For example, Russel describes his fall from social grace when rumors begin to circulate about his being gay as follows: "Over the past few weeks, I'd been exploring the Land of the Popular, and the Landscape of Love. . . . I'd gone from the Borderlands of Respectability, to the Land of the Popular, and now to Outcast Island, also known as Brian's lunch table" (pp. 195–196). In addition, it is notable that the gay teen characters' movement is restricted by the gossip and suspicions of their classmates, such that they are forced to convene or "hook up" in dark, abandoned, and out-of-the-way places, like old warehouses, the woods, and the park at night. And the irony of forming a geography club in order to claim a legitimate meeting space is not lost on cultural geographers of difference.

To be sure, there are academic geographers whose scholarly research focuses on sexuality and space, and in particular on gay and straight spaces, and how mainstream straight spaces can be subverted or "queered" through acts, words, and coded cues that construct a shared symbolic subculture. Gill Valentine and David Bell's edited collection *Mapping Desire* (1995) was an early publication in the field of queer geography. Geographer Michael Brown has examined Hartinger's novel in "A geographer reads *Geography Club*: spatial metaphor and metonym in textual/sexual space," in *Cultural Geographies* 13, 3 (2006): 313–339.

The Gay and Lesbian Atlas (2004), by Gary J. Gates and Jason Ost, utilizes census data to map demographic patterns of gays and lesbians in the United States.

Author Brent Hartinger says that the main character in the book, Russel Middlebrook, is fashioned after himself and his own experiences in high school. On his Web site, http://www.brenthartinger.com, the author provides a wealth of advice for teens who think they may be gay, or who simply want to know more about the author. Hartinger describes going to a Catholic high school, which was "even worse than I'd expected." "I think it's absolutely criminal," writes Hartinger, "that gay kids are still forced to spend their adolescent years feeling as lonely, and as freakish, as I did then." Apparently, things have not changed much, for in 2005 the University Place School District in Tacoma, Washington – which also happens to be Hartinger's hometown – banned *Geography Club* from school library shelves. In an interview[1], Hartinger says that while the School District's official objection was that the book romanticizes a relationship begun on-line (where Russel's relationship with closeted jock Kevin Land begins), the PTA of one school objected to the book "because it would 'turn straight kids into homosexuals'."

Brent Hartinger lives in Tacoma, Washington, with his partner, who is also a writer. Hartinger has authored many novels, including sequels to *Geography Club*, titled *The Order of the Poison Oak* (2005) and *Split Screen: Attack of the Soul-sucking Brain Zombies/Bride of the Soul-sucking Brain Zombies* (2007). Hartinger also writes screenplays, and teaches writing at Vermont College.

NOTE

1   See http://cynthialeitichsmith.blogspot.com/2005/11/author-interview-brent-hartinger-on.html

---

But I did talk to them again, the very next day. We met after school deep in the stacks of the library. I wasn't sure whose idea it had been – I'd got the message from Min, who'd got an E-mail from Terese – but it was the perfect place to get together. If you're looking for solitude, a high school library is one of the best places to go, especially in the two hours after classes. And if anyone saw us, we could always pretend we weren't together, that we all just happened to be looking for a book in the same aisle at the same time.

Right before the meeting, I'd been wondering how it would feel. Would it be comfortable and real, like the pizza parlor? Or would it feel stilted and embarrassing, like the school cafeteria?

The second Min and I turned the corner and saw the faces of the others, I knew the answer. I felt that little swell of excitement like when you know you're about to set the top score on a well-used video game. Being one of the Nerdy Intellectuals I mentioned earlier, I generally like libraries anyway – I love the clean, heady musk of ink and paper and carpet glue. But I'd never been exhilarated in a library before. I was even glad to see Ike.

We all nodded our hellos, but I didn't look at Kevin, because I was thinking about what Min had said two days earlier, about him maybe liking me.

"Sorry about yesterday at lunch," Terese said, whispering.

Kevin and Ike mumbled their apologies too.

"It wasn't you guys' fault," I whispered back, and Min nodded. "It was no one's fault."

"Did your friends say anything?" Min asked Terese.

"Yeah," she said. "They wondered what was up."

"What'd you tell them?"

"That we were thinking about starting a club." She shrugged. "I'm a pretty good liar."

"What kind of club?" I said.

"I didn't say. I just changed the subject." So I wasn't the only one who avoided questions by changing the subject. Maybe this was another thing we all had in common.

"This is so stupid!" Ike said. "We shouldn't have to hide like this, like political dissidents or whatever. Why can't we be seen together like normal people?"

As if in answer, Candy Moon walked by the end of the aisle. I thought I saw her slowing down ever so slightly. Suddenly, this didn't seem like such a good meeting place after all. Five people in the same aisle was a pretty big coincidence.

"Damn," Kevin said, whispering again. "I think I had a bad idea, our coming here."

This meeting had been Kevin's idea? But why hadn't he E-mailed me directly? Did it mean Min was wrong, and Kevin didn't like me after all? Or did it mean just the opposite, that Min was right and he *did* like me, but that he was too shy to do anything about it? (Kevin Land shy? That was a laugh.)

"What we need," Min spoke softly, "is someplace to meet where no one'll see us."

"We could go back to the pizza place," Ike said. "We could meet there after school."

"No," Terese said. "Sooner or later, someone'll see us. It's too close to school. The team goes there for pizza."

"Then some other restaurant," I said.

"I don't know," Kevin said. "Most nights, I got practice. It'd have to be close by. But like Terese said, if it's close to campus, we're gonna run into someone."

"The woods?" Terese said. There was this big forested area on the other side of the track field.

"Too cold and wet," Ike said.

"Wait a minute," Min said suddenly. "What Terese said. Why not start a real club?"

"Huh?" I said.

"You know," she said. "An after-school club. Don't they let you use a classroom? I mean, if you fill out the right forms?"

"What kind of club?" Terese said. She sounded suspicious. "You mean like a gay–straight alliance?" I'd heard about gay–straight alliances at other schools. Other big-city schools, that is. There were no gay–straight alliances in our town, maybe not even in our entire state, and there weren't going to be any anytime soon. If Reverend Blowhard could get so worked up over something as innocent as a teacher talking about contraceptives in a health class, it wasn't hard to imagine what he and his cadre of concerned parents would do over the existence of a gay–straight alliance at the local high school. The mushroom cloud would be visible for miles around.

"Well," Min said, "we don't need to tell anyone that's what kind of club it is. We'll just say it's a club."

"You have to," Ike said. "You have to say exactly what you are. They can't deny any club, not as long as you follow all the rules. My friends and me were going to start an Earth First! chapter, and Rall wasn't going to let us." (Remember, Mr. Rall was the school principal.) "But then Gladstein – he was our faculty advisor – he told Rall we'd sue if he didn't let us. Oh yeah, and you have to have a faculty advisor too."

No one said anything. We just thought about all the new information.

"Well, the first part is easy," I said. "We just make something up. We'll tell them it's a chess club when it's really just us."

"But what about the faculty advisor?" Ike said. "I mean, they'd be right there with us."

"Mr. Kephart," Min said.

We all looked at her.

"He's the most uninvolved teacher in the whole school! Two fifty-one in the afternoon, he's gone. If we ask him to be our advisor, the last we'll see of him is when he signs our application form."

"You think he'll do it?" Terese said.

"He will if I tell him he doesn't have to come to any of the meetings."

I felt a smile breaking out on my face. But at the same time, I saw movement at the end of the aisle. I turned to see Heather Chen staring right at us. Terese, Min, and I all snatched books from the shelves. It looked incredibly phony and probably made us look even more suspicious in Heather's eyes. When I looked back, she was gone.

"It's time to wrap this thing up," Min whispered, quieter than ever. "Are we all agreed about starting a club?"

"Hold on." Ike was barely whispering too. "There's still one more problem. If we start a club, it has to be open to every student in the school. That's the policy."

"Too bad we *can't* say it's a gay club," Terese said. "That'd keep everyone away." It was a joke, but it didn't sound like one, because she sounded so bitter.

Kevin hadn't said anything in a while, and I figured it was because he'd changed his mind and now he didn't want anything to do with this club thing. Or me.

So I was surprised when his face suddenly lit up, and he whispered, "I got it! We just choose a club that's so boring, nobody would ever in a million years join it!" He thought for a second. "We could call it the Geography Club!"

We all considered this. This time, I saw smiles break out all around.

The Geography Club, I thought. No high school students in their right minds would ever join that.

In other words, it was perfect!

"Trish Baskin's hot for you," Gunnar said to me. It was the following Saturday, and Gunnar and I were playing racquetball on a court at the Y. I didn't completely suck at racquetball (that's my modest way of saying I was really pretty good). But Gunnar had said what he'd said about Trish Baskin right before his serve, so I had to wait until we finished the rally to ask him what the hell he meant. Of course, he won the point, but only because I was distracted.

"What?" I said.

"What what?" he said.

"What about Trish Baskin?" Our voices were echoing in the close confines of the brightly lit court.

"She does," Gunnar said. "I heard it from someone who knows. You like her?"

Like her? I thought. I barely even *knew* her. Oh, and then there was the small matter of my being queer as a three-dollar bill.

"She's okay," I said. She'd been in my geometry class the year before. She was sort of the mousy type, with this whispery voice and narrow shoulders and a streaked haircut that she'd probably had to be talked into getting. "Go ahead and serve."

"Well, she really likes you," Gunnar said, right before hitting the ball again. But I wasn't distracted this time, so I pounded it right past him and took back the serve.

We kept playing, and I noticed that Gunnar seemed quiet. Unlike Min, he wasn't particularly competitive, so I doubted he was thinking about the game. No, something else was going on here.

"Hey," he said a few minutes later, when he won back the serve.

I faced him, wiping my face with the sweat towel that had been hanging from my pocket. He said, "Remember what we talked about a couple of weeks ago?"

I had absolutely no idea what he was talking about. "What do you mean?"

"You know. About my getting a girlfriend?"

Now I remembered. But he'd talked about that a zillion times before, so I still wasn't exactly sure where he was going with this.

"Yeah?" I said.

"Well, I think I got one. A girlfriend, I mean."

"Yeah? That's great! Who is it? Why didn't you tell me?" I was genuinely happy for him, in part because I wouldn't have to listen to him moan on and on about not having a girlfriend anymore.

"Kimberly Peterson."

"Gunnar, that's fantastic!" I said. "I'm really happy for you." I'd never actually spoken to Kimberly, but I'd seen her around school. She had long blond hair. That was about all I remembered, but keep in mind I don't exactly have a photographic memory when it comes to girls.

"Well, she's not really my girlfriend yet," Gunnar said. "But she did agree to go out with me."

"Well . . ." I tried to think of something positive to say. "I'm sure she'll like you once she gets to know you. Then she *will* be your girlfriend."

"Yeah," Gunnar said, tight lipped, and I knew there was something he still wasn't telling me. Whatever it was, I had a bad feeling about it.

"Go ahead and serve," I said, and he did. I won the rally, but it was just plain luck. Now we were both distracted.

"Trish Baskin's *really* hot for you," Gunnar said. "She's friends with Kimberly. That's who told me."

I faced him. "Gunnar. What's going on?"

He was suddenly fascinated by the strings on his racquet. "Remember when we talked about my getting a girlfriend?"

I nodded.

"And remember when you promised you'd do anything to help me out?"

I nodded again, even though I didn't remember promising to do *anything* exactly.

"Well, Kimberly did agree to go out with me. But only on one condition."

"Oh," I said, and instantly I knew what was going on. "Gunnar, *no!*"

"Russ, why not? It'd only be one date!" Kimberly had agreed to go out with Gunnar only if I agreed to go out with her friend Trish, in case you haven't figured that out already.

"Gunnar!" My voice really echoed. I hadn't meant to yell.

"You said you'd do anything to help me!"

I was about to tell him exactly what I was thinking – that I hadn't said I'd do "anything" to help him. And even if I had, this wasn't what I'd meant! I'd meant driving with him to the mall so he could pick out a tux for the prom.

"Please, Russ. You know how important this is to me. Besides, it's just one date. What's the big deal?"

The big deal was I wanted to be dating Kevin Land! But I couldn't tell Gunnar that. Of course, putting up a big fuss about one little date with Trish Baskin was the next best thing to telling him. It was exactly the sort of thing that would make him suspicious.

I sighed. "It's a double, right? You and Kimberly and me and Trish?"

"Definitely!"

"When? Next weekend?"

Gunnar nodded. "Saturday."

I hesitated a second longer, just to make him squirm a little. Then I said, "Okay, I'll do it."

"Oh, thank God!" he said, far more relieved than he should've been.

"Gunnar," I said.

He still couldn't keep his eyes off those strings. "Yeah?"

"You already told Kimberly I'd do it, didn't you?" He looked up at me at last, a tiny smile on his lips.

"Maybe."

"Gunnar!"

But at least he had the decency to look properly ashamed about it, so I decided to let it slide.

"A date with a girl, huh?" Min said the next day, when I met her for a walk in the park. "That should be one hot and heavy evening."

"It sure wasn't my idea," I said. "You know her?"

"Trish? No. But I went to camp with Kimberly when I was eight. She used to eat paste, if that's any help. What does Gunnar like about her?"

"The fact that she has two X chromosomes." Min knew how much Gunnar wanted a girlfriend, and she laughed at my joke, which always made me feel good. You had to be pretty smart to make Min laugh.

"So," I said. "You excited about the Geography Club?"

"Yeah," Min said. But I noticed she'd suddenly stopped laughing, or even smiling.

"Where do you want to walk?" Min said, looking out over the hills of grass and bare trees. Winter was almost over, but there was a tinge in the breeze that hinted of a chill yet to come, like the smell of gunpowder in the air the week before the Fourth of July. The ground beneath our feet was cold and hard.

"Let's head for the Children's Peace Park," I said. This was a little garden on the other side of the park. There were shrubs and flowers, and in the middle of it all, there were these six painted wooden cutouts of the children of the world all holding hands. It was extremely stupid, but they'd put it up years before, when the Olympic torch had passed through town.

"What's up?" I said as we walked.

"Nothing," she said. She shrugged. "Terese."

"What?"

"I don't know. It's stupid. It's just . . ." She shivered, pulling her coat tighter around her neck. Min was chilly. "We got together last night at our warehouse."

"Yeah?"

"And things felt different."

I looked over at her. "What do you mean?"

She thought for a second. "Remember how I said Terese and I get together only in that warehouse?" I nodded. "No one ever saw us together, no one even knew about us. When we were together, it always felt like our own little world. This perfect, special place that only we could get to. It was like it wasn't quite real."

I nodded again, but secretly I was a little jealous. It sounded wonderful.

"But last night, it felt different," Min said.

"Because people know about you now? Kevin, Ike, and me?"

"I don't know. I guess. Nothing's really changed. But we got together at the warehouse last night, and it felt different. I still love her and everything. But it felt awkward. Like she wasn't quite the same person I remembered. Like we turned on the overhead lights in the warehouse, and we could see everything clearly for the first time, but nothing looked like we thought it did. Everything was messy. I liked it the way it was. I don't want light in that room."

"No one's going to tell," I said. "If you're thinking one of us is going to tell people about you guys, we won't." I don't know why I believed this so strongly – I barely knew Kevin and Ike. But I *did* believe it, as much as I'd ever believed anything. They'd never tell anyone about Min and Terese, and neither would I.

"It's not that," Min said. But she didn't say anything else, which made me think she didn't know what it was exactly. Finally, she said, "It's just a feeling."

"Maybe you just have to get used to it," I said. "People know now. I guess that makes it feel more real. But maybe once you get used to that, you'll go back to feeling the way you did before. Or maybe it'll be different, but better."

"Maybe," Min said, but I could tell she didn't believe me.

"You want to forget the idea of the club?"

"No. What's done is done." She said this hesitantly, but I was relieved anyway. Without Terese and Min, there was no Geography Club. And with no Geography Club, there was no Kevin and me.

"But?" I said.

"But I can't shake this feeling that something bad is going to happen."

I thought, Something bad to you and Terese, or something bad to the whole Geography Club? But I didn't ask this, because it seemed rude to be thinking of myself. Still, I couldn't help but remember what had happened the last time one of us had felt just a partial feeling of impending doom.

On that happy note, we reached the Children's Peace Park. It looked incredibly cheesy, just like I remembered. The painted wooden cutouts were all these horrible ethnic stereotypes of the children of the world. But it had been changed since I'd seen it last. Someone had taken a black marker and drawn tits on the wooden cutout of the Polynesian girl in the grass skirt, and they'd given the grinning, sombrero-wearing Mexican boy a hard-on. But the rosy-cheeked Eskimo boy had it worst. They'd pulled him off his base, kicked him in half, and knocked both pieces clean out of the garden.

"Nice," Min said.

"Yeah," I said, now shivering myself, and not just from the cold.

PART EIGHT

Culture as Resource

Courtesy of Tim Oakes

INTRODUCTION TO PART EIGHT

As pointed out in the introduction to Part One of the *Reader*, the concept of culture has – in very general terms – moved from a descriptor of the *outcomes* of certain environmental or social processes to an increasingly central variable *explaining* social phenomena. One consequence of this increasing "centrality of culture," as Stuart Hall (see also p. 264) calls it, is the fact that there has been a marriage of culture with fields of study, such as economics, that were previously thought to have little if anything to do with culture.[1] It has thus become common to hear about such things as "management culture," "enterprise culture," "commercial culture," and so on, among the many other expanded uses of the term. In analyzing this expansion, Hall points out that one should not necessarily take this expansion of culture to mean that economic or business activities are now more cultural than before. As will be discussed below, it may be true that capitalist development today values "cultural" (that is, symbolic, meaning-oriented) resources more than in the past, but the more fundamental point is that culture serves as an *analytical resource* today in ways that it did not in the past. The traditional field of political economy, in which economic relations are understood in terms of their political implications, has been increasingly infiltrated by culture, such that today we speak more and more of investigating a *cultural economy*.

As Paul du Gay has observed, the term "cultural economy" signifies a break with political economy in allocating great importance to meaning in the conduct of economic life. Standard political economy tends to conceive of economic processes and practices as "things in themselves," with "objective" meanings, and people become the simple bearers of these processes and practices, rather than meaning-producing subjects. But meaning is always produced at economic sites and through economic processes. "The most important point to note about our term 'cultural economy' is therefore the crucial importance it allots to language, representation and meaning – to 'culture' – for understanding the conduct of economic life and the construction of economic identities."[2] Thus, terms like "business culture" and "commercial culture" refer to the meanings produced by business and commercial practices.

However, cultural economy *also* indicates the increasing significance of the culture industries, such as global entertainment corporations like Sony and Disney, in the economic value chains of global capitalism. *And* it refers to the growing aestheticization of products, the generating of desire for products by inscribing them with particular meanings through advertising, design, and marketing. So, while it is important to understand that economic activities have *always* had a cultural dimension – and that the analytical strategies of political economy have only recently begun to take this into account – there has also been a noticeable shift in the "substance" of the economy, with symbolic, intangible, image-oriented resources being increasingly valued as part of production and consumption.

We can speak of *culture as a resource*, then, in two distinct senses of the term. On the one hand, culture is an *analytical resource* for interpreting economic-related activities. That is, the "expansion" of culture has been a conceptual or intellectual expansion, and culture is now increasingly used as a tool in novel analytical ways. This is the general sense of the "cultural turn" in the social sciences, discussed in the Introduction to this *Reader*. This approach tends to view culture in some variation of a semiotic perspective (see Geertz, p. 29). Culture, in this sense, is like a language, a system of meanings, and as such can be brought to bear on the study of any number of human phenomena, since all human phenomena have some kind of meaning-oriented dimension to them.

On the other hand, *culture is a resource mobilized* for specific economic, developmental, or governance objectives. This second sense of culture as resource represents an empirical version of the cultural turn. That is, once we start thinking more about the cultural dimensions of economic activity, we realize that changes in those cultural dimensions can affect changes in economic activities. The same presumably applies to other social or even regulatory activities. Thus, culture has become increasingly central to the *strategies* of economic development and local boosterism. This much is apparent enough, for example, in Zukin's chapter in this part (see p. 431). But this is now a very different conception of culture than the semiotic version laid out in the above paragraph. As an instrumental or "expedient" resource (see Yúdice, p. 422), culture is not a symbolic system but a kind of performance, or a particular practice, the *meaning* of which is less important than the *uses* to which a particular performance or practice is put. Such an approach to culture echoes the "non-representational" versions encountered in the selections by Latham (p. 232), Jones and Cloke (p. 68), and Bull (p. 194).

Careful readers will probably realize, however, that these two versions of "culture as resource" are not mutually exclusive. In the selections that follow, both a *culture as analytical resource* approach and a *culture as expedient resource* approach ought to be apparent. For example, McDowell and Court's article displays both a meaning-oriented analysis of the representations of merchant banker bodies and an expedient-oriented analysis of how bankers "perform" their bodies in order to meet certain expectations, goals, or comply with dominant norms. And Jackson's article on commercial cultures (see p. 413) seeks to demonstrate both the ways meaning-oriented innovations in clothing styles impact economic processes *and* the ways clothing production cannot be fully understood by focusing solely on a political economy analysis of commodity chains.

The idea of "culture as resource" does not assume any single conception of culture, but rather is meant to highlight the ways the idea of culture has been extended into fields far beyond those conventionally thought to belong to cultural geography. And this of course is our main point. Cultural geography is not a firmly bounded field of study, but follows the twists and turns that culture itself takes as it changes and adopts according to changing social conditions.

NOTES

1 S. Hall, "The Centrality of Culture: Notes on the Cultural Revolutions of our Time," in K. Thompson (ed.) *Media and Cultural Regulation* (1997).
2 P. du Gay, "Introduction," in P. du Gay (ed.) *Production of Culture/Cultures of Production* (1997), pp. 4–5.

"Commercial Cultures: Transcending the Cultural and the Economic"

from *Progress in Human Geography* 26, 1 (2002): 3-18

Peter Jackson

Editors' introduction

One of the first questions often asked concerning the notion of "cultural economy" deals with the political implications of the term. If cultural economy signifies a break with political economy in allocating importance to meaning in the conduct of economic life, how do the politics of culture reformulate the understandings of social power relations articulated in political economy? Another way of raising this question is to consider the fact that cultural studies of economic processes tend to focus on behaviors associated with consumption, since that's where much of the "meaning-making" in economic decisions occurs. Thus, the politics of the cultural economy might be viewed primarily as a politics of consumption. If we suppose that this is true, then we might want to know whether or not a cultural economy analysis has anything useful to say about the social relations revolving around production.

For example, there has been much written from the perspective of political economy that has sought to expose the exploitative nature of Nike's global system of production, distribution, and marketing. The concept of a *commodity chain* has been central to such critiques. Here, the idea is to demonstrate that each link of an extended chain of production and consumption – i.e. links between resource producers and suppliers, various manufacturers, traders and shippers, wholesalers, and retailers – involves specific (and usually unequal) social relations, surplus value extractions, and institutional provisions structuring the economic exchanges in ways that are often exploitative or otherwise unjust. But how might the use of culture as an analytic resource impinge on this critique? When you buy a pair of Nikes for $100, does the *meaning* of that purchase make any difference to the assembly-line worker in Indonesia who earns mere pennies from that purchase? Do the complex cultures of fashion – which of course can be extremely political in terms of their subversion of dominant norms – make any difference in the process of production?

Many political economists would say not. Indeed, Marx himself argued that any meaning associated with consumption was but a "fetish," or a "veil" that masked or hid the true exploitative relations of production underlying the commodity. The Marxist term *commodity fetishism* identifies this idea by acknowledging that people do create meaning in their encounter with commodities, but that such meaning should never distract us from the material realities of the capitalist production system. And while this perspective remains incisive in its ability to identify processes of labor exploitation and other widespread injustices underlying economic processes, many scholars have remained dissatisfied with the kind of black-and-white simplicity of this kind of Marxist political economy.

If anything, then, cultural economy's break with political economy reflects this dissatisfaction. Peter Jackson's article on commercial cultures is itself inspired by this dissatisfaction with a narrow political economy understanding of phenomena like the success of Nike. Originally delivered as the "Progress in Human Geography" Lecture at the 2001 RGS–IBG Annual Conference in Plymouth, Jackson's article explores how a cultural economy approach might offer alternatives to traditional commodity chain analysis. Jackson's starting point is to question the dualisms with which we commonly categorize and structure our understanding of the social world, such as culture–economy, local–global, and consumption–production. He argues that such terms should not be *reified*, that is, assumed to express actual social formation, but instead achieve their meaning only in relational terms. Thus, for instance, consumption and production are suggested by Jackson to be mutually constituted, meaning that the significance of production depends upon claiming that it is *not* consumption, and vice versa. But in real life, these cannot be so neatly separated into discrete activities.

Insisting that the world is more complicated and messier, Jackson argues that commodity chain analysis fails to capture the ways people invest meaning in their purchases, the ways design and marketing reframe commodities, the ways agents of cultural innovation play a key role in both production and consumption. The commodity chain, Jackson argues, is a *linear model*, whereas the economy functions much more like a network of multiple and often circular connections. He argues that "notions of circuitry and interconnection have more to offer than linear constructs of modernization or globalization that posit some kind of simple transition from a traditional to a more highly commodified system of exchange." After presenting some examples of empirical work that suggest some of this "circuitry and interconnection," Jackson raises some questions regarding the political implications of rethinking linear models of economic processes.

Does a cultural analysis dull the critical political edge of commodity chain analysis? Is there really a progressive politics of consumption? Such questions are necessary, Jackson admits, but answering them should not invoke an academic division of labor between "materialist" studies of labor exploitation on the production side and "cultural" studies of meaning-oriented behavior on the consumption side. "Consumption" need not be restricted to the point of purchase, nor should the "production" be restricted to the factory floor. How do poor and marginalized people consume? What meanings are created in both consumption *and* production practices? And in what ways does creative innovation influence production networks in far-flung places? A cultural analysis, Jackson suggests, opens the door to these kinds of questions that can deepen our understanding of economic processes, and these cannot help but have significant political implications.

Peter Jackson is Professor of Geography at the University of Sheffield. A prolific cultural and social geographer, Jackson has published widely on consumption and popular culture, diversity, identity, and social difference. He is also the author of the classic introduction to cultural geography *Maps of Meaning* (1989). Recent publications include a co-authored article, "Mobilising the 'commodity chain' concept in the politics of food and farming," in *Journal of Rural Studies* 22 (2006): 129–141, and the co-edited volume *Making Sense of Men's Magazines* (2001).

Other significant work that would serve as a good background for material discussed in Jackson's article would include Sharon Zukin's *Point of Purchase: How Shopping Changed American Culture* (2003), Paul du Gay and Michael Pryke's edited collection *Cultural Economy: Cultural Analysis and Commercial Life* (2002), Paul du Gay's edited volume *Production of Culture/Cultures of Production* (1997), Scott Lash and John Urry's *Economies of Signs and Space* (1994), Ash Amin and Nigel Thrift's edited collection *The Blackwell Cultural Economy Reader* (2003), and Robert Sack's *Place, Modernity, and the Consumer's World* (1992).

INTRODUCTION

Over the last few years, 'progress' in human geography has been marked by repeated calls for a convergence between 'the economic' and 'the cultural'.

... Nor are these [calls] purely rhetorical (although I think it could fairly be said that calls to transcend the 'great divide' between the cultural and the economic have significantly outnumbered empirically grounded studies that demonstrate the difference that such a move would make in practice). There is, however, a growing number of examples of the kind of work I have in mind. A promising start was made in the early 1980s with Sharon Zukin's study of the intersection of culture and capital in the creation of New York City's real-estate market for luxury 'loft living', an approach she has subsequently elaborated in relation to the wider 'cultures of cities', including the key mediating role of the 'critical infrastructure' in the city's public culture of museums and restaurants [see p. 431]. Similar lines of enquiry have been pursued in Paul Du Gay's study [*Consumption and Identity at Work*, 1996] of the collaborative manufacture of 'enterprise culture' among retail workers and consumers in 1980s Britain, and in Linda McDowell's work on the gendered performance of culturally approved workplace identities in the City of London [see p. 457] – to name just a few of my recent favourites.

What, then, can an exploration of commercial culture add to these promising debates? At first sight, it might seem that the juxtaposition of 'culture' and 'commerce' in my title is a doomed attempt to bring together two irreconcilable ways of seeing the world. After all, 'culture' is traditionally associated with meaning and creativity, with works of the imagination and aesthetic practices that are far removed from the pursuit of economic profit. 'Commerce', on the other hand, has traditionally been regarded by social scientists with disdain, signalling a vulgar and materialistic world devoid of morality, where human agency is subordinated to the logic of capital. To quote from Raymond Williams' *Keywords*: while 'commerce' has retained a fairly neutral inflection, negative associations began to attach to the idea of 'commercialism' from around the mid-nineteenth century as 'a system which puts financial profit before any other consideration'. The perceived decline of the 'industrial spirit' in England dates from around the same time as observers began to associate industry and commerce with vulgarity.

The themes that are addressed in this paper concerning the economies, practices and spaces of contemporary commercial culture clearly have a wider resonance in terms of what Linda McDowell has recently described as the 'awkward relationship' between the cultural and the economic. Following McDowell, my argument seeks to challenge the kind of dualistic thinking that separates production

from consumption, the local from the global, or culture from economy – by emphasizing the mutual constitution of these very terms and investigating their fundamental inseparability. So, for example, the marketing of something as mundane as a jar of coffee draws on a language of seduction and sensuality, culturally encoded through references to the exotic and the erotic. In the advertisements to which I am referring here, for Cap Colombie and Alta Rica coffee, the makers Nestlé invite us to breathe an 'aroma that softly, subtly catches the attention, then charms into wilful submission': 'Voluptuous, dark and full-bodied,' with an 'aroma of passion . . . Alta Rica, we bask in your glory.' The commodity form is imbued with the sexual aura of the 'Latin' women whose bodies adorn these billboard and magazine advertisements. We are even asked to imagine that a multinational company like Nestlé has a soul: 'From the Heart of Latin America. And the soul of Nescafé.' We have become so used to this kind of commercial blandishment that we are almost immune to its absurd hyperbole ('basking in the glory' of a cup of instant coffee).

While examples like these demonstrate that cultural meanings are regularly appropriated for commercial ends, I also want to argue, conversely, that the apparently rational calculus of the market is inescapably embedded in a range of cultural processes. I am not seeking to 'reduce' the cultural to the economic (or vice versa), or to show that either side of the equation is more significant than the other. Rather, by subjecting a range of commercial cultures to theoretical reflection and empirical scrutiny, I want to try to demonstrate the value of an approach that transcends conventional dualisms between 'the cultural' and 'the economic', drawing out the links between production and consumption and making connections between a variety of scales from the local to the global while simultaneously blurring the boundaries between academic disciplines. This is an agenda of such heroic (some would say foolhardy) proportions that it can only be achieved (if at all) through specific examples and, for further illustrations, I shall be drawing on my own current research (with Claire Dwyer and Phil Crang) on contemporary transnational commodity cultures.

The intellectual agenda I am pursuing here follows recent changes in the commercial world itself, which was transformed during and after the

Thatcher years. Thus it has become commonplace to speak of a 'retail revolution' in Britain as three or four supermarket chains have come to account for an ever greater proportion of the nation's shopping expenditure. As corner shops have gone out of business, new stores have sprung up in locations that were once regarded as off limits to retail capital, like airports and railway stations. Changes to Britain's commercial landscape are, of course, much more widespread than the changing retail geography of the high street. Banks and other financial institutions now regularly sponsor opera performances and TV series, while football shirts are emblazoned with the team's sponsors (famously subverted by Robbie Fowler's parody of Calvin Klein in his support for the striking Liverpool DocKers). As these examples (which could easily be multiplied) suggest, the mutual implication of culture and commerce is clearly grounded in changes in the material world as well as in scholarly fashion.

To pursue my argument, I want to consider a number of different models for bringing the cultural and the economic into closer dialogue, seeking to privilege neither term but to 'deepen' our understanding of their mutual entanglement through some specific examples. My argument involves a move from linear commodity chains to more complex circuits and networks as a way of subverting dualistic thinking and unsettling the kind of linear logic that sees consumption at one end of a chain that begins with an equally abstracted notion of production. This emphasis on networks and circuits is not designed to demonstrate complexity for its own sake but to suggest new modes of understanding and new possibilities for intervention in what can sometimes seem an all-encompassing 'consumer culture' where every act of resistance is immediately recuperated by the market in successive rounds of commodification.

A 'SIMPLE' COMMODITY CHAIN ANALYSIS

Let us begin, then, with a simple commodity chain analysis, taken from Robert Goldman and Stephen Papson's recent study of *Nike Culture: the sign of the swoosh* (1998). The book argues that Nike have succeeded commercially through the commodification of sport, turning a 'parity product' (where there is actually very little difference between Nike sports shoes and those of their competitors) into a hugely successful brand, commanding over 40 per cent of the market share and grossing over $9 billion a year. Their commercial success is, of course, based on the exploitation of foreign labour – with Pakistani children reputedly paid as little as six cents an hour to sew footballs, and labour costs accounting for less than $3 of the retail price of a pair of $80 trainers.

Selling a global brand through marketing that appeals to local tastes (in the words of Nike Vice-Chairman Richard Donahue) involves a range of corporate strategies including the breaking up of production, subcontracting and outsourcing, with virtually no production done in-house and with an extremely high proportion of the company's budget devoted to advertising and marketing. This 'buyer-driven commodity chain' involves some 18,000 retail accounts throughout the USA plus a mix of independent distributors, licensees and subsidiaries in approximately 110 countries around the world. The production process itself is now more likely to be located in Indonesia, China and Vietnam than in Japan or South Korea as Nike, like their competitors, are constantly searching out lower wage regions. Critics of multinational firms like Nike have often expressed their role as to unmask or unveil the exploitative labour conditions and social relations involved along the commodity chain. Yet, when one tries to represent the company's supply networks, subcontractors, marketing and distribution systems diagrammatically, the notion of a simple single-stranded 'commodity chain' scarcely does justice to the complexity of the processes involved.

'UNVEILING' THE COMMODITY FETISH

It could, of course, be argued that Nike's strategy of spatially dispersed flexible production has helped shield them from external criticism. In 1997, for example, the company employed former US Ambassador Andrew Young to tour its Asian manufacturing facilities, championing Nike's Code of Conduct and their severing of links with a number of factories that were paying below the minimum wage or operating excessive working hours. Responding to public criticism of exploitative wages,

unacceptable working conditions and the harass-
ment of women workers, the company claimed that
their jobs were prized locally in comparison with
other available work, using a form of cultural relativism to justify their economic practices. These
contested relations of production are, of course,
nowhere to be seen in the company's advertising
campaigns or in their lavish Niketown retail stores.
Here, the emphasis is on the individual athlete,
personified through 'stars' like Michael Jordan
and André Agassi, with the admonition to all of their
customers to fulfil their true potential (Just Do It).
As such, the company's ideology could clearly
be represented as a classic case of commodity
fetishism, where identical shoes without the Nike
logo would be much less desirable. Yet there is
something hollow about the call to 'unveil' the
commodity fetish as though the provision of such
knowledge would automatically lead to widespread
shifts in consumer behaviour or to significant
changes in working conditions at the point of production. There is little evidence to suggest that commercial culture works in this way, notwithstanding
the success of specific consumer boycotts for products such as Coors beer or Nestlé dairy products.

David Harvey has, of course, written passionately
about the need to reveal the 'hidden geographies'
of production that lie masked on the supermarket
shelves: the fingerprints of exploitation that are
rendered invisible by the commodity form [see
"Between space and time: reflections on the geographical imagination" in *Annals of the Association
of American Geographers* 80, 418–434]. Our job,
according to Harvey, is to 'lift the veil on this geographical and social ignorance', 'tracing back' and
'revealing' what lies 'embedded' within the social
relations of contemporary consumption. Robert
Sack [in *Place, Modernity, and the Consumer's
World*, 1992] employs a similar metaphor to talk
about the history of extraction, manufacture and
distribution being 'virtually obliterated' when the
finished product is presented to the customer,
while Martyn Lee [in *Consumer Culture Reborn: the
Cultural Politics of Consumption*, 1993] asks why
commodities show 'no manifest trace' of the
labour that was invested in them during production,
calling on academics to 'reveal' the 'concealed'
exploitation of labour that lies behind the 'mask' of
the commodity form. . . . However, there is something unsatisfactory about this call for unveiling and

unmasking the commodity fetish, not least its
subtle privileging of academic knowledge over
the popular wisdom of everyday life. It shows
little respect for the political judgement or moral
integrity of ordinary consumers to represent them
as so easily duped by the manipulative forces of
contemporary capitalism. It also runs counter to all
the empirical evidence from media and cultural studies that emphasizes the agency of audiences to read
media messages in an increasingly knowledgeable
way. For these reasons and others, recent work from
a variety of theoretical perspectives has begun to
move from an analysis of commodity chains to the
less linear logic of circuits and networks.

[. . .]

REFASHIONING CULTURAL IDENTITIES?

There has, to date, been relatively little work on
British-Asian fashion, with the important exceptions
of Nasreen Khan, Parminder Bhachu and Emma
Tarlo whose work I will now briefly consider.
Nasreen Khan (1992) provides a useful introduction
to the recent cultural history of Asian women's dress,
from *burqah* to Bloggs, beginning with the hostile
reception of Zandra Rhodes' collection of ripped
saris, commissioned by the Indian government
and shown in New Delhi in 1982 [see "Asian
women's dress: from burqah to Bloggs: changing
clothes for changing times," in J. Ash and E.
Wilson, eds., *Chic Trills: a Fashion Reader*, 1992]. The
collection caused widespread offence in defiling a
garment that had become the hallmark of Indian
women. Khan goes on to illustrate the widespread
politicization of dress in South Asia, including
Benazir Bhutto's adoption of the (Islamic-inspired)
dupatta (headscarf or shawl), the growing popularity of *bindi* forehead decorations in the West
and the general revival of 'ethnic' fashion among
middle-class urban-educated consumers in India
and throughout the diaspora. Khan argues that this
'ethnic revival' was the result of a multitude of small
ventures by women, from 'suitcase collections'
and similar modest beginnings in people's attics and
garages to the development of more commercial
boutiques and designer labels.

Khan charts the initial ambivalence of many
British-Asian women to 'traditional' styles of
dressing and the tendency to adopt different

clothes for different settings: 'British on the streets, Asian at home.' During the early years of immigration, she argues, 'Asian' clothes occupied a private or secret place for many young people. By the early 1980s, however, businesses like Variety Silk House in Wembley had begun to cater for the local British-Asian community and visitors to the subcontinent were returning to Britain with evidence of the creative fashion explosion in India and Pakistan, no longer regarded as the bastion of tradition in contrast to a stereotypically 'modern' west. Designers like Geeta Sarin and firms like Libas and Egg sold their clothes through catalogues and via upmarket stores in Belgravia and Kensington, tapping a more affluent east–west market rather than the localized 'Asian' market in places like Ealing, Southall and Wembley. Meanwhile, as Claire Dwyer's research has shown, such 'traditional' items of clothing as the veil have been substantially reworked in the construction of new identities among young British Muslim women [see "Veiled meanings: young British Muslim women and the negotiation of differences," *Gender, Place and Culture* 6, 1999: 5–26].

Parminder Bhachu's account [in "Dangerous design: Asian women and the new landscapes of fashion," in A. Oakley and J. Mitchell, eds., *Who's Afraid of Feminism?* 1998] also emphasizes the agency of women in forging new identities within an increasingly global market place:

> [diaspora Asian women] have used global commodities and consumer products to create new local interpretations of cultural identity . . . patterns [which] emerge from their sophisticated command of the symbolic and political economies in which they are located. (p. 189)

Her work focuses, in particular, on the commodification of the salwaar-kameez (or Punjabi suit) and its entanglement within the commercialization of the wedding economy. She shows, for example, how the dowry system has escalated among Sikh women in Britain so it is not uncommon now for dowries to include over fifty items of clothing as well as household goods, luxury consumer items and gold ornaments. Bhachu interprets the rise of the ready-to-wear salwaar-kameez as a highly charged piece of clothing: an inscription of ethnic pride through which Sikh women, in

particular, were able to express their opposition to the 'Hindu' sari following the Indian army's action at the Amritsar temple. While the media have focused on 'western' appropriations of 'Asian' dress – such as Jemima Goldsmith's wedding outfits in 1995 or the clothes that the late Princess Diana wore on her visit to Pakistan in 1996 – Bhachu insists that cultural creativity also flows in the other direction. So, for example:

> during 1994 and 1995 many metropolitan Asian women wore the top half of the salwaar-kameez with a full body stocking and with Doc Marten or thick platform shoes . . . Mohicanized and punkized salwaar-kameez outfits have also been worn by Asian women in the last decade. The whole gamut of current styles in vogue – from punky to funky to grunge to baggy hip-hop – can be seen in the interpretation of salwaar-kameez by diaspora women. (pp. 196–197)

The activities of these cultural intermediaries, Bhachu argues, have opened up new spaces, generating new landscapes and ethnicities, new consumer styles and material economies, representing 'the subversive outcomes of the shared cultural geographies of British women in the 1990s'. Bhachu emphasizes the symbolic importance of the salwaar-kameez as 'reflective of the stitching and suturing of many terrains and textures in which Asian women are situated' and of the active negotiation of new cultural forms by British-Asian entrepreneurs who continuously reformulate their 'ethnic' traditions through the filters of their British class and local cultures.

It is no coincidence that Parminder Bhachu adopts a highly spatialized vocabulary to describe these processes of cultural creativity and commercial innovation. Her own biography is highly transnational, describing herself as a European woman of East African descent who lived for many years in Britain before taking up permanent residence in the USA. From her own experience of multiple migrations, she speaks authoritatively of the local specificity of consumption styles, the creation of new spaces and notions of citizenship and the new landscapes of transnational Asian fashion, arguing that conditions of social and economic marginality have produced some extremely powerful arenas of cultural creativity.

The contemporary resonances of Parminder Bhachu's work can be put in longer historical perspective by reference to the work of Emma Tarlo [in *Clothing Matters: Dress and Identity in India*, 1996]. Tarlo argues that dress has always played an active process in the forging of social identities in India, from the Nationalist Movement's support for swadeshi (Indian-made) clothing, through long-felt concern about the decline of the Indian hand-loom industry (first noted in 1880), to the recent revival of traditional Indian craftwork, design and embroidery, promoted by the Indian Handicrafts Board.

Tarlo brings this historical perspective to bear on her anthropological study of the recent development of an urban 'fashion village' for upper-middle-class Indian consumers in South Delhi. Hauz Khas, the case-study village, provides one particular instance of a more widespread 'ethnic revival' during the late 1980s, when urban women from the educated élite were 'returning' to the kind of clothing that rural women were themselves in the process of rejecting. During Tarlo's fieldwork in 1989, the village was dominated by a style that Tarlo describes as 'ethnic chic', actively promoted by the local Creative Arts Village Association. This commercial revival of so-called 'ethnic' style was spearheaded by designers such as Bina Ramani who was born in India but had lived abroad for twenty-five years in London, San Francisco and New York, working for Christian Dior, Givenchy and other top designers, later supplying Liberty and Harvey Nichols with her own 'classic' designs. In setting up her store in the urban village of Hauz Khas, Ramani claimed to have seen India 'with foreign eyes' – as demonstrated by her cringe-inducing admiration for 'those rural women in their fabulous and colourful garments'. The fragile foundations of this 'ethnic revival' soon became apparent, however, as local people moved to get a share of the village's commercial success. Indeed, the villagers whose innocence Bina Ramani celebrated soon demonstrated their own shrewd business judgement, undermining the village's aesthetic appeal in the process and driving its exclusive image inexorably downmarket.

Since Tarlo completed her fieldwork, Hauz Khas has been transformed into a state-of-the-art shopping complex. 'Ethnic chic' has been replaced by 'global' fashions (skimpy black lycra and platform shoes) reflecting the influence of cable television and investment from Non-resident Indians living abroad. Understanding the evolution of Hauz Khas requires a longer historical timeframe that includes the upper-middle-class appeal of 'European' dress, the Nationalist 'return' to khadi (hand-woven cloth produced from hand-spun yarn), the process of post-Independence modernization and the revival of so-called 'ethnic chic'.

Here again, then, I would suggest that notions of circuitry and interconnection have more to offer than linear constructs of modernization or globalization that posit some kind of simple transition from a traditional to a more highly commodified system of exchange.

One final twist in this tale was the invention during the late 1990s of another version of 'ethnic' or 'Asian chic', this time promoted within the British media, who detected a moment when it was suddenly 'cool to be Asian'. This latest version of 'Asian chic' appears to have been born out of the coincidence of several interrelated phenomena including the commercial success of bands like Cornershop (whose album 'Brimful of Asha' reached number one in the British charts in 1998). Meanwhile, Madonna's album 'Ray of Light' was re-mixed by Talvin Singh as Madonna herself took up yoga and began painting her hands with *mehndi* (henna dye). Fashion designers like Dries van Noten, Rifat Ozbek, Vivienne Tam and Dolce and Gabbana all included Indian fabrics and embroidery in their collections, while supermodels and film stars like Naomi Campbell and Kate Winslet all appeared in saris and with *bindis* painted on their foreheads. 'The Asian invasion,' as Sheryll Garrett described it in *The Sunday Times* (under the inevitable headline 'Who's sari now?', 23 August 1998), 'is heading this way: in the charts, on the catwalk, even on the best-dressed cushions.' Nothing was immune to the trend it seemed as *Wallpaper*, the coolest of style magazines, Garratt reported, had gone 'urban-turban', signalling 'the commercialization of anything Asian'.

The *Independent on Sunday* ran a similar feature a few months earlier ('British, Asian and hip', 1 March 1998) about the 'mainstreaming' of so-called second-generation Asian culture compared to the economic and social marginalization of their immigrant parents. The same range of cultural phenomena were noted, including Cornershop ('breaking the ethnic mould of British pop'), the radio and television comedy show *Goodness Gracious Me*, London's Anokha nightclub and the (now-defunct) *Second Generation* style magazine. The optimism of this piece, with its emphasis on 'leapfrogging the cultural divide' and not needing to compromise artistic integrity in order to reach a white British audience, contrasts strongly with the much greater reserve expressed in another article on 'Asian cool' published in the same newspaper just nine months later (6 December 1998). The article started with the now-predictable range of examples including Madonna's penchant for saris, David Beckham's sarongs, the popularity of *mehndi* tattoos, Talvin Singh, Cornershop and the Asian Dub Foundation. A different message emerged, however, in the latter part of the article, signalled by the subheading: 'Hands off our culture: If it's been in this year, it's probably been Asian. But has the appropriation of all things Eastern gone too far?' According to the journalist, Hettie Judah:

> The message is clear: when white people adopt Asian fashions, deck their houses out in Asian fabrics and furniture and mix samples of Asian instruments into their music, they embody mainstream fashion. When Asians make music, theatre or film, their work is classified as underground or fringe.

The article included a quote from the musician and composer Nitin Sawhney lamenting the 'colonial arrogance' of contemporary western attitudes to Asia, redolent of a much longer history of Orientalist fascination for all things eastern, sold as off-the-peg profundity, a panacea for the a-spirituality of western capitalism. Sawhney complained about the trivialization and fetishization of Asian culture, about the superficial level of understanding (illustrated by Madonna's unknowing appropriation of a priest's insignia at the MTV awards or more recently by the controversy over David Beckham's tattoo) and the perception that Asian culture needs to be represented by white people before it is 'accessible' to the rest of the world.

Contemporary commercial cultures are full of such ambiguities, where models of ethnic authenticity and essentialist constructions of identity are no longer tenable as guides to the complexities of cultural borrowing. Rather than casting these issues in terms of a stark opposition between the negative associations of cultural appropriation and an equally uncritical celebration of the positive potential of cultural hybridity, I want to conclude by exploring the politically contested middle ground, where cultural cannibalism and economic exploitation rub shoulders with the emergence of more critical forms of multiculturalism.

As our own research suggests, the agents of cultural innovation – be they 'ethnic entrepreneurs', cultural intermediaries or 'ordinary consumers' – exhibit a higher degree of reflexivity than they are often credited with in studies of cultural appropriation. While the agency of small-scale designers is undoubtedly circumscribed, neither are they merely dupes of an economic system that is entirely beyond their knowledge or control. For some clothing firms, like One BC in Nottingham's Lace Market, for example, 'Asian' designs are just one aspect of their cultural repertoire which also includes playful allusions to traditional 'British' dress (bowler hats) or ironic references to 'Cowboys and Asians': 'Clint Eastwood meets Ghandi, a label and an attitude . . . Fugitives from the law of averages.' Indian companies, like The Bombay Store in Mumbai, are equally capable of playing these subversive cultural games, with garment labels that read 'You're fed up with Nike, Reebok and Tommy Hilfiger . . . you're looking for an intelligent gift . . . you're an "alternative" person . . . you'd like to improve the Indian economy . . . you'd like to make us richer than we already are . . .'. As so often, of course, the use of irony is double-edged and the hint of subversiveness in these examples is (however knowingly) subordinated to the imperatives of the market. What looks like 'resistance' at one moment is rapidly recuperated by the market at the next moment as 'consumer culture' engages in another round of commodification.

[. . .]

CONCLUSION: COMMODIFYING DIFFERENCE

Just over ten years ago, Jonathan Rutherford argued [in *Identity: Community, Culture and Difference*, 1990] that capital had fallen in love with difference. Advertising, he claimed, thrived on selling things that enhance our sense of uniqueness and individuality: 'From World Music to exotic holidays in Third World locations, ethnic TV dinners to Peruvian knitted hats, cultural difference *sells*.' From such arguments a kind of cultural pessimism developed, with critics such as bell hooks detecting a cannibalistic tendency within contemporary commodity culture. In a now-familiar passage [from *Black Looks: Race and Representation*, 1992], hooks argued that 'ethnicity becomes spice, seasoning that can liven up the dull dish that is mainstream white culture', suggesting that the relationship between the 'ethnic' and the 'mainstream' was purely parasitic. According to hooks, communities of resistance had been replaced by communities of consumption, with commodification stripping the signs of difference of their political integrity and cultural meaning. Sharon Zukin [see p. 431] warns similarly of the aestheticization of difference, while Deborah Root [in *Cannibal Culture: Art, Appropriation, and the Commodification of Difference*, 1996) has characterized the wider tendency within western society towards a kind of 'cannibal culture' where the aestheticization of difference leads to a romanticization of violence. These arguments are rhetorically powerful, but tend to gloss over the wide range of meanings that can be attached to the commodification of difference. While 'eating the Other' may be an expression of power and privilege (in some circumstances), it may (in other circumstances) provide an entrée to more critical forms of multiculturalism. To move in that direction requires us to identify the many ways in which power is distributed along the chains and through the networks that we describe and analyse. It requires us to examine more closely the complexities of the production process, the politics of representation and the practices of consumption, rather than simply inferring these in some abstract, a priori way.

Each of the metaphors employed in this paper has its own political implications: 'chains' have their weak links, 'circuits' can be broken and 'networks' suggest a more diffused model of how power is distributed. This can, as I have argued, leave our analysis open to the process of recuperation where, as each time difference is recognized and acknowledged, it is immediately subject to new rounds of commodification and exploitation. There is a danger, then, that in replacing linear models with more complex understandings of cultural change we may simply be playing into the hands of the market. We might also, though, be opening up new lines of fracture, new possibilities for more equal social relations to be forged. These are issues to be struggled over, to (re)theorize and work through empirically. For me, at least, that is what is at stake in attempting to transcend the cultural and the economic, as I have attempted to illustrate here through an exploration of contemporary commercial culture.

"The Expediency of Culture"

from *The Expediency of Culture:
Uses of Culture in the Global Era* (2003)

George Yúdice

Editors' introduction

When considering the relationship between "globalization" and "culture," a typical line of scholarly critique posits the increasing infiltration of the cultural field by the commodity form of capital. That is, globalization is often seen to transform local cultural forms into commodities to be sold, for example, as tourist experiences or village crafts. Even something as intangible as "indigenous wisdom" has found ways to be packaged and sold in the marketplaces of post-industrial societies. Yet, for George Yúdice, the "commodification of culture" barely begins to describe the ways culture has been transformed in the global era. Referring to culture as resource, Yúdice seeks to capture something different than the simple colonization of the cultural field by capital. As he puts it, "Culture as resource is seen as a way of providing social welfare and quality of life in the context of diminishing public resources and the withdrawal of the state from the guarantees of the good life."

Culture, in other words, has taken on the role of *governance*. And while this in itself is not new – Yúdice points out nineteenth century and Cold War precedents – the "resourcing" of culture has expanded in concert with what many have identified as a neoliberal agenda of reducing state welfare, freeing up the marketplace, and promoting public–private partnerships in areas that were formally the exclusive realm of state management, such as the provision of public utilities, education, security, or welfare. Culture is no longer deployed as an ideological tool for inculcating "civilized" norms of behavior or the virtues of individual freedom. Rather, culture is now thought capable of helping to solve social dysfunctions – crime, substance abuse, racism, intolerance – all the while forming a major realm of capital accumulation.

Yúdice's argument is complex and theoretically charged. While pointing out the ways that culture has become a resource, he is also arguing that our definition of culture – typically thought of as a "way of life" (see Raymond Williams, p. 15) or a "signifying system" (Geertz, p. 29) (see also the introduction to Part One) – requires significant adjustment. The content of culture (that is, Williams's *structure of feeling* or Geertz's *webs of significance*) has been "hollowed out." This is what Yúdice means when he claims that culture is no longer "transcendent," that it is no longer "beyond interest." Culture is now firmly linked to political agendas, and thus its *content* is less important than its *utility* in achieving certain political (or economic or social) objectives. "Politics trumps the content of culture."

Such an approach, then, entails a *new episteme*. This is a term introduced in the introduction to Part Two of the *Reader*. Michel Foucault, in *The Order of Things* (1966/1970) used this term to describe the commonsense assumptions that provided the basis for the kinds of knowledges and discourses that were possible during a particular historical period. Foucault outlined three distinct epistemes: the Renaissance, based on resemblance; the classic period, based on representation; and the modern period, based on structuralism. To these Yúdice proposes a fourth episteme of *performativity* to characterize the global era and cultural expediency (see Latham and McDowell and Court, pp. 68 and 457). As he argues,

> Performativity is based on the assumption that the maintenance of the status quo (i.e. the reproduction of social hierarchies of race, gender, sexuality) is achieved by repeatedly performing norms. Every day we rehearse the rituals of conformity in the media of dress, gesture, gaze, and verbal interaction within the purview of the workplace, the school, the church, the government office. But repetition is never exact; people, particularly those with a will to disidentify or "transgress," do not fail to repeat, they just "fail to repeat loyally."

Drawing on Judith Butler's extensions of Foucault's thinking, Yúdice sees culture as a field of performance, in which a set of particular cultural norms – certain rituals, beliefs, activities, dress, and so on – are repeatedly performed because of the expediency that such norms have acquired in the social context of global neoliberalism. By invoking performance, Yúdice is indicating that there is nothing *inherent* linking a given culture to a particular set of practices – just as Butler claimed gender norms (such as the traits associated with "masculinity" and "femininity") are not given but must be repeatedly enacted in order to become "real." There is no inherent content to culture, he argues, but a performance of norms according to the utility to which culture has been put as a resource.

Perhaps the most significant utility in this regard is the rights and entitlements of citizenship that are accorded to recognized culture groups in many liberal states, a topic discussed in greater detail in the *Reader*'s opening Introduction in the context of the headscarf issue in France. The rise of the welfare state in the 1960s, Yúdice argues, helped bring this about, shifting the entitlements of citizenship from individual to group-based. "In this view, so long as you can assert that you have a culture (a distinctive set of beliefs and practices), you have legitimate grounds for enfranchisement." Further, "In our era, *claims* to difference and culture are expedient insofar as they presumably lead to the empowerment of a community."

Ultimately, then, Yúdice's argument critiques multiculturalism and cultural citizenship as much as it does neoliberalism. His linking of Foucault and performance with the realm of cultural policy and political economy has been anticipated in geography by Clive Barnett ["Culture, geography, and the arts of government," *Environment and Planning D: Society and Space* 19, 2001], who argues that the "cultural turn" in geography must do more than simply bring the perspective of political economy to bear on the field of culture, but more significantly must view culture as an important field of governance. In more general terms, to the extent that geographers have invoked the relationship between culture and political economy, they have tended to adopt some version of the Italian Marxist Antonio Gramsci's understanding of "cultural hegemony." The work of Barnett and Yúdice suggest that Foucauldian ideas of "governmentality" might serve as a useful alternative approach more suitable to the current global era.

George Yúdice is Professor of American Studies and Spanish and Portuguese at New York University. Co-author of *Cultural Policy* (2002), and editor of *The Challenge of Cultural Policy* (forthcoming), Yúdice has written widely on literature, art, and culture in the United States and Latin America.

■ ■ ■ ■ ■ ■ ■

But it is culture – not raw technology alone – that will determine whether the United States retains its status as the pre-eminent Internet nation. (Sever Lohr, "Welcome to the Internet, the First Global Colony")

CULTURE AS RESOURCE

I argue in this book that the role of culture has expanded in an unprecedented way into the political and economic at the same time that conventional notions of culture largely have been emptied out. I do not focus on the content of culture – that is, the model of uplift (following Schiller or Arnold) or distinction (following Bourdieu) that it offered in its traditional acceptations, or more recently its anthropologization as a whole way of life (Williams), according to which it is recognized that everyone's culture has value. Instead, I approach the question of culture in our period, characterized as one of accelerated globalization, as a *resource*. . . . [C]ulture is increasingly wielded as a resource for both sociopolitical and economic amelioration, that is, for increasing participation in this era of waning political involvement, conflicts over citizenship, and the rise of what Jeremy Rifkin [*The Age of Access*, 2000] has called "cultural capitalism." The immaterialization characteristic of many new sources of economic growth (e.g., intellectual property rights as defined by the General Agreement on Tariffs and Trade [GATT] and the World Trade Organization) and the increasing share of the world trade by symbolic goods (movies, TV programs, music, tourism, etc.) have given the cultural sphere greater protagonism than at any other moment in the history of modernity. It could be argued that culture has simply become a pretext for sociopolitical amelioration and economic growth, but even if that were the case, the proliferation of such arguments, in those fora provided by local culture and development projects as well as by UNESCO, the World Bank, and the so-called globalized civil society of international foundations and NGOs, has operated a transformation in what we understand by the notion of culture and what we do in its name.

The relation between cultural and political spheres or cultural and economic spheres is not new. On the one hand, culture is the medium in which the public sphere emerges in the eighteenth century; as Foucauldian and cultural studies scholars have argued, it became a means to internalize social control (i.e., via discipline and governmentality) throughout the nineteenth and twentieth centuries. Tony Bennett [*The Birth of the Museum*, 1995], for example, has demonstrated that culture provided not only ideological uplift, according to which people were gauged to have human worth, but also a material inscription in forms of behavior: people's behavior was transformed by the physical requirements involved in moving through schools and museums (ways of walking, dressing, talking, etc.). Also well studied are the political uses of culture to promote a particular ideology, for clientelist purposes or for currying favor in foreign relations, as evidenced in the advancement of proletarian culture by the Soviet Commissariat of Enlightenment, the clientelist sponsorship of muralism by the Mexican state in the 1920s and 1930s, or the currying of influence in foreign relations, as in the United States' Good Neighbor and cold war cultural policies.

Also on the economic front, nineteenth-century Europe saw the increasing subjection of the artist and the writer to the commercial imperative. In this context, and with the emergence of new technologies such as lithography, photography, film, and sound recording, some theorists and critics came to define art in contradistinction to the commercial. In his famous 1938 essay, "On the Fetish-Character in Music and the Regression of Listening," Theodor Adorno rejected the political-economic basis of the new mass media, which turned the engagement with art away from its use-value and toward the "fetish character of commodities." In the first half of the twentieth century, Adorno could define art as the process through which the individual gains freedom by externalizing himself, in contrast to the philistine "who craves art for what he can get out of it." Today it is nearly impossible to find public statements that do not recruit instrumentalized art and culture, whether to better social conditions, as in the creation of multicultural tolerance and civic participation though UNESCO-like advocacy for cultural citizenship and cultural rights, or to spur economic growth through urban cultural development projects and the concomitant proliferation of museums for cultural tourism, epitomized by the increasing number of Guggenheim franchises.

To illustrate the extent to which this is the case, consider *American Canvas*, a 1997 report of the National Endowment for the Arts (NEA) on the place of arts and culture in U.S. society:

No longer restricted solely to the sanctioned arenas of culture, the arts would be literally suffused throughout the civic structure, finding a home in a variety of community service and economic development activities – from youth programs and crime prevention to job training and race relations – far afield from the traditional aesthetic functions of the arts. This extended role for culture can also be seen in the many new partners that arts organizations have taken on in recent years, with school districts, parks and recreation departments, convention and visitor bureaus, chambers of commerce, and a host of social welfare agencies all serving to highlight the utilitarian aspects of the arts in contemporary society.

This expanded role for culture is due in part to the reduction of direct subvention of all social services, culture included, by the state, thus requiring a new legitimation strategy in the post-Fordist and the post-civil rights era in the United States. Advocacy for the centrality of culture in solving social problems is not new, but it took different forms in the past, such as the ideological (re)production of proper citizens (whether bourgeois, proletarian, or national). Although there have long been art therapy programs for the mentally ill and for the incarcerated, culture more generally was not regarded as a proper therapy for such social dysfunctions as racism and genocide. Nor was it considered, historically, an incentive for economic growth. Why the turn to a legitimation based on utility?

There are, I think, two main reasons. Globalization has pluralized the contacts among diverse peoples and facilitated migrations, thus problematizing the use of culture as a national expedient. Additionally, in the United States, the end of the cold war pulled the legitimizing rug out from under a belief in artistic freedom, and with it unconditional support for the arts, as a major marker of difference with respect to the Soviet Union. Of course, this politically motivated sponsorship of freedom was fundamental in giving certain artistic styles (jazz, modern dance, abstract expressionism) the shot in the arm needed for "New York to steal the idea of modern art" from Paris, according to Serge Guilbaut [*How New York Stole the Idea of Modern Art*, 1983].

Without cold war legitimation, there is no holding back utilitarian arguments in the United States. Art has completely folded into an expanded conception of culture that can solve problems, including job creation. Its purpose is to lend a hand in the reduction of expenditures and at the same time help maintain the level of state intervention for the stability of capitalism.

Because almost all actors in the cultural sphere have latched on to this strategy, culture is no longer experienced, valued, or understood as transcendent. And insofar as this is the case, appeals to culture are no longer tied to this strategy. The culture wars, for example, take the form they do in a context in which art and culture are seen as fundamentally interested . . . Conservatives and liberals are not willing to give each other the benefit of the doubt that art is beyond interest . . . As conservatives began to exercise more influence in the 1980s and 1990s, this basic belief in the interested character of art and culture was expressed by eliminating entitlements and redistributive programs bequeathed by Johnson's Great Society and the civil rights legacy, which benefit marginalized groups. Many of these programs were legitimized by claims that the needs of these groups were premised on cultural difference, which had to be taken as a deciding factor in the distribution of recognition and resources. Conservatives, on the other hand, saw these differences as incapacities or moral flaws (e.g., the "culture of poverty" attributed to racial minorities or the libertinism of gay and lesbian sexual preferences and practices) that rendered these groups ineligible for public resources.

But this move to reduce state expenditures, which might seem like the death knell of the nonprofit arts and cultural activities, is actually their condition of continued possibility. The arts and culture sector is now claiming that it can solve the United States' problems: enhance education, salve racial strife, help reverse urban blight through cultural tourism, create jobs, reduce crime, and perhaps even make a profit. This reorientation of the arts is being brought about by arts administrators. . . .

CULTURAL DEVELOPMENT

. . . [A]s powerful institutions like the European Union, the World Bank, the Inter-American Development Bank (IADB), and the major international foundations begin to understand culture as a crucial sphere for investment, it is increasingly treated like any other resource. James D. Wolfensohn, president of the World Bank, in his keynote address at the international conference "Culture Counts: Financing, Resources, and the

Economics of Culture in Sustainable Development" (October 1999), folds culture into the Bank's policies as an instrument for human development. He stresses a "holistic view of development" that focuses on community empowerment of the poor so that they may hold on to – sustain – those assets that enable them to cope with "trauma and loss," stave off "social disconnectedness," "maintain self-esteem," and also provide material resources. He writes, "There are development dimensions of culture. Physical and expressive culture is an undervalued resource in developing countries. It can earn income, through tourism, crafts, and other cultural enterprises . . . Heritage gives value. Part of our joint challenge is to analyze the local and national returns on investments which restore and draw value from cultural heritage – whether it is built or living cultural expression, such as indigenous music, theater, crafts."

Now consider the lending strategy of the IADB in the cultural sphere. According to one Bank official [Elcior Santana's remarks at a conference in Bellagio, Italy, in 1999], "Given economic orthodoxy throughout the world, the old model of state public support for culture is dead. The new models consist of partnerships with the public sector and with international financial institutions, particularly the Multilateral Development Banks (MDBs) like the World Bank and the Inter-American Development Bank." The turn to cultural capital is part of the history of recognition of shortcomings in investment for physical capital in the 1960s, human capital in the 1980s, and social capital in the 1990s. Each new notion of capital was devised as a way of ameliorating some of the failures of development according to the preceding framework. The concept of social capital was operationalized in the MDEs, taking the social fabric into consideration in their development projects. This concept also ensued from the recognition that although economic returns have been substantial in the 1990s, inequality has increased exponentially. The trickle-down premise of neoliberal economic theory has not been confirmed. Consequently, there has been a turn to investment in civil society, and culture as its prime animator.

According to Elcior Santana [of the IADB], empirical examples suggest that there is force to this argument. For example, Villa El Salvador in Peru showed an impressive increase in social indicators

in its near thirty years of existence. In 1971, homeless people invaded Lima and the government relocated them to a semidesert-like area. Twenty years later they comprised a city of eighty-one hundred people with some of the best social indicators in the country. Illiteracy declined from an index of 5.8 to 3.8, infant mortality was reduced to a lower than average rate of 67 per 1,000, and registration in basic education grew to a better than average 98 percent. The variable that explains this, according to Santana, is culture, which enables the consolidation of citizenship founded on active participation of the population. The majority of the people came from the highlands of Peru and maintained their indigenous cultural customs, communal work, and solidarity, which provided those characteristics that lead to development. Santana compared these characteristics to the civic and cultural traditions that, according to Robert Putnam [*Making Democracy Work*, 1993], enabled a northern Italian region to prosper. Consequently, if it could be shown, he added, that culture produces the patterns of trust, cooperation, and social interaction that result in a more vigorous economy, more democratic and effective government, and fewer social problems, then MDBs will be likely to invest in cultural development projects.

[. . .]

THE CULTURAL ECONOMY

. . . Artistic trends such as multiculturalism that emphasize social justice (perhaps understood no more broadly than equal visual representation in public spheres) and initiatives to promote sociopolitical and economic utility have been fused into the notion of what I call the "cultural economy" and what [Prime Minister Tony] Blair's New Labourite rhetoric dubbed the "creative economy." Also marketed at home and to the world as "Cool Britannia," this creative economy includes both a sociopolitical agenda, particularly the protagonism of multiculturalism as embodied in the work of the so-called young British artists, as well as an economic agenda, that is, the belief that the creativity provided by this new generation transformed London into "the creative hub for trends in music, fashion, art and design." Applying the logic that a creative environment begets innovation, hip London culture was touted as the foundation for the so-called new economy based on "content provision," which is supposed to be the engine of accumulation. This premise is quite widespread, with U.S. rhetoric of a "new economy" and British hype about the "creative economy" echoed in New Zealand's "HOT Nation," Scotland's "Create in Scotland," and Canada's "A Sense of Place, a Sense of Being.". . .

. . . Culture is increasingly being invoked not only as an engine of capital development, as evidenced by the ad nauseam repetition that the audiovisual industry is second only to the aerospace industry in the United States. Some have even argued that culture has transformed into the very logic of contemporary capitalism, a transformation that [according to Jeremy Rifkin] "already is challenging many of our most basic assumptions about what constitutes human society." This culturalization of the economy has not occurred naturally, of course; it has been carefully coordinated via agreements on trade and intellectual property, such as GATT and the WTO, laws controlling the movement of mental and manual labor (i.e., immigration laws), and so on. In other words, the new phase of economic growth, the cultural economy, is also political economy . . .

The culturalization of the so-called new economy, based on cultural and mental labor – or better yet, on the expropriation of the value of cultural and mental labor – has, with the aid of new communications and informatics technology, become the basis of a new division of labor. To the degree that communications enable services and independent producers to be located almost anywhere on earth, this is also a new international division of *cultural* labor, necessary for fostering innovation and creating content. Culturalization is also political economy, for the U.S. government has been a central actor in ensuring that the nation can maintain its domination of the new economy. For example, the report on Intellectual Property and the National Information Infrastructure of the White House Information Infrastructure Task Force (IITF) recommended bolstering copyright regimes so that content provision would ensure U.S. dominance in the new economy: "All the computers, telephones, scanners, printers, switches, routers, wires, cables, networks and satellites in the world," the task force argues, "will not create a successful national information infrastructure (NII) if there is no content. What will drive the NII is the content moving through it": information and entertainment resources; access to the world's cultural resources; new product innovation; greater variety in cultural consumption.

More traditional activities such as cultural tourism and arts development are also facilitating the transformation of postindustrial cities. The most sensational example is the Guggenheim Museum in Bilbao, which is serving as a model for the franchising of museums in other parts of the world, such as Rio de Janeiro and Lyons . . . Another postindustrial city that turned to culture to revive the economy is Peekskill, New York. Reasoning that "artists are a kind of pilot fish for gentrification," the city council created an arts district and offered incentives, such as cheap loft space, so that artists would relocate there from New York City.

These initiatives also have a downside, for, as in classic instances of gentrification, they tend to displace residents . . . The turn to the "creative economy" evidently favors the professional-managerial class, even as it trades on the rhetoric of multicultural inclusion . . . Culturalization, then, is also based on the mobilization and management of populations, particularly the "life-enhancing" marginal populations who nourish the innovation of the "creators." This means a marriage of culture-as-vernacular practices, notions of community, and economic development. . . .

CULTURAL CITIZENSHIP

Cultural rights include the freedom to engage in cultural activity, to speak one's language of choice, to teach one's language and culture to one's children, to identify with the cultural communities of one's choice, to discover the whole range of cultures that constitute world heritage, to gain knowledge of human rights, to have an education, to be free from being represented without consent or from having one's cultural space used for publicity, and to gain public provision to safeguard these rights. However, as [Filibek] puts it, cultural rights are the "Cinderellas of the human rights family" because their definition is still ambiguous: the full range of what is to be included in "culture" is not clear, nor is it easy to reconcile universal applicability with cultural relativism. Moreover, even though cultural rights refer to collectivities, the individual rights of members of such collectivities have priority, at least in international covenants. Consequently, cultural rights are not universally accepted and in most cases are not justifiable, unlike economic rights, whose status is firmly entrenched in international law . . .

Nevertheless, some justifiable rights overlap with cultural rights, as in the case of the right to information. Yet, how that right is exercised is dependent on cultural context. As Javier Perez de Cuéllar, president of the World Commission on Culture and Development, observes in his introduction to the UNESCO report *Our Creative Diversity* (1996), "Economic and political rights cannot be realized separately from social and cultural rights."

[. . .]

It is this notion of culture that underpins the concept of cultural citizenship as developed by Renato Rosaldo in the late 1980s [*Culture and Truth*, 1989]. At odds with conventional notions of citizenship, which emphasize universal, albeit formal, applicability of political rights to all members of a nation, Rosaldo posited that cultural citizenship entailed that groups of people bound together by shared social, cultural, and/or physical features should not be excluded from participation in the public spheres of a given polity on the basis of those features. In a juridical context that enables litigation against exclusion and a cultural-political ethos that eschews marginalizing the "nonnormative" (considered as such from the perspective of the "mainstream"), culture serves as the ground or warrant for making "claim[s} to rights in the public sphere." Because culture is what "create[s] space where people feel 'safe' and 'at home,' where they feel a sense of belonging and membership," it is, according to this view, a necessary condition for citizenship . . .

Consequently, if democracy is to be fostered, public spheres in which deliberation on questions of the public good is held must be permeable to different cultures. The relativist strain in anthropological theory, according to which "communal culture" as an ensemble of ideas and values provides the individual with identity, is mobilized here for political ends. Culture is thus more than an anchoring ensemble of ideas and values. It is, according to Flores and Benmayor [*Latino Cultural Citizenship*, 1997], premised on difference, which functions as a *resource*. The content of culture recedes in importance as the *usefulness of the claim to difference* as a warrant gains legitimacy. The result is that *politics* trumps the content of culture . . .

[. . .]

A NEW EPISTEME?

It is at this juncture that I would like to propose the notion of *performativity* as the mode, beyond instrumentality, in which the social is increasingly practiced. . . .

The expediency of culture underpins performativity as the fundamental logic of social life today. First, globalization has accelerated the transformation of everything into resource. Second, the specific transformation of culture into resource epitomizes the emergence of a new episteme, in the Foucauldian sense. Third, this transformation should not be understood as a manifestation of "mere politics" . . .

Culture and globalization. It has been argued that under conditions of globalization, difference rather than homogenization infuses the prevailing logic of accumulation. Globalization, a process of economic expansion datable from sixteenth-century European exploration and conquest and of modernization, produces encounters of diverse traditions such that "cultures can no longer be examined as if they were islands in an archipelago." The recently published *World Culture Report 1998: Culture, Creativity and Markets* attempts to map out the coordinates of this greater cultural complexity and how it might be harnessed, "creatively," for greater development and democracy.

Discourses on globalization, however, have less sanguine precedents. It was not so long ago that the economic and mediatic global reach of the United States and Western Europe was characterized as cultural imperialism. Exponents of this view endeavored to unveil the will to power that subtended the reverence for Western high art, the concealment of power differentials in celebrations of the common humanity shared by all peoples as promoted in much anthropological work, and the brainwashing of the entire globe by Hollywood. . . .

The cultural imperialism argument has been criticized for three main reasons. In the first place, it has overlooked the subordination of internal minorities that takes place within the nationalism of developing countries as they gird themselves to stave off the symbolic aggression of imperial powers. Second, migrations and diasporic movements generated by global processes have complicated the unity presumed to exist in the nation; belonging may be infra- or supranational. Third, and relatedly, the exchange of ideas, information, knowledge, and labor "multiplies the number of permutations and in the process creates new ways of life, new cultures" often premised on elements from one culture sampled into another, such as the rap music that black Brazilian youth incorporate into their own antiracist projects. It is no longer viable to argue that such hybrid cultures are inauthentic.

These arguments suggest that there is an expedient relation between globalization and culture in the sense that there is a fit or a *suitability* between them. Globalization involves the (mostly commercial and informatic) dissemination of symbolic processes that increasingly drive economics and politics. Malcolm Waters [*Globalization*, 1995] bases his entire study of globalization on this first sense of expediency: "The theorem that guides the argument of this book is that: *material exchanges localize; political exchanges internationalize; and symbolic exchanges globalize.* It follows that the globalization of human society is contingent on the extent to which cultural arrangements are effective relative to economic and political arrangements. We can expect the economy and the polity to be globalized to the extent that they are culturalized."

From culture as resource to politics. As argued above, culture is expedient as a resource for attaining an end. Culture as a resource is a principal component of what might be characterized as a postmodern episteme. In *The Order of Things* (1973), Foucault sketches out three different and discontinuous modalities of relation between thought and world, or epistemes, that enable the various fields of knowledge in each given era. In each era, knowledge is organized, according to Foucault, by a series of fundamental operative rules. The Renaissance or sixteenth-century episteme is based

on resemblance, the mode by which language relates words and the signatures that mark things. Knowledge consisted of relating, through interpretation, the different forms of language so as to "restor[e] the great, unbroken plain of words and things." The classical episteme of the seventeenth and eighteenth centuries consisted of the representation and classification of all entities according to the principles of order and measurement. It is this episteme that Borges caricatures in his image of the Chinese encyclopedia, cited by Foucault as his inspiration for thinking its obverse, the heteroclite. With the rise of the modern episteme, which Foucault locates at the turn of the eighteenth and nineteenth centuries, representation is no longer adequate for the examination of concerns with life, the organic, and history. This inadequacy in turn implies a depth or a "density withdrawn into itself" in which "what matters is no longer identities, distinctive characters, permanent tables with all their possible paths and routes, but great hidden forces developed on the basis of their primitive and inaccessible nucleus, origin, causality, and history." These hidden forces are analogous in Foucault's account to what remains concealed in Heidegger's account of modern technology. Modern knowledge thus consists of unveiling the primary processes (the infrastructure, the unconscious) that lurk in the depths, beneath the surface: manifestations of ideology, personality, and the social.

[. . .]

I would like to extend Foucault's archaeological periodization and propose a fourth episteme based on a relationship between words and world that draws on the previous epistemes – resemblance, representation, and historicity – yet recombines them in a way that accounts for the constitutive force of signs. Some [i.e. Baudrillard, *Simulations*, 1983] have characterized this constitutive force as simulation, that is, an effect of reality premised on the "precession of the model": "Facts no longer have any trajectory of their own, they arise at the intersection of the models." I prefer the term performativity, which refers to the processes by which identities and the entities of social reality are constituted by repeated approximations of models (i.e., the normative) as well as by those "remainders" ("constitutive exclusions") that fall short. As I explained above, to the degree that globalization brings different cultures into contact with each other, it escalates the questioning of norms and thus abets performativity.

[. . .]

"*Mere politics.*" Expediency in this sense refers to what is, according to the *Oxford English Dictionary*, "merely politic (esp. with regard to self-interest) to the neglect of what is just or right." I would like to modify this understanding of expediency, for it implies that there is a notion of right that exists outside of the play of interests. A performative understanding of the expediency of culture, in contrast, focuses on the strategies implied in any invocation of culture, any invention of tradition, in relation to some purpose or goal. That there is an end is what makes it possible to speak of culture as a resource. For example, the debate over Rigoberta Menchú's alleged exaggeration [in *I, Rigoberta Menchú: an Indian Woman of Guatemala*, 1984], and in some cases fabrication, of the events narrated in her testimonio turns on the productive role that culture performs. Those who . . . argue that she has distorted the truth for her own ends, for her self-interest, see her testimonio as expedient in the negative sense . . . Those who defend her . . . argue that she altered the facts of the events to make her narrative more compelling and thus to be more persuasive in attracting attention to the plight of her people. In both cases, however, there is a calculation of interest being made; and in both cases, culture is being invoked as a resource for determining the value of an action, in this case, a speech act, a testimonio.

Some readers might assume that my brief précis of the Rigoberta Menchú case entails a negative view of the instrumentalization of culture, as if the truth hovered somewhere among the various accounts, attacks, and counterattacks. My own view is that it is not possible not to make recourse to culture as a resource. Consequently, cultural analysis necessarily entails taking a position, even in those cases where the writer seeks objectivity or transcendence. But such a position need not be a normative one, based on right and wrong. Foucault rejected such moralism in the last phase of his work, positing instead an ethical basis for practice. Ethics, Foucault argued, did not entail a teleological foundation, such as is usually attributed to utilitarianism. His notion of the care of the self emphasized the active role of the subject in his or her own process of constitution. There is a compatibility between this notion of the care of the self and performativity, for Foucault's ethics entails a reflexive practice of self-management vis-à-vis models (or what Bakhtin called "voices" and "perspectives") imposed by a given society or cultural formation. Bakhtin's notion of the author may serve as a prototype of Foucault's performative ethics, since the author is an orchestration of others' "voices," an appropriation that consists of "populating those 'voices' with his or her own intentions, with his or her own accent." He or she who practices care of the self must also forge his or her freedom by working through the "models that he finds in his culture and are proposed, suggested, imposed upon him by his culture, his society, and his social group." . . .

"Whose Culture? Whose City?"

from *The Cultures of Cities* (1995)

Sharon Zukin

Editors' introduction

Sharon Zukin's book, *The Cultures of Cities*, from which this excerpt is taken, is appropriately titled in at least two ways. On the one hand, cities are home to countless groups of people who share distinct ways of life. These may be ethnic or racially based, but they may also be occupational, class, age, or sexuality-based cultures. On the other hand, Zukin's analysis relates to several different ways of defining culture. Conceptually, then, there are different cultures that inform different interpretations of cities. As she points out at the opening of the selection, culture in the sense of refined art, literature, theater, dance, and cuisine, has always been a fundamental part of the urban experience. While the upper classes have long been suspicious of the "hedonism" and baseness that often mark urban life, cities have also provided "high culture" as an antidote to this crass vision of the city. But this is just one sense in which culture plays a role in our understanding of urban life.

In addition, Zukin's work examines culture as a instrument of urban governance. Culture, she argues, is also "a powerful means of controlling cities," of regulating spaces of inclusion and exclusion, and of conditioning collective memory by determining what part of the urban landscape gets preserved as heritage or reconstructed as spectacle. This sense of culture echoes Yúdice's approach (see p. 422) to culture as an expedient resource mobilized for specific social objectives. This is not a culture of high art or literature. In fact, as Yúdice argues, the content of this culture ("high" or "low," "refined" or "popular") is far less important than its utility in achieving a certain social goal.

Culture is also a set of images or symbols used to sell the city to outsiders, to encourage them to visit, spend their money, invest, or relocate their residence or business. While this sense of culture also views it as a resource, the broader framework is to see cities developing a "symbolic economy" as a new means of accumulating footloose capital.

Finally, culture in Zukin's analysis is also a terrain of struggle, in the sense originally proposed by the Italian Marxist Antonio Gramsci. Gramsci argued that dominant classes maintained their power not simply by "controlling the means of production," as Marx would have it, but by *controlling culture*, by exerting *cultural hegemony*. Struggles by subalterns against the dominant classes, then, were as much struggles over the cultural field as they were struggles over relations of production. In Zukin's work, this struggle occurs between the dominant crafters of the "symbolic economy" (city governments, boosters, real estate developers, corporations) and local communities, neighborhoods, activists, and other grass-roots groups. The symbolic economy typically entails public–private partnerships. As governments relinquish part of their provision of welfare and other public goods to the private sector, culture is now deployed to help fill in the gaps. The spaces in which this deployment occurs are often "semi-public" in that they are privately owned or managed. And such spaces are increasingly being substituted for the traditional public spaces in which, according to Zukin, a democratic and grass-roots urban culture thrives.

The Cultures of Cities complements a broad swath of work in cultural geography that focuses on the political economy of the culture industries, and of "selling places." See, for instance, Kearns and Philo's *Selling Places: The City as Cultural Capital* (1993) and Allen Scott's *The Cultural Economy of Cities* (2000). Yet, as argued in the introduction to this section of the *Reader*, the cultural economy involves more than bringing a political-economic focus to bear on the production of culture. For Zukin, it also means recognizing that the value of economic goods is increasingly dominated by symbolic or cultural properties, and that Disney is a more appropriate model of economic value chains than Ford. There remains debate, however, over the extent to which there really has been a "symbolic turn" in the economy. Daniel Miller ("The unintended political economy," in P. du Gay and M. Pryke, eds., *Cultural Economy: Cultural Analysis and Commercial Life*, 2002, pp. 166–184), for example, has argued that there is no convincing evidence of such a shift and that the assertion that the economy is now more cultural than before represents "a sleight of hand through which a shift in academic emphasis is presupposed to reflect a shift in the world that these academics are describing."

Sharon Zukin is Broeklundian Professor of Sociology at City University of New York. She is the author of *Loft Living: Culture and Capital in Urban Change* (1989), *Landscapes of Power: From Detroit to Disney World* (1991), and *Point of Purchase: How Shopping Changed American Culture* (2003). Her work has long been influential among urban, economic, social, and cultural geographers.

■ ■ ■ ■ ■ ■

Cities are often criticized because they represent the basest instincts of human society. They are built versions of Leviathan and Mammon, mapping the power of the bureaucratic machine or the social pressures of money. We who live in cities like to think of "culture" as the antidote to this crass vision. The Acropolis of the urban art museum or concert hall, the trendy art gallery and café, restaurants that fuse ethnic traditions into culinary logos – cultural activities are supposed to lift us out of the mire of our everyday lives and into the sacred spaces of ritualized pleasures.

Yet culture is also a powerful means of controlling cities. As a source of images and memories, it symbolizes "who belongs" in specific places. As a set of architectural themes, it plays a leading role in urban redevelopment strategies based on historic preservation or local "heritage." With the disappearance of local manufacturing industries and periodic crises in government and finance, culture is more and more the business of cities – the basis of their tourist attractions and their unique, competitive edge. The growth of cultural consumption (of art, food, fashion, music, tourism) and the industries that cater to it fuels the city's symbolic economy, its visible ability to produce both symbols and space.

In recent years, culture has also become a more explicit site of conflicts over social differences and urban fears. Large numbers of new immigrants and ethnic minorities have put pressure on public institutions, from schools to political parties, to deal with their individual demands. Such high culture institutions as art museums and symphony orchestras have been driven to expand and diversify their offerings to appeal to a broader public. These pressures, broadly speaking, are both ethnic and aesthetic. By creating policies and ideologies of "multiculturalism," they have forced public institutions to change.

On a different level, city boosters increasingly compete for tourist dollars and financial investments by bolstering the city's image as a center of cultural innovation, including restaurants, avant garde performances, and architectural design. These cultural strategies of redevelopment have fewer critics than multiculturalism. But they often pit the self-interest of real estate developers, politicians, and expansion-minded cultural institutions against grassroots pressures from local communities.

At the same time, strangers mingling in public space and fears of violent crime have inspired the growth of private police forces, gated and barred communities, and a movement to design public spaces for maximum surveillance. These, too, are a source of contemporary urban culture. If one way of dealing with the material inequalities of city life has been to aestheticize diversity, another way has been to aestheticize fear.

Controlling the various cultures of cities suggests the possibility of controlling all sorts of urban ills, from violence and hate crime to economic decline. That this is an illusion has been amply shown by battles over multiculturalism and its warring factions – ethnic politics and urban riots. Yet the cultural power to create an image, to frame a vision, of the city has become more important as publics have become more mobile and diverse, and traditional institutions – both social classes and political parties – have become less relevant mechanisms of expressing identity. Those who create images stamp a collective identity. Whether they are media corporations like the Disney Company, art museums, or politicians, they are developing new spaces for public cultures. Significant public spaces of the late nineteenth and early twentieth century – such as Central Park, the Broadway theater district, and the top of the Empire State Building – have been joined by Disney World, Bryant Park, and the entertainment-based retail shops of Sony Plaza. By accepting these spaces without questioning their representations of urban life, we risk succumbing to a visually seductive, privatized public culture.

THE SYMBOLIC ECONOMY

Anyone who walks through midtown Manhattan comes face to face with the symbolic economy (see map of Manhattan). A significant number of new public spaces owe their particular shape and form to the intertwining of cultural symbols and entrepreneurial capital.

▪ The AT&T Building, whose Chippendale roof was a much criticized icon of postmodern architecture, has been sold to the Japanese entertainment giant Sony; the formerly open public areas at street level have been enclosed as retail stores and transformed into Sony plaza. Each store sells Sony products: video cameras in one shop, clothes and accessories related to

performers under contract to Sony's music or film division in another. Sony's interactive science museum features the opportunity to get hands-on experience with Sony video equipment. Sony had to get the city government's approval both to enclose these stores and set them up for retail shopping, for the original agreement to build the office tower had depended on providing *public* space. Critics charged that retail stores are not public space, and even the city planning commissioners admitted they were perplexed by the question. "In return for the retail space," the chairman of the local community board said, "we would like to hold Sony to the original understanding to create a peaceful refuge, which certainly didn't include corporate banners and a television monitor." "We like it," the president of Sony Plaza replied. The banners "are seen as art and bring warmth and color to the space" (*New York Times*, January 30, 1994).

▪ Two blocks away, André Emmerich, a leading contemporary art dealer, rented an empty storefront in a former bank branch to show three huge abstract canvases by the painter Al Held. Entitled *Harry, If I Told You, Would You Know?* the group of paintings was exhibited in raw space, amid falling plaster, peeling paint, exposed wires, and unfinished floors, and passersby viewed the exhibit from the street through large plate glass windows. The work of art was certainly for sale, yet it was displayed as if it were a free, public good; and it would never have been there had the storefront been rented by a more usual commercial tenant.

▪ On 42nd Street, across from my office, Bryant Park is considered one of the most successful public spaces to be created in New York City in recent years. After a period of decline, disuse, and daily occupation by vagrants and drug dealers, the park was taken over by a not-for-profit business association of local property owners and their major corporate tenants, called the Bryant Park Restoration Corporation. This group redesigned the park and organized daylong programs of cultural events; they renovated the kiosks and installed new food services; they hired a phalanx of private security guards. All this attracted nearby office workers, both women and men, who make the park a lively midday gathering place, as it had been prior

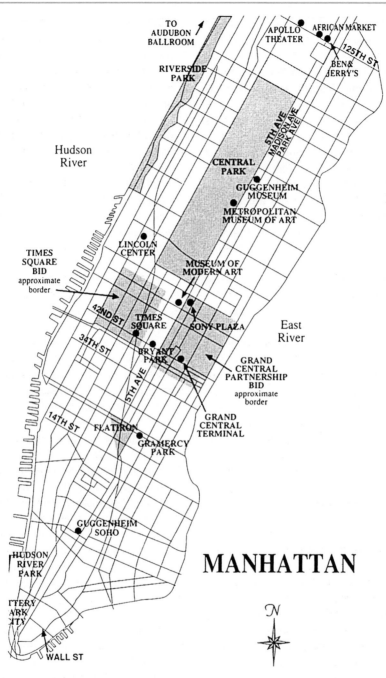

Figure 1 Manhattan

to the mid 1970s – a public park under private control.

Building a city depends on how people combine the traditional economic factors of land, labor, and capital. But it also depends on how they manipulate symbolic languages of exclusion and entitlement. The look and feel of cities reflect decisions about what – and who – should be visible and what should not, on concepts of order and disorder, and on uses of aesthetic power. In this primal sense, the city has always had a symbolic economy. Modern cities also owe their existence to a second, more abstract symbolic economy devised by [what Molotch called] "place entrepreneurs," officials and investors whose ability to deal with the symbols of growth yields "real" results in real estate development, new businesses, and jobs.

Related to this entrepreneurial activity is a third, traditional symbolic economy of city advocates and business elites who, through a combination of philanthropy, civic pride, and desire to establish their identity as a patrician class, build the majestic art museums, parks, and architectural complexes that represent a world-class city. What is new about the symbolic economy since the 1970s is its symbiosis of image and product, the scope and scale of selling images on a national and even a global level, and the role of the symbolic economy in speaking for, or representing, the city.

In the 1970s and 1980s, the symbolic economy rose to prominence against a background of industrial decline and financial speculation. The metamorphosis of American-made products into Mexican blue jeans, Japanese autos, and East Asian computers emptied the factories where those goods had been made. Companies that were the largest employers in their communities went out of business or were bought and restructured by takeover artists.

The entrepreneurial edge of the economy shifted toward deal making and selling investments and toward those creative products that could not easily be reproduced elsewhere. Product design – creating the look of a thing – was said to show economic genius. Hollywood film studios and media empires were bought and sold and bought again. In the 1990s, with the harnessing of new computer-based technologies to marketing campaigns, the "information superhighway" promised to join companies to consumers in a Manichean embrace of technology and entertainment. "The entertainment industry is now the driving force for new technology, as defense used to be," the CEO of a U.S. software company said.

The growth of the symbolic economy in finance, media, and entertainment may not change the way entrepreneurs do business. But it has already forced the growth of towns and cities, created a vast new work force, and changed the way consumers and employees think. In the early 1990s, employment in "entertainment and recreation" in the United States grew slightly more than in health care and six times more than in the auto industry. The facilities where these employees work – hotels, restaurants, expanses of new construction and undeveloped land – are more than just workplaces. They reshape geography and ecology; they are places of creation and transformation.

The Disney Company, for example, makes films and distributes them from Hollywood. It runs a television channel and sells commercial spinoffs, such as toys, books, and videos, from a national network of stores. Disney is also a real estate developer in Anaheim, Orlando, France, and Japan and the proposed developer of a theme park in Virginia and a hotel and theme park in Times Square. Moreover, as an employer, Disney has redefined work roles. Proposing a model for change in the emerging service economy, Disney has shifted from the white-collar worker described by C. Wright Mills in the 1950s to a new chameleon of "flexible" tasks. The planners at its corporate headquarters are "imagineers"; the costumed crowd-handlers at its theme parks are "cast members." Disney suggests that the symbolic economy is more than just the sum of the services it provides. The symbolic economy unifies material practices of finance, labor, art, performance, and design.

The prominence of culture industries also inspires a new language dealing with difference. It offers a coded means of discrimination, an undertone to the dominant discourse of democratization. Styles that develop on the streets are cycled through mass media, especially fashion and "urban music" magazines and MTV, where, divorced from their social context, they become images of cool. On urban billboards advertising designer perfumes or jeans, they are recycled to the streets, where they become a provocation, breeding imitation and even violence. The beachheads of designer stores, from Armani to A/X, from Ralph Lauren to Polo, are fiercely parodied for the "props" of fashion-conscious teenagers in inner city ghettos. The cacophany of demands for justice is translated into a coherent demand for jeans. Claims for public space by culture industries inspire the counterpolitics of display in late twentieth century urban riots.

The symbolic economy recycles real estate as it does designer clothes. Visual display matters in

American and European cities today, because the identities of places are established by sites of delectation. The sensual display of fruit at an urban farmers' market or gourmet food store puts a neighborhood "on the map" of visual delights and reclaims it for gentrification. A sidewalk cafe takes back the street from casual workers and homeless people. In Bryant Park, enormous white tents and a canopied walkway set the scene for spring and fall showings of New York fashion designers. Twice a year, the park is filled by the fashion media, paparazzi, store buyers, and supermodels doing the business of culture and reclaiming Bryant Park as a vital, important place. We New Yorkers become willing participants in the drama of the fashion business. As cultural consumers, we are drawn into the interrelated production of symbols and space.

Mass suburbanization since the 1950s has made it unreasonable to expect that most middle-class men and women will want to live in cities. But developing small places within the city as sites of visual delectation creates urban oases where everyone *appears* to be middle-class. In the fronts of the restaurants or stores, at least, consumers are strolling, looking, eating, drinking, sometimes speaking English and sometimes not. In the back regions, an ethnic division of labor guarantees that immigrant workers are preparing food and cleaning up.

This is not just a game of representations: developing the city's symbolic economy involves recycling workers, sorting people in housing markets, luring investment, and negotiating political claims for public goods and ethnic promotion. Cities from New York to Los Angeles and Miami seem to thrive by developing small districts around specific themes. Whether it is Times Square or el Calle Ocho, a commercial or an "ethnic" district, the narrative web spun by the symbolic economy around a specific place relies on a vision of cultural consumption and a social and an ethnic division of labor.

As cities and societies place greater emphasis on visualization, the Disney Company and art museums play more prominent roles in defining public culture. I am speaking, first, of public culture as a process of negotiating images that are accepted by large numbers of people. In this sense, culture industries and cultural institutions have stepped into the vacuum left by government. At least since the 1970s debacles of Watergate and the Vietnam War, through Irangate in the 1980s and the confessions of politicians in the 1990s, government has lacked the basic credibility to define the core values of a common culture. On the local level, most mayors and other elected officials have been too busy clearing budget deficits and dealing with constituents' complaints about crime and schools to project a common image. The "vision thing," as George Bush called it, has been supplied by religious leaders from Jerry Falwell to Jesse Jackson and by those institutions whose visual resources permit or even require them to capitalize on culture.

I also see public culture as socially constructed on the micro-level. It is produced by the many social encounters that make up daily life in the streets, shops, and parks – the spaces in which we experience public life in cities. The right to be in these spaces, to use them in certain ways, to invest them with a sense of our selves and our communities – to claim them as ours and to be claimed in turn by them – make up a constantly changing public culture. People with economic and political power have the greatest opportunity to shape public culture by controlling the building of the city's public spaces in stone and concrete. Yet public space is inherently democratic. The question of who can occupy public space, and so define an image of the city, is open-ended.

Talking about the cultures of cities in purely visual terms does not do justice to the material practices of politics and economics that create a symbolic economy. But neither does a strictly political-economic approach suggest the subtle powers of visual and spatial strategies of social differentiation. As I suggested in *Landscapes of Power* (1991), the rise of the cities' symbolic economy is rooted in two long-term changes – the economic decline of cities compared to suburban and nonurban spaces and the expansion of abstract financial speculation – and in such short-term factors, dating from the 1970s and 1980s, as new mass immigration, the growth of cultural consumption, and the marketing of identity politics. This is an inclusive, structural, and materialist view. If I am right, we cannot speak about cities today without understanding:

- how cities use culture as an economic base,
- how capitalizing on culture spills over into the privatization and militarization of public space, and
- how the power of culture is related to the aesthetics of fear.

CULTURE AS AN ECONOMIC BASE

Suppose we turn the old Marxist relation between a society's base and its superstructure on its head and think of culture as a way of producing basic goods. In fact, culture supplies the basic information – including symbols, patterns, and meaning – for nearly all the service industries. In our debased contemporary vocabulary, the word *culture* has become an abstraction for any economic activity that does not create material products like steel, cars, or computers. Stretching the term is a legacy of the advertising revolution of the early twentieth century and the more recent escalation in political image making. Because culture is a system for producing symbols, every attempt to get people to buy a product becomes a culture industry. The sociologist Daniel Bell used to tell a joke about a circus employee whose job it was to follow the elephant and clean up after it; when asked, she said her job was in "the entertainment business." Today, she might say she was in "the culture industry." Culture is intertwined with capital and identity in the city's production systems.

From one point of view, cultural institutions establish a competitive advantage over other cities for attracting new businesses and corporate elites. Culture suggests the coherence and consistency of a brand name product. Like any commodity, "cultural" landscape has the possibility of generating other commodities. Historically, of course, the arrow of causality goes the other way. Only an economic surplus – sufficient to fund sacrifices for the temple, Michelangelos for the chapel, and bequests to art museums in the wills of robber barons – generates culture. But in American and European cities during the 1970s, culture became more of an instrument in the entrepreneurial strategies of local governments and business alliances. In the shift to a post-postwar economy, who could build the biggest modern art museum suggested the vitality of the financial sector. Who could turn the waterfront from docklands rubble to parks and marinas suggested the possibilities for expansion of the managerial and professional corps. This was probably as rational a response as any to the unbeatable isolationist challenge of suburban industrial parks and office campuses. The city, such planners and developers as James Rouse believed, would counter the visual homogeneity of the suburbs by playing the card of aesthetic diversity.

Yet culture also suggests a labor force that is well suited to the revolution of diminished expectations that began in the 1960s. In contrast to high-rolling rappers and rockers, "high" cultural producers are supposed to live on the margins; and the incomes of most visual artists, art curators, actors, writers, and musicians suggest they must be used to deprivation. A widespread appreciation of culture does not really temper the work force's demands. But, in contrast to workers in other industries, artists are flexible on job tasks and work hours, do not always join labor unions, and present a docile or even "cultured" persona. These qualities make them, like immigrants, desirable employees in service industries. Dissatisfaction with menial and dead-end jobs does not boil over into protest because their "real" identity comes from an activity outside the job.

[. . .]

CULTURE AS A MEANS OF FRAMING SPACE

For several hundred years, visual representations of cities have "sold" urban growth. Images, from early maps to picture postcards, have not simply reflected real city spaces; instead, they have been imaginative reconstructions – from specific points of view – of a city's monumentality. The development of visual media in the twentieth century made photography and movies the most important cultural means of framing urban space, at least until the 1970s. Since then, as the surrealism of *King Kong* shifted to that of *Blade Runner* and redevelopment came to focus on consumption activities, the material landscape itself – the buildings, parks, and streets – has become the city's most important visual representation. Indeed, in *Blade Runner*, the modern urban landscape is used as a cult object. Far more than King Kong's perch on the Empire State Building, *Blade Runner's* use of the Bradbury Building, an early twentieth century office building in downtown Los Angeles that has been preserved and lovingly restored, emphasizes the city's material landscape as a visual backdrop for a new high-tech, global society . . .

[. . .]

E
I
G
H
T

More common forms of visual re-presentation in all cities connect cultural activities and populist images in festivals, sports stadiums, and shopping centers. While these may simply be minimized as "loss leaders" supporting new office construction, they should also be understood as producing space for a symbolic economy. In the 1960s, new or restored urban shopping centers from Boston to Seattle copied suburban shopping malls by developing clean space according to a visually coherent theme. To the surprise of some urban planners, they actually thrived. No longer did the city's dream world of commercial culture relate to the bourgeois culture of the old downtown or the patrician culture of art museums and public buildings. Instead, urban commercial culture became "entertainment," aimed at attracting a mobile public of cultural consumers. This altered the public culture of the city.

Linking public culture to commercial cultures has important implications for social identity and social control. Preserving an ecology of images often takes a connoisseur's view of the past, re-reading the legible practices of social class discrimination and financial speculation by reshaping the city's collective memory. Boston's Faneuil Hall, South Street Seaport in New York, Harborplace in Baltimore, and London's Tobacco Wharf make the waterfront of older cities into a consumers' playground, far safer for tourists and cultural consumers than the closed worlds of wholesale fish and vegetable dealers and longshoremen . . .

[. . .]

Culture can also be used to frame, and humanize, the space of real estate development. Cultural producers who supply art (and sell "interpretation") are sought because they legitimize the appropriation of space. Office buildings are not just monumentalized by height and facades, they are given a human face by video artists' screen installations and public concerts. Every well-

designed downtown has a mixed-use shopping center and a nearby artists' quarter. Sometimes it seems that every derelict factory district or waterfront has been converted into one of those sites of visual delectation – a themed shopping space for seasonal produce, cooking equipment, restaurants, art galleries, and an aquarium. . . .

[. . .]

So the symbolic economy features two parallel production systems that are crucial to a city's material life: the *production of space*, with its synergy of capital investment and cultural meanings, and the *production of symbols*, which constructs both a currency of commercial exchange and a language of social identity. Every effort to rearrange space in the city is also an attempt at visual re-presentation. Raising property values, which remains a goal of most urban elites, requires imposing a new point of view. But negotiating whose point of view and the costs of imposing it create problems for public culture.

Creating a public culture involves both shaping public space for social interaction and constructing a visual representation of the city. Who occupies public space is often decided by negotiations over physical security, cultural identity, and social and geographical community. These issues have been at the core of urban anxieties for hundreds of years. They are significant today, however, because of the complexity and diversity of urban populations. Today the stakes of cultural reorganization are most visible in three basic shifts in the sources of cultural identity:

- from local to global images,
- from public to private institutions, and
- from ethnically and racially homogeneous communities to those that are more diverse.

These rather abstract concepts have a concrete impact on framing urban public space.

"The Invention of Regional Culture"

from R. Lee and J. Wills (eds)
Geographies of Economies (1997)

Meric Gertler

Editors' introduction

It has been observed that as a network of capitalist production and commercial trade increasingly wraps the globe, nation-states no longer provide the basic spatial framework for economic geography that they once did. It has become easier, in other words, for localities and regions *within* nation-states to "link up" directly with global networks and bypass the national scale altogether. Thus, globalization is said to have brought about a new kind of regionalization. Geographers like Michael Storper (*The Regional World*, 1992) have argued that sub-national regions – such as California's Silicon Valley, the "Third Italy," or Germany's Baden-Württemberg – are at the center of flexible, knowledge-based production systems. And it is these types of systems that are best suited to succeed in the global economy.

The success of some of these regional economies compared to others has typically been attributed to specific characteristic within those regions themselves. Thus, we see an odd revival of the traditional "culture region" in the guise of a "cultural turn" in economic geography. That is to say, culture often becomes an explanatory variable accounting for the successes or failures of regional economies within the networks of global capitalism. There are many variations of this view of culture's role in the development of regional economies. One of the most well-known accounts is Michael Porter's *Clusters and the New Economics of Competition* (2002). Sharon Zukin (see p. 431) has argued that localities turn to promoting "culture industries" in order to sell themselves as distinctive places (for instance, via tourism, recreation, or entertainment), since symbolic resources have become key sites of capital accumulation in post-industrial societies. Others, like Annalee Saxenian (in *Regional Advantage: Culture and Competition in Silicon Valley and Route 128*, 1994), have sought to identify for cultivation the kinds of regional cultural patterns and behaviors that lead to an agglomeration of successful businesses. Culture, in these terms, becomes a mysterious kind of resource that some places are lucky to have but that other places might somehow also cultivate. Indeed, "cultural regulation" has been identified as a key policy front in the promotion of regional economic development.

In the selection below, Meric Gertler argues that this turn toward culture as an explanatory variable mystifies the actual regulatory and institutional processes going on. Moreover, he argues that these regulatory and institutional processes often operate at national scales, thus tempering the idea that globalization has brought about a new era of economic regionalism. At the outset, it appears that Gertler might be echoing Don Mitchell's (see p. 11) argument that "there is no such thing as culture," only social processes the political implications of which are too often "naturalized" by appealing to cultural explanations. But Gertler is more interested in the interaction between cultural forces and social practices. His argument is less a denial of the existence of regional cultural distinctions than an analysis of how such distinctions are conditioned by social practices such as labor-market regulations and norms of institutional behavior.

While Gertler is an economic geographer interrogating the "cultural turn" in economic geography, his work is important for cultural geography because it provides a critical analysis of cultural patterns as *produced* by certain social assemblages that are relatively easy to identify. While, as Yúdice (see p. 422) points out, culture is increasingly regarded as a resource capable of achieving certain social ends, Gertler's argument flips this equation on its head. His argument implies that success in the global economy depends less on the promotion of place-specific culture than on the development of particular kinds of regulatory and institutional frameworks at the national scale. In some respects the contrasting arguments of Gertler and Yúdice reveal very different understandings of culture. But they probably share a conviction that treating culture as a resource shifts development policy away from a more important focus on broader social systems of regulation and governance.

A Professor of Geography at the University of Toronto, Meric Gertler conducts research on technology production, regional and national innovation systems, regional economic development planning, and the political economy of technological change in North America and Europe. He is the author of numerous journal articles and book chapters, including a contribution on the cultural-economic geography of production in Anderson, Domosh, Pile and Thrift's *Handbook of Cultural Geography* (2003). Gertler is also co-editor of *The New Industrial Geography: Regions, Regulation, and Institutions* (1999) and *Innovation and Social Learning: Institutional Adaptation in an Era of Technological Change* (2002).

INTRODUCTION

'Culture' has re-entered the lexicon of the economic disciplines with a prominence not seen for some time. With the growing interest in the social nature of production systems, signified by the use of terms such as 'industrial networks', 'industrial districts', and especially the 'new social economy' and 'socio-economics', a new significance has been ascribed to socio-cultural context. Hence, in emerging production systems in which the social division of labour is recognized as being of increasing importance, social and cultural characteristics have begun to figure prominently in the work of economic geographers, industrial economists, political economists and management theorists. The inter-firm relations which have come to dominate the 'new competition' are said to be based increasingly upon non-market forms of interaction bound by trust, in which cultural commonality between co-operating and transacting partners is seen as an advantage.

Extending this line of thinking, a position which is gaining currency in the growing volume of work on the adoption and propagation of network relations between firms holds that Anglo-American 'business culture' is not favourably predisposed to the idea of co-operation. The claim is made that the ethic of rugged individualism and the rhetoric of 'dog-eat-dog' competition are so strong as to discourage Canadian, American or British firms from participating in (or reaping the full benefits from) inter-firm co-operation and collaboration. The converse of this argument is that particular national or regional cultures are inherently more predisposed to co-operation, or that cultural ties are coming to dominate all others in shaping the emerging alliances and partnerships between businesses.

A related argument is the idea that certain cultural traits – for example, the 'traditional' Japanese values of dedication to education and hard work, devotion to higher authorities such as one's employer, and a sense of social cohesion, or the 'typical' German predisposition towards all things technical and complex – explain the success with which particular national economies have adopted post-Fordist production methods, including new forms of complex production technologies and modes of workplace organization *within* the individual firm. Conversely, the absence of these traits in other cultures explains the failures of their own indigenous firms to adopt these new practices with the same degree of success.

Hence, a new variable has entered the debate on regional and national competitiveness, and the prescriptions for policy flow directly from the diagnosis: in the absence of a naturally inherited manufacturing culture, the state must attempt to create or 'manufacture' one, by exhorting firms to 'co-operate to compete' – that is, to change firms' behaviour by convincing them that it is in their own best economic interest to co-operate with other firms. Furthermore, to help them along in this process, the state should train individuals to act as 'brokers', to bring reluctant firms together by helping them recognize complementarities they may share with other firms (normally within the same region).

I wish to argue in this chapter that the role of culture in this debate has not been adequately specified, on either a theoretical or an empirical level, and that it needs to be thoroughly rethought. I hope to demonstrate the need to examine the process by which industrial cultures – whether at the level of the workplace, the region or the nation – are themselves constructed by social practices. For the purposes of this analysis, I shall focus on one small but significant part of this process, by examining the role of economic institutions and regulatory frameworks – primarily public ones, and operating at both the regional *and* national scale – in shaping practices, customs, norms of economic behaviour and even what appear to be individual traits. In so doing, I hope to take issue with the notion that certain regional or national cultures are somehow more naturally predisposed to engaging successfully in post-Fordist manufacturing activities than others – that their success is due to their naturally endowed 'manufacturing culture'. I also wish to assess critically the recent arguments that such manufacturing cultures can themselves be readily manufactured in particular places where they were previously undeveloped. Here, I shall argue that the process by which industrial practices are produced is more complex, involving forces of regulation operating not only at the level of the individual workplace, corporation, community or region, but also at the spatial scale of the nation-state.

At the same time, and somewhat ironically, I shall argue that this process is considerably more transparent than the current literature would have us believe. When economic analysts resort to 'cultural' influences to explain the behaviour of managers,

firms and workers, this is normally tantamount to an admission of ignorance. It is as if the processes at work arise from some timeless, primordial traits whose formation mystifies and confounds understanding. Instead, I shall attempt to demonstrate that the motivations underlying many of these practices within and between individual firms can be seen to arise quite directly from the structure of the macroregulatory environment in which these entities function. In doing so then, I hope to begin the process of demystification in our study of contemporary economic relations. I also wish to argue that those prescriptions for regional economic renewal which focus on the need to correct the dysfunctional behavioural tendencies and culturally shaped attitudes of individual firms, managers and workers are usually based on a mis-diagnosis of the problem. Consequently, these prescriptions are misdirected at changing only the attributes and traits of managers and workers, when they should also focus on the broader, systemic characteristics of the regulatory environment.

[...]

CULTURE AND REGIONAL SYSTEMS OF INNOVATION AND PRODUCTION

Integral to the claim that the nature of capitalist competition has shifted in the late twentieth century is the key idea that systems of innovation and production have become more social in nature. This assertion has two distinct but related components. First, production systems are coming to be characterized by a more finely articulated social division of labor, achieved through the process of vertical disintegration of large firms and the growing use of various forms of outsourcing, including subcontracting to smaller supplier firms. This externalization of the production process is said to offer the chief advantage of agility in meeting the needs of ever more rapidly changing and fragmented markets. As market demands shift qualitatively, producers are able to respond more effectively in such 'open' systems because (a) they can more readily absorb the innovative ideas of supplier firms to help them devise new products and improvements, and (b) they can rework their sources of supply to match the particular attributes of the 'product of the moment', in both cases drawing upon the rich resources of a large collection of suppliers.

The second component is that, as individual firms come to rely more heavily on their relations and exchanges with other firms, *non-market* forms of interaction become more important. Viewed in terms of the Williamsonian continuum between public markets and private hierarchies, much of the interesting action is seen to be taking place in the middle ground: relations are social, but are increasingly buttressed by trust. In particular, as Harrison ["Industrial districts: old wine in new bottles?" *Regional Studies* 26, 1992] has pointed out, for these innovative production systems to function properly, firms must develop a considerable degree of interdependence on one another (including surrendering proprietary information) but will do so only when a relationship of trust has been established. Such relations are more likely to arise when firms interact with one another directly and repeatedly over time, as they are more likely to do when they are located in the same region. However . . . this interaction takes place through informal as well as formal mechanisms, and is reinforced by shared histories and cultures.

. . . This interaction is said to be especially important at times when technological development crosses the threshold to a new paradigm. When the technology in question is particularly complex, expensive, and subject to rapid change, then 'closeness' – in both a physical and cultural sense – is crucial to successful interaction leading to effective innovation. Spatial proximity facilitates the easy, frequent face-to-face contact necessary for the exchange of detailed technical information. Cultural commonality further reinforces this link, since it is easier for producer and user to understand one another at deeper levels of meaning.

[Michael] Storper [*The Regional World: Territorial Development in the Global Economy*, 1992] . . . describes a phenomenon he dubs 'product-based technological learning' or PBTL, which he observes to be occurring most commonly in dynamic, subnational agglomerations he calls 'technology districts'. This phenomenon, according to Storper, is underpinned crucially by what he refers to as 'conventions', which 'structure the participation of agents' in such districts. Furthermore, these conventions are 'territorially bounded' and serve to 'define the qualitative basis of the external

economies of PBTL systems'. Storper's concept of conventions is rich and multifaceted. In essence, they amount to a set of acknowledged and shared rules 'that mobilize resources and regulate interactions so as to make PBTL possible', and that create 'localized expectations' and 'preference structures' concerning concepts such as time horizons, payoff points, etc.' . . .

More recently, Saxenian [*Regional Advantage: Culture and Competition in Silicon Valley and Route 128*, 1994] has introduced culture as a key variable in her analysis of the reasons for the widely diverging performances of two regions producing innovative products such as semiconductors and personal computers: California's Silicon Valley and Massachusetts's Route 128 . . . In attempting to explain Silicon Valley's continued technological success and the failure of Route 128, despite the fact that they competed in the same product markets and as recently as the 1970s boasted comparable levels of economic activity, Saxenian attributes causality to the divergent 'industrial systems' that characterized these two different regions. This regional industrial system is said to have three closely interconnected dimensions: 'local institutions and culture, industrial structure, and corporate organization'. The first of these elements is described as follows:

Regional institutions include public and private organizations such as universities, business associations, and local governments, as well as the many less formal hobbyist clubs, professional societies, and other forums that create and sustain regular patterns of social interaction in a region. These institutions *shape and are shaped by* the local culture, the shared understandings and practices that unify a community and define everything from labor market behavior to attitudes toward risk-taking. (p. 8; emphasis added)

Note here that the 'culture' at work is explicitly local or regional in character. Furthermore, while Saxenian draws attention to the reflexive interaction between regional institutions and regional culture, national institutions and culture do not figure in this discussion.

More recently still, Kanter ["Thriving locally in the global economy," *Harvard Business Review*,

September–October 1995] has extended the application of cultural ideas to the process of regional economic growth by arguing that communities that wish to serve as successful destinations for foreign direct investment need to create, among other things, a local culture of collaboration. As she puts it, 'In addition to the physical infrastructure that supports daily life and work – roads, subways, sewers, electricity, and communications systems – communities need an infrastructure for collaboration to solve problems and create the future.' According to Kanter, this infrastructure is largely informal in nature, with the chambers of commerce in rapidly growing communities such as Spartanburg and Greenville in South Carolina acting to provide the 'social glue' that fosters cooperative action and joint learning in the region.

[. . .]

MACHINERY CULTURE: THE SOCIAL CONSTRUCTION OF 'OVERENGINEERED'

. . . Since 1991, I have been studying the process by which manufacturers in Ontario have acquired and implemented their new process technologies. The study has focused on the relationship between these Canadian 'user' firms and the companies which produce these technologies for them ('producers'). Given the somewhat underdeveloped state of the Canadian advanced machinery industry . . . many of the leading producers of advanced manufacturing technologies in use in the Canadian plants are now found in Japan, Germany, Italy, Sweden and other European and Asian countries. After surveying and interviewing a selection of these users in Ontario, I was able to identify the sources of advanced machinery and equipment used by these firms. In a subsequent phase of the study, I then visited and interviewed a sample of producers of these technologies in Germany, one of the leading offshore sources for such production systems. In addition, I conducted interviews with a number of representatives and suppliers of these foreign machinery producers, residing in Canada and serving as intermediaries between overseas producers and local users . . .

Generally speaking, the findings indicate that many users in Canada continue to experience significant problems of implementation and operation

long after the installation is completed. Hence, even after being given time to 'work out the kinks' and 'move along the learning curve', users (including some large, relatively sophisticated operations with deep financial resources and in-house technical staff) have had a difficult time achieving effective implementation. The machinery and systems once installed, failed to live up to the user's expectations (or the salesperson's claims) for product flexibility, speed of production and change-over, quality, ease of use and reliability. Breakdowns and malfunctions were frequent and downtimes were lengthy and disruptive. In general, the returns from such costly and difficult investments were often disappointing. Furthermore, and crucially (given the theme of this chapter), these problems seem to have been particularly likely to arise (and to be especially acute) when the technology in question originated in 'far-off places' such as Germany, Japan and many other overseas sources.

When users (or producers) were asked to explain the reasons for and sources of these difficulties they pointed first to the minor but significant complications introduced when trying to carry out communications and transactions involving complex technical subjects (including both the initial specification of technology requirements and the subsequent problem-solving and 'trouble shooting' procedures) over long distances. These relate to the delays introduced by intervening time zones, the difficulties of technical problem-solving without face-to-face contact (despite the widespread use of information and telecommunication technologies to connect users to producers), and problems of comprehension which may arise owing to differences of language . . .

However, subsequent discussions revealed that a deeper source of these problems lies in the fundamental differences in expectations, characteristic workplace practices and norms, managerial routines, transactional behaviours, and understandings of key concepts such as 'technology' itself – in short, what appear to be substantially different industrial or business cultures in Canada and Germany respectively. Indeed, interviewees on both sides of the Atlantic readily identified differences in 'culture' or 'mentality' (a term used far more frequently by German respondents than by Canadian ones) as the root of their problems in dealing with one another. Nevertheless, notwithstanding this diagnosis, what

I have been able to show, by examining some specific instances in which these differences have become salient, is that underlying these apparently cultural gaps are *fundamentally different regulatory regimes and institutional structures* which are themselves *instrumental in reproducing these 'cultural' differences*. Presented below is a sample of two specific symptoms – expressed as differences in expectations, attitudes, accepted business customs and practices – which have led to misunderstandings, disappointments, conflict and, in extreme cases, termination of the relationship between machinery producer and user.

Maintenance

One of the clearest differences to emerge from this study was in the contrasting practices of German and North American users regarding machinery and equipment maintenance. German producers remarked (usually with disbelief and more than a little disdain) that North American industrial culture did not seem to assign much value to the importance of regular, preventive maintenance. As a consequence, production systems in Canadian and American plants would, in the view of the producers, fail with predictably greater frequency. This stood in sharp distinction to the dominant practice in German plants, where not only managers but also the operators themselves would maintain and service the machinery on a regular basis. More than one German producer commented on how, in their German customers' plants, the operators were 'married to' or 'owned' their machines, and would lavish attention upon them. In the words of one German manager, 'German workers . . . have the feeling, "that is my machine, and I am responsible for it"'.

Machine complexity and ease of operation

Canadian users complained that the production systems supplied by German producers were considerably more difficult to operate effectively than they had been led to believe at the time of sale. They may have held this impression despite having travelled to another user's plant (often in Germany) to observe the operation of a similar system in real time before deciding to make the purchase. A frequently heard comment (both from those users that did buy German machinery and those that did not) was that German technology was 'overengineered' and 'too complex'. This was usually accompanied by remarks to indicate that this was due to cultural traits that predisposed German producers and users to overly complicated technical solutions. The German producers had difficulty knowing how to regard such complaints, since they were aware that similar systems worked perfectly well and with little difficulty in the plants of their German customers. Instead, in the face of criticism from users that producers were 'rigid', 'unbending', or trying to 'dictate' inappropriate technical solutions to their precise production problems, the German firms would tend to place the blame with the user, accusing it of not doing enough training of its workers and managers, or of investing insufficient attention and resources in maintenance ('the problem must be yours').

As I have indicated above, the distinctive differences between German and Canadian practices were most frequently comprehended and described by those interviewed as arising from cultural dissimilarities. Indeed, the two sets of characteristic practices, expectations, attitudes and norms documented above *might themselves be viewed as constituent parts of distinct industrial and business cultures*. However, this diagnosis begs the obvious question, namely: how are such differences produced? More to the point, if one accepts that 'culture' (industrial or otherwise) is not some natural, prior, unchanging and inherited whole, then how does it interact with contemporary social practices in its own production and reproduction? One way of answering this is to set these cultural characteristics within their broader social and political context, by examining their relationship to readily identifiable institutional and regulatory features. Given that many aspects of this context *also* differ markedly between Germany and the Anglo-American economies, it should come as no surprise that these larger, background differences might play a role.

In fact, I would argue that the differences observed above can be linked quite directly to the nature of social institutions which regulate capital markets and business finance, labour markets, labour relations and the employment relations of user firms. Beginning with the issue of sharply divergent maintenance practices, much of this can be explained by examining the enduring differences in capital market structures in the two countries, which create marked differences in time horizons between the German and North American machinery users. Canada and the USA have created business environments, based on the classic Anglo-American system of public capital markets for equity investments, in which there is a strong division between financial and industrial capital. Shareholders usually exert significant power, creating strong pressures to produce short-term returns on investment. In contrast, German businesses raise the bulk of their equity capital through private investments. In a system in which financial institutions and industrial firms are closely linked, and in which (as a result of the labour relations institutions described below) a broad array of stakeholders (including workers and unions) are routinely represented on boards of directors, investment objectives are longer-term. The pursuit of short run returns is tempered by sources of capital which are patient or 'quiet', and by a stronger voice in favour of social returns, resulting from the direct representation of workers on the managing boards of many larger German firms. As a consequence, German industrial firms have considerably more latitude to wait longer periods of time for investments to bear fruit, explaining their considerably longer managerial time horizons, relative to their North American counterparts.

Hence, the stark differences in maintenance practices can now be understood as arising, at least in part, from the structure of industrial investment finance and the institutions shaping capital markets. When investment capital is acquired on terms that are so strongly skewed in favour of quarterly returns, it should come as no surprise that Canadian (or American) users treat their capital equipment in a manner consistent with the prevailing truncated time horizons. When their decision-making horizon stops at two to three years and their expectation is that a machine will be in active service only this long, it is understandable that managers will undervalue regular expenditures for the purpose of longer-term machine and system maintenance.

This tendency is further reinforced by sharp distinctions between the German and North American institutions and systems of regulation shaping labour markets and the employment relation. One of the most distinctive features of the German economy is its system of labour relations based on the principle of 'co-determination'. Under this system, workers – both directly through firm-based 'works councils' and indirectly through national unions – have a significant and institutionalized role in many aspects of the firm's decision-making, including training, technology acquisition and implementation, and day-to-day operations. Furthermore, and as a result of labour's institutionalized power, there are serious curbs on employers' ability to fire or lay off workers. Instead, the system works to encourage a stable employment relation characterized by long length of employment tenure and the active use of internal labour market practices to manage firms' personnel needs. Furthermore, with a much greater degree of centralization of wage determination, and strong concordance between wages in union and non-union workplaces, inter-firm competition based on wages is held in check.

All of this stands in sharp contrast to the Anglo-American norm, where employment relations are far less stable over the long term, where employers make far more extensive use of external labour market practices (hiring and firing), leading to the high turnover rates discussed earlier. Furthermore, apart from some key sectors such as automotive assembly, unionization rates are low and (at least in the USA) declining, as is labour's power in the workplace in general. As a result, the degree of inter-firm variation in wages and working conditions is significantly greater than in Germany, and employers are encouraged to view labour cost as one of the chief dimensions of inter-firm competition.

These fundamental differences in the institutional and regulatory framework surrounding employment play a large role in producing the practices and attitudes documented earlier and described so frequently as being cultural in origin. Hence, it should not be surprising that North American workers do not develop the same sense of 'ownership' of their machinery as was seen to be the case in Germany, and do not engage in the same kind of lavish maintenance behaviour that the German producers so admired in the practices of their domestic customers. Furthermore, when you have a system in which machine operators are much more likely to participate in the decision to purchase the machinery in the first place (including the process of deciding on technical specifications), this is a powerful force in the development of the sense of 'ownership' of a machine that was referred to earlier.

[. . .]

CULTURE, INSTITUTIONS AND INDUSTRIAL PRACTICES: IMPLICATIONS FOR THEORY AND POLICY

I have endeavoured to show how the traits and attitudes we commonly understand as being part and parcel of inherited cultures are themselves produced and reproduced over time by day-to-day practices that are strongly conditioned by surrounding social institutions and regulatory regimes. Hence, we can see that workplace practices, attitudes and norms in the use of advanced machinery in Germany or North America do themselves constitute distinctive industrial cultures – but ones which are actively shaped by the prevailing macro-regulatory context. By demonstrating the impact of the institutional setting on the formation of industrial culture, I have hoped to convey something of the perils arising from the more prevalent approach to the question of culture's influence in national and regional economic systems. The argument advanced here implies strongly that the very practices we take as signifiers of distinct cultures are themselves influenced by a set of institutions constituted outside the individual firm. Moreover, in the story told here, the institutions that seem to matter most are largely *national* in origin.

This implies that we as analysts need to be much more careful in our use of cultural concepts to 'explain' differences in the performance of local or regional production systems. Culture is not a static, analytically prior concept, which 'produces' these differences. To a very significant extent, it is the outcome of regulatory forces emanating from a set of socially constructed institutions for the governance of investment and the use of labor. A further implication is that what we have sometimes taken to be organic, *sui generis* behavior – among, say, the artisanal firms of the Third Italy or the mechanical engineering firms of Baden-Württemberg – is to an important extent strongly

consistent with the overarching national system of regulation (akin to what Nelson and others have referred to as national systems of innovation; *see* Nelson, 1993). Therefore, it is much easier to understand how the technical excellence of German engineering firms is *produced rather than simply 'inherited'* when one examines the broader social context within which these firms operate.

[. . .]

These insights provide both an optimistic and a pessimistic prospect for regional development policy. On the up side, they demystify the hitherto murky origins of successful economic systems, showing how they can in fact be produced by deliberate state action. On the down side, those policy-makers who would wish to intervene *solely* at the regional scale (or, for that matter, at the level of the individual firm) in order to alter industrial practices will be discouraged to know that their initiatives will be somewhat futile in the absence of generally supportive (or at least, not actively antithetical) national regulatory features. . . .

This raises another issue of significance for theory and policy: namely, the relative importance of regional versus national institutions in the production of favourable industrial practices. It is clear from the preceding analysis that the most telling and significant differences between the German machinery producers and their Ontario customers originate from national-level distinctions: in systems of labour market regulation, in training systems, in industrial relations, and even in the systems of industrial finance and capital markets. As such, the arguments in this chapter stand in marked contrast to much recent work in economic geography and related disciplines (such as that of Storper and Saxenian reviewed earlier) which has accorded causal significance to *regional* institutions. Indeed, so little of the difficulty arising in this bilateral relationship appears to be regional in origin that it is worth reflecting on this issue at greater length.

[. . .]

Hence, it remains important for economic geographers, other social scientists and policy-makers to appreciate the importance of nation-state institutions in creating the enabling, accommodative space within which particular regional growth phenomena may arise. In this sense, then, we can understand the spatial construction of industrial practices as occurring through the interaction of local, national and subnational regulatory forces as well as corporate strategy. However, it is equally important to consider the provenance of the very institutions which we have implicated as having so much power to shape corporate and regional practices. Just as it is crucial to espouse a dynamic conception of culture, so too is it important not to treat institutions as if they were 'carved in stone' or inherited from on high. Indeed, it is likely that the relationship between institutions and practices is fundamentally dialectical in nature, with the latter possessing the potential to reshape the former over time. . . .

"Destination Museum"

from *Destination Culture: Tourism, Museums, and Heritage* (1998)

Barbara Kirshenblatt-Gimblett

Editors' introduction

Tim Oakes (*Tourism and Modernity in China*, 1998) has observed that ethnic villages in China's interior tend to model their tourism plans on the ethnic culture theme parks that have sprung up in many of China's urban centers. The irony of course is that while theme parks are modeling their displays on "original" villages, those villages are turning themselves into theme parks. Such villages, we might observe, become exhibits of themselves, or "open air museums." Jean Baudrillard called this kind of thing a *precession of simulacra*, in which the model or exhibit of something (i.e. its representation) precedes the thing itself. He believed this to be the norm in our postmodern age. And while it may be debatable the extent to which such a "precession of simulacra" characterizes contemporary postindustrial societies (Baudrillard has his share of critics), the tourism and heritage industries certainly suggest that he is on to something. As Barbara Kirshenblatt-Gimblett observes in the selection below, "Increasingly, we travel to actual destinations to experience virtual places." And sometimes, those "virtual places" are themselves simulations of the actual destinations in which they are located.

Kirshenblatt-Gimblett notes, for example, the plans to build a "Gatwick Airport theme park" inside Gatwick Airport. Then there was the controversy over plans to build a "Key West World" theme park in Orlando, Florida, just seven hours' drive away from the actual Key West. This not surprisingly resulted in a few humorous reactions in the press, as observers noticed the irony of Florida building theme park exhibits of itself. In *The New York Times Magazine*, David Ives wrote a whimsical article entitled "Welcome to World World" (1995). Playing on the absurd lengths to which the "precession of simulacra" can be taken, Ives proposed several more theme parks, such as "Mall World" (a " 'mall' with shops, fast-food restaurants and ficus" that will look exactly like a "mall with shops, fast-food restaurants and ficus" but will charge admission and give every visitor a button "identifying him or her as a 'customer'") and "Walt Disney World World" (a replica of Disney World within Disney World "for visitors of Disney World who don't have time to do all of Disney World").

Kirshenblatt-Gimblett's chapter "Destination Museum", from *Destination Culture*, considers this question of what happens when places are put on display for tourists. What happens when a place becomes a "sight to be seen"? What happens when a village, or city, or country becomes a museum? Culture, in this context, becomes a resource for the exhibition of a place. And as culture gets reinvented according to the demands of cultural display, the tourism–heritage–museum industry is involved in the production of new cultural geographies. The project of selling places, of making them distinctive, of inventing an organic culture region, a new cultural geography – that project, Kirshenblatt-Gimblett claims, is museological. Whole places are being treated as exhibits of themselves.

There seem to be two slightly different interpretive takes on this kind of process. On the one hand, there is the question of authenticity. At what point do we lose sight of a clear distinction between the "authentic original" and its simulation? At what point, as Baudrillard would have it, does the simulation or exhibit start to seem more "real" than the original it supposedly references? At what point does the "original" cease to become an important referent at all? Are tourists happier experiencing Key West as a sanitized theme park instead of going to the island itself? And does asking this question betray an elitist condescension toward the hapless and easily fooled tourist? Does it betray an attempt to reinforce an elitist notion of culture?

On the other hand, there is the approach that views such landscapes of exhibition and display as "landscapes of power". In her influential book of the same title, Sharon Zukin argues that such landscapes enable the power of social control to be masked by the seductions of leisure and the naturalizations of culture. In *Colonizing Egypt* (1988), Timothy Mitchell argued that the spate of "world exhibitions" in nineteenth and early twentieth century Europe were part of larger *episteme* in which the world was known via its representation (see also the introduction to Part Two of the *Reader*). As also noted in the selection by Yúdice (see p. 422), the term episteme was used by Foucault, in *The Order of Things* (1966/1970), to describe the commonsense assumptions that provided the basis for the kinds of knowledges and discourses that were possible during a particular historical period. For Mitchell, this meant that nineteenth century Europeans understood the world primarily as an exhibition, objectified and framed before them to gaze upon. "Outside the world exhibition," he wrote, "one encountered not the real world but only further models and representations of the real."

But such a view of the world can lead to a perverse logic in the interests of heritage preservation and cultural display. Kirshenblatt-Gimblett notes the example of Burmese plans to forcibly remove Padaung people from the mountains and into a "model village" for display during "Visit Myanmar Year." And Timothy Mitchell (*Rule of Experts*, 2002) documents the case in Egypt of the Gurna villagers' eviction in order to make way for heritage. Mitchell writes, "The Gurnawis were to be treated as ignorant, uncivilized, and incapable of preserving their own architecture heritage. Only by seeing them in this way would the architect have an opportunity to intervene, presenting himself as the rediscoverer of a local heritage that the locals themselves no longer recognized or knew how to value." Such cases make the issues of power and social justice fundamental to any consideration of what happens when places seek to put themselves on display.

Barbara Kirshenblatt-Gimblett is University Professor and Professor of Performance Studies at the Tisch School of the Arts, New York University. Her many books and essays on tourism, heritage, exhibitions, Jewish culture, food, and aesthetics include *Image before my Eyes: A Photographic History of Jewish Life in Poland, 1864–1939* (reissued 1995), *They Called Me Mayer July: Painted Memories of a Jewish Childhood in Poland before the Holocaust* (2007), and the co-edited volume *Museum Frictions: Public Cultures/Global Transformations* (2007).

Related works on the politics of cultural display, heritage, and tourism include: Michael Herzfeld's *A Place in History* (1991), Sharon Macdonald's *The Politics of Display* (1998), and Joy Hendry's *The Empire Strikes Back* (2000).

■ ■ ■ ■ ■ ■

When Gatwick Airport's theme park opens in 1998, visitors for whom the experience of actual travel is no longer enough will be taking "a tour through baggage, security and emergency facilities, a mock control tower where visitors can have a go at landing planes and a 'white knuckle' ride through a replica of a baggage handling system." The very trials and tribulations of travel are becoming attractions in their own right through principles that have long connected tourism and museums.

Whole countries market themselves as "the world's largest open air museum." Deep in this marketing ploy for Turkey is the unnerving insight that tourism may beat museums at their own game by enabling travelers to encounter "some of the most stunning, intact, works of art and architecture anywhere. Such as St. Sophia, the Blue Mosque and the sumptuous Topkapi Palace" and to experience them in situ, before they have been dismantled and shipped off to a museum. The Bikini Islands is developing an atomic theme park in the areas devastated by nuclear testing. The U.S. National Park Service characterizes the ships and bombs at the bottom of a Bikini Island lagoon as an "unmodified museum of the dawn of the era of the atomic bomb." Such promotions promise an experience that is more real, more immediate, or more complete, whether they deliver an actuality (Gatwick Airport) or a virtuality (Gatwick Airport theme park) – or both at the very same place.

Immersion in a world other than one's own is a form of transport, whether one travels twenty-six hours from Europe to New Zealand, strolls from Samoa to Fiji within the virtual space of the Polynesian Cultural Center in Hawaii, or crosses the road separating Chinatown from Little Italy in Manhattan. What is most ordinary in the context of the destination becomes a source of fascination for the visitor – cows being milked on a farm, the subway in Mexico City during rush hour, outdoor barbers in Nairobi, the etiquette of bathing in Japan. Once it is a sight to be seen, the life world becomes a museum of itself.

Tourism needs destinations, and museums are premier attractions. Museums are not only destinations on an itinerary: they are also nodes in a network of attractions that form the recreational geography of a region and, increasingly, the globe. Museums, by whatever name, are also an integral part of natural, historical, and cultural sites. Such facilities orient the visitor to Napier's art deco district, the Waitomo Caves, and the Waitakere rain forest, in New Zealand. Some businesses establish full-fledged museums devoted to their own history (Atlanta's World of Coca-Cola) or the history of their product (Toronto's Bata Shoe Museum). Museums are also events on a calendar. Blockbuster exhibitions are known in the trade as event tourism.

Museums have long served as surrogates for travel, a particularly important role before the advent of mass tourism. They have from their inception preserved souvenirs of travel, as evidenced in their collections of plants, animals, minerals, and examples of the arts and industries of the world's cultures. While the museum collection itself is an undrawn map of all the places from which the materials have come, the floor plan, which determines where people walk, also delineates conceptual paths through what becomes a virtual space of travel.

Exhibiting artifacts from far and wide, museums have attempted from an early date to reconstruct the places from which these things were brought. The habitat group, period room, and re-created village bring a site otherwise removed in space or time to the visitor. During the nineteenth century, exhibitions delivered to one's door a world already made smaller by the railroad and steamship. Panoramas featured virtual grand tours and simulated the sound and motion of trains and ships and the atmospheric effects of storms at sea. A guide lectured and otherwise entertained these would-be travelers. Such shows were celebrated in their own day as substitutes for travel that might be even better than actually going to the place depicted. As one commentator explained in *Blackwood's Magazine* (1824), panoramas were a painless form of travel:

> Panoramas are among the happiest contrivances for saving time and expense in this age of contrivances, What cost a couple of hundred pounds and half a year a century ago, now costs a shilling and a summary manner. The affair is settled in a quarter of an hour. The mountain or the sea, the classic vale or the ancient city, is transported to us on the wings of the wind . . . If we have not the waters of the Lake of Geneva, and the bricks and mortar of the little Greek town, tangible by our hands, we have them tangible by the eye – the fullest impression that could be purchased, by our being parched, passported, plundered, starved, and stenched, for 1,200 miles east and by south, could not be fuller than the work of Messrs Parker's and Burford's brushes. The scene is absolutely alive, vivid, and true; we feel all but the breeze, and hear all but the dashing of the wave.

Viewers might prefer the panorama of Naples to Naples itself because it is "even more pleasant to look upon in Leicester Square, than is the reality with all its abominations of tyranny, licentiousness, poverty, and dirt" [see Ralph Hyde, *Panoramania! The Art and Entertainment of the "All Embracing" View*, 1988].

Furthermore, not everyone could travel, and for them panoramas and dioramas were, in the words of Charles Dickens, a "mode of conveyance." Mr. Booley's travel account in *Household Words* (1850) turns out to be based on a panorama – "all my modes of conveyance have been pictorial." The panorama's value, in Booley's words, lay in its ability to convey "the results of actual experience, to those who are unable to obtain such experiences for themselves." In addition, the panorama might convey "aspects of soil and climate . . . with a completeness and truthfulness not always to be gained from a visit to the scene itself." Displaced by cinema and amusement parks by the end of the century, this exhibition tradition can be found today in the atavism of museum dioramas, the futurism of IMAX projection, the special effects of rides like Back to the Future at Universal Studios, and hi-tech panoramas at the Museum of Sydney. Museums continue to enact transformations in perception linked to *the* technologies and practices of travel.

Museums now also serve as literal travel agents and organize exclusive tours to distant places. Travel with a Purpose tours, many of them led by curators from the Powerhouse Museum in Sydney, focus on ecotourism and the arts. These tours are intended to be "informative expeditions into other cultures for those of us not interested in poolside tourism experiences"; during the last months of 1994, groups went to Bhutan, France, Nepal, and India. The cost of the tour includes a donation to one of the sponsors – World Wide Fund for Nature – and to the Powerhouse Museum.

Instead of waiting for the tourists to come to them, museums are going to the tourists. Thanks to an exhibition program inaugurated in 1980 by the San Francisco Airport Commission, more than one person making a connecting flight in 1996 stepped off the motorized walkway to stroll through a display of kitchen equipment and tableware from the Ritz Collection at the California Academy of Sciences or slowed down for an exhibition of vintage ukuleles from the collection of Akira Tseumara in an otherwise bleak corridor.

Museums are even reproducing the protocols of travel. Visitors can purchase a Museums Passport to more than 190 museums in Queensland, Australia, get their documents stamped as they complete each visit, and save the passport as a souvenir. The American Museum of Natural History in New York, to celebrate its twelve anniversary in 1995,

thematized visits to its galleries as an expedition comparable to those the museum once sponsored to collect the specimens on display. *Expedition Passport*, available at two Base Camps in the building, welcomes the young visitor, the primary audience for this booklet:

> Most explorers travel to far-off places, but your journey will take place right in the footsteps of those scientists who have travelled the world and who have brought back many of the treasures you will see today. On this expedition, you can move back in time to the Age of Dinosaurs, You can touch a meteorite as old as the solar system. You can see a young Chinese woman on the way to her wedding. You can visit the woodlands, savannahs, and mountain regions of Africa. You can even shrink to the size of an ant. A great adventure lies before you today: To begin, turn the page.

At field stations in live galleries, visitors get their "passports" stamped.

Such tropes form an archive of historical understandings that go uncontested. Their playfulness insulates them front the very critiques that destabilized celebrations of the Columbus Quincentenary and that have brought museums themselves to task for their historic role in grand projects of discovery and conquest. Marketing a troubled history that glorifies colonial adventure and a repudiated anthropology of primitivism, tourism provides a safe haven for these ideas. A 1987 Iberia Airlines promotion began, "With 100 tours to choose from, Spain is once again open to invasion," and added tourists to a list that included Phoenicians, Greeks, Romans, and Visigoths – "Get ready for a vacation that's destined to go down in history."

[. . .]

[W]hat is the fate of the "museum product," however it is defined, in today's tourism economy? The presumption in some quarters is that visitors are no longer interested in the quiet contemplation of objects in a cathedral of culture. They want to have an "experience." Museums worry that they will be bypassed as boring, dusty places, as spaces of death – dead animals, dead plants, defunct things. This is why Te Papa Tongarewa, The Museum of New Zealand, in its Wellington Visitors' Center, has made a preemptive strike, first anticipating the negative image of the museum as a solemn place, "somewhere you have to whisper like [in] a church" and are not allowed to touch old things in glass cases. Then it tells the visitor that "[w]e are re-imagining the term 'Museum'," as a place alive, exciting, and unique – exactly what tourism markets. The flyer announcing Te Papa defines the museum experience as "an amazing adventure – one in which all New Zealanders are travellers," for "[t]he Museum is going to take us on a journey." The destination is collective self-understanding. Museums engaged in the task of imagining the nation must define its location, a responsibility that has repercussions beyond the journey within its walls.

Even as museums model themselves on tourism – the promise of "experience" indexes the immediacy of travel – the industry in parts of the world like New Zealand and Australia has been slow to develop "cultural tourism." Most tourism in these relatively young states is based on nature and the rest on purpose-built tourist attractions. There are several reasons for this emphasis.

There is the problem of how to define the uniqueness of a destination the better to market it in a competitive industry. What makes *this* place different? Australia and New Zealand have tended to identify their uniqueness as tourist destinations with the indigenous and to identify culture with the places from which settlers came. Yet, despite a high rate of endemism, their difference from other places is not natural but cultural; that is, difference is produced, not found. For Anthony Trollope, writing in 1873, "the great drawback to New Zealand – or I should more properly say to travelling in New Zealand – comes from the feeling that after crossing the world and journeying over so many miles, you have not at all succeeded in getting away from England. When you have arrived there you are, as it were, next door to your own house, and yet you have a two months' barrier between yourself and your home." Identifying New Zealand's specificity with unique aspects of its natural endowments is a cultural practice. Judging from Trollope's observation, it is not an obvious one. More than a century later, the information pamphlet in a Dunedin motel room keeps alive the idea that "[p]acked into this small country is seemingly a piece of every part of the world. England's countryside, Norway's fjords, Switzerland's Alps, Canada's lakes,

Oregon's coast, and Hawaiian beaches are but a few of the similarities one may find while travelling around this South Pacific gem."

Tourism can be taken as a barometer, and it operates as an instrument, of local and national self-understanding. As Christopher Wood, art historian and founder of Australians Studying Abroad, commented, "[I]n trying to package itself to attract a burgeoning new class of curious and sophisticated travellers, Australia is in a real sense having to invent itself . . . What we're doing, if you like, is creating a whole new cultural geography based on things other people want to learn about; making Australia into a bounded place with a vast typology of things to see." That process is museological.

New Zealand tourism projects an imagined landscape that segments the history of the country into three hermetic compartments. The nature story stops with the coming of people. The indigenous story stops with the coming of Europeans. And the Europeans (and later immigrants) have until recently not been convinced that their story is very interesting. The divided consciousness of settler societies, with one foot here and the other there, is registered in the very history of tourism. Where tourists once travelled all the way from Europe only to arrive in "Europe," today they disembark in "the world's oldest land," according to *Welcome to Australia*, the guest information book at the Brisbane Hilton. The map of Australia found there features flora, fauna, sports, Uluru (aboriginal name for Ayers Rock), aborigines, and a few buildings – in other words, natural attractions, indigenous people, and sports.

[. . .]

Consider the Queensland Government Cultural Statement: "The Business of Culture" will promote "what makes Queensland culture distinctive – our social history and heritage, our Indigenous cultures and natural environment, our quality products, regions and many diverse cultures." Or *Destination New Zealand*'s proposition: "while our cultural heritage can be presented as 'entertainment' in the hubs, it can be experienced as 'lifestyle' in the regions." This formulation elides several notions of culture: culture as lived practice, culture as heritage, and the culture industry. It also raises several questions. How does a way of life become "heritage"? How does heritage become an industry? And what happens to the life world in the process?

There is a reciprocity, a recursiveness, between the exhibition of the world and the world as exhibition of itself. Museums, through their exhibitions, create "an effect called the real world." That effect is one of tourism's most valuable assets. But, it is not enough, from the industry's perspective, to open the bus and release tourists into the lifespace of their destination – the "real world," available everywhere, always open, and free of charge. The industry prefers the world as a picture of itself – the picture window, cultural precinct, and formal performance.

First, model villages and performing troupes are transportable. Maori cultural performances were exported to Australia and England during the 1860s and to the Festival of Empire Celebrations in England in 1911. Tourists to Bali today can see performances related to those created for international expositions in the course of the last hundred years and specially during the thirties in Paris. Second, designated precincts are more profitable than the lifespace because they "add value" to it. Controlled access to all areas makes it possible to charge a fee. Third, model villages and cultural concerts are more manageable and less intrusive on the lifespace, hence less destructive of it.

The appeal of the lifespace is its high resolution, its vividness and immediacy. One problem with the lifespace is its low density, the dead space between attractions. A second problem is saturation: as they increase in number, tourists fill the space and displace what drew them to it in the first place. To address the saturation issue, the industry markets exclusive *sites* to high-end tourists, thereby generating more revenue from fewer visitors. This is the promise of the empty beach. This is the message of photographs that show the site, but not the tourists.

To address the density issue, the industry develops linkages among sites in a region to form "heritage corridors" arid itineraries that link sites in a region. *The International Express: A Guide to Ethnic Communities along the 7 Train* provides reasons to get off at *every* stop on the route:

The #7 train passes above so many ethnic and immigrant communities on its seven-mile route through northwest Queens [New York City] that it has been dubbed The International Express. We invite you to experience it yourself.

Get off in Sunnyside, spend an evening at a Spanish theater and a night at a Romanian disco; get off in Woodside, rent a Thai video and strike up a conversation at an Irish pub; get off in Jackson Heights, visit an Indian sari shop and dance at a Colombian night club . . .

Or, the industry designs cultural precincts like Brisbane's riverside district, which will provide "a showcase for the finest performers, artworkers and the State's cultural heritage, integrated with food, shopping, and other exciting lifestyle experiences," a convention and exhibition center and a casino: "It will be a model cultural tourism concept that will promote an integrated lifestyle and local cultural experience." In this way the district brings "free, inherent and natural resources" or "incidental resources from various industries" within the scope of the tourism industry proper.

Purpose-built tourist attractions like the Polynesian Cultural Center in Hawaii, where you can experience the Cook Islands, the Marquesas. Samoa, and Fiji all in one spot, are not only dense, they also insulate from tourists the lifespace represented there, while controlling its representation and bringing it firmly within the industry. Guides to the site are Pacific people who have converted to Mormonism, many of them students at Brigham Young University. Through their performance of a way of life they no longer live, made safe for display by that very fact, they also exhibit their conversion.

Theme parks achieve the highest density of all – the whole world within a few acres, often in places that have nothing else to draw tourists. The parks generally stand in an arbitrary relation to the sites where they are built, since fantasy has no fixed geographic location. Nor do recreations. New Yorkers who visit the New York–New York resort in Las Vegas may wonder why they left home. In Central Florida, orange groves and swamps have been displaced by highways, motels, and restaurants that serve the 34 million tourists who "visit Orlando each year to see the world," or rather the "world's showcases" – including, soon, a representation of Key West, a rival destination near by. Key West World "will distill the essence of the tiny island into a land-locked five-acre theme village" at Sea World, just seven hours away from Key West itself. The park is to offer charm without crime and "introduce guests to the island's 'fascinating inhabitants' as well as to its subtropical ecosystem." Is the theme park competition or free advertising for Key West itself? One pundit has proposed that Key West create an "Orlando World," in which visitors would "park in gigantic parking lots, ride trams to the main gate, purchase tickets and spend the remainder of the day standing in an enormous, nonmoving line," after which "they would buy ugly T-shirts, get back on the trams, spend an hour or so trying to find their cars, then spend the rest of the evening driving around trying to decide which one of 317 Sizzler restaurants to eat dinner in." In a word, a theme park of a theme park – all infrastructure, low density, dead space.

When these same tourists return home, they may well discover that the places they left have themselves become destinations. Small towns in Britain have become so popular that they are turning visitors away. *The Age* reported in 1994 that "'Town Full' is a sign of the times." Three million guests visit 30,000 hosts at Windsor, where the ratio is one hundred to one and even higher in the peak season. So resentful are the locals in areas such as Bath that residents have been known to turn hoses on open-top buses. Tourists, it is said, are spoiling the towns for each other and making them uninhabitable for residents, who are fed up with congestion, pollution, and erosion of the sites themselves. Some towns, conceding that they cannot keep tourists away, are drafting "visitor managemnent plans." Others, like Cambridge, are refusing to promote themselves at all.

[. . .]

Heritage, in this context, is the transvaluation of the obsolete, the mistaken, the outmoded, the dead, and the defunct. Heritage is created through a process of exhibition (as knowledge, as performance, as museum display). Exhibition endows heritage thus conceived with a second life. This process reveals the political economy of display in museums and in cultural tourism more generally . . .

[. . .]

HERITAGE IS A "VALUE ADDED" INDUSTRY

Heritage adds value to existing assets that have either ceased to be viable (subsistence lifestyles,

obsolete technologies, abandoned mines, the evidence of past disasters) or that never were economically productive because an area is too hot, too cold, too wet, or too remote or that operate outside the realm of profit because they are "free, inherent and natural resources" or inalienable possessions. Heritage organizations ensure that places and practices in danger of disappearing because they are no longer occupied or functioning or valued will survive. It does this by adding the value of pastness, exhibition, difference, and, where possible, indigeneity.

The Value of the Past

"The past is a foreign country" thanks to the heritage industry. The notion of time travel is explicit in invitations to "[t]ake a trip through history" (Taranaki Heritage Trail) or "walk down memory lane" (Howick Historical Village), both in New Zealand. The very term "historic" can be taken as an indication of obsolescence: no calls can be placed from the "Historic Telephone Box" on the Heritage Trail in Palmerston North. It is enshrined by the City Corporation with the words, "This is a protected building," but its windows now display real estate listings for Harcourts, a business older than the box. Harcourts, which has been operating since 1888, is not on the Heritage Trail.

The Value of Exhibition

Heritage and tourism are collaborative industries, heritage converting locations into destinations and tourism making them economically viable as exhibits of themselves. Locations become museums of themselves within a tourism economy. Once sites, buildings, objects, technologies, or ways of life can no longer sustain themselves as they formerly did, they "survive" – they are made economically viable – as representations of themselves. Heritage projects in Pennsylvania address the massive deindustrialization of the state – by one estimate, "65 percent of land zoned for industrial use lies abandoned" – by providing new uses for derelict buildings and jobs for unemployed industrial workers, who serve as guides to their former lives as miners and steelworkers, to what has become industrial heritage.

Dying economies stage their own rebirth as displays of what they once were, sometimes before the body is cold. In the former East Germany, tourism is stepping in where the heavy industry encouraged by the Communist regime is in decline. Thuringia is selling the good old days of Luther and Goethe by featuring its medieval castles, Renaissance town hall, and churches. Just north of Berlin, on a former army base, "the bad old days" are the subject of a museum and theme park. The museum will present the political and social history of East Germany; the theme park will re-create Communist life there. "Clerks and shopkeepers will be surly and unhelpful. The only products for sale will be those that were available in East Germany."

[…]

The Value of Difference

To compete for tourists, a location must become a destination. To compete with each other, destinations must be distinguishable, which is why the tourism industry requires the production of difference. It is not in the interest of remote destinations that one should arrive in a place indistinguishable from the place one left or from any of a thousand other destinations competing for market share. The Queensland Government Cultural Statement recognizes this all too well when, under the heading "The Business of Culture," it states that "[t]he Government will expect the subsidized arts sector to ensure the cost effective delivery of distinctive Queensland cultural products and services to the State's audiences." It is about "profiting from difference," as the report put it, and benefiting from the "spillover effect" of "a positive Queensland image."

[…]

HERITAGE IS PRODUCED THROUGH A PROCESS THAT FORECLOSES WHAT IS SHOWN

Exhibition is instrumental in the foreclosing of what is shown. The destruction of cultural forms under the pretext of preservation has precedents in the Protestant Reformation, the French Revolution,

the formation of colonial empires, the emergence of nation-states, and the reform of Judaism in the nineteenth century, to mention but a few cases. Utopian longings notwithstanding, the world imagined under the banner of heritage is a battlefield. Which is not to say that all combat waged there is equally bloody, or that the terms of the conflict are the same.

[…]

The tourist stands at the edge of an open grave, not with spade in hand to bury old traditions but with a pen to record them.

The process of negating cultural practices reverses itself once it has succeeded in archaizing the "errors"; indeed, through a process of archaizing, which is a mode of cultural production, the repudiated is transvalued as heritage. The very term "folklore" marks a transformation of errors into archaisms and their transvaluation once they are safe for collection, preservation, exhibition, study, amid even nostalgia and revival. How safe is another matter. In the words of John Comaroff, "[F]olklore … is one of the most dangerous words in the English language" because it often obscures "a highly unreflective populism," or worse, in the case of Splendid China theme park in Florida.

Documentation and exhibition are implicated in the disappearance of what they show, whether intended to induce disgust in those still internalizing the new norms; justify genocide, as the Nazis intended their planned exhibition of an extinct race to do; or demonstrate improvement, in the case of a sanitized Maori model village. …

A recent effort of Burmese authorities to relocate "long-necked" minority women from their homes in eastern Burma to Rangoon to live in a model-village tourist attraction implicates exhibition in the disappearance of what it shows. A Burmese opposition group protested the forced removal of "ethnic minority people from more than two hundred villages in Thandaung township in the hills of northern Karen state," including "members of the Padaung ethnic group whose women put metal rings around their necks giving them a 'long-necked' look." Some of them "will be forced to live in a model village, which is being built near Rangoon in time for next year's 'Visit Myanmar Year'" and is described by the dissidents as an "ethnic human zoo." This is not the first time that "Padaung people have been promoted as tourist attractions," nor is it the first time that human exhibits have been featured in zoos.

According to a plan for the "New Luxor," "the 100,000 residents of Qurna, currently living above and among ancient tombs, will ultimately be relocated from this archeological zone to Al-Taref." To encourage tourists to stay longer – if not for a thousand and one nights, then for "Six Egyptian Nights" – developers plan a golf course and "a model village that portrays aspects of Egyptian life – Pharaonic, Bedouin, Nubian and rural cultures."

Bushmen, "routed almost out of existence" by early settlers and now few in number, were expelled from Kalahari Gemsbok Park in 1970, because "management decided that tourists did riot like seeing hungry-looking Bushmen. The tribesmen's lack of materialism made them unreliable, many employers say, and they were eating too many animals." Twenty years later forty Bushmen have been brought from a shantytown to the Kagga Kammna Game Park north of Cape Town, where tourists can view them for $7.00 ($1.50 of the fee goes to the Bushmen) . . .

[. . .]

Like museums, tourism is predicated on dislocation – on moving people and, for that matter, sites from one place to another. Take Luxor – Luxor Las Vegas, that is:

> Luxor Las Vegas, which opened on October 15, [1993], is a 30-story pyramid encased in 11 acres of glass. The hotel's Egyptian theme is reflected in the decor of its 2,526 rooms and 100,000 square foot casino. Guests travel by boat along the River Nile from the registration desk to the elevators, which climb the pyramid at a 39-degree angle. Other features include an obelisk that projects a laser light show in the pyramid's central atrium; seven themed restaurants, and an entertainment complex offering high-tech interactive "adventures" into the

past, present, and future. Double rooms at the Luxor, 3900 Las Vegas Boulevard South, are $59 to $99.

Is getting to and from the registration desk to the elevators by boat along the river Nile any stranger than squeezing the Temple of Dendur into the Metropolitan Museum of Art in New York? Any stranger than traveling to Luxor, Egypt, itself? Travel Plans International promises a cruise up "the legendary Nile in a craft that surpasses even Cleopatra's barge of burnished gold. . . . It is a yacht-like 44-passenger vessel carefully chosen for its luxuriously intimate appointments. Each cabin provides panoramic views through picture windows as well as the convenience and comforts of private showers, individual climate control, and television." What *Travel Plans International* (1988) does not tell you is that several years later "[t]ourism in Luxor has all but ended because of violence." Islamic militants were planting bombs in Pharaonic monuments, both to drive out tourists and to wipe out traces of idolatry.

Go to Las Vegas, experience Egypt. Go to Stockholm, experience all of Sweden – at the Skansen open-air museum. Go to Elancourt, outside Paris, and experience the glories of France-Miniature – including scale models of the Arc de Triomphe, the Cathedral of Notre Dame, and the Alps. Stay in the Acapulco Motel in Auckland or the Sahara Guesthouse and Motel in Dunedin or the fully generic Heritage Motor Inn, in faux Tudor, in Rotorua. In Christchurch, at Orana Park, where African cheetahs, rhino, and giraffes roam, "The Serengeti Restaurant offers brilliant views over the African Plains," just twenty-five minutes from the heart of the city. The International Antarctic Center invites you to "Experience Antarctica Right Here" – "It's better than being there."

Increasingly, we travel to actual destinations to experience virtual places . . .

"Performing Work: Bodily Representations in Merchant Banks"

from *Environment and Planning D: Society and Space* 12 (1994): 727–750

Linda McDowell and Gill Court

Editors' introduction

In Part One we were introduced to Alan Latham's approach to culture as a kind of embodied practice, or "performance," as opposed to the "web of meaning" suggested by Clifford Geertz (see pp. 60 and 20). While Geertz argued that people interpret the meaning of their world semiotically – that is, through signs and symbols – Latham's work suggested that for many people meaning comes about in ways that don't necessarily involve conscious reflection. People may derive meaning less from creating and interpreting symbols around them than from their embodied movements, senses, and encounters. The metaphor of performance has become a common way of rethinking culture according to a less cognitively oriented approach to culture. George Yúdice (see p. 422), for example, uses the idea of "performativity" to capture a more instrumental and expedient approach to culture.

Similarly, Linda McDowell and Gill Court draw on the metaphor of performance to suggest a way of understanding merchant banking culture that focuses on embodied practices of bankers. They are also interested in how representations of those embodied practices in the media reflect and inform dominant understandings of masculinity and femininity. Such understandings, in turn, provide the models that individual "performances" of masculinity and femininity either conform to or deviate from. This idea of *performance* is drawn primarily from the work of Judith Butler, whose approach is outlined most clearly in *Gender Trouble* (1990), and which also informs Yúdice's approach. Butler argues that "gender ought not to be constructed as a stable identity or locus of agency from which various acts follow; rather, gender is an identity tenuously constituted in time, instituted in an exterior space through a *stylized repetition of acts*." Thus, if gender is constituted by its repeated performance, rather than by any innate quality, it is also subject to performances that fail to repeat the norms, or "models," loyally. In this way, some bodies transgress the norms of masculinity or femininity, and, over time, such norms are also subject to change.

McDowell and Court's focus is thus on embodiment and bodily representation. Along with scholars such as Gill Valentine (see p. 395), Peter Jackson (see p. 413), Robyn Longhurst (see p. 388), and David Bell, McDowell and Court were among the earliest geographers to take seriously the body as a constitutive element of culture. They note, however, that most of this work focuses on bodies that transgress the norms, rather than "serious" bodies in suits at work. Such bodies tend to be "invisible" in that they are just "normal." McDowell and Court's research, however, suggests that the "serious" bodies of merchant bankers have their own distinct performative dimensions.

While this article reveals the ways a feminist approach to performance can inform geographies of work, it also represents a productive intersection between cultural and economic geography. The "cultural turn" in economic geography can be thought as consisting of two related lines of inquiry. On the one hand we see a focus on the rising importance of a *symbolic economy* in which images, meanings, and experiences are increasing central to economic value chains (see Zukin, p. 431). On the other hand, we see new interpretations of economic activities that focus on their discursive, symbolic, and (in the case of McDowell and Court) embodied qualities. Thus, echoing the argument made also by Peter Jackson (p. 413), culture and economy should not represent two mutually exclusive categories of social life. Instead, as McDowell and Court show, economic activities have a deeply cultural dimension, and can be subject to the same kind of cultural analysis that traditionally might have only been contemplated in some far-flung village among "traditional natives."

Linda McDowell, Professor of Human Geography at St. John's College, Oxford University, has been at the forefront of feminist geographies of contemporary social and economic change. Along with many journal articles, she is author or editor of several books, including *Capital Culture* (1997), *Gender, Identity and Place* (1999) and *Redundant Masculinites? Employment Change and White Working Class Youth* (2003). *Hard Labour* (2005) examines the post-war lives of European women migrant labourers in Britain.

There is now an increasing amount of work in cultural geography on the performance and practice of culture, including two special issues of *Environment and Planning D: Society and Space* 18: 4–5 (2000). Nigel Thrift has written numerous pieces on the subject, such as "The still point" in *Geographies of Resistance*, edited by Pile and Keith (1997) and "Afterwords" in *Environment and Planning D: Society and Space* 18:2 (2000). Additional perspectives can be found in Gillian Rose's "Performing space," in *Human Geography Today*, edited by Massey, Allen, and Sarre (1999), Catherine Nash's "Performativity in practice: some recent work in cultural geography" in *Progress in Human Geography* 24 (2000), and Jon May's "A little taste of something exotic: the imaginative geographies of everyday life geography" in *Geography* 81 (1996). Research focusing on the culture and performance of work includes Chris Gibson's "Cultures at work," in *Social and Cultural Geography* 4, 2 (2003) and Phil Crang's "'It's showtime!' On the workplace geographies of display in a restaurant in South East England" in *Environment and Planning D: Society and Space* 12 (1994).

INTRODUCTION

Labour, unlike other commodities, has a human embodiment that cannot long be denied. (David Smith, *The Apartheid City and Beyond* 1992, pages 6–7)

. . . it could be argued that men don't have any bodies at all. Look at the magazines! Magazines for women have women's bodies on the covers, magazines for men have women's bodies on the cover. When men appear on the covers of magazines, it's magazines about money, or the world news. Invasions, rocket launches, political coups, interest rates, elections, medical breakthroughs. *Reality.* Not *entertainment.* Such magazines only show the heads, the unsmiling heads, the talking heads, the decision making heads, and maybe a little glimpse, a coy flash of suit. How do we know there's a body, under all that discreet pinstriped tailoring? We don't, and maybe there isn't. What does this lead us to suppose? That women are bodies with heads attached, and that men are heads with bodies attached? Or not, depending?' (Margaret Atwood, *Good Bones* 1992, pages 80–81)

And, moreover, whatever the brain might do when the professions were open to it, the body remained? (Virginia Woolf, *Three Guineas*, 1977, page 10)

In this paper, the consequences of these statements are examined – Smith's argument that *all* workers are embodied, and Atwood's and Woolf's recognition that embodiment is gender differentiated – for understanding the ways in which women's experiences as waged workers differ from those of men. Our argument is based on empirical work in the world of high finance – a preeminently serious world, at least at first sight, in which disembodied pinstripe suits are the major actors. Our major sources are of two different types: first, the representations of bankers that are conveyed through the printed medium, especially the financial pages of broadsheet newspapers but also through publicity material produced by merchant banks, and second, images of self and the ways in which these are constructed as part of everyday social relations between colleagues and between bankers and clients. This second source is derived from detailed interviews with men and women merchant bankers currently employed in the City of London. The aim is to provide a sociocultural reading of economic practices in order to make, as [Erica] Schoenberger recently suggested [at the AAG conference in Atlanta, 1993], "a small and partial start to untangling the complexities arising from the fact that corporations are run by real people". As she argues, "in order to understand corporate strategies we need to understand something about corporate strategists" and, as we argue here, to understand the effectiveness of policies to enhance workplace equality, we need to know how workers relate to each other. Although ethnographic analyses of economic institutions are not yet common, at least in the geographic literature, a number of geographers have recently argued for social analyses of economic institutions. Although the initial intention of the research from which this paper is drawn was to undertake a full-scale ethnography, observing, if not actually participating in, everyday social relations in the range of departments that constitute merchant banks, questions of sensitivity and confidentiality precluded open access. In the event a range of interviews were held, sometimes in the participant's office or workspace but more often in the more neutral space of a meeting room. However, long interviews elicited rich details about the social practices within banks.

Material social practices, however, are deeply imbued with and are undertaken within the context of a set of cultural and symbolic meanings. The world of merchant banking and the composition of the key social actors in the City are saturated with symbolic significance. The 'old' world of the City of London with its distinctive built environment reflecting the nineteenth century expansion is paralleled by class-specific and gender-specific images of bankers – the white, male, and bourgeois world of the public school, elite universities, and masculine clubs. These are juxtaposed with contemporary images of 'fast' money and slick operators in the new cut-throat deregulated City represented in films such as *Wall Street* or the UK television serial *Capital City*. These images exert an influence not only on the popular imagination and representations of the City but also affect social practices, from recruitment to relations with clients. It is this relationship that we address in this paper. In an attempt to deconstruct the distinction between 'image' and 'reality', we draw on ideas from social theory, and from the recent productive interactions between feminist theories and deconstruction in particular, in which objects and everyday social practices are constituted and may be read as texts or narratives.

Our aim is to develop what Game [in *Undoing the Social: Towards a Deconstructive Sociology*, 1991] has termed a materialist semiotics – that is, an understanding of meaning as both temporal and embodied. As Game argues, 'this approach to meaning breaks with distinctions between representation and the real, text and context, theory and practice' (page x). The idea that reality is fictitious or that the fictitious is real is not yet a common notion within economic geography, where feminist and other 'critical' approaches have tended to remain within the discursive and theoretical constraints of the mainstream subject. In contrast, the recent work of feminist theorists in many disciplines, and in certain subareas of geography, has been richly informed by interdisciplinary perspectives. Drawing on literary, cultural, and film theory and on developments in French philosophy, researchers have posed questions about the meaning and representation of social practices. In contrast, for most economic geographers, like the sociologists whom Game addresses, their subject matter is 'facts – social reality, the empirical – and theory, and the correspondence between these . . . [P]ractice is conceived of as representation of the real' (page 4). However, in areas of the geographical discipline more influenced by the humanities and recent feminisms, in studies of landscape by cultural geographers, for example, reading the landscape as text is becoming an accepted approach. However, the ways in which representation, meanings, and practice are interconnected in economic institutions have not been explored in detail, although [Nigel] Thrift has recently made an argument for the analysis of money markets in general and the City in particular 'as socially constructed institutions and dealing in money as a social and cultural affair'. Here, we explore a particular link between representation and social practice seen through a gender lens, showing how meaning and everyday social behaviours are connected in merchant banks. We argue that themes such as desire, subjectivity, and the body, which may seem unfamiliar to economic geographers, are an important part of an understanding of the ways in which economic institutions operate.

[. . .]

NARRATIVES OF POWER: READING CITY BODIES

Within the last two decades or so, culminating in the mid to late 1980s, a noticeable shift has taken place in the representation of the world of money, and in particular the City of London. A world that was pre-eminently both serious and inaccessible has become the subject of novels, plays, and films, a theme in advertising, and the object of press attention ranging from praise to excoriation as financial scandals become more common, or at least more visible. This emphasis on fictional representations is paralleled by shifts in the nature of 'economic reality' as the commodities dealt with in the new financial world themselves are increasingly fictional entities: futures trading (in commodities that do not yet exist) and junk bonds, for example, invisible earnings, something called credit, which, as Grace points out, is really debt ["Business, pleasure, narrative: the folktale in our times," in R. Diprose and R. Ferrell, eds., *Cartographies: Poststructuralism and the Mapping of Bodies and Spaces*, 1991]. Although the City has long dealt with variants of these commodities, the element of fiction seems heightened in recent years as extreme losses and gains predominate. For example, in 1993 the British Government 'wasted' millions from the reserves trying to prop up the pound, at the same time as the financier–speculator–philanthropist Georg Soros made millions of dollars by astute currency deals. As Grace suggests, 'Clearly, then, a world of high fiction is observable, a daily soap opera, full of the

most extreme occurrences. Everyday economic life has become a fiction of terrifying realism, a horror scenario with such convincing special effects that, at times, you really feel you too are there, in the middle of it.' The metaphorical or fictitious nature of money is well recognised. Roberts recently emphasised this in her recent work on offshore financial flows in which not only money but space is dubbed fictitious ["Fictitious capital, fictitious spaces: the geography of offshore financial flows," in S. Corbridge *et al.*, eds., *Money, Power and Space*, 1994]. As she argued, 'the realisation that money only works because people believe in it underscores the fictitious quality of money'. It therefore seems particularly appropriate to treat the financial sector as if it literally were a fiction in the sense of a constructed narrative.

This fictional world is one . . . in which heroic but flawed individuals struggle against extreme forces. It is not insignificant that Corbridge, Thrift, and Martin chose to introduce their new book [*Money, Power and Space*, 1994] with Harvey's words 'Love and money may make the world go round . . . but love of money provides the raw energy at the centre of the whirlwind'. This combination of sex and greed with elemental forces is reflected in press coverage of the affairs of the City. In the financial and business pages of the serious press, photographs of key individuals in the world of money are now a common feature. The photograph is an important part of the way of presenting the images of power and influence that are so strongly associated with money and in presenting an image of a tragic hero engaged in a struggle against elemental forces . . .

A number of cultural critics . . . have suggested that the visual images of powerful men that appear in popular cultural representations and on the business pages might be interpreted in terms of a cinematic or televisual folk drama, in which the forces of good struggle with the forces of evil . . . Drawing on these arguments, on an analysis of the British press undertaken over a two-year period from autumn 1991, and on a reading of taped interviews with men and women in a variety of professional positions in merchant banks, we present below a materialist semiotics of the different characters, bodily norms, and gendered performances in one part of the British financial sector in the early 1990s.

EMBODIED BANKERS: CORPOREAL REPRESENTATIONS AND PRACTICES

The *dramatis personae* that peopled the financial world of Britain in the early 1990s, present in the press and in the offices of corporate financiers and in the trading and dealing rooms, include not only the patriarch, as hero or villain, the desirable princess, and tough, but still smart and sexy, older women, but also the young prince, who may or may not be a pretender. Only time will tell whether he might be sufficiently worthy to inherit his father's powerful position. Although the patriarchal figures are closest to the disembodied masculinity valorised in bureaucratic organisations, the bodily performance of younger men challenges this idealisation.

MULTIPLE MASCULINITIES: PATRIARCHS AND PRINCES

In the critical discourses about sexuality and organisations, it has been argued that the term 'man' has been used as an unmarked universal category to stand in for humanity in general. Organisational practices that purport to be nondiscriminatory are in fact based on assumptions that are masculine. The ideal, disembodied, rational worker is in fact a man against whom the embodied woman appears as an inferior 'other'. These arguments have been critical in the challenge to the privileges that accrue to men as a consequence of the implicit male bias. However, the central assumption of 'men' or 'masculinity' as a single oppositional category itself needs deconstruction. In different circumstances, a particular hegemonic version of masculinity is dominant and this construction positions not only women but alternative styles of masculinity as inferior or inadequate. For men, too, there are a range of gender performances or ways of 'doing gender' in the workplace. Our analysis of press photography and other images of male bankers and material about presentation of self and social interaction in the workplace revealed different versions of masculinity in merchant banks.

What we have termed a patriarchal masculinity perhaps conforms most closely to traditional stereotypical representations of bankers. This masculinity is class specific and is perhaps the least

conscious performance. None of the men whom we interviewed who fell into this category referred to notions of image, of performance or masquerade, or to selling themselves, unlike all the other categories of respondents. These men are sober and industrious, with a solid background of a good family and a public school behind them. In certain cases they are members of a dynasty of bankers whose names are commemorated in merchant banks today . . .

What has changed to some extent is the class basis of the City, the associated hegemonic masculinity and ways of doing business. The rapid recruitment in the 1980s resulted in a wider class composition and a more open way of doing business. As a respondent commented, 'in merchant banking it was a boys' club. I think it's changed considerably in the last ten years' . . .

What we might term the new City of the last decade or so is dominated by an alternative version of masculinity – the slick young pretender or prince. This character is an international figure (rather than a representative of a particular bourgeois version of Englishness), distinguished by his youthful appearance, his energy, activity, and virility. His characteristic site is the floor of the stock exchange, in the midst of a chaotic spectacle. The spectacle of hysterical traders is familiar from recent stock market 'adjustments' – Black Monday in 1987, Black Wednesday in September 1992, and the demise of the ERM in 1993.

These images [of hysterical traders] – which emphasise the embodiment, the sheer physical exuberance of the characters – challenge the notion of the masculine worker as a rational disembodied being. From the taped interviews, an interesting narrative representation of the qualities needed for successful workplace performance in this arena of banking emerged. It is a representation that is also both classed and gendered, but in different ways from the patriarchal representatives of the banking world. In this performance there is an emphasis on the 'natural' characteristics of successful traders and dealers, and on the unity, rather than the dualism, of mind and body.

In the view of respondents from other parts of the banking world, these young men are 'natural mathematicians, good with figures'; their chief characteristic is 'basic raw intelligence'. They are 'barrow boys, natural sellers. If they weren't selling bonds, they'd be selling fruit and vegetables in the East End.' 'Traders and salesmen are born not made. They used to equate it with selling apples off the barrow, if you could do that you could trade.' They 'come to the bank at eighteenth, no training, all natural ability, straight from school, highly numerate, amazingly quick minds'.

These recorded views are themselves fictional, based on a stereotypical figure. However, as a respondent remarked, 'the trading floor's much wider in fact in regard of social group nowadays' and all but one of the respondents whom we interviewed from this area of the banking world was a graduate, differing from the corporate finance divisions only in the fact that somewhat less prestigious universities had been attended.

The characteristics required for success in dealing, trading, and selling are less class based than gender specific. They are those attributes conventionally associated with masculinity. As respondents explained, 'you have to be tough and ruthless to succeed' and 'there seems to be an incredible need to bite everybody's head off and knock them out of the way and trample on their heads'.

In an essentialist reading of human nature, women are assumed to lack these attributes, which come naturally to men. This was a common assumption among our interviewees. Thus, 'it doesn't come naturally to women to shout down phones', or 'women aren't tough enough; it's not in their natures', and 'an aggressive male dealer – well that's how dealers are; an aggressive woman – it's not natural'. And, as this respondent rather condescendingly explained, even if women act tough, it does not benefit them: 'There's a certain female type in the City; trying to be men, wanting to show themselves as that much more aggressive. I find it quite sad, really. It more often backfires with colleagues than helps.'

[. . .]

These men revealed themselves in the course of the interviews to be aware of the importance of their bodies and bodily discipline, almost all of them referring to dress, style, and their weight. Weight seemed particular significant. . . . In an image-conscious, class-conscious, and youthful occupation like merchant banking, weight (or rather lack of it) is a crucial element of success. Our respondents were astonishingly physically uniform – the majority of them of average height and suit-able weight for that height. That our respondents were aware of this is clear from one protestation: 'We are not all clones, you know.'

For many men the maintenance of a sleek body required considerable effort and expenditure. Over a third of the respondents worked out on a regular basis and almost a half were actively involved in sporting activities . . . 'I tell people in my team to look after themselves, sort out their BO or weight. You have got to look good,' argued a male interviewee. And, as he explained, using revealing metaphors, self-improvement is possible, if not limitless: 'It's no good putting someone in a frock coat if they should be in a donkey jacket, but you can make yourself look better with hard work.'

Clearly, corporeality has a materiality that matters, and not all applicants to enter the world of high finance are given the option. Bodily size may be a barrier to entry. There were no obviously overweight people among those whom we interviewed and, as one respondent candidly admitted, 'If someone was very fat or ugly it would make a difference.' Reflecting on a recent job applicant who was, according to this respondent, three or four stone overweight, he continued, 'the fact that he was very large is going to weigh on the client's mind . . . We don't recruit physical stereotypes but we are selling a service and if people don't want to buy the service from that person . . . well.'

PRINCESSES, HONORARY MEN, AND FEMININE MASQUERADE

What about images and representations of women? How do women present themselves and construct a workplace performance in these two worlds of banking when they are clearly neither patriarchs nor princes?

Photographs of women so rarely appear on the business pages that when they do their presence is marked. The style of photography and the location of the woman immediately marks out the difference from representations of powerful men. An emphasis either on the whole body or on a close-up shot of the face rather than a head and shoulders businesslike shot is usual . . . As Grace argued, it is desire not authority that structures these images.

A second significant feature of images of women in the financial sector is the fact that these women are often pictured either without an identifiable location or in their homes rather than in their offices in the bank, let alone on the trading floor. Countering the images of energy, activity, and virility, women are represented as passive and domestic, as private rather than public and so out of place in the public arena of work. . . . Although the femininity of these women is emphasised, they are also women 'of a certain age', smart, confident, knowledgeable, and experienced. Indeed, these women are no longer 'princesses', nor even mistresses. . . .

Grace suggests that, as women have entered powerful positions in the workplace, 'a certain anxiety seems to have emerged and a displacement of woman as representation of desire and pleasure has begun to take place. Increasingly, the sex objects are men, partly in terms of their bodily attributes . . . but partly because of what also might be called their mind attributes; the bright young executive or bond dealer, who outsmarts the competition.'

[. . .]

The overall impression given by representations of women on the business pages is that they are 'the other', objects of desire, at least until they reach the age of menopause and become 'honorary men'. As Rodgers argued in the context of her work on women in the British Parliament ["Women's space in a men's house: the British House of Commons," in S. Ardener, ed., *Women and Space: Ground Rules and Social Maps*, 1981], older women who have established their position with difficulty are often overtly hostile to younger women who display attributes of embodied femininity, such as pregnancy, too openly. She suggests that 'Women whose success has been geared to the male construct have discarded the symbols by which they would be anchored into the traditional domain of domesticity and nature. They fear that if one of their women colleagues openly combines the public symbols with the female domestic ones, they themselves will be at risk of being seen as the women, which, on some levels they, of course, are. Their position in the dominant category is after all a tenuous one" (pages 60–61).

Rodgers's argument parallels that of other feminist theorists who have argued that women, to be successful in masculinist organisations, adopt a workplace performance that constructs them as honorary men. . . . This contention was supported by comments from many of the junior women we interviewed who emphasised that, through their clothes and their attitudes, they attempted to conform to the dominant image in the part of the bank in which they worked. Thus women told us they felt they were accepted as 'one of the boys' or that they adopted, in the capital markets and corporate finance departments, a masculine disguise: 'I always wear a suit' or 'I try to look neutral, dark colours; a jacket is essential.' 'For most of the time I am an honorary man. They [her male colleagues] do treat me like an honorary male and that's what I prefer. It means that I can see the way they look on women. If I go out for a drink with them, then they will comment on anything that walks past in a short skirt, things that friends wouldn't say if I was there. I guess I'd rather be an honorary man than be on the other side.'

However, a significant number of women, some of them junior but especially the more senior women, were adamant that masquerading as a man was impossible. In the succinct words of one of the tiny number of women directors, 'I'll never be a man as well as a man is.' Another woman elaborated on the problems: 'It's difficult, even demeaning, to try to be one of the boys. Don't play a man at his own game because I think quite frankly you'll fail if you try to do that. You are not a man.'

Indeed, many women suggested that overt femininity conferred advantages, realising that, as Bordo (1993) argues [*Unbearable Weight: Feminism, Western Culture and the Body*, 1993], ' "feminine" decorativeness may function "subversively" in professional contexts which are dominated by highly masculinist norms (such as academia)' (page 193). Thus a respondent suggested that she deliberately adopts different stereotypical images of femininity to confuse her male colleagues: 'Sometimes I'll chose the "executive bimbo" look; at other times . . . it's easiest if I look as if I'll blend into the background . . . I do sometimes dress quite consciously because you've got to have some fun in life, and sometimes wearing a leather skirt to work is fun because you know they [her male peers and superiors] can't cope with it.' Another rather older women told us, 'I just use what I am. I don't turn up disguised as an executive bimbo because I'm not really of an age to do that. Some women do use that image. Some women are disguised as bimbos and it turns out there's a first class brain hammering away underneath all that.' Metaphors of performance, disguise, and masquerade were commonly used by women respondents. As one reported, 'There's different ways to skin a cat, different roles to adopt.'

It became clear, however, through the analysis of the transcripts of male respondents, that many men felt uneasy about the performance of their women colleagues, whether they acted as 'men' or as 'women'. Attempting a masculine parody, whether through dress or behaviour, tends, as a number of respondents intimated, to be counterproductive . . .

But neither is femininity costless for women professionals. Indeed there was a common feeling among the men whom we interviewed that femininity conferred unfair advantages on women, particularly in interactions with clients: the key element of 'selling' work, whatever the context. Here, too, the discursive construction of feminine attributes as natural was noticeable. Women were regarded as unfair competition because 'women are good at getting on with people; people tell them things', and 'women may have a natural advantage, as the majority of clients are men, and clearly their PR skills and general warmth of approach is much better than a man's'.

Or, as another male respondent reported somewhat ingenuously, putting words into the mouth of an anonymous female colleague: 'I never really thought of this, but she said, "Sometimes I can use my female skills to get things that you couldn't get". There's a definite advantage to being a woman. Being feminine, even slightly sugar-coated, can be a great advantage both within the bank and with clients, because girls [the word he used] can manage to strike up an almost instant rapport, you know, with their director and their clients.' And, what clearly worried him, 'I think it will influence the choice of promotion to a degree, as long as it's somebody who has the other skills.'

UNEASY SEDUCTION

The unstated implication of these comments is that women use their sexuality and femininity to seduce their male clients, albeit not in the literal sense. Women respondents were far more outspoken about the ways in which interactions between themselves and their clients were based on the manipulation of conventional heterosexual norms of attraction between the opposite sexes. The language in the following quotations is that of flirtation. 'You need flair . . . I'm no good with a client I am not interested in or he in me . . . if there is no spark in a relationship, you just can't turn it on.' 'I have the ability to listen and make polite noises. I gain clients' confidence. It's a different way of doing things.' And, most explicitly, 'Women seduce their clients, not literally. I'm quite certain it's done that way.'

However, many women also enunciated a certain feeling of unease about their performance, unhappy in their adoption and exploitation of a parodic femininity which they found demeaning. 'If you are an attractive woman in this environment it can help on the male side of things. Frankly you have to learn to use all your assets and swallow your pride sometimes because in some form or other, obviously not in the literal sense, but in some form or other, it can be a form of prostitution of your sex . . . and you, hmm, and you . . . you have to learn to use that' . . . Many of the women whom we interviewed not only explicitly used the language of performance to describe everyday social interaction in the workplace but also suggested that their workplace persona was unreal. They talked about 'building up a shell', of 'adopting a different sense of myself', of 'not using my real personality'. Thus the unease that men expressed about women in merchant banking is paralleled by unease among women themselves.

The extent to which this uncertainty affects women's performance and their promotion prospects remains unclear. However, it is clear that men retain their hold on the key positions in merchant banks. Only 11 per cent of professional positions in merchant banks in the City of London as a whole are held by women. Although there are clearly a number of reasons for this low representation, including previous gender differences in educational levels and in recruitment practices, media representations of women as 'other' and everyday social interactions based on norms of masculinity, be they the disembodied male of conventional organisation theory and bureaucratic practice or the variants of masculine performance outlined here, undoubtedly are part of the reason for women's lack of success in penetrating the inner sanctums of financial power.

CONCLUSIONS

It is clear that the gendered identities of the employees whom we interviewed are an essential element of the overall service that they provide for clients. In merchant banking, as in other interactive service occupations, corporeality – in the threefold sense of anatomical sex, gender identity, and gender performance – is a crucial part of selling a service, in this case monetary advice. It is also clear that, with the important proviso that they must be within an exclusively heterosexual scenario, there are several gender performances available that are acceptable and appropriate fictions in particular circumstances. The disembodied ideal of the male bureaucrat in which rational advice was constructed as a cerebral product, purportedly unconnected to the specific embodiment of the purveyor of that advice, has been displaced in the contemporary world of high finance. . . .

The movement towards work as a performance, in which individualistic criteria of appearance, personality, panache, style, and deliberate self-presentation are increasingly emphasised, has begun to undermine the notions of worth and achievement that used to typify bureaucracies. The net result is that the relative evaluation of individuals against universalistic criteria is increasingly difficult. While not denying an iota of the pertinent critique of the implicit masculinist nature of the old attributes valued by purportedly universalistic assessment and evaluation schemes, we find it ironic that, just as women are gaining access to formal examinations and professional credentialisation, the formal criteria of access to positions of power and status are becoming less valued. However, rather than rueing a time that is past, it is clear that those concerned with the position of women in the workplace must look beyond the distributional conception of justice that lies behind conventional equal opportunities policies. . . .

COPYRIGHT INFORMATION

PART THREE   LANDSCAPE

PART FOUR   NATURE

PART FIVE   IDENTITY AND PLACE IN A GLOBAL CONTEXT

PART SIX   HOME AND AWAY

PART SEVEN   DIFFERENCE

PART EIGHT   CULTURE AS RESOURCE

Index